ANATOMY OF THE
DICOTYLEDONS

ANATOMY OF THE DICOTYLEDONS

SECOND EDITION

VOLUME I
SYSTEMATIC ANATOMY OF LEAF AND STEM,
WITH A BRIEF HISTORY OF THE SUBJECT

BY

C. R. METCALFE

Honorary Research Associate of the Royal Botanic Gardens, Kew
and formerly keeper of the Jodrell Laboratory.

AND

L. CHALK

formerly head of the Section of Wood Structure,
Commonwealth Forestry Institute, Oxford.

With contributions from
Katherine Esau, Leo J. Hickley, Richard A. Howard, William L. Theobald,
Joseph L. Krahulik, Reed C. Rollins, Hazel P. Wilkinson, and David F. Cutler

CLARENDON PRESS · OXFORD
1979

Oxford University Press, Walton Street, Oxford OX2 6DP
OXFORD LONDON GLASGOW
NEW YORK TORONTO MELBOURNE WELLINGTON
KUALA LUMPUR SINGAPORE JAKARTA HONG KONG TOKYO
DELHI BOMBAY CALCUTTA MADRAS KARACHI
IBADAN NAIROBI DAR ES SALAAM CAPE TOWN

Published in the United States by Oxford University Press, Inc., New York.

British Library Cataloguing in Publication Data
Metcalfe, Charles Russell
    Anatomy of the dicotyledons.
    Vol.1: Systematic anatomy of the leaf and stem,
    with a brief history of the subject.
    — 2nd ed.
    1. Dicotyledons — Anatomy
    I. Title   II. Chalk, Laurence
    583'.04'4   QK495.A12   78-40745
    ISBN 0-19-854383-2

Set by Hope Services, Abingdon
Printed in Great Britain at The Pitman Press, Bath

PREFACE

During the 29 years that have elapsed since the appearance of the first edition of this work a large amount of new and interesting material has been published. Various changes in the format have been introduced in order to make it easier to incorporate the new information. Apart from technical changes such as the adoption of a larger page with the text in two columns as a means of economizing space, there has been a major change in the character and purpose of the introduction, which in the first edition occupied only about 50 pages. The introduction, in its new guise, virtually amounts to a textbook of systematic anatomy as applied to the dicotyledons. Well-known textbooks of general plant anatomy already exist, and indeed we frequently refer to these very worthy publications. The present introduction differs in emphasizing those characters that are of taxonomic or diagnostic significance. It also indicates the various lines of inquiry, such as the structure of the phloem and of the petiole, that have now become part of the broadened concept of taxonomy to which we are becoming increasingly accustomed.

In the course of assembling material for the introductory survey, it has become apparent that there is so much that must be included that it has been found preferable to have two relatively short, general survey volumes instead of only one. Volume I is devoted mainly to leaves, stems, and roots, but it contains frequent references to the structure of woody plants. The subject of wood structure will be fully elaborated by L.C. in Volume II. We envisage, however, that Volumes I and II should form a closely knit pair in which the subject-matter is so closely woven that neither volume can be effectively used without frequent reference to the other. Important topics such as ecological anatomy, taxonomy, and phylogeny are covered in Volume II, together with the relationship of systematic anatomy to related disciplines such as chemotaxonomy.

Subsequent volumes will contain descriptions of the individual families and these will be described in Professor Takhtajan's revised sequence of dicotyledonous families which he has kindly provided as a contribution to Volume II of this work. This is a slightly modified version of the sequence already published in Takhtajan's *Flowering plants*, of which the authorized translation by C. Jeffrey was published in 1969. There was much discussion before we decided to follow this system, but we were strongly influenced by the fact that Takhtajan's scheme for the families has the great merit of being among those that harmonize reasonably with the anatomical features of the plants, so far as they are known at present.

Each volume will include a comprehensive bibliography of the literature references mentioned in the text, together with some others recommended for further reading.

The sequence of chapters in Volumes I and II begins with the history of the subject, followed by the structure of leaf, stem, and root. The component cells and tissues are described, and also the patterns of cellular organization that are to be seen in the various organs that go to make up the plant body. In order to relate the histological aspects of taxonomy to the external characters that are the traditional criteria used by herbarium taxonomists, various topics that are on the borderline between the two approaches have also been briefly included. These range from external morphology in the traditional sense to trichomes and leaf architecture.

It is sometimes said that the anatomy of the vegetative organs of angiosperms is already so well known that it is no longer worth while to pursue the subject any further. It is true that the study of histology is now a long-established discipline which started with the invention of the compound light microscope and has kept pace with the many improvements in its performance. For this reason there has been a steady accumulation of descriptive information, the sheer volume of which is now great enough to give the false impression that our knowledge is already complete. It is true that many of the basic concepts about cells and tissues remain unchanged. On the other hand, many new techniques for making histological preparations have been introduced, and a whole new era in plant histology has been made possible by the invention and development of the electron microscope; the scanning electron microscope in particular is adding greatly to our knowledge of plant surfaces. The significance of the scanning electron microscope in taxonomic studies is pointed out by Dr. D.F. Cutler in Part III of our first chapter. Nevertheless even the histological details that have been observed and recorded by using the older techniques are very incomplete. This is partly because the number of species so far examined represents only a very small fraction of the total number that are accepted as valid, but also because views have changed over the years concerning the significance and importance of the characters to be described. Another difficulty is that only dried herbarium material of many of the taxa has hitherto been available for histological examination, often giving a distorted or misleading appearance under the microscope. This all goes to show that, far from having an exhausted field of study, we have only touched the fringe of the subject of systematic anatomy.

When we embarked on our first edition (Metcalfe and Chalk 1950) the subject of systematic anatomy was fairly sharply divided into wood structure on the one hand and the structure of leaf, stem, and root on the other. At that particular point in the history of botany, wood structure had never received adequate attention and this defect was made good as far as it could be by L.C. who, with the assistance of his colleagues, was entirely responsible for dealing with the structure of wood in the form of specimens that were large enough to be recognized as timber by foresters and wood technologists. L.C.'s endeavours came at a propitious moment, for vast collections of authentically named wood specimens were pouring into Oxford from Commonwealth countries, and he had the very generous co-operation of Professor Samuel J. Record of Yale University in the U.S.A. who was collecting and studying wood specimens on a worldwide basis, especially from South America. New methods of cutting sections of wood had recently been introduced to Oxford from the U.S.A., and contact with the U.S.A. was also promoted when Dr. Margaret Chattaway visited Yale University to collect histological details of woods that were not available at Oxford. The great contribution made by L.C. is fully recognized in many countries, now that sufficient time has elapsed for his descriptions and ideas to diffuse amongst botanists around the world.

We feel, however, that the real value of our first edition was the fact that we agreed to work together, so that the book covered the whole field of systematic anatomy. C.R.M. dealt with all aspects of systematic anatomy apart from wood structure, and for his original work the unrivalled collections of living specimens at Kew were invaluable. Supplementary material from the Museums of Economic Botany at Kew and also from the Kew Herbarium was also used. Special collections of material were also sent to Kew for our investigations.

Since the first edition appeared in 1950 there has been a considerable change in the attitude of our taxonomic colleagues to the use of histological characters. Some taxonomists were then reluctant to use microscopical characters at all, but with the passage of time it has now become generally recognized that characters made visible with the compound microscope can be just as valuable as macroscopic characters in taxonomic studies.

It has already been mentioned that an important innovation in our second edition is the broadening of the subject-matter by special contributions from new authors. Taking them in alphabetical sequence they are as follows:

(1) Emeritus Professor Katherine Esau of the University of California at Santa Barbara, has contributed a valuable section of the taxonomic significance of phloem characters. In writing this contribution Professor Esau has received assistance from her colleague Professor Vernon I. Cheadle, who placed some of his microscope slides at her disposal. Professor Esau's writings on phloem are already familiar throughout the botanical world, and her contribution on the taxonomic characters of phloem is very welcome.

(2) Dr. Leo J. Hickey, of the Division of Paleobotany at the Smithsonian Institution at Washington D.C., has provided a contribution on leaf architecture. Various attempts have previously been made to use the vascular pattern of the leaf and characters of the leaf outline and margin for taxonomic purposes. Attempts to use these characters have, however, been only partially successful owing to a lack of precisely defined characters. Leo Hickey and his collaborators have for some time been engaged on large-scale investigations of this problem. Hickey's contribution, which constitutes chapter 4, is a modified version of an important article which he published in the *American Journal of Botany*. It is clear that leaf architecture could become a very important component of systematic anatomy, and Hickey's contribution is therefore very greatly to be welcomed.

(3) Professor Richard A. Howard was until recently the distinguished Director of the Arnold Arboretum of Harvard University and is now Professor of Dendrology at the University. It may surprise many that over a long period of years he has made a spare-time study of the structure of petioles and nodes, the results of which are here presented in chapters 8 and 9. Professor Howard and C.R.M. have also collaborated in writing a brief account of the external morphology in Chapter 3.

(4) Professor William Theobald, with guidance from Professor R. C. Rollins of the Grey Herbarium

at Harvard University and the assistance of his former colleague Joseph L. Krahulik, has produced a novel classification of trichomes. Joseph Krahulik is now at Honolulu Community College, Hawaii. Professor Theobald, some of whose early research was undertaken at the Jodrell Laboratory, held a succession of academic posts in various parts of the U.S.A. while his contribution on trichomes was being written. He is currently Director of the Pacific Tropical Botanical Garden on Lawai, one of the Hawaiian Islands.

(5) Our final contributor to Volume I is Hazel P. Wilkinson. Dr. Wilkinson is particularly well qualified to write about the diagnostic characters presented by stomata, hydathodes, extra-floral nectaries, and other structures to be seen on, or forming part of, the plant surface. Her Ph.D. studies at London University, on the leaf surface characters of certain members of the Anacardiaceae make an admirable introduction to the more broadly based treatment of dicotyledons in general which she has contributed, under the general title of 'The plant surface' in Chapter 10. For some of her investigations she has made use of the scanning electron microscope.

If we may anticipate the appearance of our second volume, the preparation of which is already well advanced, readers may like to know that we shall have a specialist contribution from Dr. J. B. Harborne of Reading University to show how chemotaxonomy and the findings of histologists frequently go hand in hand. Dr. E. Chenery has very kindly placed at our disposal many of his data relating to the taxonomic value of aluminium accumulation.

Although many of the illustrations in the first edition have been repeated there are others that are new. Most of the new illustrations, both line drawings and photomicrographs, are original. Photomicrographs are used much more freely in this edition than in the first.

As with the first edition we gratefully acknowledge the wholehearted co-operation of our many colleagues and friends who have assisted in different ways. Once again we have made use of the slide collections at our respective institutions. At Kew C.R.M. continued for some time to have the support of Mr. F. Richardson, who was the chief anatomical technician at the Jodrell Laboratory until Mr. Richardson retired in 1975. At the Jodrell Laboratory there is also a very comprehensive card index to the literature. This was initiated and for many years maintained by C.R.M., but it has now been greatly expanded and made more generally useful by Miss Mary Gregory. Miss Gregory's index has been very valuable indeed in dealing with the bibliographical side of the work.

Dr. Hazel Wilkinson was employed at Kew as C.R.M.'s assistant by the Bentham Moxon Trustees, who also contributed towards the cost of typing. We are greatly indebted to the Trustees for their interest and support.

A further word of thanks is due to our outside contributors who have already been mentioned individually above. We feel that they have collectively formed an invaluable team and we are greatly indebted to them. It should be added that Professor Esau wishes it to be made known that the research and bibliographic work upon which the contribution on the phloem tissue is based were conducted with the aid of grants from the National Science Foundation, U.S.A. The phloem material for study and illustrations was generously provided by Professor Vernon I. Cheadle and some of the photographs were taken by Dr. Clyde L. Calvin.

Kew   C.R.M.
Oxford   L.C.
January 1979

ACKNOWLEDGEMENTS

The authors have much pleasure in thanking all those who have contributed in many different ways to this volume. A special word of appreciation is due to those new authors who have broadened the scope of the work by contributing their own chapters. Others, not only at Kew, but in many different countries have been involved in very rewarding discussions and correspondence, thereby adding substantially to the value of the volume. The very great help given by Miss Mary Gregory, bibliographer at the Jodrell Laboratory, in reading the proofs has been quite invaluable.

The authors also thank those authors and publishers who have so willingly given permission for illustrations in journals and other publications to be reproduced. Their names in alphabetical sequence are as follows: The publishers of *Annales des Sciences Naturelles*; Dr. F. Amelunzen; Professor Tyge W. Böcher and D.B. Lyshede; Dr. R. Crawford and the publishers of *Flora*; Dr. C.L. Cristóbal, M.M. Arbo, and the Editor of *Darwiniana*; Dr. E. Hausemann and Professor A. Frey-Wyssling; Dr. David H. Janzen; Dr. H. Reule and the Editors of *Flora*; Dr. Thomas L. Rost, Dr. Caroline Sargent; Dr. C.A. Stace and the Trustees of the British Museum (Natural History). The authors of the individual illustrations are indicated in the underlines.

CONTENTS

1

HISTORY OF SYSTEMATIC ANATOMY

PART I : GENERAL ANATOMY

C.R. METCALFE

Early investigators

It is not intended in this historical review to deal with the history of plant anatomy as a whole, but rather to trace more specifically the manner in which systematic anatomy has developed out of the general corpus of anatomical knowledge. Those who wish for a broader picture of plant anatomy are referred to the works cited on p.00.

The serious study of plant structure began with the investigations made by Nehemiah Grew (1641–1712) in England and Marcello Malpighi (1628–94) in Italy. Slightly before this time Robert Hooke (1605–1703), who was appointed 'Curator of Experiments' to the Royal Society in 1665, published his *Micrographia* (1665). This is a pioneer work of great importance, but it does not have much bearing on our present discussion because Hooke was only casually interested in plant structure. Anton van Loeuwenhoeck, a Dutch contemporary of Hooke, was likewise an extremely skilled but sporadic microscopist. On the botanical side he studied crystals in *Iris* and pitted vessels in secondary wood.

Nehemiah Grew, like Robert Hooke, was supported by the Royal Society and in 1672 subscriptions amounting to fifty pounds were collected from amongst the Fellows to enable him to hold a Curatorship for the anatomy of plants. Grew, who was initially a medical man, started to investigate plant structure while still in his twenties. The results of his morphological and histological investigations were published in three books (Grew 1672, 1675, 1682), of which the third entitled *The anatomy of plants* is the best known and most important. His studies dealt both with the naked-eye morphology of plants and with their cellular structure. His work confirmed the cellular structure of the plant body; he demonstrated the existence of vessels in wood, fibres in bark and parenchyma in the pith and cortex of the stem. Indeed the term parenchyma was first used by Grew. In stems Grew recognized 'the skin' (epidermis), 'the cortical body' (cortex), 'the ligneous body' (xylem), 'the pith' and 'the inserted pieces' (rays).

Grew's work was illustrated very effectively and it is indeed very remarkable that he gained so much

reliable factual information. Unfortunately his efforts to discover how structure is related to physiological function were far less satisfactory. This was probably because, with his medical background, he was obsessed with the idea that there must in plants be a pulsating physiological action corresponding to the heartbeat of animals.

Grew clearly demonstrated that each kind of plant that he investigated had its own distinctive structure by means of which it can be recognized. This in fact amounted to a pioneer essay in systematic anatomy. Indeed in Grew's writings the bifurcation of histology into taxonomic and physiological lines of investigation became apparent for the first time and both of these lines of enquiry have become the principal branches into which the subject is divided today.

Modern anatomical investigations also include ontogenetic and developmental studies which are considerably broadening our anatomical horizons. Greater magnifications, especially those that have resulted from the use of the electron microscope, are also clarifying much that was previously obscure. How much these approaches will eventually add to our understanding of structure in relation to botanical classification is still a matter of conjecture, because the proportion of the world's angiosperm flora that has been studied ontogenetically or with the electron microscope is still far too small to enable reliable conclusions to be drawn. However this may be, it is of considerable interest that the division of anatomy into the physiological, including the ecological, aspects on the one hand and taxonomic aspects on the other were already foreshadowed in the writings of Grew and Malpighi.

Marcello Malpighi, Grew's equally distinguished contemporary, was also a medical man who turned his attention to plant anatomy. His views were published in his *Anatome plantarum* (Malpighi 1675). A manuscript of this great work was sent by Malpighi to Grew in the hope that the Royal Society would arrange for it to be published. This indeed happened, but only after some delay, a circumstance which led some botanists to believe that Grew's results were merely copied from Malpighi. A careful perusal of Birch's

History of the Royal Society (1660-87), as well as the writings of Agnes Arber (1941 *a–c*) and W. Carruthers (1902) shows, however, that Grew was a man of the highest integrity and there are no grounds for the view that his results were second-hand. Indeed, Grew was at pains to tell Fellows of the Royal Society when Malpighi was in advance of him, as for example when Malpighi demonstrated the occurrence of spiral thickenings in vessels. Malpighi and Grew in fact appear to have had a high regard for each other. It must, however, have been difficult for them to communicate with each other because Malpighi could not read English and he was less able than Grew to express himself in Latin. The anatomical data recorded by Grew and Malpighi show a close agreement. There were, however, some topics such as the spiral thickenings in vessels which have already been mentioned, and the structure of stomata, especially those of *Nerium oleander* which occur in crypts, which were more fully understood by Malpighi than by Grew.

Malpighi and Grew had no students to perpetuate the high initial standards which they established for the study of plant anatomy. In consequence interest in the subject was minimal for the next hundred years after they died. There was, however, a somewhat colourful personality named John Hill (1716-75) who combined an interest in wood structure with being a playwright. In 1770 he published *The construction of timber*. He also designed and illustrated in his book an instrument which he described as a 'cutting engine', which was a primitive form of microtome which is said to have produced sections 1/2000 inch thick 'when the knife was sharp'.

From von Mohl to D.H. Scott

Interest in plant structure now passed to Continental Europe, but even here the standard of work was not very high until we come to the time of Hugo von Mohl (1805-72). He was Professor of Botany at Tubingen and author of *Principles of the anatomy and physiology of the vegetable cell* (1852) and other works. He was one of the first to devote attention to the contents of cells, rather than to the cell walls; he recognized protoplasm for the first time and called it sarcode. He also demonstrated that vascular bundles consist of xylem and phloem.

One reason for von Mohl's success was that he realized the need for microscopists to illustrate their work with their own drawings rather than to rely on professional artists who had no real understanding of

their own concerning plant structure. The employment of professional delineators had become fashionable partly in the mistaken belief that an anatomist could not be relied upon to make an unbiased drawing of what he saw under the microscope.

Von Mohl's work came at the outset of a time when the study of classical plant anatomy reached its peak during the second half of the nineteenth century and the beginning of the twentieth century. Many of the most famous anatomists of all time flourished during this period. One can recall such well-known names as those of Mirbel (1776-1834), Schleiden (1804-81), Nägeli (1817-91), Hartig (1805-80), Vesque (1848-95), Radlkofer (1829-1927), Haberlandt (1854-1945), not to mention Julius von Sachs (1832-97) and Anton de Bary (1831-88) through whose influence, as we shall see, the study of plant structure was reintroduced into England. It is beyond our scope to follow in detail the contributions that were made by all of these and other botanists of that period who were also very distinguished.

In pursuing the subsequent history of systematic anatomy we may start with Radlkofer (1829-1927), who was Professor of Botany at Munich. He was particularly interested in the Sapindaceae and his work on *Serjania* included one of the first serious attempts to make use of anatomical characters in classifying the members of an angiosperm genus. His anatomical data for the Sapindaceae are summarized in Engler and Prantl's *Die natürlichen Pflanzenfamilien* (Radlkofer 1896) and in Engler's *Das Pflanzenreich* (Radlkofer 1934). His importance in relation to this present book becomes even more apparent when we remember that Hans Solereder (1860-1920), who became Professor of Botany at Erlangen, was not only a pupil of Radlkofer, but also author of the great work of which an English translation was published in 1908 under the title of *Systematic anatomy of the dicotyledons* (Solereder 1908). Solereder's book formed the starting-point for the first edition of this present work.

While these important developments were taking place on the European Continent, a group of then youthful botanists in England, who found the teaching of botany in British universities in those days to be dull and uninspiring, were attracted to the laboratories of Julius von Sachs at Wurzburg and of Anton de Bary at Strasbourg. For an excellent account of these developments the reader is referred to Frederic Orpen Bower's *Sixty years of botany in Britain (1875-1935)* published in 1938 which gives the impressions of an eyewitness. One of the key personalities who was involved in these activities was Dukinfield Henry Scott, whose article entitled 'German reminiscences of the early eighties' (Scott 1925) makes interesting reading. S.H. Vines (1925), who like Scott also studied under von Sachs at Wurzburg, has published his recollections about life at the Botanisches Institut during this period. In their accounts both Scott and Vines make it clear that in those days students had a very long and onerous working day. They used microscopes that to modern eyes would seem very primitive, and microtomes were almost unknown. It is evident that von Sachs and de Bary were outstanding teachers whose influence on the study of plant anatomy was profound.

On his return to England, Scott held several botanical posts in the University of London before he accepted the honorary Keepership of the Jodrell Laboratory at the Royal Botanic Gardens Kew in 1892. He held this position with great distinction until 1906. F.W. Oliver (1935), who was one of the Jodrell workers at the time, has recorded his impressions of the congenial atmosphere that prevailed in the laboratory.

From D.H. Scott to the present time

Although Scott became best known as a very distinguished palaeobotanist, his influence on the development of systematic anatomy was also profound. It was at Scott's instigation that the great work by Solereder, already mentioned on p.00, was translated into English with the very capable assistance of F.E. Fritsch and L.A. Boodle. When the English translation was undertaken Scott made some additions of his own to the text, so that there are some points of difference between the German text and the English translation. It was because so much new knowledge had become available since Solereder's time that the first edition of this present work was written by Metcalfe and Chalk, a task which began in the 1930s some thirty years after the appearance of the English translation of Solereder's *magnum opus*.

It is a notable fact that although Solereder's studies were unavoidably incomplete they were seldom inaccurate; indeed this applies to most of the anatomical investigations made at the end of the nineteenth century and the beginning of the twentieth century. The incompleteness of the factual data may on occasions have led to erroneous conclusions, but when the first edition of this present work was written very few alterations had to be made concerning the facts

presented by Solereder. Systematic anatomy was, however, still in its infancy in Solereder's time, as was pointed out by D.H. Scott (1889). Scott rightly emphasized that one cannot draw sound taxonomic conclusions by examining sections of a few herbarium specimens. On the other hand he also indicated that many students of systematic anatomy have destroyed their own chances of making good progress by becoming too greatly absorbed in detail. Both of these comments are as valid today as they were when Scott made them, and he did well in urging research workers to view their own investigations with a well-developed sense of perspective.

The events so far described here and the personalities who have been mentioned have been selected so as to trace the history of systematic anatomy from the days of Nehemiah Grew and Marcello Malpighi until the present time. It is to be hoped, however, that our readers will realize that many other botanists have also been involved in developing the subject. Indeed at the present time, although systematic anatomy constitutes only a relatively small part of botanical science, the subject is being actively pursued in many different countries. The important contribution from German plant anatomists has already been made abundantly clear. For further details concerning the German work, readers should consult particularly the works by von Sachs (1906), Green (1914), and Schmucker and Linnemann (1951). Although the contribution from our French colleagues is less directly connected with the reintroduction of plant anatomy to England from the European continent at the end of the nineteenth century, it nevertheless constitutes an important component of anatomical lore. Readers particularly interested in the French contribution are referred to the very informative article by Hocquette (1954).

The study of plant anatomy was introduced to the U.S.A. comparatively recently, one of the best known of the pioneer workers being Jeffrey (1866–1952). I.W. Bailey was one of the most notable of Jeffrey's students and his publications on many aspects of plant anatomy constitute a landmark in the subject in the U.S.A. This is due to the integrity of Bailey as a man, to the care with which his work was done, and to his profound influence as a teacher of research students who have so ably carried forward the traditions which Bailey established. Happily many of these students are still active workers today. Some of Bailey's more important publications have been collected together in book form at the instigation of his students (Bailey 1954). All this, and much more besides, has been very well presented by Sherwin Carlquist (1969).

A strong tradition for research in plant anatomy, much of it with a taxonomic bias, has also been established in India. P. Maheshwari (1904–66) who was Professor of Botany at Delhi, played an outstanding part in promoting the study of comparative anatomy which he recognized as being closely linked to his own special field of embryology. An historical account of these Indian developments (1939–50) has come from Chowdhury, Rao, and Mitra.

Although plant anatomy has been and is being actively studied in other countries besides those that have just been mentioned, the writer is unaware of any historical surveys describing the development of the subject in any one of them. In conclusion it may be mentioned that there is a specially long tradition for anatomical investigation in the Netherlands, and that Scandinavian, Swiss, and Japanese botanists have added in no small measure to our knowledge of the subject. In recent years there has been an increasing spate of publications on various aspects of plant anatomy from the U.S.S.R.

It is sometimes suggested that we already know so much about the classical aspects of plant anatomy that there is no need to pursue the subject any further. On the other hand there are other botanists who bewail the lack of investigators in the field of plant anatomy. The present revision has shown two things. Firstly, that so many angiosperms in the world remain unexamined histologically that there is good reason for all available workers to press on with this task. The diversity of structure is as great as the diversity of taxa and for this reason there is little likelihood

that future exercises in systematic anatomy will be purely repetitive. The second main point is that the present output of anatomical publications that are either mainly concerned with or which have a bearing on taxonomy is so great and the journals and books in which they are published are so scattered that very few botanists anywhere can be fully aware of their existence and, still less, fully informed about their content.

Literature cited

Publications relating to the botanists concerned

Arber 1941*a,b,c*; Birch 1660; Bower 1938; Carlquist 1969; Carruthers 1902; Chowdhury, Rao, and Mitra (No date on the reprint available to the writer); Green 1914; Hocquette 1954; Oliver 1935; Radlkofer 1896, 1934; Sachs 1906; Scott, D.H. 1889, 1925; Vines 1925.

Publications dealing with anatomical structure

Bailey 1954; Grew 1672, 1675, 1682; Hill, J. 1770; Hooke 1665; Link 1840; Malpighi 1675, 1679; Mohl 1852; Schmucker and Linnemann 1951; Solereder 1899, 1908.

Suggestion for further reading

Metcalfe in Baas, Bolton, and Catling (eds.) 1976: history of the Jodrell Laboratory as a centre for systematic anatomy.

PART II : THE HISTORY OF WOOD ANATOMY

L. CHALK AND C.R. METCALFE

There is no clear line of demarcation between the structure of wood and the general histology of the whole plant. Nevertheless we find today that wood structure is a more-or-less self-contained discipline, which is studied for the most part by foresters and wood technologists in institutions that are separate from the plant science laboratories, formerly known as departments of botany, where most general studies on plant structure are or have been undertaken. This tendency for studies in wood anatomy to be segregated from anatomical studies dealing with other plant tissues has led wood anatomists to develop their own special modes of description and vocabularies of descriptive terms. Moreover the study of wood struc-

ture has become largely an applied science, devoted to investigating the relationship between structure and the working properties of commercial timbers and the accurate identification of timber specimens. Wood anatomy also plays an essential part in archaeology, palaeobotany, and forensic science. It was perhaps inevitable that increasing specialization was bound to occur during the historical development of the study of plant structure. The separation has led to great advances in wood technology and for this reason it is to be welcomed, but the reduced contact between the fundamental and applied sides of anatomy has also had its weaknesses. Fortunately there seems to be an increasing tendency for the two branches to come together again. This is at least partly because of a greatly increased realization that the study of taxonomy (including phylogeny) cannot effectively develop any further if it is confined to herbarium studies of the restricted type that became traditional before the importance of medium and high magnifications, as well as of experimental investigations and other approaches, were fully appreciated by taxonomists.

The history of wood anatomy shows how it gradually came to be recognized that the naming of a reference wood specimen cannot be accepted as authentic unless there is a voucher herbarium specimen from the same tree. This one might think would be a simple matter to arrange, but there are more difficulties about the correlation of wood specimens and herbarium vouchers than is sometimes appreciated. In the first place elementary mistakes arise when labels are attached to the wrong specimens. This type of fault is usually relatively easy to detect because the structure of the wood and the identity of the herbarium specimens are usually different enough for it to be at once obvious that something is wrong. However, it is not always easy to ensure that there is effective collaboration between the forester who collected the material, the wood technologist, and the appropriate taxonomic expert in the herbarium. Even when these difficulties have been overcome there may be differences of opinion among taxonomists concerning the precise taxonomic designation of the herbarium vouchers. Then again, with the passage of time, taxonomic opinions may change, thereby causing further problems for the wood anatomist. It can never be assumed that the names given to herbarium specimens are fixed and immutable. It follows that the names of the wood specimens must also vary with the ebb and flow of taxonomic change. The most that the wood anatomist can hope for is a reasonable period of stability in naming, and that taxonomists will avoid making name changes without good reason.

In spite of these difficulties and problems it is evident that reference collections of woods backed by herbarium vouchers serve as a tool which is essential to the wood anatomist and herbarium taxonomist alike. The two classes of material must be kept together, and administrators and others who think of breaking up or disposing of wood collections should be actively discouraged from doing so. It has always been taken as axiomatic by taxonomists that it is essential to maintain herbaria complete and in good working order, especially when the specimens have been critically examined by well-known experts. Wood collections

clearly merit the same respect.

The older wood collections are often worth preserving even when they were assembled before the days when the value of voucher specimens was fully appreciated. During the nineteenth century many wood specimens from the tropics and the southern hemisphere were sent to various European countries when the economic plant products from these areas were being listed and investigated for the first time. Many of the specimens became incorporated in our national museums, such as the museums of economic botany at Kew, which were initiated by the first Director, Sir William Hooker. Because these specimens are not supported by herbarium voucher specimens, the belief has grown up that they must be wrongly named or at least unreliable. No doubt this is to some extent a valid contention, but it must be remembered that the specimens were for the most part collected by botanists of repute who were as well informed as anybody in their day about the botanical identity of the woods in which they were interested. Most of these specimens are correctly named at least to the genus level, except where the concepts of the genera have undergone great changes since the time when the specimens were collected. Where species names have remained stable the older wood specimens may also be correctly named to the species level as well. This opinion of the older Kew wood specimens has been formed as the result of one of us (C.R.M.) having worked with the samples in question for many years (see Metcalfe ∙ in Baas, Bolton, and Catling 1976). Mistakes have no doubt been made, but it is very easy to decry the work of our predecessors who were frequently better informed than we are apt to believe.

As we shall see later in this present discussion, there has been a tendency for historically important wood collections to be stored in unsuitable places where they have received inadequate curatorial treatment. Wood specimens rapidly deteriorate in such circumstances, and much damage has been done. This is especially evident from two publications by Stern (1957, 1973). The earlier of these two communications is a world-wide guide to institutional wood collections. The list was based on information received in reply to a questionnaire sent out by Stern himself to the authorities in charge of all the wood collections that were then known to him. The questionnaire was drawn up with the assistance of some of the leading authorities on wood structure at the time. Stern's 1973 article, which refers to the principal wood collections in the U.S.A., is a document of the greatest historical interest because it gives particulars of the circumstances in which the individual collections were initiated. He also comments on the number of specimens in each collection, and on the effectiveness with which they were curated. The largest collection then was, and still is, at the Forest Products Laboratory at Madison, Wisconsin, with 100 000 specimens. (This includes specimens previously in the Field Museum.)

Size alone is not the only criterion by which the value of wood collections should be judged, and in Stern's opinion the Record Memorial Collection, which at the time was still at Yale University, was given pride of place. The nucleus of this collection already existed in 1901, but the specimens were largely destroyed by fire. A fresh start was made in 1905, but it was not until S.J. Record was appointed Professor of Forest Products in 1910 that the collection began to expand rapidly. By 1928 it contained 1000 specimens, and when Record died in 1945 the number had risen to 40 000. Eventually the 55 000 specimens in the collection were transferred to the Forest Products Laboratory at Madison where they are still housed. The value of the collection was due mainly to Record's personal supervision in assembling the samples, and to the notes about many of them that he published.

Of particular interest in our present context is the liaison that was established between Record and Chalk, who had initiated a fine collection of permanent microscope slides for use in conjunction with the well-known and increasingly important collection of wood specimens at the Imperial Forestry Institute at Oxford (England). In the 1920s there was a marked improvement in the technique for preparing wood sections through a more complete realization of what could be achieved by using the sledge microtome. At the same time improved methods of filing the slides had been devised and so the way had been opened up for developing reference collections of microscope slides, of which the one at Oxford was a particularly notable example. It was in these circumstances we are told by Stern: 'In return for specimens of wood, Chalk arranged to have permanent slides prepared of them which were returned to Record. These slides formed the basis of the large collection associated with the woods at Yale.'

At that time B.J. Rendle, together with his colleagues on the staff of the Forest Products Research Laboratory, which was subsequently established at Princes Risborough, was sharing laboratory accommodation with the staff of the newly formed Imperial Forestry Institute. The staffs of both departments had temporary accommodation at Oxford in the old Forestry School building and in Keble Road.

Whilst the staffs of the F.P.R.L. and the Imperial Forestry Institute were in such close contact, Professor C.C. Forsaith of Syracuse University spent a sabbatical year at Oxford. Forsaith's visit had a marked influence on the study of wood structure at both institutions because of his knowledge of how to organize a wood collection and of the then most recent techniques of cutting sections of wood.

The similar methods which were developed at the F.P.R.L. and Imperial Forestry Institute led to remarkably close collaboration between the two departments in subsequent years. Since Oxford and Princes Risborough are only some 20 miles apart, effective co-operation between the two departments was easy to maintain.

It may be noted in passing that it was Chalk's slide collection at Oxford which served as a model for the reference collection of microscope slides at the Jodrell Laboratory at Kew. However, the Kew collection was extended to include slides of other tissues besides wood.

I.W. Bailey's work on wood structure at Harvard University is so well known that it is not surprising that the specimens which he assembled are of particular interest. As one of us (L.C.) points out, I.W. Bailey insisted on the certification of all wood specimens used for research purposes by voucher specimens in the herbarium. In addition to this he recognized the extent of the structural variation that occurs in different parts of an individual tree and also in different trees belonging to the same species. This variation makes it necessary to examine many more than a single specimen of each species under investigation. By means of his collections Bailey and his pupils were able to carry out researches that revolutionized many ideas then current about the significance of the wood characters and of plant development generally. His work on the significance of the length of the cambial initial as a measure of phylogenetic status is well known (Bailey and Tupper 1918). Bailey's pupils extended this work to cover many of the tissues that enter into the composition of wood, and this subject is discussed more fully in Vol. II. It is perhaps not always realized that Bailey also pointed out that the sizes of cambial cells and the methods by which they divide are reflected in fossilized woods as well as in present-day timbers.

Stern, like Chalk, recognized the scientific importance of Bailey's collections, and it is encouraging to learn from an article by Wetmore and Barghoorn (1974) how Bailey's specimens have been arranged in a modest but well-appointed laboratory specially designed and constructed to take them. One of us (C.R.M.) had an opportunity to see the Bailey specimens in their new surroundings shortly before the new laboratory was officially opened and it was very encouraging to see how much practical common sense had been applied to the new arrangement. The specimens, both modern and fossilized, were there, with microscopes and other laboratory facilities close at hand to enable them to be examined.

Although this account refers mainly to some of the American and British wood collections, it must be realized that there are important collections in many other countries as well. For further particulars the reader is referred to Stern's 1957 and 1978 articles and to recent numbers of the *Bulletin of the International Association of Wood Anatomists*. A notable example of a young but promising department devoted to the study of wood structure (and of systematic anatomy in general) is the one at the Rijksherbarium at the University of Leiden (Holland) which is being developed by Dr. P. Baas and his associates.

After the First World War there was a great expansion of forest services in the tropics. These developments afforded much better opportunities for the collection of authentically named wood samples. It was during this period of active expansion that the International Association of Wood Anatomists came into being. An organization was felt to be needed to ensure that those working in the field of wood structure could share each others' interests and research problems. There was also the need for a glossary of descriptive terms that could be generally accepted. Finally it was hoped that a new association would facilitate the exchange of specimens.

The first steps which led to the formation of the International Association of Wood Anatomists were taken by L. Chalk and B.J. Rendle (both of whom, as we have already seen, were at that time working at Oxford) and S.J. Record of Yale University. Record not only controlled the important wood collection at Yale to which we have already referred, but he had recently started *Tropical Woods*, a technical journal devoted to the furtherance of the knowledge of tropical woods and forests. This journal, started in 1925, served as a means of communication between wood anatomists until Record died in 1945. Quite naturally *Tropical Woods* served as a medium for the exchange of ideas about the formation of an international body to study wood (Record 1930–45), and the first formal meetings to discuss the proposition took place at

Cambridge (England) during the International Botanical Congress held there in 1930. These discussions led to the setting up of an organizing committee of which the members were to report to the Congrès international du bois et de la sylviculture that was to be held in Paris from 1 to 5 July 1931. The association, now known to all wood anatomists as the I.A.W.A., was formally constituted on 2 July 1931 with members from 16 countries. Since its inception the fortunes of the I.A.W.A. have fluctuated, but there can be no doubt of its continuing value and indeed it is true to say that it was never more active than it is today. Evidence of this is afforded by the present high standard of the I.A.W.A. Bulletin, currently published in the Netherlands from the Rijksherbarium at Leiden. In 1975 the I.A.W.A. membership was spread over 36 countries, which is a considerable advance on the initial sixteen (*I.A.W.A. Bulletin* 1975, 2, p.36). Probably the association's most important achievement has been to publish an international glossary of terms used in describing woods, which is well known and widely used (*Trop. Woods* No. 107 (1957)).

There are now numerous well-known reference books dealing with the timbers from individual countries and geographical areas. These will be mentioned repeatedly throughout this second edition and there is no need to consider them in detail here. A few that have become landmarks in the historical development of the subject may be briefly mentioned. In *A manual of Indian timbers* by Gamble (1902) the first really comprehensive attempt was made to describe the timbers of the Indian Subcontinent. When this book is compared with more modern publications there are obvious weaknesses, chiefly because of the inadequate section-cutting techniques that were in use at the time. In consequence it was eventually succeeded by Pearson and Brown's *Commercial timbers of India* (1932) which was followed by Chowhury and Ghosh (1958, 1963) and Rao and Purkayastha (1972). In Australia there was Baker's *Hardwoods of Australia* (1919). Amongst other Australian wood anatomists the well-known names of Dadswell, Wardrop, and Chattaway must be mentioned, and also that of Patel from New Zealand.

One of the most outstanding treatises concerning the woods from a definite geographical area is that on Indonesian woods in six volumes by Moll and Janssonius (1906–36). The work was based on 2400 specimens representing 1070 species of 380 genera from 80 families. This account of Indonesian woods is particularly noteworthy for the accuracy and detail of the descriptions compiled by Janssonius. The descriptions were quite outstanding at the time and are still models of descriptive wood anatomy. Janssonius (1952) later published a key to Javanese woods as a supplement.

A particularly high descriptive standard was also maintained in *Forest trees and timbers of the British Empire* edited by Chalk and Burtt Davy (1932-9). Four parts of this important work, all dealing with timbers from different parts of Africa, were published, after which it is greatly to be regretted that the publication was discontinued. If this work had been completed it would have stood out as a monumental reference book on Commonwealth timbers.

In spite of the large amount of work that has been devoted to the structure of wood, only a small proportion of the woody plants that exist today have as yet been investigated. Brazier (1968) has pointed out that the factor limiting the wider use of wood anatomy as an aid in taxonomy is the relatively small amount of material available for study, compared with the vast collections of herbarium specimens that are deposited in our national herbaria. This points to the need for us to give up the commonly held belief that our knowledge of wood structure is already complete.

In a more recent communication Brazier (1975) has surveyed research activity in wood anatomy since 1900, 'with special reference to the pattern of development in (1) descriptive and comparative anatomy, (2) structure in relation to working properties and (3) the influence of growth on wood structure'.

Throughout the history of wood anatomy, ecological influences have been investigated rather spasmodically, but in recent years the ecological approach has received fresh attention. One important example is provided by the recent publications of Carlquist, and readers should note especially his *Ecological strategies of xylem evolution* (Carlquist 1975). While fully accepting the views of Bailey and his followers concerning the phylogenetic significance of the dimensions and the nature of the pitting in xylem elements (see Vol. II), Carlquist nevertheless feels that environmental and physiological factors must have some influence on the structure of the cells that enter into the composition of wood. Other workers, particularly in Holland, have also expressed similar views. We shall revert to this topic more fully later in this present work. The subject is mentioned here mainly to emphasize that a return to the ecological and physiological approaches to the study of wood structure is a recent historical development.

The scanning electron microscope, usually referred to as the s.e.m., is assuming increasing importance in the study of wood structure. It is a far cry from the days when timber identification was based solely on characters made visible with a simple hand lens to the revelation of the structure of, for example, vestured pits by the s.e.m. However, let it not be supposed that the s.e.m. will replace the simple hand lens or the compound microscope completely. Each has its place in our studies and they will probably remain complementary for many years to come.

We are aware that the above history of the study of wood anatomy is far from being fully comprehensive. For example, the many interesting investigations that have taken place or are in progress in such areas as China, Japan, Taiwan, or the U.S.S.R. have not been mentioned. We do not feel, however, that we have studied the historical details of these researches in sufficient depth to provide a more adequate account. This is largely due to language problems, and it is to be hoped that, with the passage of time, these difficulties will be overcome. We are also aware of the work in progress in South American countries. These investigations are of particular interest because of the many species of which the wood structure is being or has been described for the first time.

Reverting to the Indian subcontinent, together with Burma and Sri Lanka, we must not forget Chowdhury's (1968) important history of the study of wood structure in these areas. Chowdhury's studies of growth rings in tropical trees (Chowdhury 1964) also constitute an important contribution to the history of wood anatomy. There are many other regional surveys of timber structure which have some content of historical interest. These will be mentioned elsewhere in this treatise, and a few suggestions for further reading are given at the end of this chapter.

To revert to the theme with which this discussion started: it would seem that the facts of wood structure will gradually be incorporated in the general corpus of histological knowledge. It is now generally agreed that histological details, when carefully selected, can be accepted as taxonomic characters, and in this way our knowledge of the structure of wood will become part of the warp and woof of the fabric of which taxonomy consists. Current developments seem at least to be pointing in this direction.

Literature cited

Bailey and Tupper 1918; Baker, R.T. 1920; Brazier 1968, 1975; Carlquist 1975; Chalk and Burtt Davy (eds) 1932–9; Chowdhury 1964, 1968; Chowdhury and Ghosh 1958, 1963; Gamble 1902; *International Association of Wood Anatomists, Bulletin* 1975; Janssonius 1952; Metcalfe *in* Baas, Bolton, and Catling (eds) 1976; Moll and Janssonius 1906–34; Pearson and Brown 1932; Rao, K.R. and Purkayastha 1972; Record 1930–45; Stern 1957, 1973, 1978; Wetmore and Barghoorn 1974.

Suggestions for further reading

Brazier 1975: a general history of wood anatomy;
Desch 1941, 1954: Malayan timbers;
International Wood Collector's Society Bulletin (ed. Frost);

Kanehira 1921*a*: Japanese woods;
— 1921*b*: Formosan woods;
— 1926: woods of the Japanese Empire;
Record 1934*a*: wood anatomy and taxonomy;
— 1934*b*: identification of North American timbers;
Record and Hess 1943: New World timbers;
Rehder 1911–18: The Bradley bibliography — this contains numerous early anatomical references in chronological sequence;
Stern 1967: wood collections of the world;
— 1973: future uses of wood collections;
— 1976: multiple uses of wood collections;
— wood collections of the world;
Tortorelli 1956: Argentinian woods;
Wetmore, Barghoorn, and Stern 1974: rejuvenation of systematic wood anatomy by the wood collection at Harvard University.

PART III : THE SCANNING ELECTRON MICROSCOPE IN RECENT SYSTEMATIC PLANT ANATOMY

D.F. CUTLER

The scanning electron microscope (s.e.m.) is an ideal instrument for examining surfaces. Because it is relatively easy to operate and has a magnification range of about 10× to 200 000×, it has been used in a wide variety of studies. In systematic plant anatomy, it is not so much the resolving power (currently about 7 nm) or the high magnifications that make the s.e.m. such a useful tool, as the depth of field obtainable. This is about 500 times better than that of the light microscope. Consequently images can give a very good impression of the three-dimensional nature of a specimen. If stereo pairs of photographs are prepared and correctly viewed, this effect is very marked.

Preparation of botanical material is quite straightforward. Some types of material can be viewed while still alive, for example the shoot apices of many species and leaf surfaces of the more mesophytic plants. However, these quickly deteriorate in the vacuum, and photographs must be taken within about 5 minutes, before the specimen has time to collapse. Most plant material is much more suitable for observation in the s.e.m. after pre-treatment, which in its simplest form involves normal fixation in FAA (a mixture of formalin, alcohol, and acetic acid), selection of the area or section to be viewed, and dehydration through an alcohol series to air. The last stage is carefully conducted by keeping the specimen flat on a microscope slide inside a desiccator. Alternatively, the fresh specimen may be freeze-dried or, preferably, dried by using the 'critical point technique' in special apparatus. The dry specimen is stuck to a stub appropriate to the microscope, with an electrically conducting glue or tape. Alternatively a conducting paste may be used. A very thin film of conducting material is then deposited on the surface of the specimen. Gold, gold–palladium, or either of these preceded by a carbon coat are the usual treatments.

It has sometimes been said that little new has been revealed by the s.e.m., and that it is possible with intelligent use of high-quality optical microscopes for the trained observer to see adequately all the features of importance to the systematic anatomist. There is an element of truth in this view, particularly if a darkfield, epi-illuminating light microscope is used. However, photographs taken on even the best optical equipment cannot match those from an s.e.m. for depth of field. Also, features at the limits of resolution of the light microscope can be readily seen with the s.e.m. In addition the s.e.m. is normally used to view only the surface of a specimen. I will show later how this can be a great advantage in the study of some leaf surfaces. Above all, it is the ease with which even the untrained observer can interpret photographs taken with the s.e.m. which makes it such an important tool for the plant anatomist. Certainly students find the three-dimensional, 'solid' appearance of material such as secondary xylem very easy to understand. They can then look at the thin, flat sections on their microscope slides with renewed interest and comprehension.

Secondary xylem

Apart from the ease with which the interrelationships between cell types can be observed, the main benefit in using the s.e.m. in comparative studies of secondary xylem is the clarity by which structures such as vestured pits can be observed. These are rather obscure at the best of times when seen in the light microscope, but the s.e.m. has made it possible to distinguish readily between the different types of vesturing. Other features of wall ornamentation can also be seen readily, such as, for example, warts and intercellular connections in the intercellular spaces. Spiral (helical) thickenings are also quickly distinguishable from splits in the vessel element or tracheid walls. Perforation plates are also remarkably clear. It is often not necessary to cut wood surfaces for viewing but, if this has to be done, it is best to use a new safety razor blade. Simply splitting a specimen in the correct planes can produce very informative surfaces. Portions of dry twigs from herbarium specimens when split need only to be stuck to a stub and coated before viewing.

A secondary application is found in the identification of small carbonized fragments of wood, and mineralized wood from archaeological sites. Characters which can be very difficult to see by other means are often well displayed in the s.e.m.

Leaf surfaces

In some genera hairs are very valuable as diagnostic or taxonomic characters (see Chapter 5). Many hairs are rigid enough to be seen on leaves which have been fixed and dehydrated through an alcohol series. Glandular hairs are often best seen in living material, or specimens prepared in the critical point apparatus. The depth of field of the s.e.m. at low magnification is used to full advantage on such material. At higher magnifications, details of warty hairs can be seen.

Stomata with complex cuticular flanges make good subjects for study in the s.e.m. They are readily interpreted in a much shorter time than is necessary when using the light microscope. Often it is better to use the light microscope to study subsidiary cells, since these may not show clearly in the s.e.m. (see Plates 2 and 3).

Patterns in the cuticle, or cuticle/outer cell wall are often of interest to the systematic anatomist. In some genera, e.g. *Aloe*, to take an example from amongst the monocotyledons, the cuticle-to-cell-wall interface is not clear cut, but short strands of wall material penetrate the cuticle and vice versa. When viewed from the surface in the light microscope this zone appears granular and obscures the cuticular pattern. Since the s.e.m. at low accelerating voltages produces an image of the leaf surface with little beam penetration, a clear picture is obtained and cuticular patterns are quite distinct (see also pp. 140–56 and Plates 4 and 6).

Wax on leaf surfaces can also be studied in the s.e.m. The three-dimensional form of crystals or particles shows well. It is suspected that some waxes which have a low melting point may not be stable in hot sun and consequently more than one type of crystal could appear in those species which have such waxes. However it is probable that most waxes are not of this type. While doubt persists, caution should be exercised about the use of wax types for taxonomic purposes, (see also pp. 158–61 and Plates 9 and 10).

Sections

Thick sections will often stick to s.e.m. stubs without the use of additional adhesives, if dried on from absolute alcohol. Certain cell types, e.g. transfer cells (see p. 73), tracheary cells in nodal complexes (see chapter 8), stomata (see pp. 97–116), can be more readily studied by this method than with the light microscope. However, for most taxonomic purposes little advantage is gained and the light microscope is perfectly adequate for systematic studies.

Apical meristems

The apical meristems of many plants are remarkably robust, and can be studied both alive and dried in the s.e.m. Comparative, systematic studies of apical meristems are not common, but there is a great potential for such research. Also, developmental studies on apices can be clarified by using the s.e.m.

Seeds and Pollen

Although outside the scope of this volume, it must be recorded that the s.e.m. is a very useful tool in the study of seeds and pollen for systematic, developmental and physiological studies, and in the study of breeding systems.

The wide range of application of the s.e.m. in systematic anatomy is evident. It may not be so obvious but it is equally true that the s.e.m. is of most systematic value when it is used in conjunction with optical microscopy and sometimes transmission electron microscopy. In fact it sometimes needs to be used in conjunction with the hand lens. It can happen that new techniques and newly designed instruments are used for the sole purpose of showing them off. This does not apply to the s.e.m., which has now been in use long enough for its value to the taxonomist, and in particular to the systematic anatomist, to be fully appreciated.

2

PURPOSES OF SYSTEMATIC ANATOMY

C.R. METCALFE

The chief aim of those who study systematic anatomy is to relate the structure, particularly of the vegetative organs, to the taxonomic classification of the plants in which the characters are exemplified. The angiosperms show such a wealth of external morphological characters of proved taxonomic value, which can be readily observed with the naked eye or simple hand-lens, that taxonomists have not always been attracted by the rather time-consuming processes that histological investigation demands. It is evident, however, that the numerous problems in taxonomy that still remain unsolved are causing the traditional methods of herbarium taxonomists to be supplemented to an increasing extent by laboratory disciplines such as anatomy, cytology, palynology, chemotaxonomy, etc. Indeed the broadening of the concept of what taxonomic investigation should entail is one of the outstanding developments at the present time.

Of the supplementary disciplines that accompany herbarium botany, the study of the histology of the vegetative organs is one of the oldest. The main problem about it today is that although numerous anatomical investigations have been undertaken during a long period of years, and many more are currently in progress, broad surveys of the angiosperms in histological terms are still very incomplete. In the first edition of this present work as well as in the previous studies by Solereder (1908) an attempt was made to bring together and summarize what was known of the subject at the times when these books were written. Since then new publications have been so numerous that a fresh attempt to obtain an overall picture of the anatomy of the vegetative organs as a contribution to the classification of the angiosperms is badly needed. This synthetic approach, which is the main aim of this work, is difficult to achieve, partly because the factual data that have to be taken into account are so numerous, scattered throughout a considerable number of journals, and written in so many languages that there can be very few botanists who are thoroughly familiar with them all. Secondly, histological characters vary in their taxonomic importance and significance as one passes from family to family. This last fact should cause no surprise since the same principle has, for many years, been known to apply in using the traditional external morphological characters in taxonomy. In view of the numerous gaps in our factual knowledge about the structure of flowering plants it is unfortunate that there are some botanists who misunderstand the present position sufficiently to believe that further anatomical investigations along traditional lines will be unrewarding. Here I would agree that our knowledge of the fundamental principles of plant construction as revealed by the light microscope appears to be relatively complete, but on the other hand there is still much that remains unknown about the various ways in which the organization of cells and tissues is expressed in passing from one taxon to another. Little further progress can be made towards a broad understanding of angiosperm taxonomy unless and until the gaps in our histological knowledge have been filled in. The fulfilment of this ambition is one of the main and most urgent purposes for which research in systematic anatomy is needed at the present time.

The applications of systematic anatomy in solving practical problems have so often been stressed that there is no need to repeat them in detail. It is evident that any exercise that involves the identification of vegetable material when it is in a fragmentary condition, partly decomposed, or when reproductive organs are not available, can be achieved only by resorting to the methods of comparative histology. This applies in confirming the identification of economic plant products ranging from timbers to foodstuffs as well as crude drugs of vegetable origin and natural plant fibres. Adulterants and substitutes can also be detected. Passing to subterranean parts of the plant, it may be noted that the correct identification of tree roots can often be achieved by microscopical investigation. This is important when, for example, field drains or water mains are obstructed by roots of an undetermined species in the surrounding vegetation or where the growth of tree roots is causing damage to the foundations of buildings. Then again there are the often intricate problems of establishing the identity of archaeological and palaeobotanical specimens. Finally it should be remembered that there are numerous problems in forensic science that can be solved by histological methods. It is not our primary purpose in this work to discuss these applied problems in detail, but rather to survey some of the principles that underlie their solution. It is nevertheless hoped that the attention of students and research workers will be drawn to the fact that the principles of systematic anatomy should be used to solve these practical problems. My personal experience leads me to believe that students today often remain unaware of the value of practical applications of systematic anatomy. This is to be regretted at a time when there is so much interest, especially among the younger botanists, in solving practical problems.

Before leaving this initial consideration of the usefulness of systematic anatomy it may be as well to remind ourselves that plant structure is produced by biochemical and physiological change. Since plant structure is taxonomically diverse, this at least opens the question of whether biochemical and physiological processes and pathways may not themselves be more diverse than has sometimes been supposed. It has so often been implied in textbooks that there is a great uniformity of physiological and biochemical processes throughout the whole range of the angiosperms. No doubt this is true in a general way, but one cannot fail to ask oneself whether the structural diversity in the plant kingdom may provide evidence in concrete form that the physiology and biochemistry that gives rise to this structural diversity may itself lack uniformity. The relationship between structure and function is very intimate and it seems not unlikely that the diversity of structure that is portrayed in this present work should direct attention to taxonomic groups in which metabolic diversity can be expected. For example, it is a fairly recent discovery that photosynthesis in leaves with 'Kranz' structure (see p. 72)

takes place along a different pathway from that in leaves in which 'Kranz' structure does not exist. Here we have definite evidence of a correlation between the arrangement of the photosynthetic cells in the leaf and the manner in which photosynthesis takes place. Furthermore it should be noted that 'Kranz' structure has a precise but limited taxonomic distribution among the dicotyledons. We might go on and enquire whether the presence of laticiferous secretory canals or other structures that are of restricted occurrence among the dicotyledons may not similarly provide evidence of metabolic diversity. This seems to me to be a matter of fundamental interest, for it provides a focal point at which physiological and anatomical studies meet. During the latter part of the last century there was no clear dividing line between the study of structure and the study of physiology, and the recognition of this mutual dependence of the two aspects of the same subject was one of the strengths of botanical investigation during that period. Since then, investigations of structure and function have tended to go their separate ways. Fortunately the study of the physiology of cells and tissues has become fashionable again and it is to be hoped that the contents of the present and future volumes of this present work will be useful to those who are interested in function, as well as to those who are concerned with structure and classification.

Suggestions for further reading

Bailey 1953a; Metcalfe 1953, 1954.

3

EXTERNAL MORPHOLOGY
PART I
C.R. METCALFE

It is clearly beyond the scope of this work to cover in full the great range of differences in external morphology exhibited by the dicotyledons. On the other hand without a preliminary discussion of certain aspects of plant form it is difficult to appreciate the full significance of some of the histological features with which the systematic anatomist has to deal. For this reason the following topics are discussed: (1) Trees, shrubs, and herbs; (2) Axis and leaves; (3) The leaf; (4) Stipules; (5) Glands; (6) Phyllodes; and (7) Epiphyllous inflorescences.

Trees, shrubs, and herbs

Although the distinction between trees, shrubs, and herbs is well understood in a popular sense, the differences between these categories become less obvious when the morphology of the plants is taken carefully into consideration. In a general way trees are much larger than shrubs but both groups share the common character of being woody. Woodiness, however, is a character that is difficult to define. It depends partly on cambial activity being sufficiently prolonged to produce enough secondary xylem to justify the classification of the plants concerned as trees or shrubs. However, if, for example, a Baobab (*Adansonia digitata*) is compared with a well-grown ash (*Fraxinus excelsior*) the Baobab with its grotesquely thickened trunk stands out in marked contrast to the more stately trunk of the ash. The precise definition of a herb is even more elusive. In a general sense it can be taken to mean a plant that is not woody. By some a herb is envisaged as a plant that has no persistent parts above soil level, but this is unsatisfactory because it does not take perennial herbs into account. Moreover there are some plants that are referred to as woody herbs.

Raunkiaer (1934) realizing that the popular terms herb, shrub, and tree lack any precise meanings, put forward a classification for the growth forms of angiosperms on a more logical basis. Raunkiaer's classification rests on differences in the duration of the life of the shoots and on the position in relation to ground level of the resting buds and their mode of

protection. He recognized five basic growth forms. These are: **phanerophytes** in which the resting buds are well above the surface of the ground; **chamaephytes** with resting buds very much closer to the soil surface; **hemicryptophytes** with resting buds at soil level; **cryptophytes** (also known as geophytes) with resting buds below soil level; and **therophytes** which persist only as embryos enclosed in seeds. The first four growth forms are illustrated in Fig. 3.1.

The growth form of trees, which for many years has tended to be ignored, is now receiving renewed attention. Readers who are interested in this topic should consult the important book by Hallé and Oldeman (1970). In a more recent article Hallé (1971) draws attention to the architectural diversity presented by members of the Euphorbiaceae and contrasts this with the architectural poorness of the Myristicaceae. The same author (Hallé 1974) has also described the architecture of trees in the Rain Forest of Morobe district in New Guinea. A detailed discussion of this subject is outside the scope of this present work, but it is clearly evident that it is desirable to take the architecture of trees into account when studying the systematic anatomy of the plants concerned. The architecture of plants is genetically controlled and cannot be ignored (see also Hallé, Oldeman, and Tomlinson 1978).

An important point for the systematic anatomist to remember is that whilst there are many families such as the Caryophyllaceae and Brassicaceae in which herbs predominate, and others, such as the Magnoliaceae or Fagaceae that are predominantly woody, there are few or no families that are wholly herbaceous or wholly woody. This means that the herbaceous or woody habit is not, in itself, a character of fundamental taxonomic importance, a fact that is generally acknowledged and accepted by most botanists.

Axis and leaves

The popular concept of the aerial part of the plant body of most dicotyledons is that it consists of a flower- and leaf-bearing axis (stem). The stem is typically cylindrical and variously branched. At ground

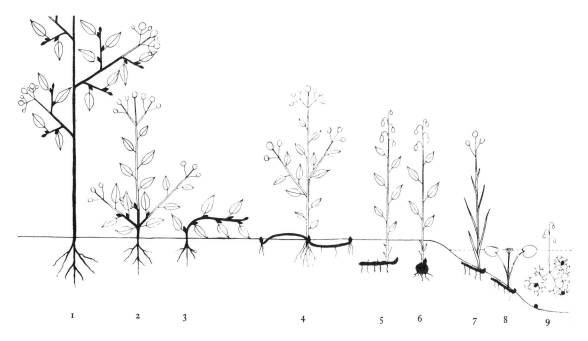

Fig. 3.1. Diagram of the first four types of life-form: phanerophytes (1), chamaephytes (2–3), hemicryptophytes (4), and cryptophytes (5–9). The parts of the plant which die in the unfavourable season are unshaded; the persistent axes with the surviving buds are black. Proceeding from phanerophytes on the left and going farther and farther to the right, it is seen how the plants enjoy progressively better protection during the unfavourable season, the surviving buds being attached lower and lower. In chamaephytes the buds are on the surface of the ground (2 and 3), in hemicryptophytes they are in the soil-surface (4), and lastly in genuine cryptophytes (5 and 6) the buds are actually in the soil, or at the bottom of the water in helophytes (7), and hydrophytes (8–9). (From Raunkiaer, C. (1934), *The life forms of plants and statistical plant geography*, Clarendon Press, Oxford.)

level the basal part of the stem gradually passes over into the subterranean part of the axis (root) which serves to attach the plant to the ground and to take up water and mineral foodstuffs from the soil. The stem and the leaves which it bears are generally so arranged that the foliage is effectively exposed to the light and air.

It is generally acknowledged that the distinction between stem and root is morphologically more fundamental than that which exists between stem and leaves. Nevertheless, organs that combine some of the characteristics of stems and roots are known to occur. An interesting example is afforded by the axillary tubercles of *Ranunculus ficaria* (Metcalfe 1938a). It is, however, pointed out by Howard (1974) and by previous authors mentioned in his paper, that the distinction between leaf and stem is not always clearly defined and it is debated amongst morphologists whether it is justifiable to regard leaves and stems as fundamentally distinct organs.

Howard (1974, Summary and conclusions) brings together the various definitions and interpretations of the leaf that have been proposed at one time or another. For example the leaf has sometimes been interpreted as an enation or protuberance from the stem which obtained its vascular supply from the stem. An alternative is to view the leaf as a modification in one way or another of the stem system itself. C. De Candolle (1868) regarded the leaf as a flattened stem in which the adaxial part has atrophied. Arber (1941d) concluded that 'the leaf is a partial shoot . . . revealing an inherent urge toward becoming a whole shoot, but never actually attaining this goal, since radial symmetry and the power of apical growth and self-reproduction are curtailed or inhibited.' The telome theory (Wilson 1942) suggested that the primitive stem system consisted of dichotomously branched, terete axes which became flattened by broadening to give rise to leaves. The broadening was either, to take the more familiar view first, from a single axis, or it could result from the fusion of axes (Croizat 1973, Footnote 8). Howard also refers to various authors (Lignier 1888; Morvillez 1919; Maekawa 1952; Croizat 1960) who have interpreted the modern leaf as a compound structure formed from an assembly of components. Howard also draws attention to the widely differing views concerning the morphological appearance of the primitive leaf that have been put forward by such authors as Sinnott and Bailey (1914) and Corner (1949). Sinnott and Bailey pictured the primitive leaf as simple, palmately veined, probably three-lobed and associated with a trilacunar node. Majumdar (1957) likewise regards simple leaves as more primitive than compound leaves. Corner, on the other hand, suggested that the primitive leaf was both massive and pinnately compound.

Howard concludes that discussion of these topics calls for the 'abandonment of modern botanical semantics, and especially the current concepts of the leaf as a fundamental structure or a unit of the plant.'

'The "leaf" of our current thinking', he says, 'may well have multiple possibilities of origin.'

Whilst readily conceding the validity of the view expressed by Howard, and other authors quoted by him, that the stem, the node and the leaf constitute a morphological continuum, the distinction between leaf and axis is maintained in this present work as a matter of descriptive convenience. There is a general understanding of what is meant by stems and leaves and it seems easiest to handle the vast mass of descriptive data concerning them under these familiar terms. The leaves of most plants, as understood in this popular, everyday sense, consist of a flattened blade (lamina) which is either attached to the stem at a node to give a sessile leaf, or a leaf stalk (petiole) is interpolated between the lamina and the stem if the leaf is petiolate. These considerations are elaborated in the following extract from Howard's (1974) article. His account of vascular structure of petioles will be found on pp. 88–96.

PART II

RICHARD A. HOWARD

Leaf morphology

A simple definition of a leaf proves to be a difficult exercise, due to the extreme diversity of morphology. Leaves are generally described as axillary appendages to the stem, dorsiventrally flattened, with a restricted terminal growth, an interpolated petiole, and primarily a photosynthetic function. There are many exceptions to all of these descriptive phrases. Even the relationship of the leaf to the stem remains a philosophic question, unanswered morphologically. Are the leaf and the stem independent structures? Are leaves of common phylogenetic origin, or do leaves of the major taxa of the plant kingdom have diverse origins? Foster (1949) stated, and many authors have repeated the statement that 'it is difficult on both theoretical as well as practical grounds to demarcate the leaf from the stem'. Leaves and stems are actually integrated portions of the shoot system without sharp boundaries.

The mature leaf may be simple or compound. Pinnate and palmate compound leaves are recognized in several degrees of division and modification. Leaf blades may be entire, lobed, diplophyllous, or peltate. The distinction between a pinnately compound leaf and a pinnately lobed leaf may be extremely difficult to ascertain.

The leaf may be sessile or with a petiole and the petioles of different taxa vary considerably in length. The leaf may also have a pulvinus, consisting of flexible tissue, at the base of the petiole, and there is sometimes a second pulvinus at the petiolar apex. When a pulvinus is present it is normally of greater diameter than the petiole above it. The flexible nature of its tissue enables it to produce nastic movements which are a well-known feature of the leaves of many tropical plants (Satter and Galston 1973). The pulvinus has a large amount of cortical tissue, a relatively small amount of supportive tissue, and at the same time it lacks any cambial activity and contains no sclerenchyma. The leaf base may subtend a bud, or nearly or completely surround the bud. Alternatively it may extend laterally and thus be sheathing in nature.

Abscission of the leaf may be by means of a layer or layers of cells developed at the base of the pulvinus, within the pulvinus, or at its apex. In simple leaves the lamina may abscise, leaving the petiole to become detached later, or the whole leaf may be abscised at one time. In pinnately compound leaves the leaflets may abscise before the petiole or rachis, or all may fall as a unit. Decompound leaves (i.e. leaves that are several times divided or compounded), may abscise in successive parts, the ultimate divisions falling first, then successively more proximate divisions, until finally the basal petiolar portion abscises from the stem. The nature of these 'units' of the leaf has received little practical or theoretical consideration beyond the original definition of them by C. De Candolle (1868), who classified leaves as unimerous, dimerous, trimerous etc., to polymerous. The areas of articulation, joints or pulvini, have been noted and classified by Funke (1929). Needless to say, the pattern of vascular tissue may be distinctive in these zones. No consideration of the pulvinus development appears in ontogenetic literature.

Most leaves are dorsiventrally flattened, and they may show differences in the structure of the cuticle or of the pubescence and stomata on the two surfaces. Many examples of terete leaves are also well known. There are also many leaves that have been described as unifacial. A classification of unifacial leaves of Australian plants was proposed by Flachs (1916). He recognized 'aquifacial' (laterally flattened), 'unifacial' 'bifacial', and 'transitional' types. The classification was based mainly on the orientation of the bundles.

The abrupt transition from a laminar leaf of *Hakea* (Proteaceae) to the dissected leaf with terete segments is referred to by Troll (1939). *Franklandia fucifolia* (Proteaceae) is commonly cited as an example of foliar dichotomous divisions in terete or unifacial leaves.

Duration of leaves

Kraus (1880) offers one of the early references to the persistence of leaves. Some conifers he reported as retaining their needles for up to 12 years. Dicotyledons often retain their leaves for 2 to 3 years in many species of *Ilex*, *Fagus*, and *Rhododendron*, with persistence of 5 years in *Buxus* and *Hakea*. Hallé (1967) reported the tropical *Schumanniophyton* with leaf retention of 4 years, while I (Howard 1969) observed *Trichilia* (Meliaceae) having pinnate leaves of 4 years' duration. Pease (1917) found that the relatively smaller leaves of *Chimaphila* and *Pachystima* were retained in some cases for 8 years.

The petiole of leaves more than one year old usually shows secondary xylem, with some evidence of annual increments. Attempts to estimate the volume of annual xylem production, or to correlate the amount of xylem in leaves of comparable age, were not successful.

Samantarai and Kabi (1953) observed the development of secondary xylem in the petiole of rooted leaves of *Amaranthus gangeticus*, *Chenopodium album*, and *Ipomoea batatas*. Secondary growth in *Amaranthus* was initiated outside the original vascular bundles by accessory cambia in distinct arcs.

Leaf as a continuum

There is continuity of the vascular supply from the stem to the apex of the leaf. The nature of its variations, as revealed in transverse sections, within the length, are presented in the existing literature in only a very few papers (Acqua 1887; Swamy and Bailey 1949; Nakazawa 1956; Yamazaki 1965; Schofield 1968; Sugiyama 1972).

For convenience one may consider the traces as departing from the stem and entering the leaf. The individual traces may separate ordinarily from the principal vascular system, and be seen in the cortex only immediately below the leaf gap or, in some species, in many varying numbers of internodes below the leaf attachment. The description of a transverse section through an internode taken immediately below the attachment of the leaf base to the stem may include information concerning the number and nature of the leaf traces. The presence or absence of cortical and/or medullary bundles and of fibres or sclereids may also be noted.

At the node in the classical anatomical sense, that is, the area of leaf gaps, one determines the size and nature of the traces, cathodic or anodic[1] precocity, and the attitude or path of departure of all traces which may be on a gradual slope or abruptly at right-angles to the axis. Only occasionally will traces depart from above the petiole base and curve backwards, even running down the stem cortex (de Fraine 1913; Fahn 1963). Lateral traces, if present, usually pursue an abrupt, nearly horizontal path and, when sheathing stipules are present, may produce branch traces which enter the stipular sheath.

The basal parts of sessile leaves often show the same arrangement of bundles as there is in the pulvinus. Bearing in mind, as has already been pointed out, that the petiole is a portion of the leaf which is interpolated between the blade and the leaf base during development, in pinnately compound leaves it is represented by the portion of the leaf between the leaf base and the first leaflet or pair of leaflets. The remainder of the axis which bears leaflets in a pinnately compound leaf is referred to as the rachis.

In palmately compound leaves, the petiole is between the leaf base and the area of origin of the leaflets. Leaflets of pinnately or palmately compound leaves may have petiolules.

[1] For the meaning of 'cathodic' and 'anodic' in an anatomical sense, see Benzing (1967a).

Transverse sections, taken successively from the base of the petiole to its apex, reveal variations in the number, position, and arrangement of the traces (Fig. 8.1, p. 80, and Figs. 9.1–5, pp. 91–4). The lower end of the petiole is supplied with a characteristic number of traces, but successive sections reveal that these may remain independent, fuse, divide, or be rearranged, with changes in orientation. The petiole vasculature is most complex at the geographic middle of the petiole where there is the greatest amount of secondary tissue, if a cambium has developed. Collenchyma or sclerenchyma also reaches its maximum development in the cortex and pith of the petiole in sections taken at the same middle level. Near the upper end of the petiole, the vascular pattern which has developed from the basal orientation may be undone in an exact reversal of its formation. Thus in some simple leaves the upper pulvinus shows a vascular pattern comparable with that of the lower pulvinus. In other leaves which do not have an upper pulvinus, the vascular pattern of the petiole continues into the midrib until interrupted by the departure of the primary veins.

Dormer (1972) concluded that a common feature of leaves in which separate strands exist in the petiole is that these should be linked by a massive collar or bridge of vascular tissue in the leaf base. This collar is often situated at the upper end of a stipular or sheathing region. He adds that 'in many compound leaves a similar type of cross connection may be associated with the attachment of each pinna'.

Palmately compound leaves usually have a plexus of tissues at the apex of the petiole, and it appears that each petiolule receives a supply of vascular tissues from each of the original traces. Palmately compound leaves have been associated with unilacunar, trilacunar, or multilacunar nodes.

Peltate leaves may have a relatively simple vascular pattern at the apex of the petiole, or one of great complexity.

In leaves in which the vascular pattern is simplified in the upper pulvinus, the vascular pattern is again restored within the base of the leaf blade. The pattern of vascular tissues in the midrib is interrupted by the departure of the primary veins in a palmate or pinnate fashion.

The distinctions between deeply lobed leaves and pinnately compound or palmately compound leaves are arbitrary. Normally, if a leaflet is attenuated at the base to a petiolule which develops an abscission layer, the leaf is considered as compound. Compound leaves may be associated with unilacunar nodes,

although most appear to be associated with trilacunar nodes. In general, a leaflet of a pinnately compound leaf receives its vascular supply from branches of the median and one lateral trace.

Diplophyllous leaves may vary considerably in the amount of vascular tissue which enters the secondary laminae.

The amount of vascular tissue in the midrib of the leaf or leaflet is greatest at the base of the blade, and is progressively diminished as veins are produced. Normally, some vascular tissue, usually from the median trace, continues to the ultimate apex of the leaf blade. In some pinnately compound leaves the ultimate terminal leaflet may abort, yet a vascular supply is evident in the rudiment. The mucro of the fused leaflets of *Bauhinia* is vascularized by the median trace.

Stipules

Although the presence or absence of the stipules, as well as their form, has long been used as a diagnostic character in plant taxonomy, the morphological nature and origin of the stipule remains obscure. The stipule is commonly defined as an appendage or pair of appendages at the base of the leaf. The stipules may be paired or single, equal or unequal in size, large, foliaceous, and persistent or reduced to a mere protuberance or a small, readily deciduous scale. Some stipules have become modified to thorns or tendrils. Stipules are often represented by a stipular sheath, and may be below the point of attachment of the leaf, well above the point of attachment of the leaf, associated with the leaf, or free from it. Alternatively stipules may be axillary (Philipson 1968) to the leaf, on both sides of the leaf but attached to the stem, opposite the leaf, or at varying levels of the pulvinus or petiole. Sinnott and Bailey (1914) believed that the stipules were primarily associated with trilacunar nodes, and many authors have indicated that most stipules are vascularized from the lateral traces of a trilacunar node.

A fundamental but still unresolved question is whether the stipules are part of the leaf, or independent structures. Guédès (1972) apparently believes the stipules are comparable to portions of the leaf blade or leaflets, for he states: 'a petiole generally occurs because of intense intercalary growth either below the lowest pair of leaflets and the leaf is exstipulate or between the first pair of leaflets from below and the second one. The lower pair (of leaflets) is then isolated near the leaf insertion and becomes a pair of proto-stipules. It rises to the state of true stipules when proleptic development is obtained. . . . The petiole can also develop above several pairs of leaflets and there are then several pairs of stipules.'

Regel (1843), Agardh (1850), Clos (1879), and Tyler (1897) have all suggested the independence of stipules from the foliage leaf in nature as well as in development. More recently Croizat (1940) proposed that the two stipules are basic structures, and between them a 'dab' of meristem is found. Croizat refers to this meristem as a primary nerving centre which may develop into a foliage leaf.

Examples of the seemingly independent development of stipules and the leaf are many. A shoot developing from a dormant bud may produce successively cataphylls, lobed stipular-like structures, similar lobed structures with a nonpetioled lamina, and finally a normal petiolate leaf with basal stipules (Furuya 1953). Many authors have used the illustration of *Ribes* to show the retarded development of the leaf blade in successive stem appendages. Here the bud scales are clearly stipules, with the petiole and abortive leaf blade represented by a ridge, or a ridge and a small appendage. Fahn (1967, p. 194) gives an illustration of bud scales of *Vitis vinifera* where each pair of stipular scales has an included leaf. Large stipules with a rudiment of a leaf attached protect the buds in species of *Magnolia*. Large foliaceous stipules are often precociously developed in the Fabaceae. The seasonal growth of *Ilex laevigata* ends abruptly in transition from the normal foliage leaves with two minute gland-like basal stipules to an organ of three structures, the outer two comparable in size and shape to the stipules of the preceding leaf, and the middle one no larger in size, and of the same shape but in theory the primordium of the foliage leaf.

Traces which vascularize the stipules may be either the complete lateral traces (Furuya 1953) or branches of the lateral traces of a trilacunar node (Sinnott and Bailey 1914). When the leaves are opposite at the node, e.g., in many Rubiaceae, the stipules of adjoining sides of the opposite leaves may be variously united, sometimes with a bifid apex to the connecting tissue, or the connecting tissue may taper to a single point (Howard 1970). The vascular supply to these conspicuous interpetiolar stipules may come from the sides of the single arc-like trace of each leaf. The branch bundles run horizontally or at an angle into the stipules, and may remain independent. In multilacunar nodes the leaf may be vascularized primarily by the centrally located vascular bundles, with small

traces departing from the lateral nodes, running horizontally to the leaf base while giving rise to vertical bundles which vascularize the stipules. In some sheathing stipules (Ozenda 1949; Sugiyama 1972), the bundle or bundles most distant from the median trace may enter the stipules without connections with other bundles or with the leaf itself (Fig. 2, 1*i*). Sugiyama (1972) reports that the bundle opposite the median trace in *Magnolia virginiana* may branch while remaining free and be non-contributory to the leaf itself.

The bundles of stipular sheaths which are above and free from the leaf, as in *Platanus*, may be vascularized by traces completely unassociated with any vascular supply to the leaf. So, too, with stipules which are borne on the side of the stem opposite the leaf scar (*Ricinus*), or borne below the leaf (Cunoniaceae).

A further note on the vascular structure of stipules will be found on p. 81.

Glands

The presence and nature of foliar glands (see also Chapter 10) has been recognized as a useful taxonomic character (Gregory 1915; Dorsey and Weiss 1920). Zimmermann (1932) has given an extensive systematic survey and classification of extrafloral nectaries. Schnell, Cusset, and Quenum (1963) also supplied a broad survey of foliar glands with a classification, with suggestions on the phylogenetic origin of glands and with some anatomical details.

Glands can occur on the pulvinus, stipules, petiole, or blade in various positions. They can be sessile or stalked, and with or without vascular tissue. Bernhard (1964) found the petiolar glands in certain Euphorbiaceae to be present in the primordium, and compared the glands to other lobes of the lamina. Dorsey and Weiss (1920) also considered the gland to be the equivalent of a portion of the blade, and Messager (1886) earlier had proposed the origin of the gland as an abortive laminar structure. Schnell, Cusset, and Quenum (1963) regard the glands as non-foliarized 'éléments foliaires', or the homologues of lobes of the lamina. Cusset (1970) later proposed the term 'métamère' for the 'article foliaire' in a connotation differing from that used earlier by De Candolle. He suggested the primitive leaf was a single métamère glandularized at the apex, while the foliar leaf of most plants is to be regarded as a compound product of several métamères with the glands persisting or lost in the evolutionary process.

The petiolar glands may be associated with simple or complex petiolar vascular patterns arising from unilacunar or trilacunar nodes. The majority of vascularized glands receive their vascular supply from the lateral traces or their derivatives, as sequential lateral traces or rib traces. The very large glands of *Pithecellobium* obtain a vascular supply from the adaxial bundles of the complex petiole vasculature (Elias 1972).

Phyllodes

The term phyllode literally means leaflike, and has been applied in plant morphology in a variety of ways to flattened photosynthetic petioles, to rachides of compound leaves which have lost their leaflets, and to a quantity of dissected leaves with terete segments. A flattened leaflike stem may be termed a platyclade, a phylloclade, or a cladode.

The phyllodes of species of *Acacia* may be genetically without leaflets, or in maturation fail to develop leaflets, or lose the leaflets by abscission at various stages of development. Boke's (1940) study of the laterally (vertically) flattened phyllodes of several species of *Acacia* revealed fundamental differences in development, including the dominance of the adaxial meristem in producing the flattened organ. A short pulvinus and stipules were also found to be present. The orientation of the vascular bundles simulated the pattern present in the petiole of other leaves. Boke concluded that the phyllode was homologous to a petiole-rachis of a pinnately compound leaf. The initial three vascular traces in the phyllode are supplemented with interpolated bundles. Slade (1952) reported a similar interpolation of traces in cladodes of several New Zealand brooms. Peters (1912) recognized three types of phyllodes in *Acacia* which he termed 'Platentypus,' 'Binsentypus' (rushlike), and 'Übergangstypus' (transitional types). Phyllodes in species of *Oxalis*, e.g. *O. fruticosa, O. ptychochala*, represent the petiole of a palmately compound leaf which has lost the leaflets in development. (See also Metcalfe 1933.)

Articulated leaves have been described and illustrated for several families (Troll 1939). The 'Gelenkknoten' of *Polyscias* species are clearly in the position of leaflets lost phylogenetically or ontogenetically from terete rachides, since one or more foliage leaflets are present. Species of *Citrus* occasionally show a similar articulated form. Troll (1939) illustrated the articulated phyllode of *Phyllarthron* (Bignoniaceae). Saha (1952) studied *Phyllarthron commorense* and concluded that the 'leaves are simple, petiolate, with segmented blades'. He denied the existence of articulation within a leaf, ignoring the frequent occurrence of such segments in the Bignoniaceae. *Phyllarthron* should be re-examined to determine if it is not a compound leaf which has lost or not developed pinnae.

Kaplan (1970b) has studied the development of the 'rachis-leaves' in two genera of the Apiaceae, and concluded that these terete leaves with septa are the equivalent of the rachis of a pinnately compound leaf with the appendages reduced, the transformed pinnae functioning as hydathodes.

Terete leaves have often been described as unifacial leaves. A cross section of the leaf may reveal that the bundles are arranged in a circle with the phloem of each trace orientated to the periphery. Stomata may have a uniform distribution around the leaf surface, instead of being limited to the abaxial surface as in most dorsiventrally flattened leaves. Flach's (1916) classification of unifacial leaves has already been mentioned (p. 17).

Epiphyllous inflorescences

Flowers may be borne on leaves, singly, or in inflorescences. They have been described as epipetiolous, epiphyllous or hypophyllous according to their position. An epirachial type should be recognized for compound leaves (Fig. 3.2(e)). Johnson (1958) has supplied a listing of genera and families in which epiphyllous inflorescences are known to occur. A slightly modified version of this list is to be found on p. 190. C. De Candolle (1890), after studying the vascular structure of the petiole and the blade, concluded instead that the epiphyllous inflorescence was of foliar origin. Stork (1956) debated whether the flattened structures bearing flowers might be termed 'cladophylls'. Melville (1962) regarded the epiphyllous inflorescence as evidence of his proposed primitive 'gonophyll'.

The hypophyllous inflorescence of *Erythrochiton hypophyllanthus* (Rutaceae) (Fig. 3.2(i)) is apparently unique in the Angiospermae in bearing a flower on the lower side of the lamina (Engler 1897). Melville (1962) stated that 'above the insertion of the inflorescence there is nothing unusual in the structure of the midrib. Below the flowers, the midrib bundle lies above and quite distinct from the vascular strands supplying the inflorescence. It is evident that in *Erythrochiton* the inflorescence in truly adnate to the lower surface of the leaf.' No specimens are available to me for further examination of the vascular supply within the petiole to determine the origin of the traces supplying the inflorescence. Whether these are axillary to the leaf bearing the inflorescence or are from the subjacent node is not known.

Epipetiolar inflorescences are reported at the base of the petiole, along the petiole, or at the apex of the petiole. Johnson (1958) has shown that the epipetiolar inflorescence of *Turnera ulmifolia* (Fig. 3.2(d)) originates as a branch trace and becomes associated in the petiole, but is not fused with the vascular supply of the leaf. He concluded that 'the epiphyllous inflorescence in *Turnera* is axillary in origin, and through subsequent growth in the foliar buttress is displaced to a petiolar position ... 10 out of 13 species of *Turnera* examined have epiphyllous inflorescences which originated in the axil of a foliar primordium.'

Several inflorescences are shown arising from the longer petioles of *Mocquerysia multiflora* (Flacourtiaceae) by Letouzey, Hallé, and Cusset (1969), but no data on the vascular structure of the petiole are given.

Barth (1896) studied various species of the Dichapetalaceae in which petiole-borne inflorescences (Fig. 3.2(a and h)) are present (fertile leaves) or absent (sterile leaves). In the petioles of fertile leaves he found a stem-like vascular structure, the adaxial portions of which formed the vascular supply of the inflorescence. Such leaves may have an axillary bud. In some sterile leaves the petioles also had an adaxial vascular supply comparable to that of the fertile leaves. In others, the petiole had only a simple arc of xylem and a bud in the axil. Additional species were intermediate between the two extremes. Barth accepted the idea that the epiphyllous inflorescence is a congenital accrescence of the petiole with the floral peduncle; and although the bundles are of foliar origin, concluded that the inflorescence is nothing more than one bud of several which might be produced in the axil. Stork (1956) concluded from his studies that 'various degrees of coalescence prevail in species of Dichapetalaceae where the inflorescences are borne on the petioles'. To Melville (1962) 'the sequence of vascular structures to be observed in the Chailletiaceae provides no evidence for adnation ... but rather, it shows the last stages in the disengagement of the fertile branch from its leaf.'

Dickinson and Sattler (1974) studied the development of the epiphyllous inflorescences of *Phyllonoma integerrima*, and reported that the inflorescence primordium is initiated on the adaxial side of the leaf primordium, and above an axillary bud. They concluded that neither the inception nor the procam-

bial supply of the inflorescence provided evidence of congenital fusion of inflorescence and leaf. (See the note on De Candolle's investigation of *Phyllonoma* below.)

The inflorescences borne on leaf blades have been reported for simple leaves with the sole exception of *Chisocheton pohlianus* (Meliaceae) (Fig. 3.2(e)), which may bear clusters of flowers in the axils of the pinnae of a pinnately compound leaf. Harms (1917) illustrated a specimen with flowers borne in the axils of three pairs of leaflets. His illustration also shows an axillary bud at the base of the leaf, and immature pinnae at the apex. Contrasting is another shoot illustrated with an axillary racemose inflorescence and a normal pinnately compound leaf. No anatomical data are offered by Harms on the relationship of the vascular supply of the epirachial inflorescence to that of the rachis of the leaf blades. Melville (1962) refers to 'a new species of *Chisocheton*' in which sections of the rachis were cut to reveal a 'stem-like structure below the insertion of the inflorescence and leaf-like structure above'. Knowing the complex vascular patterns found in the rachis of leaves of the Meliaceae, this statement by Melville is unrevealing.

Flowers that are borne on the leaf blade are generally associated with the midrib, or proliferation of it, in pinnately veined leaves (Fig. 3.2(g)). Possible exceptions are found in *Peperomia foliiflora*, in which the inflorescences arise from the base of diverging palmate venation. No material is available for anatomical investigation. Data are also wanting for the epiphyllous flower borne on palmately veined leaves of *Begonia prolifera* and *Begonia sinuata* cited by Johnson (1958).

The flowers may be at one of several positions on the blade, being near the middle of the length of the blade or along the extreme portions of it, even at the ultimate apex. Letouzey, Hallé, and Cusset (1969) concluded that the relative position of the inflorescence may vary and is without taxonomic significance. They further explain the presence of a large bract subtending the epiphyllous inflorescence of *Phylloclinium bracteatum* as an example of hyperfoliarization.

C. De Candolle's (1890) study of the vascular supply to the epiphyllous inflorescence of *Polycardia* (Celastraceae) (Fig. 3.2(f)), *Phyllonoma* (Fig. 3.2(c)) (Dulongiaceae), and *Helwingia* (Fig. 3.2(b)) (variously assigned to Cornaceae, Helwingiaceae, and Araliaceae), formed the basis for his conclusion that the inflorescences were foliar in origin and not concrescent branches. Melville (1962) stated: 'In all of these the vascular system of the petiole, where it left the branch,

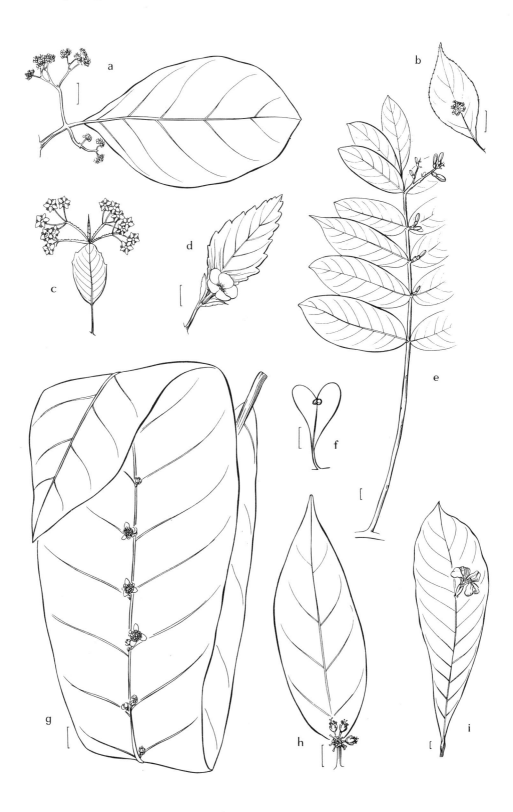

was an open arc which further up arched round and formed a complete circle, as in a stem ... In all of these examples where the midrib extends beyond the inflorescence its trace in section has the open arc appearance, which is very common in Angiosperm leaves.' Johnson (1958) reached the same conclusion. Watari (1939) was not aware of De Candolle's work and supported the contrasting idea of Thouvenin (1890) that the traces for the flowering branch were fused with the petiole and lamina. He cited species of *Saxifraga, Chrysosplenium,* and *Vahlia,* in which branch traces fuse with the foliar traces in a comparable manner.

With the exception of the study by Johnson (1958) on *Turnera* and *Helwingia,* the majority of the anatomical data available in the literature do not make any reference to the nodal origin of the vascular supply of the epiphyllous inflorescences.

The Dichapetalaceae, Turneraceae and Flacourtiaceae have trilacunar nodes. The Celastraceae has a unilacunar node. *Helwingia,* which as already noted is variously assigned to several families, has a unilacunar node, which is exceptional in the Cornaceae and most unusual in the Araliaceae with which Melville (1962) associates the genus. *Phyllonoma,* at one time placed in the Saxifragaceae, which has trilacunar or unilacunar nodes, is now associated with the Escalloniaceae (Melville 1962) with unilacunar nodes. It has also been placed in the monotypic Dulongiaceae (Airy-Shaw, in Willis 1966). *Phyllonoma* is reported by Stork (1956) to have stipules and it is the only genus with stipules and an epiphyllous inflorescence.

Literature cited

Acqua 1887; Agardh 1850; Arber 1941*d*; Barth 1896; Benzing 1967*a*; Bernhard 1964; Boke 1940; Candolle C. De, 1868, 1890; Clos 1879; Corner 1949; Croizat 1940, 1960, 1973; Cusset 1965, 1970; Dickinson and Sattler 1974; Dormer 1972; Dorsey and Weiss 1920; Elias 1972; Engler 1897; Fahn 1963, 1967; Flachs 1916; Foster 1949; Fraine, de 1913; Funke 1929; Furuya 1953; Gregory 1915; Guédès 1972*b*; Hallé 1967, 1971, 1974; Hallé and Oldeman 1970; Hallé, Aldeman, and Tomlinson 1978; Harms 1917; Howard 1969, 1970, 1974; Johnson 1958; Kaplan 1970b; Kraus 1880; Letouzey, Hallé, and Cusset 1969; Lignier 1888; Maekawa 1952; Majumdar 1957; Melville 1962; Messager 1886; Metcalfe 1933, 1938*a*; Morvillez 1919; Nakazawa 1956; Ozenda 1949; Pease 1917; Peters 1912; Philipson 1968; Raunkiaer 1934; Regel 1843; Saha 1952; Samantarai and Kabi 1953; Satter and Galston 1973; Schnell, Cusset, and Quenum 1963; Schofield 1968; Sinnott and Bailey 1914; Slade 1952; Stork 1956; Sugiyama 1972; Swamy and Bailey 1949; Thouvenin 1890; Troll 1939; Tyler 1897; Watari 1939; Willis 1966; Wilson 1942; Yamazaki 1965; Zimmermann 1932.

Suggestions for further reading

Eckhardt 1957: critique of the 'New morphology', see Thomas 1932 and Tomlinson and Gill 1973.

Emberger 1952: the axis is interpreted as the basic organ in vascular plants from which roots and leaves arose as modifications.

Fahn and Rachmilevitz 1970: autoradiography of nectar secretion.

Gill and Tomlinson 1971: in *Rhizophora mangle,* although growth is influenced by season and environment, the basic control is endogenous.

Lawrence 1952: co-operative research needed owing to expanding basis of taxonomy.

Markgraf 1964: a general discussion concerning the morphological interpretation of plant organs.

Fig. 3.2. **Epipetiolar, epiphyllous, epirachial, and hypophyllous infloresences.** (a) *Dichapetalum latifolium* (Dichapetalaceae), inflorescences from petiole; (b) *Helwingia japonica* (Cornaceae), epiphyllous cluster of flowers, plants dioecious; (c) *Phyllonoma laticuspis* (Dulongiaceae), epiphyllous inflorescences; (d) *Turnera ulmifolia* (Turneraceae), single flower with bracts borne on the petiole; (e) *Chisocheton pohlianus* (Meliaceae), epirachial flowers and racemes from single pinnately compound leaf; (f) *Polycardia phyllanthoides* (Celastraceae), terminal epiphyllous flowers; (g) *Phyllobotryum spathulatum* (Flacourtiacea) (redrawn from Hooker's Icones **14**: Plate 1353. 1881, and herbarium material), flowers borne on upper surface of midrib (epiphyllous); (h) *Tapura latifolia* (Dichapetalaceae), flowers borne here at apex of petiole, but may also be on the lamina; (i) *Erythrochiton hypophyllanthus* (Rutaceae) (redrawn from Engler, Engler, and Prantl. *Nat. Pflanzenfam.* III. 4: 96. fig. 96 F. 1896), flower borne on abaxial surface of lamina (hypophyllous). Scale markers are 1 cm. (From Howard 1974.)

Rachmilevitz and Fahn 1975: floral nectary of *Tropaeolum*.

Sattler 1966: root, shoot, phyllome, caulome, and epiphyllous branch should not be regarded as mutually exclusive organs; they all merge into one another.

Thomas 1932: the old morphology and the new; this should be read in conjunction with Eckhardt's critique (1957).

Tomlinson and Gill 1972: external form of trees.

Troll 1932: a discussion of the morphological relationship of the peltate and scutiform leaves to other types of leaf.

Uhlarz 1975: glandular stipules of *Euphorbia*.

Weberling 1955a: lower part of the leaf primordium (Unterblatt) of dicotyledons in general.

— 1955b: stipules of Coriariaceae.

— 1956: rudimentary stipules of Myrtales.

— 1957a: stipules of Lecythidaceae and Sonneratiaceae.

— 1957b: Unterblatt of Balsaminaceae and Plum-baginaceae.

— 1957c: stipules of Caprifoliaceae.

— 1963: stipules of Geissolomataceae, Myrtaceae (*Heteropyxis*) and Penaeaceae.

— 1966a: stipules of *Kania* (Myrtaceae)

— 1966b: stipules of *Heptacodium* (Caprifoliaceae)

— 1967: taxonomic value of stipules.

— 1968a: stipules; rudimentary (Cyrillaceae, Lythraceae).

— 1968b: stipules; rudimentary (Resedaceae).

— 1970a: stipules; Piperales.

— 1970b: stipules; Polygonaceae.

— 1970c: stipular thorns of *Zanthoxylum*, *Fagara* and *Acanthopanax*.

— 1974a,b 1975: relationship of stipules to sheathing lobes.

— 1976: pseudostipules of Sapindaceae.

— Weberling and Leenhouts 1965 [1966]: stipules in Anacardiaceae, Burseraceae, Meliaceae, Sapindaceae, Simaroubaceae).

4

A REVISED CLASSIFICATION OF THE ARCHITECTURE OF DICOTYLEDONOUS LEAVES

LEO J. HICKEY

In this chapter Dr. Hickey presents an outline of the characters on which he classifies leaves in terms of leaf architecture. With further co-operation from Dr. Hickey it is hoped in future volumes of this work to present brief summaries of the leaf architectural characters for as many of the families as possible. [*Editor*]

Introduction

Leaf morphology has remained a virtually unexploited tool for systematic studies of the dicotyledons. This has resulted, in part, from the lack of a comprehensive and unambiguous classification of leaf morphological features and from a belief that such features respond in a relatively plastic manner to environmental forces. In 1973 I published a standardized classification of leaf architectural features of the dicotyledons which developed as an outgrowth of my efforts to find reliable criteria for identifying fossil angiosperm leaf floras. This system was based in part on two earlier schemes: that of Turrill, first published by Lee (1948), which established the most simple yet precisely defined shape classes for leaves; and that of von Ettingshausen (1861) which classified the patterns of leaf venation. In adapting these earlier schemes I retained or added terms which a rigorous evaluation of both cleared leaves and herbarium specimens in a present total of approximately 1850 genera in 192 dicot families showed to be of importance either from a descriptive or systematic standpoint.

In this chapter I now present an updated version of my 1973 treatment, modified to include a classification of compound leaf configurations and a detailed examination of the architectural elements of marginal teeth. This latter has proven of particular importance in leaf systematic studies (Hickey and Wolfe 1975).

As used here, the term 'leaf architecture' denotes the position and form of the elements constituting the outward expression of leaf structure. These include venation pattern, marginal configuration, leaf shape, and gland position. Architecture is the aspect of morphology which applies to the spatial configuration and coordination of those elements making up part of a plant without regard to histology, function, origin, or homology.

The essential justification for this classification is the fact that dicotyledonous leaves possess consistent and recognizable patterns of leaf architectural organization at all levels from the subclass to the species (Hickey and Wolfe 1975). Thus, the Magnoliidae (*sensu* Takhtajan 1969 and Cronquist 1968) have basically simple leaves while the Rosidae have basically pinnately compound leaves. At the family level, most members of the Theaceae exhibit serrations of a characteristic shape, with glandular setae. They also have pinnate, camptodromous venation with elongated intercostal areas and large, loosely organized, irregular areoles. Character sets for families or genera sometimes overlap, e.g. in the Lauraceae, making a positive assignment impossible. Sets of characters proving significant for recognizing one taxon may be completely different from those distinguishing another, although, in general, leaf organization (simple vs. compound), configuration of the major veins, and marginal features, particularly the tooth morphology, are especially valuable.

Numerous taxa of dicotyledons contain more than one basic pattern of leaf architecture. This is true of taxa that are believed to be polyphyletic such as the Subclass Hamamelididae, as it is currently recognized by Takhtajan (1969) and Cronquist (1968). The Hamamelididae includes both simple, actinodromous leaves and pinnately compound blades. In other taxa, variations from the basic pattern appear in those genera or species in environments more extreme than the mean to which the group is adapted. For example, most of the Theaceae live in paratropical and subtropical forests where the majority of them possess the characteristic glandular serrate margins mentioned above. Species living in drier, more open areas (*Laplacea* and *Ternstroemia*), in high montane settings (*Eurya*), or in the lowland tropics (some *Gordonia* and *Shima* spp.) show a trend first toward loss of the serrations and finally to the loss of marginal glands. As a general rule in all groups of dicotyledons the small, coriaceous, entire-margined leaves of xeric, arctic, or alpine environments show relatively few of the characters needed for taxonomic identification.

All terms listed in the following outline are illustrated and defined except where their meaning is obvious. For the sake of simplicity the leaf architecture of simple leaves and of the leaflets of compound leaves is treated together as occurring within laminae. Although I have set limits based on observed breaks in morphological features in describing some characters such as size of primary veins and of areolation, in many cases it is not possible to avoid being somewhat arbitrary. Thus, although most acute leaf apices can be distinguished from most acuminate ones, there are a number which fall in a transitional zone. The same sort of transition is present between the imperfect actinodromous venation class and the pinnate class where the lowest pair of secondary veins is set at an angle different from those above. Such difficulties are inevitable in imposing a classification on natural features and in no way detract from the utility of doing so.

I. Leaf orientation

The basic axes of orientation in the leaf are indicated in Fig. 4.1(1) (as modified after Wolfe, written communication, 1968):

A. *Apical: toward the apex (upward).*

B. *Basal: toward the base (downward).*

C. *Exmedial: away from the axis of symmetry of the leaf.*

D. *Admedial: toward the axis of symmetry of the leaf.*

The curvature of leaf elements (margin or part thereof, sides of a serration, venation) (Fig. 4.1(2)) is referred to as:

E. *Convex: curved away from the centre of the leaf, or its axis, or (for venation in the actinodromous class) the point of origin of the primary veins.*

F. *Concave: curved toward the centre of the leaf, etc.*

II. Leaf organization

A. *Simple: consisting of a single lamina, all parts of which are connected by foliar tissue (Fig. 4.3(53)).*

B. *Compound: leaf divided into separate laminar sub-units.*

1. Pinnate: sub-units arranged along an axis (rachis).

a. Arrangement
   (i) Even: having a number of primary laminar sub-units integrally divisible by two (Fig. 4.1(3)).

(ii) Odd: having a number of primary laminar sub-units not integrally divisible by two (Fig. 4.1(4)).
Special cases:
   (a) Pinnately trifoliolate: distinguished from palmately trifolioate by the extension of the rachis beyond the lateral leaflets (Fig. 4.1(5)).
   (b) Ternate: all sub-units of a bipinnate or more highly dissected leaf rigidly opposite giving the appearance that the leaf axes branch in threes (Ranunculidae) (Fig. 4.1(6)).

b. Dissection
   (i) Once pinnate: leaflets attached to the leaf axis (rachis) either sessilely or by a petiolule (Fig. 4.1(7)).
   (ii) Bipinnate: leaf twice dissected, leaflets arranged along minor axes (rachillae) which are attached to the rachis (Fig. 4.1(8)).
   (iii) Tripinnate: leaf cut to the third degree.
   (iv) Quadrapinnate: leaf cut to the fourth degree.

2. Palmate: laminar sub-units attached at the apex of the rachis (Fig. 4.1(9)).

a. Arrangement
Special case: palmately trifoliolate rachis not extending beyond the point of attachment of the lateral leaflets but either ending in a distinctly swollen apex to which the leaflets are attached (Fig. 4.1(10)) or the leaflets all sessile (Fig. 4.1(11)).

b. Dissection
If second order dissection occurs the second and subsequent orders will be pinnate (Fig. 4.1(12)). No case is known of a bipalmate leaf.

III. Leaf shape

A. *Lamina*

1. Symmetry
a. Whole lamina
   (i) Symmetrical (Fig. 4.1(1)).
   (ii) Asymmetrical (Fig. 4.2(17)).
b. Base only
   (i) Symmetrical (Fig. 4.1(1)).
   (ii) Asymmetrical (Fig. 4.2(18)).
2. Form (modified after Stearn 1956)
a. Oblong: widest portion constituting a zone through the middle of the long axis of the leaf, margins parallel, or nearly so, within this zone (Fig. 4.2(13)). The length–width (l/w) ratios given for the subclasses are lower limits except for the last.

		l/w ratio
(i)	Linear	10:1 or more
(ii)	Lorate	6:1
(iii)	Narrow oblong	3:1
(iv)	Oblong	2:1
(v)	Wide oblong	1·5:1
(vi)	Very wide oblong	1·2:1 or less

b. Elliptic: axis of greatest width perpendicular to the approximate midpoint of the leaf axis (Fig. 4.2(14)). Margins in this and succeeding classes convex or some combination of convex and concave.

LEAF ORGANIZATION

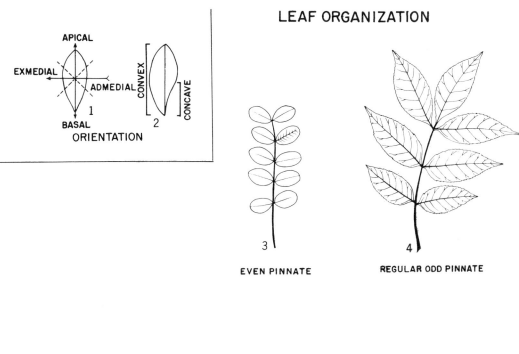

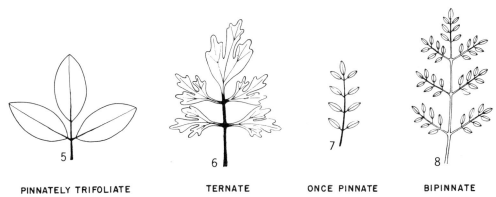

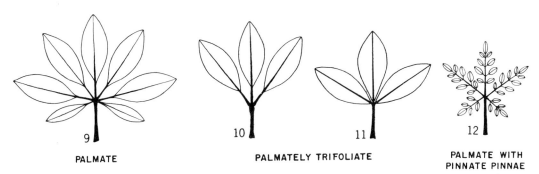

Fig. 4.1 (1–12). **Leaf architectural features** — orientation and organization of the leaf.
(From Hickey 1973.)

		l/w ratio
(i)	Very narrow elliptic	6:1 or more
(ii)	Narrow elliptic	3:1
(iii)	Elliptic	2:1
(iv)	Wide elliptic	1·5:1
(v)	Suborbiculate	1·2:1
(vi)	Orbiculate	1:1
(vii)	Oblate	0·75:1 or less

c. Ovate: axis of greatest width intersecting the leaf axis basal to the midpoint of the latter axis (Fig. 4.2(15)).

		l/w ratio
(i)	Lanceolate	3:1 or more
(ii)	Narrow ovate	2:1
(iii)	Ovate	1·5:1
(iv)	Wide ovate	1·2:1
(v)	Very wide ovate	1:1 or less

d. Obovate: axis of greatest width intersecting the long axis of the leaf apical to the midpoint of the latter axis (Fig. 4.2(16)).

		l/w ratio
(i)	Narrow oblanceolate	6:1 or more
(ii)	Oblanceolate	3:1
(iii)	Narrow obovate	2:1
(iv)	Wide obovate	1·2:1
(v)	Very wide obovate	1:1 or less

e. Special forms (including needle, awl, or scale shaped).

B. Apex: that portion of the leaf bounded by approximately the upper 25 per cent of the leaf margin.

1. Acute: straight to convex margins forming an angle of less than 90° (Fig. 4.2(19)).
2. Acuminate: tip acute, margins markedly concave, either long or short acuminate (Fig. 4.2(23)).
3. Attenuate: margins straight or only slightly concave, gradually tapering to a narrow acute apex (Fig. 4.2(27)).
4. Obtuse: straight to convex margins forming an angle of more than 90° (Fig. 4.2(20)).
5. Rounded: margins forming a smooth arc across the apex (Fig. 4.2(25)).
6. Mucronate: apex terminating in a sharp point which is a continuation of the midvein (Fig. 4.2(24)).
7. Retuse: apex slightly notched as if by removal of the termination of the midvein; internal angle of the sinus generally less than 25° (Fig. 4.2(21)).
8. Emarginate: apex broadly notched by the embayment of the leaf tissue (Fig. 4.2(22)).
9. Truncate: apex terminating abruptly as if cut, margin perpendicular to the midvein or nearly so (Fig. 4.2(26)).
10. Other.

C. Base: that portion of the leaf bounded by approximately the lower 25 per cent of the margin.

1. Acute: margins forming an angle of less than 90°.
a. Normal: base with curved margins terminating at the petiole without appreciable change in direction (Fig. 4.2(28)).
b. Cuneate: margins straight, or nearly so, and forming a 'wedge' of less than 90° (Fig. 4.2(30)).
c. Decurrent: margin extending downward along the petiole at a gradually decreasing angle to it (Fig. 4.2(32)).
2. Obtuse: margins forming an angle of more than 90°.
a. Normal: as above (Fig. 4.2(29)).
b. Cuneate: margins straight, or nearly so, and forming a 'wedge' of more than 90° (rare) (Fig. 4.2(30)).
3. Rounded: margins forming a smooth arc across the base (Fig. 4.2(31)).
4. Truncate: terminating abruptly as if cut, margin perpendicular to the midvein or nearly so (Fig. 4.2(33)).
5. Cordate: leaf base embayed in a sinus whose sides are straight or convex (Fig. 4.2(34)).
6. Auriculate to Lobate: small to large rounded projections whose inner margins (those toward the petiole) are in part concave (Fig. 4.2(35)).
7. Saggitate: with two large pointed lobes whose apices are directed downward, i.e., at an angle of 45° or less from the leaf axis (Fig. 4.2(36)).
8. Hastate: with two large pointed lobes whose apices are directed outward, i.e., at an angle of greater than 45° from the leaf axis (Fig. 4.2(37)).
9. Peltate: petiole attached within the boundaries of the leaf margin (Fig. 4.2(38)).
10. Other.

IV. Form of leaf margin

A. Entire: margin forming a smooth line or arc without noticeable projections or indentations.

B. Lobed: margin indented one quarter or more of the distance to the midvein or (where this is lacking) to the long axis of the leaf (Fig. 4.2(41)).

C. Toothed: margin having projections with pointed apices, indented less than one quarter of the distance to the midvein or long axis of the leaf (Figs. 4.2(42,43,45–7,49,50)).

1. Dentate (Fig. 4.2(42)): dentations are pointed, with axes of symmetry approximately perpendicular to the trend of the margin; as with leaf apices, these can be acute (Fig. 4.2(19)), obtuse (Fig. 4.2(20)), acuminate (Fig. 4.2(23)), attenuate (Fig. 4.2(27)), or mucronate (Fig. 4.2(24)) (definitions as under apices, above).

2. Serrate (Fig. 4.2(43)): serrations are pointed, with their axes inclined (i.e., at an oblique angle) to the trend (tangent) of the margin.
a. Apical angle
   (i) Acute: angle formed by the two sides less than 90° (typical).
   (ii) Obtuse: angle formed by the two sides greater than 90° (rare).
b. Serration type: determined by the shape of the basal side of the tooth (shown on the vertical or lettered columns of Fig. 4.2(52)) vs. the shape of the apical side (shown on the horizontal or numbered rows of Fig. 4.2(52)). Families and genera often exhibit a high degree of consistency in their possession of one or two types of serration. See Fig. 4.2(47) for orientation. The configurations used for each side of the tooth are given in Table 4.1.

D. Crenate (Fig. 4.2(44)): crenations are smoothly rounded, without a pointed apex.

E. Erose (Fig. 4.2(48)): irregular, as if chewed.

F. Revolute or enrolled (Fig. 4.2(51)): margin turned under or rolled upon itself like a scroll – applies to both entire and non-entire margins.

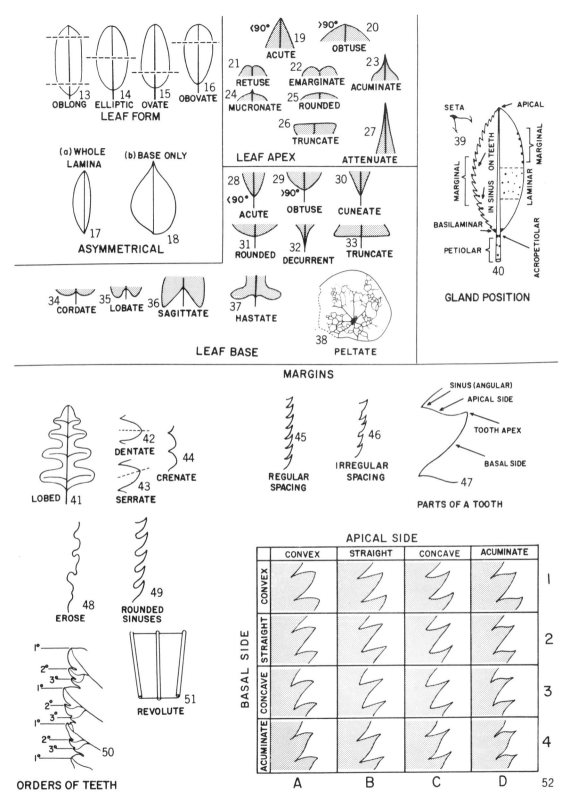

Fig. 4.2 (13–52). Leaf architectural features (*cont.*) — leaf(let) form, shape of apex and base, gland position, and marginal configuration. (From Hickey 1973.)

Table 4.1. *Serration types (see Fig. 4.2(52))*

Marginal configuration	Apical	Basal
	Column	Row
Convex	A	1
Straight	B	2
Concave	C	3
Acuminate	D	4

G. Sinuses: incisions between marginal projections of any sort − lobes, dentations, serrations, or crenations.

1. Rounded (Fig. 4.2(49)): margins of sinus meeting in a smooth curve.
2. Angular (Fig. 4.2(47)): margins of sinus meeting at a point.

H. Spacing: interval between corresponding points on the teeth or crenations.

1. Regular (Fig. 4.2(45)): interval varying by no more than 25 per cent.
2. Irregular (Fig. 4.2(46)): interval varying by more than 25 per cent.

I. Series: teeth separated into size groups.

1. Simple (Fig. 4.2(45)): teeth all of one size.
2. Compound (Fig. 4.2(50)): teeth in two or more definite size groups − double serrations, etc.

In addition to the form and spacing of the marginal teeth, their vein configuration and glandular characteristics have proven highly important in systematic determinations (Hickey and Wolfe, in preparation). These additional characteristics of tooth morphology are treated in Section IX of this outline (p. 36).

V. Leaf texture

A. Membranaceous: thin and semi-transparent, like a fine membrane.

B. Chartaceous: opaque and like writing paper.

C. Coriaceous: leathery, thick, stiff.

D. Other.

VI. Gland Position (includes nectaries, hydathodes, tanniniferous glands, etc.) (Figs. 4.2(39 and 40))

A. Petiolar: on the tissue of the petiole; includes acropetiolar at the top of the petiole.

B. Basilaminar: on the foliar tissue at the base of the blade.

C. Laminar: generally distributed on the foliar tissue.

D. Apical: on the leaf apex.

E. Marginal: distributed on the margin or marginal processes.

1. At the margin in entire margined leaves.
2. On the teeth.

a. As a glandular thickening.
b. As a glandular seta or bristle (Fig. 4.2(39)).
3. In the sinus.

VII. Petiole(ule)

A. Normal: without noticeable thickenings or other processes.

B. Inflated: thickened, includes pulvini.

C. Winged: with a narrow strip of foliar tissue on each side.

D. Absent: blade sessile, arising directly from the axis of attachment, without an intervening bladeless area of the leaf.

VIII. Types of venation
(For definitions of vein orders see p. 32).

1. Pinnate: with a single primary vein (midvein) serving as the origin for the higher order venation.
a. Craspedodromous[1]: secondary veins terminating at the margin.
 (i) Simple: all of the secondary veins and their branches terminating at the margin (Fig. 4.3(53)).
 (ii) Semicraspedodromous: secondary veins branching just within the margin, one of the branches terminating at the margin, the other joining the superadjacent secondary (Fig. 4.3(54)).
 (iii) Mixed: some of the secondary veins terminating at the margin and an approximately equal number of (usually intervening) secondaries otherwise (Fig. 4.3(55)).
b. Camptodromous: secondary veins not terminating at the margin.
 (i) Brochidodromous[2] − secondaries joined together in a series of prominent arches (Fig. 4.3(58)).
 (ii) Eucamptodromous: secondaries upturned and gradually diminishing apically inside the margin, connected to the superadjacent secondaries by a series of cross veins without forming prominent marginal loops (Fig. 4.3(59)).
 (iii) Reticulodromous: secondaries losing their identity toward the leaf margin by repeated branching into a vein reticulum (Fig. 4.3(60)).
 (iv) Cladodromous[3] − secondaries freely ramified toward the margin (Fig. 4.3(61)).
c. Hyphodromous: all but the primary vein absent, rudimentary, or concealed within a coriaceous or fleshy mesophyll (Fig. 4.3(56)).
2. Parallelodromous: two or more primary veins originating beside each other at the leaf base and running parallel to the apex where they converge (Fig. 4.3(57)).
3. Campylodromous[4] − several primary veins or their branches, originating at, or close to, a single point and running in strongly developed, recurved arches before converging toward the leaf apex. Vein pattern convergent above and below (Fig. 4.3(70)).

---

[1] From the Greek *kraspedon*, edge, border; and *dromos*, a running, course.
[2] From the Greek *brochos*, a noose.
[3] From the Greek *klados*, a branch.
[4] From the Greek *campylos*, a bend or curve.

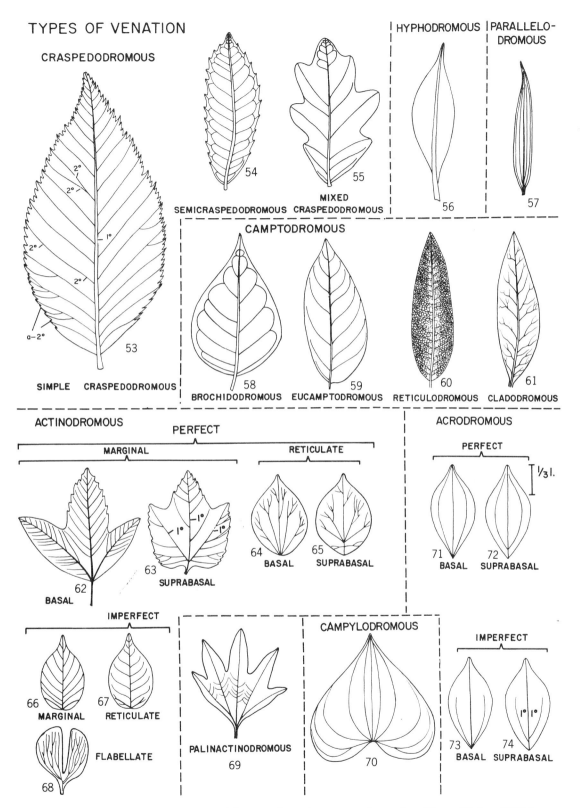

TYPES OF VENATION

CRASPEDODROMOUS

54

55

SEMICRASPEDODROMOUS

MIXED CRASPEDODROMOUS

HYPHODROMOUS

PARALLELO-DROMOUS

56

57

2°
2°
1°
2°
2°
2°
a—2°

53

SIMPLE CRASPEDODROMOUS

CAMPTODROMOUS

58

59

60

61

BROCHIDODROMOUS

EUCAMPTODROMOUS

RETICULODROMOUS

CLADODROMOUS

ACTINODROMOUS

PERFECT

MARGINAL

RETICULATE

ACRODROMOUS

PERFECT

1° 1°

62

BASAL

63

SUPRABASAL

64

BASAL

65

SUPRABASAL

71

BASAL

72

SUPRABASAL

⅓ l.

IMPERFECT

66

MARGINAL

67

RETICULATE

68

FLABELLATE

PALINACTINODROMOUS

69

CAMPYLODROMOUS

70

IMPERFECT

73

BASAL

1° 1°

74

SUPRABASAL

Fig. 4.3 (53–74). Leaf architectural features (*cont.*) — types of venation. (From Hickey 1973.)

4. Acrodromous: two or more primary or strongly developed secondary veins running in convergent arches toward the leaf apex. Arches not recurved at base (Fig. 4.3(71–74)).
a. Position.
   (i) Basal: acrodromous veins originating at the base of the leaf (Fig. 4.3(71,73)).
   (ii) Suprabasal: acrodromous veins originating some distance above the leaf base (Fig. 4.3(72,74)).
b. Development.
   (i) Perfect: acrodromous veins well developed, running at least two-thirds of the distance to the leaf apex (Fig. 4.3(71,72)).
   (ii) Imperfect; acrodromous veins running less than two-thirds of the distance to the leaf apex (Fig.4.3(73, 74)).
5. Actinodromous: three or more primary veins diverging radially from a single point (Fig. 4.3(62–7)).
6. Palinactinodromous: primaries diverging in a series of dichotomous branchings, either closely or more distantly spaced, e.g., *Platanus* (Fig. 4.3(69)).
(The following categories apply to both 5 and 6 above)
a. Position of the first point of primary vein radiation.
   (i) Basal: initial point of radiation at the leaf base (Fig. 4,3(62,64,66,67))
   (ii) Suprabasal; initial point of radiation located some distance above the leaf base (Fig.4.3(63,65,69)).
b. Development
   (i) Perfect: ramifications of the lateral actinodromous veins covering at least two-thirds of the blade area (Fig. 4.3(62–5)).
    (a) marginal: actinodromous veins reaching the margin (Fig.4.3(62,63)).
    (b) reticulate: actinodromous veins not reaching the margin (Fig. 4.3(64,65)).
   (ii) imperfect: veins originating on the lateral actinodromous primary veins covering less than two-thirds of the blade area.
    (a) marginal (Fig. 4.3(66)).
    (b) reticulate (Fig. 4.3(67)).
   (iii) Flabellate: several to many equally fine basal veins diverge radially at low angles and branch apically (Fig. 4.3(68)).

Orders of venation

Introduction

In most leaves venation is clearly differentiated into a number of size classes. Veins of a particular size class also display some degree of uniformity in their courses and patterns of distribution in relation to other classes and to the marginal features of the leaf. In practice, the objective designation of vein order is more complex than easily observed differences in thickness, course, and pattern among size classes would seem to indicate. However, the recognition of vein orders is essential in describing leaf architecture.

The primary venation of the leaf serves as the starting point for the identification of the various vein orders. Veins of the primary order are the thickest veins of the leaf and occur singly or as a medial vein accompanied by others (lateral primaries) of roughly equal thickness. These emerge from the petiole of the leaf, or the medial and lateral primaries may give rise to additional lateral primary veins above the base.

The fundamental rule for the determination of the order of a vein is its relative size at its point of origin. Where a lateral vein branch is approximately equal in width, measured just above the point of branching, to the continuation of the source vein just above the point of branching, both branches are of the same order; where the lateral vein branch is markedly finer than the continuation of its source, that branch is of a higher order. This rule is applied when tracing a vein of any order distally from its point of origin. Thus, the next set of branches of markedly smaller size than their primary source are the secondary veins, while the next finer set arising from both primary and secondary veins are designated the tertiary veins, and so on until reaching the ultimate vein order present in the leaf.

Subsidiary to the rule of size at the point of origin in the recognition of vein order is a consideration of the behaviour of a vein in relation to veins of other orders and to marginal features of the leaf blade. It is necessary to consider this additional criterion when dealing with orders having a wide range of widths or where the boundaries between size classes are not particularly sharp. Three of the most common situations where designation of vein order is difficult are:

(1) suprabasal lateral primary veins — i.e., primary veins originating as lateral branches of a primary vein above the base of the leaf;
(2) branches of secondary veins; and
(3) intersecondary veins.

Lateral primary veins frequently occur as suprabasal branches of a primary vein. In most examples, the thickness of these veins at their point of branching will be the same, or very nearly the same, as the continuation of their source. In other cases the lateral primary branches will be somewhat thinner than their primary source but still considerably thicker than the next thinner set of veins, i.e., the secondaries. Here a number of features permit the recognition of these lateral branches as primary veins. The primary branches will give rise to secondary veins whose size and behaviour is no different from secondaries arising on primary veins which originate at the base of the leaf. The primary branch may form the midvein of a leaf lobe as do the other primaries, and in all other ways its behaviour will be comparable to that of the other members of its order (Fig. 4.3(69)). However, if the thickness of a branch of the primary vein is the same as that of the typical secondary veins, it is regarded as a secondary, despite some differences in its behaviour (usually angle of origin) in relation to them.

Branches of the secondary veins ('outer secondary nerves' of von Ettingshausen, a–2° of Fig. 4.3(53)), are frequently of the same thickness at their points of branching as the continuation of the secondary veins from which they arise. In cases where the branches are slightly thinner there is no difficulty in classifying them as secondaries because of their geometric relationships. However, in series where the size of these branches gradually diminishes toward that of the third order their behaviour also gradually alters until by both criteria they have become indistinguishable from the tertiaries.

Intersecondary veins are intermediate in thickness between second and third order veins and most often occur interspersed between the secondaries of pinnate leaves (Fig. 4.4(75e)). Although their length is shorter than that of the secondary veins their course is parallel, or nearly so, to them, and they more closely resemble the secondaries in their relationship to other orders of veins.

Other difficulties in designating vein order arise as the tendency toward dichotomous branching increases in the higher vein orders. However, the behaviour of the resultant branches frequently serves to distinguish separate vein orders. In other cases the higher order venation of a leaf merges into a reticulum in which designation of vein order is impossible (Fig. 4.4(91, 92)). [Further details concerning high-order venation, vein terminations, and vein islets, etc. are presented on pp. 70–2. *Editor.*]

B. Key to orders of venation

1. *Primary veins (1°)*: the thickest vein(s) of the leaf, occurring either singly as the midvein, or as a series of veins of relatively equal thickness emerging from the petiole (Fig. 4.3(62,64)), or as branches of the medial or lateral primary veins above the base of the blade (suprabasal primaries). In the latter case, both the primary branch and the continuation of its primary source are of the same relative thickness measured just above the point of branching.
a. Size: determined midway between the leaf apex and base as the ratio of vein width (vw) to leaf width (lw).

$$\frac{vw}{lw} \times 100\% = \text{Size}.$$

    (i) Massive: > 4% (Fig. 4.3(56)).

    (ii) Stout: 2–4% (Fig. 4.3(54)).
    (iii) Moderate: 1·25–2% (Figs. 4.3(55,58,59)).
    (iv) Weak: < 1·25% (Fig. 4.3(71)).
b. Course
    (i) Straight: lacking noticeable curvature or change in course (Fig. 4.3(58)).
      (a) Unbranched: lacking ramifications of primary rank (Fig. 4.4(75)).
      (b) Branched: with one or more primary ramifications (Fig. 4.3(63)).
    (ii) Markedly curved: bent noticeably into a smooth arc (Fig. 4.3(54)).
    (iii) Sinuous: repeated smooth changes in direction of curvature (Fig. 4.4(77)).
    (iv) Zigzag: repeated angular changes in direction (Fig. 4.4(78)).
2. *Secondary veins (2°)*: the next smaller size class of veins (which arise from the primaries) and their branches (Fig. 4.4(75)) of relatively equal thickness measured just above the point of branching.
a. Angle of divergence: measured between the branch and the continuation of the source vein above the point of branching (Fig. 4.4(75)).
    (i) Acute: angle less than 80°.
      (a) Narrow: < 45°.
      (b) Moderate: 45–65°.
      (c) Wide: 65–80°.
    (ii) Right-angle or approximately so: 80–100°.
    (iii) Obtuse: > 100°.
b. Variations in angle of divergence.
    (i) Divergence angle nearly uniform (Fig. 4.4(75)): the general case, departures from which are often taxonomically useful features.
    (ii) Upper secondary veins more obtuse than lower (Fig. 4.4(79)).
    (iii) Upper secondary veins more acute than lower (Fig. 4.4(80)).
    (iv) Only lowest pair of secondary veins more acute than pairs above it (Fig. 4.4(81)).
    (v) Lower and upper secondary veins more obtuse than middle sets (Fig. 4.4(82)).
    (vi) Divergence angle more acute on one side of the leaf than on the other (Fig. 4.3(59)).
    (vii) Divergence angle varies irregularly (Fig. 4.4(83)).
c. Relative thickness of secondary veins: a measure of the width of the middle secondary veins compared to those of the primary and tertiary orders. Such relative estimates of thickness for this and succeeding vein orders are essentially a measure of the proportional reduction in width from one vein order to the next. This is a useful charcter only in cases of marked departure from the width expected in the proportional reduction series.
    (i) Thick: proportionately wide in relation to the primary and tertiary vein orders or to the secondaries in other leaves of similar size.
    (ii) Moderate: the general case.
    (iii) Fine to hair-like; proportionately narrow in relation to the primary and tertiary vein order or to the secondaries in other leaves of similar size.
d. Course — more than one of the following terms may apply.
    (i) Straight without noticeable deviation in course (Fig. 4.3(53)).
    (ii) Recurved: arching basally for a portion of its course

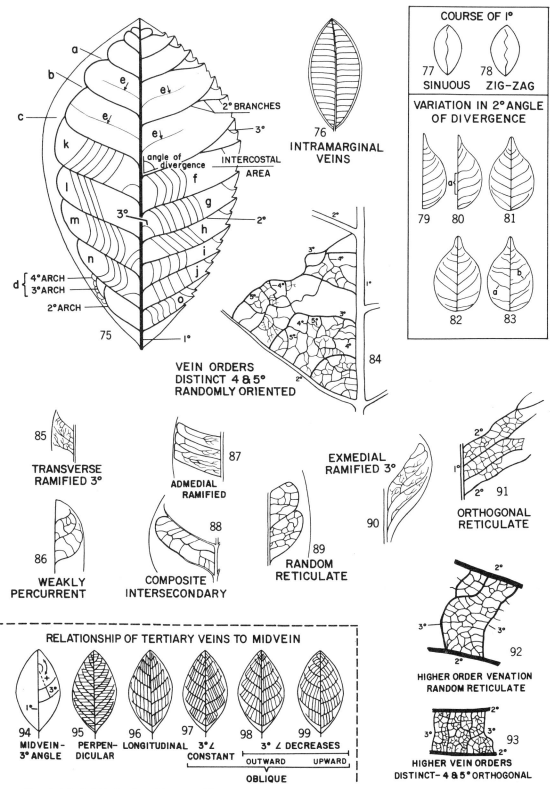

Fig. 4.4 (75-99). Leaf architectural features (*cont.*) — orders of venation and vein configuration.
(From Hickey 1973.)

(lower secondaries in Fig. 4.3(53)).

  (iii) Curved: bending in a arc.

    (a) Uniformly: arc smooth or gradually increasing in degree of curvature (Fig. 4.3(66)).

    (b) Abruptly: sharp local increase in degree of curvature (Fig. 4.3(58)).

  (iv) Sinuous: repeated smooth changes in direction of curvature (Fig. 4.4(83),a).

  (v) Zigzag: repeated angular changes in direction (Fig. 4.4(83),b).

  (vi) Unbranched: without second order ramifications (Fig. 4.3(58)).

  (vii) Branched: with one or more secondary ramifications (Fig. 4.4(75)).

  (viii) Provided with outer secondary veins: a series of secondary branches arising on the exmedial side of the secondary vein (Fig. 4.3(53 d)).

e. Behaviour of loop-forming branches (if any).

  (i) Joining superadjacent secondary at an acute angle (Fig. 4.4(75),a).

  (ii) Joining superadjacent secondary at a right-angle (Fig. 4.4(75),b).

  (iii) Joining superadjacent secondary at an obtuse angle (Fig. 4.4(75),c).

  (iv) Enclosed by secondary arches. 3° or 4° arches (Fig. 4.4(75),d).

  (v) Forming an intramarginal vein (Fig. 4.4(76)).

f. Intersecondary veins: thickness intermediate between that of the second and third order veins; generally originating from the medial primary vein, interspersed among the secondary veins, and having a course parallel, or nearly so, to them. They are of two types:

  (i) Simple: consisting of a single vein segment (Fig. 4.4 (75),e).

  (ii) Composite: made up of coalesced tertiary vein segments for over 50 per cent of its length (Fig. 4.4(88)).

g. Intramarginal vein: a vein closely paralleling the leaf margins and into which the secondary veins merge. Probably the result of the fusion and straightening of the exmedial brochidodromous secondary arch segments to form what appears to be an independent vein (Fig. 4.4(76)).

h. Intercostal areas: those portions of the leaf blade lying between the secondary veins (Foster, 1950).

3. *Tertiary veins (3°)*: the next finest branches of the secondary veins and those branches of equal thickness from the primaries.

a. Angle of origin (defined above). When the average angle of tertiary origin on the exmedial (lower) side of the secondary veins is compared with the average on the admedial (upper) side of the secondary veins, the combinations shown in Table 4.2 are possible. This trait is of diagnostic value.

As a rule, in those tertiary veins which originate on the admedial side of the secondary veins and curve to join the primary forming the midvein, the angle of tertiary vein origin on the midvein equals the angle of tertiary vein origin on the exmedial side of the secondary veins of the leaf. Departure from this rule is a taxonomically significant feature.

b. Pattern.

  (i) Ramified: tertiary veins branching into higher orders without rejoining the secondary veins (although their higher order derivatives may do so).

    (a) Exmedial: branching oriented toward the leaf axis rare (Fig. 4.4(90)).

Table 4.2 *Analysis of tertiary vein origin*

Angle of tertiary origin on the admedial (upper) side of the secondaries	Angle of tertiary origin on the the exmedial (lower) side of the secondaries		
	Acute	Right	Obtuse
Acute	AA (f)	RA (i)	OA (l)
Right	AR (g)	RR (j)	OR (m)
Obtuse	AO (h)	RO (k)	OO (n)

The lower-case letters refer to examples shown in Fig. 4.4(75).

    (b) Admedial: branching oriented toward the leaf margin (Fig. 4.4(87)).

    (c) Transverse: branching oriented across intercostal area; the commonest case (Fig. 4.4(85)).

  (ii) Reticulate: tertiary veins anastomosing with other tertiary veins or with the secondary veins.

    (a) Random reticulate: angles of anastomoses vary (Fig. 4.4(89)).

    (b) Orthogonal reticulate: angles of anastomoses predominantly right angles (Fig. 4.4(91)).

  (iii) Percurrent: tertiaries from the opposite secondaries joining.

    (a) Course

     (1) simple: unbranched (Fig. 4.4(75)).

     (2) forked: giving rise to third order ramifications (Fig. 4.4(88)).

     (3) straight: passing across the intercostal area without a noticeable change in course (Fig. 4.4 (75 h,l)).

     (4) convex: middle portion of the vein curving away from the centre of the leaf (Fig. 4.4 (75f)).

     (5) concave: middle portion of the vein curving toward the centre of the leaf (Fig. 4.4(75 n)).

     (6) sinuous: repeatedly changing direction of curvature (Fig. 4.4(86)).

     (7) retroflexed: forming a single *S*-shaped curve concave apically and convex basally (Fig. 4.4(75 o)).

     (8) recurved: curving inward from point of origin on the adaxial side of a secondary vein to terminate on the midvein of the leaf (Fig. 4.4(75)).

    (b) Relationship to midvein (Fig. 4.4(94)).

     (1) approximately at right angles (Fig. 4.4(95)).

     (2) longitudinal: approximately parallel (Fig. 4.4 (96)).

     (3) oblique: trending in an obtuse or, rarely, an acute angle to the midvein.

      [a] tertiary angle with midvein remaining approximately constant (Fig. 4.4(97)).

      [b] angle decreasing exmedially (Fig. 4.4(98)).

      [c] angle decreasing apically (Fig. 4.4(99)).

    (c) Arrangement

     (1) predominantly alternate: joining each other with an offset, i.e. an abrupt angular discontinuity such as a branch (Fig. 4.4(88,93)).

     (2) predominantly opposite: joining each other smoothly, i.e. in a straight or curved path (Fig. 4.4(75)).

(3) alternate and opposite in about equal proportions.

(4) distant: interval between veins < 3 veins/cm.

(5) close: interval between veins > 3 veins/cm.

4. *Higher-order venation*: the next finer order of veins originating from the tertiaries and those of equal size from lower order veins is known as the quaternary (4°) venation and the veins originating from these and those of equal size from lower orders are the quinternaries (5°), etc.

a. Resolution

(i) Higher order venation forming a reticulum in which vein orders cannot be distinguished (Fig. 4.4(92)).

(ii) Vein orders distinct (Fig. 4.4(84,93)).

b. Quaternary veins

(i) Size: a relative measure of the width of the quaternary veins compared to those of the third and fifth orders. Such relative estimates of thickness for this and the fifth-order veins (below) are essentially a measure of the proportional reduction in width from one vein order to the next. Notable only is any marked departure from the width expected for the fourth order (fifth order, below) veins as part of a proportional reduction series.

(a) Thick: wider than expected.

(b) Thin: narrower than expected.

(ii) Course

(a) Relatively randomly oriented (Fig.4.4(84)).

(b) Orthogonal: arising at right angles (Fig.4.4(93)). Their subsequent courses may or may not be at right-angles.

c. Quinternary veins (analysed as in b. (i) above).

(i) Size (as above).

(a) Thick

(b) Thin

(ii) Course (analysed as in b. (ii) above).

(a) Random (Fig. 4.4(84), as above).

(b) Orthogonal (Fig.4.4(93), as above).

d. Highest vein order of leaf: 3°, 4°, 5°, 6°, 7°.

e. Highest vein order showing excurrent branching: 2°, 3°, 4°, 5°, 6°.

f. Marginal ultimate venation.

(i) Incomplete: freely ending veinlets directly adjacent to the margin (Fig. 4.5(102,104)).

(ii) Looped: the major portion of the marginal ultimate venation recurved to form loops (Fig.4.5(100,101, 103)).

(iii) With a fimbrial[1] vein: higher vein orders fused into a vein running just inside the margin (fimbrial vein) (Fig.4.5(105,106)).

5. Veinlets: the freely ending ultimate veins of the leaf and veins of the same order which occasionally cross aeroles (see below) to become connected distally.

---

[1] Melville is correct in his objection that the word *fimbriate*, which I used to describe this condition in previous versions of this system, has the connotation of *fringed*, in botanical Latin, rather then *hemmed*, as in classical Latin. However, in order to maintain uniformity with published descriptions of leaf architecture using my terminology and to distinguish this type of marginal venation from the very different intramarginal vein, I will retain the word *fimbrial* for it. *Fimbrial* does not seem to have been used in the sense of fringed and thus should retain its classical meaning.

a. None (Fig. 4.5(110)).

b. Simple: without branches.

(i) Linear (Fig.4.5(111)).

(ii) Curved (Fig. 4.5(112)).

c. Branched: giving rise to ramifications by dichotomizing.

(i) Once (Fig. 4.5(113)).

(ii) Twice (Fig. 4.5(114)).

(iii) Three times (Fig. 4.5(115)), etc.

6. Areoles: the smallest areas of the leaf tissue surrounded by veins which taken together form a contiguous field over most of the area of the leaf. Thus, smaller areas occasionally formed when veinlets cross their areoles are excluded. Any order of venation in a leaf from the primary to the highest order below that of the freely ending veinlets can form one or more sides of an areole. However, only the order represented by the veinlets will intrude, or occasionally cross, the islets formed by the non-freely ending veins. The appearance and characteristics of the areoles are termed areolation.

a. Development

(i) Well developed: meshes of relatively consistent size and shape (Fig. 4.5(119)).

(ii) Imperfect: meshes of irregular shape, more or less variable in size (Fig. 4.5(118)).

(iii) Incompletely closed meshes: one or more sides of the mesh not bounded by a vein, giving rise to anomalously large meshes of highly irregular shape (Fig. 4.5(117)).

(iv) Areolation lacking; as in hyphodromous or succulent leaves. Very rarely venation simply ramifying into the intercostal spaces with no coherent shape, size or pattern to the areas surrounded by veins (Fig. 4.5(116)).

b. Arrangement

(i) Random: areoles showing no preferred orientation (Fig.4.5(118)).

(ii) Oriented: areoles having a similar alignment or pattern of arrangement within particular blocks or domains (Fig.4.5(119)).

c. Shape

(i) Triangular (Fig. 4.5(109)).

(ii) Quadrangular (Fig. 4.5(108)).

(iii) Pentagonal (Fig. 4.5(107)).

(iv) Polygonal: with more than five sides.

(v) Rounded

(vi) Irregular

d. Size: the size classes of areoles chosen appear to represent, at least partially, naturally occurring grouping.

(i) Very large: > 2 mm.

(ii) Large: 2-1 mm.

(iii) Medium: 1-0·3 mm.

(iv) Small: < 0·3 mm.

IX. Elements of tooth architecture

Since my system was first presented (Hickey 1973) it has become apparent that tooth architecture is more important than was at first realized. This has made it necessary to append the present section, in which attention is confined to diagnostically important characters such as the configuration of glands, the structure of the tooth apices, and the venation of the lateral teeth. For further details see Hickey and Wolfe (1975).

The shapes of marginal processes are dealt with under Section IV. C.

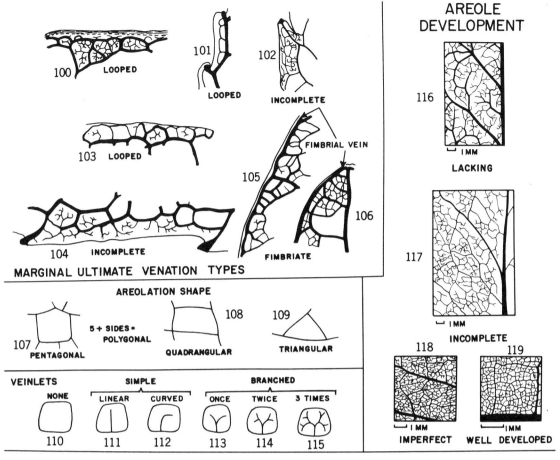

AREOLE
DEVELOPMENT

100 LOOPED
101 LOOPED
102 INCOMPLETE
103 LOOPED
104 INCOMPLETE
105 FIMBRIAL VEIN
106 FIMBRIATE

MARGINAL ULTIMATE VENATION TYPES

116 LACKING
117 INCOMPLETE
118 IMPERFECT
119 WELL DEVELOPED

AREOLATION SHAPE

107 PENTAGONAL
5 + SIDES = POLYGONAL
108 QUADRANGULAR
109 TRIANGULAR

VEINLETS

NONE	SIMPLE		BRANCHED		
	LINEAR	CURVED	ONCE	TWICE	3 TIMES
110	111	112	113	114	115

SOME FEATURES OF TOOTH ARCHITECTURE

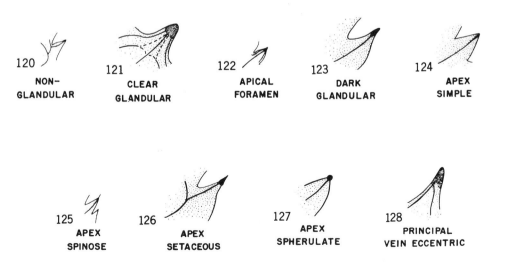

120 NON-GLANDULAR
121 CLEAR GLANDULAR
122 APICAL FORAMEN
123 DARK GLANDULAR
124 APEX SIMPLE

125 APEX SPINOSE
126 APEX SETACEOUS
127 APEX SPHERULATE
128 PRINCIPAL VEIN ECCENTRIC

Fig. 4.5 (100–28). Leaf architectural features (*cont.*) – ultimate venation, areolation, and selected elements of tooth architecture. (100–19 from Hickey 1973.)

SOME FEATURES OF TOOTH ARCHITECTURE

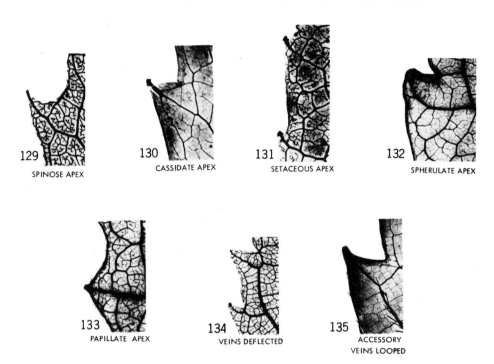

Fig. 4.6 (129–35). Selected features of tooth architecture.

A. Glandularity: recognized by concentrations of opaque material in the tooth apex or vein termination or by dense masses of cells either on the apex or lining an apical opening.

1. Non-glandular (Figs. 4.5(120), 4.6(129)).
2. Glandular
a. Clear: recognized by dense cell concentrations (Fig. 4.5 (121)) or by an apical opening (foramen) (Fig. 4.5(122)).
b. Dark: recognized by dark or opaque concentrations caused by accumulation of tannin or resins (Figs. 4.5(123), 4.6(132)).

B. Apical termination of the tooth.

1. Simple: tooth apex formed by the change in direction of the leaf margin without the additional elements listed below (Fig. 4.5(124)).
2. Spinose: principal vein of tooth projecting beyond the apex (*Castanea*) (Figs. 4.5(125), 4.6(129)).
3. Cassidate or mucronate: with an opaque, non-deciduous cap or mucro fused to the tooth (Monimiaceae) (Fig. 4.6(130)).
4. Setaceous: an opaque, deciduous bristle or cap thickened proximally and not fused firmly with the remaining tooth substance (Theaceae, Ochnaceae) (Figs. 4.5(126), 4.6(131)).
5. Spherulate: having a spherical callosity fused to the apex (*Idesia*, Salicaceae) (Figs. 4.5(127), 4.6(132)).
6. Tylate[1]: with a translucent pad of densely packed cells (hydathodal or nectariferous tissue) into which the veins

---

[1] From the Greek *tyle* – callus, pad, bolster.

run and disappear (Cucurbitaceae, Begoniaceae) (Fig. 4.5 (121,128)).
7. Papillate: having a clear, nipple-shaped, glandular apical termination (Fig. 4.6(133)).
8. Foramenate: with a foramen which broadens from the termination of the principal vein to the exterior (Rosaceae, Saxifragaceae) (Fig. 4.5(122)).

C. Principal vein configuration of the tooth: generally a secondary or tertiary vein.

1. Course of the vein
a. Central: approximately following the axis of symmetry (Figs. 4.5(121,124), 4.6(130)).
b. Eccentric: running to one side of the axis of symmetry (Figs. 4.5(128), 4.6(134)).
2. Origin of the vein
a. Direct: running straight into the tooth as a continuation of the laminar venation (Figs. 4.5(123), 4.6(130)).
b. Deflected: either arising just below the point of the marginal sinus, or the laminar vein which runs to the tooth branches equally before entering it, with one branch entering the tooth and the other running to the sinus (common in the Subclass Rosidae) (Figs. 4.5(120), 4.6(134)).

D. Accessory veins: those of higher order than the principal vein of the tooth.

1. Absent (Fig. 4.5(125)), or incomplete (Fig. 4.6(134)) or looped (Fig. 4.6(135)).

2. Present
a. Joining with the principal vein (Chloranthaceae) (Fig. 4.6 (130)).
b. Connivent with the principal vein, terminating in the glandular pad (Cucurbitaceae) (Fig. 4.5(121)) or in the apical foramen (Rosaceae) (Fig. 4.5(122)).

Postscript

Since this chapter was drafted, Melville (1976) has published an alternative method of classifying leaf architecture. He advocates that his new system should be used instead of those which Mouton (1970) and I (Hickey 1973) have already proposed and which are now further developed in this present chapter. He also objects to the earlier views of von Ettingshausen which served as a starting point for the work by Mouton and myself. Melville's objections to our proposals rest on three principal arguments:

(1) It is necessary to have a system of leaf architecture that covers all megaphyllous leaves since the time such leaves first appeared in the fossil record, rather than of only the angiosperms, or a part of them, since the Cretaceous.
(2) It is necessary to return to a Latin rather than a Greek base for the terminology of the classification, on the grounds of greater compatibility with botanical descriptive practice.
(3) It is complained that von Ettingshausen disregarded terms already in use or available for leaf venation.

Space allows only a brief outline of my reply here. However, careful examination of cleared specimens shows that the common patterns of venation that Melville recognizes in the leaf architecture of ferns, pteridosperms, and angiosperms are achieved by disregarding certain very important characters unique to angiosperm venation including the presence of:

(1) three or more orders of venation of discretely different width and behaviour,
(2) freely ending veinlets, often branching, and
(3) vein anastomoses between two or more orders of veins.

By disregarding such important differences in vein pattern or by viewing convergences in the vein pattern of reduced, xeromorphic angiosperm leaves as the result of common ancestry, Melville is able to find seeming homologies between glossopterid leaf venation and that of such advanced and divergent families as Brassicaceae, Proteaceae, Apocynaceae and even Asteraceae. In addition, recognition of the categories which he proposes frequently combines morphology with an inferred phylogeny or ontogeny; features are frequently not defined in terms of their simplest components; and important distinctions, such as between leaf shape and vein course, the veins of different orders, or between venation and areolation, are not made.

In his objection to Greek roots for terms, Dr. Melville overlooks the fact that Greek has always provided botanists with a rich store of terms — one has only to turn to Stearn's work (1966) on Botanical Latin with its extensive section on Greek syntax and transliteration to be aware of this. Finally, in regard to priority of terminology, many of the terms in use before the publication of von Ettingshausen's system were inexactly defined or ambiguously used. The advantage of von Ettingshausen's approach is that his terms form part of a unified and logically organized codex in which the meaning of the terms could be understood both from their definitions and also from their relationship to the other terms of their set. Contrary to what Melville says, however, few terms were actually employed for vein patterns before the time of von Ettingshausen. Melville himself uses only nine terms that were previously applied to venation and all of them were previously adopted by von Ettingshausen and myself. Most of the fifty or so terms that he proposes are new applications of Latin, or occasionally Greek, words to venation.

Literature cited

Cronquist 1968; Ettingshausen, von 1861; Foster 1950; Hickey 1973; Hickey and Wolfe 1975; Lee 1948; Melville 1976; Mouton 1970; Stearn 1966; Takhtajan 1969.

5

TRICHOME DESCRIPTION AND CLASSIFICATION[1]

WILLIAM L. THEOBALD, JOSEPH L. KRAHULIK, AND REED C. ROLLINS

Introduction

Trichomes have long been of considerable importance in comparative systematic investigations of angiosperms. They are frequently present, easily observable, and have often been found to have variation patterns which correlate with other features of the taxa under investigation (Cowan 1950; Cutler 1969; Heitzelman and Howard 1948; Metcalfe 1960, 1971; Metcalfe and Chalk 1950; Netolitzky, 1932; Ramayya 1972; Roe 1971; Solereder 1908; Tomlinson 1961, 1969; Uphof, Hummel, and Staesche 1962). The definition of trichomes ranges from the generalized morphological one of 'hair or bristle' (Lawrence 1951) to the somewhat more specific anatomical one of an epidermal outgrowth of diverse form, structure, and function (Esau 1965a). In most anatomical works (de Bary 1884; Esau 1965a; Foster 1949) trichomes are distinguished from emergences, which represent a combined epidermal and subepidermal outgrowth. However, as noted by Netolitzky (1932), both trichomes and emergences intergrade and only through developmental and comparative studies can one ascertain whether the origins of some outgrowths are solely epidermal, or represent outgrowths of both epidermal and subepidermal origin. Although papillate epidermal cells, as a group, are sometimes treated as morphologically distinct from trichomes, they also intergrade with them, and might best be treated as a subgroup. Their significance in comparative anatomical studies has been well documented by Metcalfe (1960, 1971) for the Gramineae and Cyperaceae. They are also discussed on pp. 148-51 of this present work. The structures described as water vesicles by Haberlandt (1914) have been treated as a specialized type of trichome by Foster (1949) and Esau (1965a), and are characteristic of some members of the Aizoaceae. Between all of the above-mentioned specialized cases there are complete series of intergrades, which must be kept in mind when undertaking any type of comparative study. Root hairs, although by definition considered to be trichomes, have been omitted from this treatment because of their relative uniformity,

their lack of relevance to date in systematic studies, and the obvious difficulties connected with their utilization.

The number of species that are completely devoid of trichomes on all parts of the plant represent a minority of the angiosperms as a whole. Many instances of the glabrous condition represent cases where the trichomes have degenerated at an early stage in their development or were lost shortly after maturation. Descriptive terms used for trichomatous surfaces (e.g. tomentose, villous, hirsute, etc.) often give a very sketchy idea of the type of trichome present, and they give no idea at all of the histological characteristics of the trichomes. As is well known, there is a great deal of intergradation in vesture types, and perhaps as a result the rather imprecise use of these terms is almost characteristic of many taxonomic works. Metcalfe and Chalk (1950) have noted that length, size, and density are far more variable in response to varied environmental conditions than are the types of trichomes, and these aspects are, therefore, of lesser value as a basis for many kinds of interpretation than the trichome type itself. The presence of a particular type of trichome, on the other hand, can frequently delimit species, genera, or even whole families (Metcalfe and Chalk 1950). It is not our intention to suggest that terms referring to the nature of the vesture, including presence or absence, be discontinued in descriptive taxonomy. Rather, it seems to us important to include, in addition, a definitive description of the trichome types involved. It is in the trichome as a morphological entity that precision of description can be achieved if the careful use of clearly defined terms is strictly followed. The style of a standard morphological description of a taxon, which has served so well in taxonomy for the last several hundred years, should now be applied to trichomes.

The adaptive value of trichomes and their possible role in plant defence are areas of investigation that have just begun to be utilized by systematists, evolutionists, and ecologists. Levin (1973) has brought together a great wealth of scattered information in this regard and has clearly and succinctly stated the case for further studies of this type. He notes,

---

[1] To be read in conjunction with the tabulated data on pp. 190-8.

It is clear that trichomes play a role in plant defense especially with regard to phytophagous insects. In numerous species there is a negative correlation between trichome density and insect feeding and oviposition responses, and the nutrition of larvae. Specialized hooked trichomes may impale adults or larvae as well. Trichomes may also complement the chemical defense of a plant by possessing glands which exude terpenes, phenolics, alkaloids or other substances which are olfactory or gustatory repellents. In essence, glandular trichomes afford an outer line of chemical defence by advertising the presence of 'noxious' compounds. In some groups of plants, protection against large mammals is achieved by the presence of stinging trichomes. Intraspecific variation for trichome type and density is known in many species, and often is clinal in accordance with ecographic parameters. The presence of such correlations does not imply that differences in predator pressure are the causal factors, although this may indeed be the case.

The absence of trichomes on plant parts where they would otherwise be expected to be present may be a simply inherited trait (Rollins 1958). It is easy to understand how the developmental sequence leading to trichome elaboration could be interrupted by simple mutation or genetic change, whereas a genetic system that codes the production of a particular trichome type on a given part of the plant must of necessity be fairly complex. Experience with the indumentum of higher plants as a taxonomic character has led to the axiom that mere presence or absence may be of trivial importance but the type of trichome present is often of considerable value for taxonomic purposes. It is clear that the axiom is soundly based.

Important use has been made of the particular types of trichomes in studies of many groups of taxa. Metcalfe and Chalk (1950) have brought attention to various types and their occurrence in particular families within the dicotyledons. In the Gramineae, Metcalfe (1960) has shown the value of a careful study of minute trichomes as well as large ones for numerous genera of this family. Tomlinson (1969) described the types within the Commelinales and Zingiberales and drew attention to their value in delimiting various groups. He also noted the diagnostic value of the trichome base in definitive studies in the Zingiberaceae. Cowan (1950), in a classic study of trichomes in the genus *Rhododendron,* illustrated and described the numerous and often elaborate types within this genus and showed their value in delimiting sub-genera and species. Heitzelman and Howard (1948), working with the Icacinaceae, have shown the value of determining types of trichomes, both relative to the number of types within a genus, and their distribution on the plant. Carolin (1971) has surveyed the trichomes found within the Goodeniaceae and questioned the taxonomic groupings made by earlier workers. At the

specific level, studies of trichomes have been found to be of value by many workers (Faust and Jones 1973; Heiser 1949; Hunter and Austin 1967; Roe 1971; Rollins 1941, 1944; Rollins and Shaw 1973; Theobald 1967a, 1967b).

The taxonomic significance of the trichome complement, or range of trichomes on the entire plant, has been emphasized by Carlquist (1958, 1959a,b, 1961) for members of the Compositae (= Asteraceae). Especially, he has been able to illustrate clearly the evolution of particular types of trichomes through comparative studies of development within and between taxa of the tribe Madiinae. On the basis of a single trichome type such conclusions probably would not have been possible and a similar understanding of the group would not have been elucidated.

Many of the difficulties with comparative studies involving trichomes centre around the rather imprecise use of terms for both the morphology and anatomy of the trichome. The term 'shaggy' gives a sketchy impression of the trichome and really conveys nothing about anatomy. Stellate hair is a descriptive term which also conveys little concrete information. So-called stellate trichomes may consist of a cluster of unicellular or multicellular trichomes (Fig. 5.1(a)). Alternatively they are sometimes multicellular (Fig. 5.1(b)) or even unicellular (Fig. 5.1(c)). Also the component cells may be papillose, smooth, thin or thick-walled, etc.

Because of the intergradations between trichome types, and also in consequence of the use of imprecise and varied morphological and histological terms, it is often difficult, from the voluminous published descriptions that exist, to determine exactly what type of trichome is being discussed, unless there is an illustration. These difficulties are often made worse when there is not enough information for valid comparisons between groups of taxa. With this in mind, the present authors have attempted in a general way to clarify problems of trichome description and to bring to the attention of taxonomists methods of analysis of trichome morphology and anatomy. A brief outline is also presented of simple methods for the preparation of slides which will provide the basis for bringing a wealth of anatomical information into the description of trichomes. No attempt has been made to describe all forms of existing trichomes, nor is it in any way intended that the outline presented will fit all existing cases. Instead the emphasis has been on the step-by-step analysis of trichomes on the basis of their morphology and anatomy, with distinctions being drawn between the over-all surface appearance the trichome complement, the gross morphological appearance of the individual trichome, and the anatomical characteristics of the particular trichome. Also, we have been interested in presenting a range of illustrations that emphasize the diversity of trichomes found among the angiosperms.

Methods for study

Both living plants and herbarium specimens are well suited as sources for studies of trichome morphology and anatomy. However, it is generally easier to work with living material, which can, if properly prepared, give a great number of characters for comparison and observation. This is especially true with respect to glands and internal cellular components. Nevertheless, herbarium material is often all that is readily available, and with proper care excellent results may be achieved. Metcalfe and Chalk (1950) have noted that care must be taken in all types of observations. For example, in the Turneraceae, there can be a substantial difference between the glandular trichomes on living material and on dried herbarium specimens.

Methods of collection and preparation will vary considerably, depending on the objectives of the collector, the nature of the material, the facilities and equipment for investigation, and the amount of time available for study. In most instances any limitations that may be met can be overcome by a selection of one or several of the methods described or referred to below. Some are relatively simple, while others may require some practice and modification depending on the nature of the material under investigation. An initial test of several methods will usually give an indication of the most suitable one for each purpose. More comprehensive works, such as Johansen (1940), Sass (1958), and Jensen (1962), give a much more detailed account of various aspects of microtechnique and explain more fully some of the procedures described below.

Preparation of material

The most common method of killing and fixation is through the use of FAA (90 ml 50 or 70 per cent ethanol, 5 ml glacial acetic acid, and 5 ml formalin). Propionic acid may be substituted for the acetic acid in this mixture, and it often gives a better preparation. Killing and fixation with various combinations of chromic acid, acetic acid, and formalin (Sass 1958, p.18; Johansen 1940, pp.42–5; Jensen 1962, p.79) generally give less shrinkage, less hardening, and better

general appearance than with FAA, but they are slower to penetrate, often require vacuum infiltration, and are more difficult to handle.

Preparation of herbarium material is simple and only requires boiling in water for approximately 15–30 minutes, or until such time as all the air is driven out and the material sinks. The addition of soap or one of the commercial wetting agents greatly facilitates the penetration of water. After cooling, the material can be stored in FAA or 50 per cent alcohol until needed.

Following fixation, the material to be examined should be transferred from the fixative (FAA) to 50 per cent alcohol for 15–30 minutes, and then placed in another change of 50 or 70 per cent alcohol prior to examination, or dehydration and embedding. In some cases a third change of alcohol may be necessary to remove all traces of the fixative. Observations can easily be made from freehand or sliding microtome sections, epidermal peels and scrapes, whole mounts, and sectioned embedded material.

Unembedded sectioning

Freehand or sliding (sledge) microtome sections can be made from material placed between pieces of pith, cork, carrot, or balsa. Sections should be cut at varying thicknesses for a complete range of mounts. At all times the material should be kept moist with 50 or 70 per cent alcohol. Sections are easily removed from the microtome knife with a camelhair brush and placed in a small dish of 50 or 70 per cent alcohol. For quick examination, sections can be run back to distilled water, transferred to a drop of glycerine on a slide, covered, and then examined. In many cases the material may be satisfactorily transferred directly to the glycerine from the original 50 per cent alcohol. The slides may be made somewhat more permanent by ringing the coverslip with paraffin wax, Duco cement, fingernail polish, or some suitable substitute. A permanent mount may be made by transferring the sections from 50 to 70, 95, and then 100 per cent alcohol with a few minutes in each concentration. A Gouch crucible is very useful for transferring large numbers of sections from dish to dish. The material is sometimes mounted directly from the 100 per cent alcohol in Euparal (Diaphane), or transferred from the absolute alcohol to xylene and then mounted in Canada Balsam, or preferably one of the artificial mounting substances such as Permount. In the latter, drying is much faster than in the Euparal mounts, but the xylene often causes shrinkage or curling of the sections.

Clearing

Problems with herbarium material that is often badly shrunken or discoloured can be overcome by treatment with 2·5–5 per cent NaOH after sectioning and rehydration with tap water. The duration of treatment varies considerably with the material. A very rapid clearing can be obtained with one of the commercial bleaching agents such as Clorox. Either full concentration or various dilutions may be used. However, care must be taken in the timing, because too lengthy treatment disintegrates some materials. Improved results are well worth some testing. The bleaching agents render the cell walls more visible, cause shrunken cells to swell to normal or near normal size, and enable a clearer staining of the cell walls. All treatments with these bleaching agents are best undertaken after sectioning, as a prior treatment often renders the tissues too soft for good sectioning unless they are embedded in paraffin wax. After bleaching, the material should be rinsed in at least two changes of water prior to examination in glycerine or it should be dehydrated in an alcohol series before permanent mounting. A well-outlined simple procedure for clearing leaves has been described by Payne (1969).

Staining procedures

These procedures may be kept relatively simple, a single stain often being adequate for many observations. If trichome surveys are to be combined with other anatomical investigations, any of the multiple staining procedures (safranin and fast green, safranin and haematoxylin, etc.) described by Jensen(1962), Johansen (1940), and Sass (1958) would be found to be useful.

Delafield's haematoxylin is an excellent general-purpose stain which requires little time and effort. Unembedded material can be transferred after sectioning and clearing directly from 50 per cent alcohol or water to the stain for 5–20 minutes. After staining, the sections should be washed in acidified water (a few drops of concentrated HCl added) for a very brief period until the material is slightly destained. This step may be omitted in many instances. The sections are then mounted temporarily in glycerine or they may be dehydrated and mounted permanently as described above.

A dual-purpose staining procedure combining safranin and Delafield's haematoxylin (Metcalfe, 1960) is also very useful. Sections are transferred from 50 to 70 per cent alcohol to a 15:1 mixture of safranin (1 per cent in 70 per cent alcohol) and Delafield's haematoxylin. After staining for 2–12 hours the material is transferred to 50 per cent alcohol (acidified, if necessary) for destaining, followed by quick changes to 70, 95, and absolute alcohol prior to mounting by the methods decribed above. The stain mixture should be made fresh for each use. The addition of a few drops of NH_4OH to the 70 per cent alcohol often increases the contrast between the safranin and haematoxylin.

Whole mounts

This method is usually successful with thin or delicate structures, including some leaves and flowers. In most cases the material should be cleared in 2·5–5 per cent NaOH, a bleaching or some other reagent prior to examination or staining. Delafield's haematoxylin in often useful.

Epidermal peels and scrapes

The ease with which these preparations can be made is almost completely dependent on the nature of the material under investigation. The methods employed for a special group such as the Gramineae (Metcalfe 1960) are basically the same as those applicable to other monocotyledons and many dicotyledons. Materials for study, usually leaves, are placed with the outer surface facing downwards on a glass plate or similar flat surface and flooded with a commercial bleaching agent. The material is then carefully scraped with a razor blade until the epidermis is reached. The bleaching agent acts as a lubricant and simultaneously softens the cell layers as they are scraped away. In some instances the material can be softened in 2·5–5 per cent NaOH prior to treatment, and the bleach may or may not have to be eliminated. In all cases practice and modification can usually lead to a successful preparation. Only a small portion of the epidermis is necessary for most observations. Subsequent staining procedures are the same as those described above.

Bailey and Nast (1945) described another method that is particularly useful with material that is easily broken. A portion of a leaf is attached to a slide with Haupt's adhesive and then scraped in the usual manner. After scraping, the slide may be stained by any of the staining procedures already outlined.

Embedded material

Embedding allows for serial sections as well as for the preparation of thin sections. Also, more refined staining procedures can be employed. However, the procedure is longer and more involved, and will not be described here. A detailed description can be found in Jensen (1962), Johansen (1940), and Sass (1958).

If serial sections are not required and the material has been found to be difficult to section on the sliding microtome, the materials can be embedded in paraffin and sectioned on the rotary or sliding microtome. This allows for thin sections and a short preparation time. After sectioning, the paraffin sections can be placed in a dish of xylene and allowed to remain there (approximately 5 minutes) until all the paraffin is removed. After a second change of xylene the sections may be rehydrated up to 50 per cent alcohol and then stained or mounted unstained, as suggested for the unembedded sections.

Scanning electron microscope

This latest tool for taxonomic research has produced both dramatic and worthwhile results with the minimum of preparation, though the expense may be prohibitive. There are no special procedures involved and Rollins and Shaw (1973), as well as others, have admirably displayed its value. (See also Chapter 1, pp. 9–11.)

Trichome description

Adequate trichome description requires the investigator to take four basic steps in the descriptive process before he attempts to interpret trichomes. A sequence of the four following procedures is involved:

I. Examination of the overall surface appearance (indumentum)
II. Investigation of the morphology of individual trichomes
III. Study of the trichome complement
IV. Histological description of trichomes

The first of these procedures is or was an essential part of past and present-day taxonomic descriptions, while the second gives a general picture of individual trichomes. The third line of enquiry gives an overall view of the range of types present, while the fourth tells us in terms that are as precise as possible the anatomical details of the trichomes. As can be seen in Figs. 5.1–5.7 there is so much diversity that written descriptions are often inadequate, and a figure is sometimes the best and only method of conveying a correct impression. Nevertheless, a combination of the morphological description of the individual trichome with histological details is the only way to build up a worthwhile classification or grouping of trichome types.

A simple example will suffice to explain the process. In the Asteraceae, *Vernonia blodgettii* Small has been described by Gleason (1922) as having leaves which are 'glabrous above, sparsely short-pubescent and frequently glandular beneath'. Faust and Jones (1973) have analysed this taxon with regard to trichome morphology, complement, and, to a limited degree, anatomy. They have observed bilobed (both surfaces), awl-shaped (upper surface), awl-shaped glandular (lower surface), and L-shaped (lower surface) leaf trichomes. Together these form the trichome complement. It is apparent from the illustrations that the trichomes are glandular or non-glandular, unicellular or multicellular, and thin-walled. The only drawback in this study was the use of the term uniseriate to describe the morphology of certain trichomes in some species. This is really an anatomical term and does not describe the shape. The trichomes in Figs. 5.2 (e–g), 5.3 (a–d,g), and 5.4 (g–i) are uniseriate and in many cases they are not at all similar.

I. Overall surface appearance (indumentum)

For centuries the nature of the indumentum or vesture has been observed and described by botanists. As with most vegetative features the terminology is very qualitative in nature and subject to varied interpretation (e.g. tomentose, pilose, stellate, etc.). Often one has to be fairly familiar with a particular group of plants or a particular author's descriptive method before one can adequately make comparisons. It is beyond the scope of the present discussion to go into an analysis of this problem, but suffice it to say that some of the terms, such as stellate, describe the nature of the indumentum as well as varied types of individual trichomes (Fig. 5.1).

II. Morphology of individual trichomes

This is probably the single most difficult aspect of trichome description and classification. Although most trichomes can be placed in one of the categories to be described, others simply defy placement. Inventing a whole new terminology might appear worth while, but it would probably be little understood or used and it is also doubtful that it would solve the problem. One of the better systems so far devised, even though it covers only a limited group, is that of Roe (1971) for *Solanum*. When this is modified slightly and combined with the much broader and

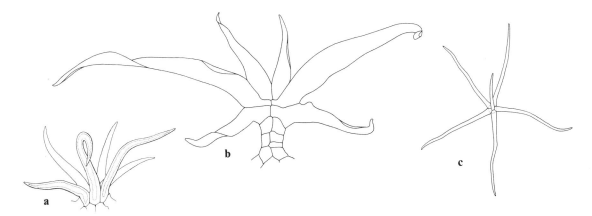

Fig. 5.1. Stellate trichomes. (a) *Humirianthera rupestris* Ducke (Icacinaceae), × 103; (b) *Miconia chrysophylla* Urban (Melastomataceae), × 378; (c) *Croton glandulosus* L. (Euphorbiaceae), × 45.

excellent coverage of Metcalfe and Chalk (1950), one does have groups which are both inclusive and flexible. There will always remain the unusual types of trichome which cannot be satisfactorily placed and in these cases it is best just to say so and possibly to indicate, with reasons, the group of trichomes showing the closest resemblance.

Although the terms glandular and non-glandular are commonly used and convenient ones for describing trichomes, they do not really tell us much about the morphology of the trichome itself. Any number of different trichomes (e.g. simple, scaly, etc.) may be either glandular or non-glandular. Therefore these terms should be used in conjunction with any of the following groups as here recognized (Table 5.1). Some of the subcategories can be combined within any one group.

Papillae. These small epidermal outgrowths are often treated as distinct from trichomes but they are for convenience placed here as a group. They can be of importance and interest and often show many varied features which may be of systematic value. Under the scanning electron microscope *Uldinia ceratocarpa* (Apiaceae) exhibits an unusual internal wall thickening (Plate 1(A)) that had been observed in an earlier study under the light microscope (Theobald 1967*b*). Unusual cuticular patterns can be seen on the papillae (Plate 1(B)) of *Tordylium apulum* (Apiaceae). Some of the smaller simple trichomes (Fig. 5.2) approach papillae and no sharp line can be drawn between them. (Papillae are also discussed in this work on pp. 148–52).

Table 5.1. *Trichome types and their major features: glandular and non-glandular*

(1) *Papillae*	(5) *Scales*
	(a) Sessile
(2) *Simple (unbranched)*	(b) Peltate
(a) Short	(c) Porrect
1. Thin	
2. Thickened	(6) *Dendritic (branched)*
(b) Long	(a) Few-branched
1. Thin	(b) Many-branched
2. Thickened (shaggy)	(c) Branching terminal
	(d) Branching medial
(3) *Two- to five-armed*	(e) Branching basal
(a) Two-armed	
1. T-shaped	(7) *Specialized types*
2. U-, V-, Y-, J-shaped	
(b) Three- to five-armed	
(4) *Stellate*	
(a) Rotate	
(b) Multiangulate	
(c) Porrect	
(d) Geminate (candelabraform)	
(e) Tufted	

Trichomes simple (unbranched). This large grouping of unicellular and multicellular trichomes is extremely common and can be conveniently subdivided into two major subgroups: short and long. It is true that the terms 'short' and 'long' lack definition, but in general the short forms (Fig. 5.2(a–f,p–r)) are here taken to mean those that give a puberulose or strigose appearance with the hand lens or to the naked eye, while those in the long category are the more common types (Fig. 5.2(g–o,s–u)) which result in such surface appearances as villose, pilose, tomentose, etc. Each of these groups can then be further subdivided into the

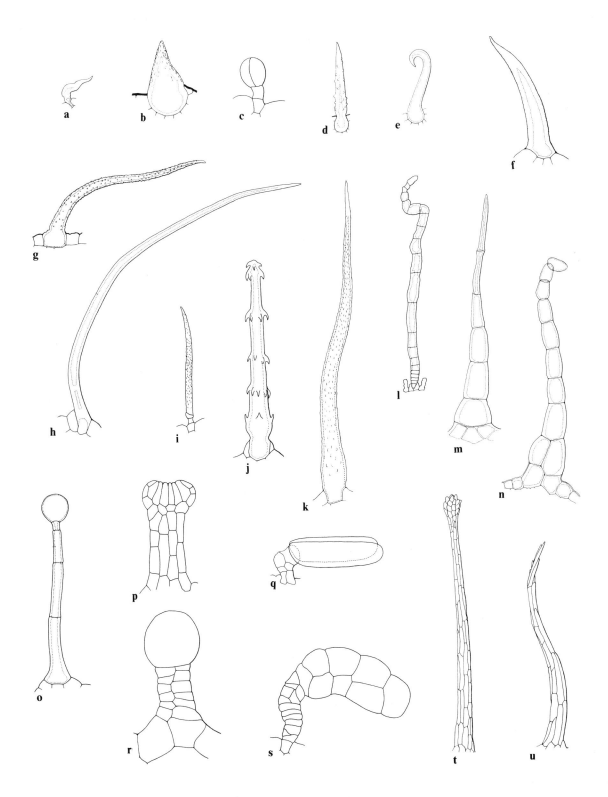

thin, or basically uniseriate forms which appear thin (Fig. 5.2(a–o)), and the thickened or multiseriate forms which appear thickened or massive (Fig. 5.2(p–u)) and have been referred to as 'shaggy' by some. This last term is used especially for the long forms. Biseriate forms intergrade but are relatively rare. Capitate forms may be found within any of these categories (Fig. 5.2 (c,o–s)) and are usually indicative of a glandular condition.

The short, simple forms intergrade with papillae and can be further subcategorized as 'prickles', which are generally short cells with swollen bases and a tapering exposed portion which ends in a point (Fig. 5.2(b)).

Trichomes, two- to five-armed. This particular group can be conveniently subdivided into the two-armed (Fig. 5.3(a–g)) and three- to five-armed types (Fig. 5.3 (h–n)). The two-armed are often diagnostic within groups and are fairly distinctive. They may be unicellular (Fig. 5.3(c)) or multicellular (Fig. 5.3(b)) and can be equal-armed (Fig. 5.3(c)), unequal-armed (Fig. 5.3(a)), horizontal or T-shaped (Fig. 5.3(a–d)), or may be variously V-shaped, U-shaped, J-shaped, or Y-shaped (Fig. 5.3(e,f)). Within a taxon they are fairly stable as to the number of arms.

The three- to five-armed forms are generally more variable and approach the stellate and dendritic types. They may be sessile (Fig. 5.3(j)) or stalked (Fig. 5.3(k)) and unicellular (Fig. 5.3(l)) or multicellular (Fig. 5.3(j)), and they are often irregular in their branching pattern. Also, within a particular taxon the number of arms may vary and small side arms are common (Fig. 5.3(m,n)).

Trichomes stellate. As can be seen in Figure 5.1 these are very variable in structure even though all are 'star-shaped'. They may be sessile (Fig. 5.1(a)) or stalked (Figs. 5.1(b); 5.4(d,g,j)) and the rays (more than five) may be in one plane (rotate) as in Fig.

5.4(a–c,e) and Plate 1(C,D), or they may be multiangulate. Often the rotate forms approach scales in appearance (Fig. 5.4(c)). The multiangulate forms are usually stalked and have their arms radiating outwards in all directions (Fig. 5.4(g–j)). However, they may be sessile (Fig. 5.1(a)). Sometimes the central ray is greatly enlarged and these hairs have been described as porrect-stellate (Roe 1971). This may occur with rotate (Fig. 5.4(d,f), or multiangulate forms.

Sometimes a series of stellate clusters may be superimposed on each other (Fig. 5.4(h,i)). These have been called geminate (Roe 1971) or candelabraform hairs and they intergrade with some dendritic types. Tufted, stellate types, as here defined, comprise those forms which are distinctly raised on a mound of epidermal and possibly sub-epidermal tissues. They are usually massive and more elaborate than the other stellate trichomes described above and they are multiangulate with rays projecting at various irregular angles (Fig. 5.5(a,b)). On occasion they may be also porrect (Fig. 5.5(a)).

Scales. These are generally flattened and either sessile (Fig. 5.6(a)) or stalked (Fig. 5.6(c,f,h,j,l,n)). They may be unicellular (Fig. 5.6(g)) or more commonly multicellular (Fig. 5.6(a–f,h–q)). In some cases they are not flattened (Fig. 5.6(c,h)) but instead narrowly convex or rounded. They are often glandular (Fig. 5.6 (c,h,j,p)). They intergrade with some of the rotate-stellate forms (Fig. 5.6(g)) and there is no simple line of demarcation between them, except that those that are divided into segments or rays to at least a depth of one-half of the radius might best be considered stellate. In some cases the scales are porrect (Fig. 5.6(n)).

Trichomes dendritic (branched). These are trichomes which branch along an extended axis. They may be unicellular (Fig. 5.7(c,d)) or multicellular (Fig. 5.7(a,b, e–k)), and uniseriate (Fig. 5.7(f)), or multiseriate

---

Fig. 5.2. Simple (unbranched) trichomes. (a) *Poraqueiba sericea* Tul. (Icacinaceae), × 38; (b) *Carrichtera annua* Prantl (Brassicaceae), × 59; (c) *Salvia occidentalis* Sw. (Lamiaceae), × 126; (d) *Pleurisanthes simpliciflora* Sleumer (Icacinaceae), × 84; (e) *Iodes philippinensis* Merrill (Icacinaceae), × 74; (f) *Heliophila pilosa* Lam. (Brassicaceae), × 126; (g) *Menonvillea cuneata* Rollins (Brassicaceae), × 162; (h) *Boronia albiflora* R. Brown (Rutaceae), × 336; (i) *Adesmia filipes* A. Gray (Fabaceae), × 118; (j) *Caiophora lateritia* Benth. (Loasaceae), × 176; (k) *Arabis auriculata* Lam. (Brassicaceae), × 294; (l) *Bellucia imperialis* Sald. and Cogn. (Melastomataceae), × 210; (m) *Olearia aculeata* (Compositae = Asteraceae), × 220; (n) *Olearia rapae* F. Brown (Asteraceae), × 220; (o) *Salvia occidentalis* Sw. (Lamiaceae), × 378; (p) *Dontostemon glandulosus* Schulz (Brassicaceae), × 105; (q) *Miconia globulifera* Naud. (Melastomataceae), × 210; (r) *Turnera diffusa* Willd. (Turneraceae) × 294; (s) *Inga hirsutissima* Rusby (Fabaceae–Mimosoideae), × 315; (t) *Andryala cheiranthifolia* Link. (Asteraceae), × 126; (u) *Rodigia commutata* Spreng. (Asteraceae), × 105.

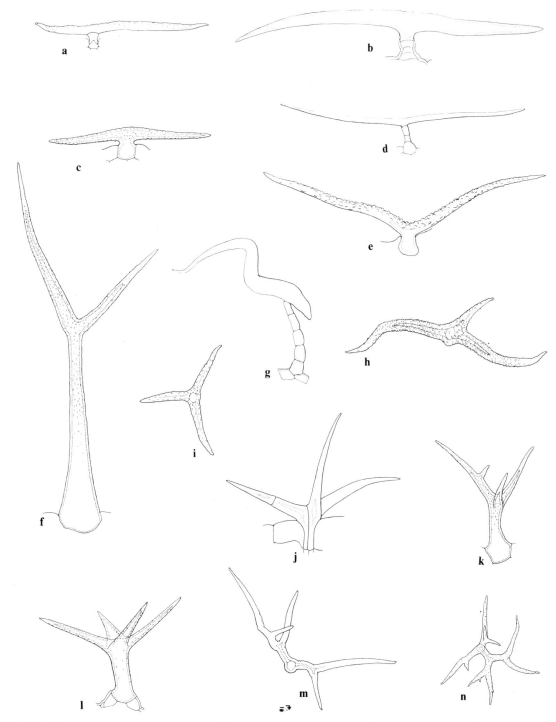

Fig. 5.3. Two- to five-armed trichomes. (a) *Astragalus canadensis* L. (Fabaceae), × 158; (b) *Olearia dentata* Moench. (Asteraceae), × 273; (c) *Arabis drummondii* A. Gray (Brassicaceae), × 158; (d) *Artemisia absinthium* L. (Asteraceae), × 210; (e) *Erysimum capitatum* Greene (Brassicaceae), × 273. (f) *Arabidopsis stricta* Hook. and Th.(Brassicaceae), × 336; (g)*Vernonia senegalensis* Less. (Asteraceae), × 189; (h) *Malcomia africana* R. Br. (Brassicaceae), × 189; (i) *Arabis divaricarpa* A. Nelson (Brassicaceae), × 63; (j) *Turnera hermannioides* Cambess. (Turneraceae), × 126; (k) *Arabis auriculata* Lam. (Brassicaceae), × 126; (l) *Arabis crucisetosa* Const. and Rollins (Brassicaceae), × 118; (m) *Cheiranthus parryoides* Kurz (Brassicaceae), × 168; (n) *Descurainia californica* Schulz (Brassicaceae), × 105.

Fig. 5.4. Stellate trichomes. (a) *Lesquerella engelmanni* S. Wats. (Brassicaceae), × 105; (b) *Alyssum corsicum* Duby (Brassicaceae), × 84; (c) *Lesquerella thamnophila* Rollins and Shaw (Brassicaceae), × 63; (d) *Croton glandulosus* L. (Euphorbiaceae), × 126; (e) *Shepherdia canadensis* Nutt. (Elaeagnaceae), × 137; (f) *Croton niveus* Jacq. (Euphorbiaceae), × 221; (g) *Olearia asterotricha* F. Muell. (Asteraceae), × 84; (h) *Olearia nernstii* (Asteraceae), × 74; (i) *Andryala cheiranthifolia* Link. (Asteraceae), × 84; (j) *Calycogonium calycopteris* Urb. (Melastomataceae), × 189.

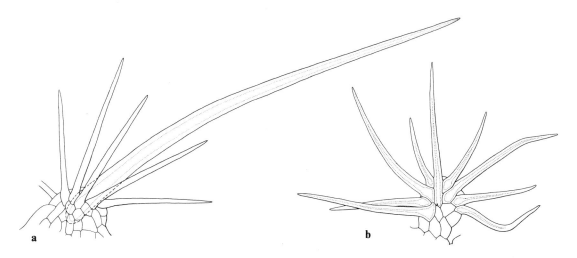

Fig. 5.5. Stellate trichomes. (a) *Piriqueta racemose* Sweet (Turneraceae), × 378; (b) *Piriqueta cistoides* Mey (Turneraceae), × 179.

(Fig. 5.7(a,b,e,g,j,k)). They may be described as branching throughout (Fig.5.7(i–k)), branching basally (Fig. 5.7(f–h)), branching medially, or branching terminally (Fig. 5.7(b–e)) or a combination thereof (Fig.5.7(h,i)). Forms such as the one shown in Fig. 5.7(h) defy proper placement and might be considered porrect-stellate. However, examination of material would show that the rays are irregular in pattern, much like most dendritic trichomes. As noted earlier, some of the candelabraform, stellate hairs and the three- to five-armed types intergrade with this group. Also, some of the larger simple hairs often have protrusions along their surface which would be an initial step towards dendritic structure.

Trichomes of specialized types. As noted earlier there are many specialized types such as stinging hairs, pearl glands, vesicular hairs, snail-shaped glands, chalk glands, mucilage hairs, calcified hairs, funnel-form hairs, etc. Most of these are well illustrated in the first edition of this present work and it is beyond the scope of the present survey to illustrate each of them.

III. Trichome complements

After accurately describing the individual trichomes on a particular taxon it is often evident that more than one type is present. When making comparisons all of these types should be taken into account. As noted by Carlquist (1961), Faust and Jones (1973), and others, these comparisons of complements can be useful within certain groups. Distribution of complements on various organs of the same taxon should also be taken into account and analysed for its value.

IV. Histological description of trichomes

Anatomical features of trichomes are numerous and in most cases it is only the lack of time or of sufficient material and facilities which delays the preparation of adequate descriptions. The following list indicates some of the common features that should be included if comparative studies are to be undertaken. They are listed in order of possible significance which is also the sequence in which they should be described.

(1) Glandular or non-glandular
(2) Unicellular or multicellular
(3) Uniseriate or multiseriate
(4) Presence of surface features — warts, papillae, striations, etc.
(5) Presence of constrictions or swollen portions of cells

Fig. 5.6. Scales. (a, b) *Miconia globulifera* Naud. (Melastomataceae), × 126; (c, d) *Pilea* sp. (Urticaceae), × 189; (e, f) *Olearia albida* Hook.f. (Asteraceae), × 74; (g) *Lesquerella schaffneri* S. Wats. (Brassicaceae), × 147; (h, i) *Ledum glandulosum* Nutt. (Ericaceae), × 252; (j, k) *Rhododendron acraium* Balf. and Smith (Ericaceae), × 162; (l, m) *Croton argyranthemus* Michx. (Euphorbiaceae), × 210; (n, o) *Phebalium squamulosum* Vent. (Rutaceae), × 221, (p) *Rhododendron carolinianum* Rehder (Ericaceae), × 252; (q) *Shepherdia canadensis* Nutt. (Elaeagnaceae), × 84.

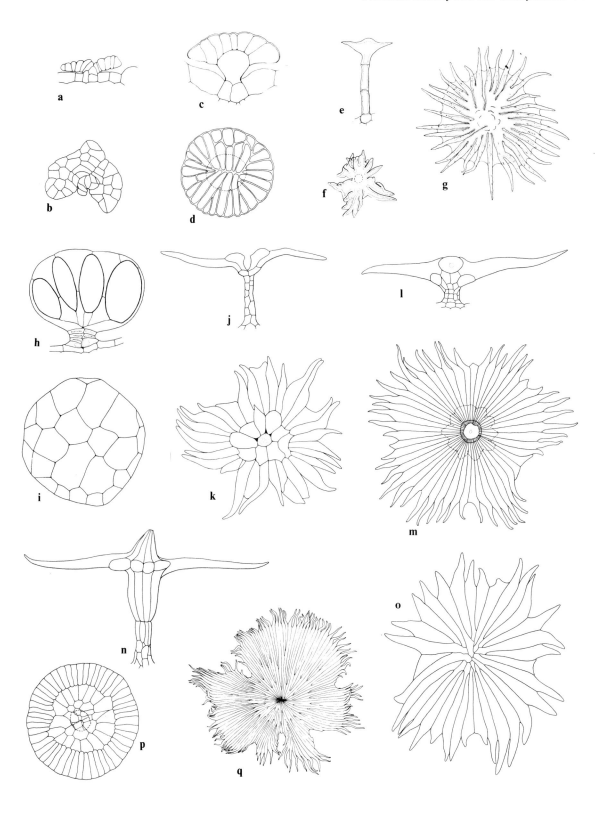

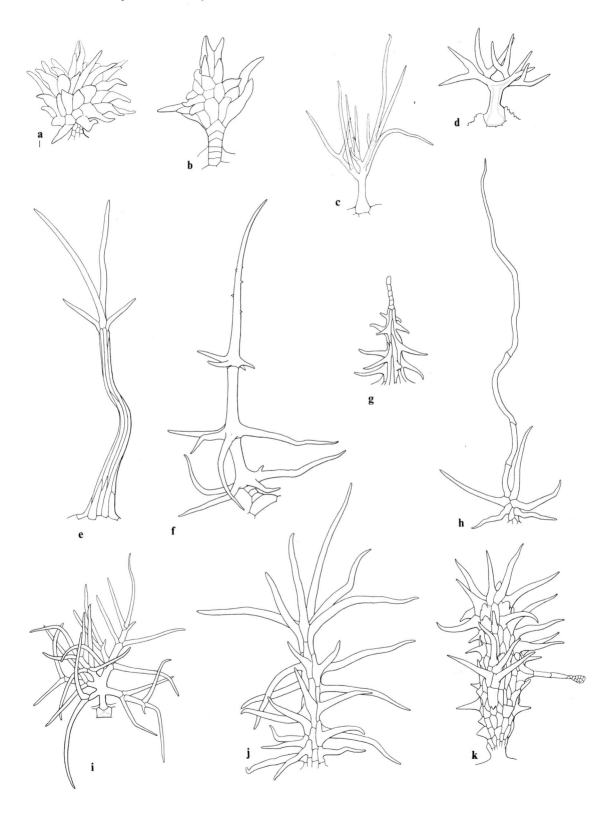

(6) Different cell types within trichome – base (foot), body, apex

(7) Differences in wall thickness, if any

(8) Lignification if evident

(9) Cuticle thickness

(10) Presence of crystals or cystoliths

(11) Wall impregnation with silica or calcium carbonate

(12) Cellular constituents

Differences in surrounding epidermal cells should also be noted, as well as the position of the trichome with regard to the surface.

Trichome classification

Our knowledge of trichomes has greatly expanded in the past twenty-five years yet we are still far from a definitive classification. This is in great part due to the infinite forms trichomes may take and our lack of knowledge regarding their ontogenetic development. Even if this were known, however, one might still not be able to classify them in a manner that would enable each type to be categorized. From surveys of plant taxa it is evident that they have had infinite independent origins and as a result no so-called 'phylogenetic' sequences can be drawn up, except on a very limited scale within certain small groups. However, certain types are more common in one taxon than another and for this reason they are of diagnostic value. Therefore a consistent pattern of description can assist a taxonomist or systematist in comparisons. Descriptions should be written in a logical clear sequence which will enable a person to picture the trichome in question.

The simple method of description given above is straightforward and undemanding in its approach. For the busy taxonomist with a wealth of material to examine it is probably as good as any that can be devised. Its great merit is that it requires only the use of a combination of simple morphological and anatomical characters to be observed and presented in a recommended sequence. Even with this simple approach, difficulties of description will inevitably arise. Two examples will suffice to illustrate this point. The trichome in Fig. 5.3(c) might be described as two-armed, T-shaped, arms equal, unicellular, stalk absent, wall minutely warty. The trichome in Fig. 5.3(b) might be described as two-armed, T-shaped, arms slightly unequal, multicellular, uniseriate, short stalked, stalk two- to three-celled, terminal cell horizontally extended, wall smooth.

Literature cited

Bailey and Nast 1945; Bary, de 1884; Carlquist 1958, 1959a,b; 1961; Carolin 1971; Cowan 1950; Cutler 1969; Esau 1965a; Faust and Jones 1973; Foster 1949; Gleason 1922; Haberlandt 1914; Heiser 1949; Heitzelman and Howard 1948; Hunter and Austin 1967; Jensen 1962; Johansen 1940; Lawrence 1951; Levin 1973; Metcalfe 1960, 1971; Metcalfe and Chalk 1950; Netolitzky 1932; Payne 1969; Ramayya 1972; Roe 1971; Rollins 1941, 1944, 1958; Rollins and Shaw 1973; Sass 1958; Solereder 1908; Theobald 1967a and b; Tomlinson 1961, 1969; Uphof, Hummel, and Staesche 1962.

Suggestions for further reading

Inamdar and Gangadhara 1975: classification of trichomes in the Cucurbitaceae.

Thurston and Lersten 1969: stinging hairs.

[The references in the above list are only a very small selection from the vast literature that exists concerning trichomes. Many further references will be recorded under the descriptions of the individual families in future volumes of this work. *Editor.*]

Fig. 5.7. Dendritic (branched) trichomes. (a) *Calycogonium domatiatum* Urb. and Ekm. (Melastomataceae), × 95; (b) *Calycogonium grisebachii* Triana (Melastomataceae), × 252; (c) *Alyssum leucadeum* Guss. (Brassicaceae), × 162; (d) *Arabis holboellii* Hornem. (Brassicaceae), × 84; (e) *Miconia guatemalensis* Cogn. (Melastomataceae), × 294; (f) *Alternanthera stellata* Uline and Bray (Amaranthaceae), × 273; (g) *Henriettea granulata* Berg (Melastomataceae), × 105; (h) *Marrubium micranthum* Boiss. and Heldr. (Lamiaceae), × 336; (i) *Lavandula vera* DC. (Lamiaceae), × 158; (j) *Mimosa furfuracea* Benth. (Fabaceae-Mimosoideae), × 231; (k) *Conostegia lindenii* Cogn. (Melastomataceae), × 105.

6

SOME BASIC TYPES OF CELLS AND TISSUES

C.R. METCALFE

Parenchyma

In the course of evolution the cells and tissues of which the plant body is composed have become specialized in various ways primarily suited to meeting their respective physiological functions. The least physiologically specialized tissue is **parenchyma** which is made up parenchymatous cells. These cells are the most important type in herbaceous plants in which the plant body is often, in fact, mainly parenchymatous. However, the 'soft' parts of woody plants, such as the leaves and young stems and to some extent the roots, include a large proportion of parenchyma. The other cells and tissues in the plant either provide mechanical support or serve as channels through which the foodstuffs and sap are conducted. These elementary facts are well known, or should be well known, to all botanists, but we must remind ourselves that the differentiation of cells and tissues is, as we have already seen, to meet physiological requirements and whether or not they provide a recognizable taxonomic pattern is of secondary significance. This present treatise is, however, primarily concerned with the diagnostic or taxonomic aspects of cellular organization, and the sum total of the evidence presented shows very clearly that the pattern of cellular organization does reflect the taxonomic affinities of plants. There is a great diversity of cellular organization in the vegetative anatomy of different angiosperm families. For the moment, we must consider very briefly the basic categories of cells that will be encountered in our search for characters that are taxonomically significant.

Parenchyma is a rather imprecise term that was first introduced by Nehemiah Grew (1682) (see p. 1). Living parenchymatous cells make up the ground tissue of the plant, and the tissue is to be found especially in positions where there is no need for cells that are morphologically specialized to meet a precise physiological need. It is a commonly expressed view that parenchymatous cells are more or less equidimensional and thin-walled. Although this is frequently true it is far from applying universally. Parenchymatous cells are often thick-walled, and their shapes range from stellate cells, such as occur in the pith of certain aquatic plants, to the variously shaped cells to be seen in the mesophyll of leaves (see pp. 66–7), or they may

be elongated. The ground tissue of stems (especially the cortex and pith) and roots, as well as the mesophyll and epidermis of leaves, provide examples of parts of plants where parenchyma is usually well developed. It also occurs as a component of bark, phloem and xylem. Indeed it may be found in almost any part of the plant. The walls of parenchymatous cells are commonly of cellulose, but they may become suberized or lignified, and sometimes the cells themselves contain secreted material such as mucilage or crystals. Very few generalized statements can be made about a tissue that is so variable, but it is so widely distributed that it is very frequently mentioned in the text of this work.

Although parenchyma has been described above as an unspecialized tissue, it nevertheless sometimes exists in specialized forms or it may acquire a secondary specialized function. The stellate cells mentioned above may very well have arisen in response to the aquatic mode of life of the plants in which they occur. Another familiar example is parenchyma in which the cells contain chlorophyll, and constitute the chlorenchyma or photosynthetic tissue which usually attains its maximum development in leaves. Parenchyma is in fact a very versatile tissue.

In speaking of parenchyma as an 'unspecialized' tissue it should be remembered that the term is not being used in the same sense as that in which 'unspecialized' is employed with reference to phylogenetic status. 'Unspecialized' in the phylogenetic sense really means primitive from the evolutionary standpoint, whereas the sense in which we are using it in this present discussion implies that cells and tissues are 'unspecialized' when, so far as we are aware, they lack any specialized physiological function. Phylogenists frequently do not regard parenchyma as 'unspecialized' in their sense. Jeffrey's (1925) view, expressed in his classical work on the origin of parenchyma in geological time, suggests that in the course of evolution, storage parenchyma has been derived from tracheids.

Collenchyma

Collenchmya (Figs. 6.1 and 6.2) is a tissue of living cells (sometimes termed **collocytes** (Roland 1967) which gives mechanical support as well as elasticity to leaves and stems. It is uncommon in roots, but it has been recorded by Duchaigne (1955) in the roots of *Diapensia* and *Vitis*, and literature citations concerning root collenchyma in other plants are also given. Collenchyma is particularly characteristic of immature tissues where the living collocytes can grow in length and so compensate for extension growth. Although collenchymatous cells sometimes remain short they are usually much more axially elongated than parenchymatous cells and they may even be comparable in length with fibre cells and, like them, sometimes have pointed ends. The cell walls often contain a high proportion of pectic material and they may be made up of alternating layers that are rich and poor respectively in pectic material. The cell walls of collocytes are also unevenly thickened, and the thickening is often particularly conspicuous, in transverse sections, at the corners of the cells. The 'thickening at the corners' actually consists of strips of thickening on the vertical walls and these strips are particularly well developed wherever 3 cells meet together. Although the walls of collocytes are usually interpreted as being of primary origin, they do, in some instances, become secondarily lignified. Duchaigne (1953,1954*a,b*,1955) refers to lignified collenchyma in certain Acanthaceae, Apiaceae, Bignoniaceae, Fabaceae, Lamiaceae, Piperaceae, and Polemoniaceae.

Collenchyma often occurs in the form of peripheral, commonly hypodermal, axial strands separated from one another by parenchyma. On the other hand it may form a continuous cylinder. It is of interest that although collenchyma in most members of the Apiaceae is in the form of strands it nevertheless forms a continuous cylinder in *Eryngium*. Although collenchyma occurs very widely amongst the flowering plants, Duchaigne (1955) says that it is specially characteristic of the petioles of the Apiaceae and the stems of Lamiaceae. The structure of the collenchyma in the petioles of celery (*Apium*) and *Heracleum sphondylium* L. is particularly well known through the investigations by Esau (1936) and Majumdar (1941).

Apart from its peripheral forms, collenchyma is sometimes also associated with vascular bundles. Duchaigne (1955) draws a sharp distinction between this perivascular collenchyma and the peripheral forms (Fig. 6.1). Perivascular collenchyma is most common on the outer faces of the vascular bundles, replacing the pericyclic fibres (see p. 000). Perivascular collenchyma sometimes borders on the xylem poles of the vascular bundles as well. Then again in still other plants in which there is a morphologically distinct

endodermis (see p. 000) the perivascular collenchyma may be situated between this layer and the vascular tissues.

Foster (1949) says that the term collenchyma was first used by Schleiden in 1839 with reference to the thick-walled, hypodermal cells in certain members of the Cactaceae. Foster, however, does not cite Schleiden's publication. Roland (1967), on the other hand, attributes the first use of the term to Link in 1837, but gives no citation. Müller (1890) after examining more than 400 angiosperms, realized that the familiar type of collenchyma, with walls thickened at the corners, is not the only kind that exists and he classified the tissue in seven more-or-less distinct types. This scheme was simplified by subsequent workers such as Majumdar (1941) and Duchaigne (1955) who pointed out that Müller's original seven types were not always sufficiently distinct to be recognized. Both Majumdar (1941) and Duchaigne (1955) recognized three types of collenchyma (Fig. 6.2). Duchaigne's scheme is as follows:

(i) **Angular collenchyma** (Fig. 6.2(a,d)) (Müller's *Eckencollenchym*) is the commonest type. This is recognized in transverse section by having cells that are thickened at the corners. According to Duchaigne, angular collenchyma occurs, for example, in the stems of *Atropa belladonna* L. and *Solanum tuberosum* L., in the petioles of *Begonia rex* Putz., *Ricinus communis* L. and *Vitis vinifera* L. Angular collenchyma also occurs in immature members of the Apiaceae and Lamiaceae.

(ii) **Tangential collenchyma** (Fig. 6.2(b)) (Müller's *Plattencollenchym*) has the thickening mainly on the inner and outer tangential walls of the cells. This type is said by Duchaigne to be relatively uncommon, but it occurs for example in *Armoracia rusticana* Gaertn., Mey. and Scherb. (*Cochlearia armoracia* L. (Brassicaceae) as well as in *Astrantia* (Apiaceae) and it is particularly well developed in *Sambucus nigra* L. (Caprifoliaceae).

(iii) **Annular collenchyma** (Fig. 6.2 (c,e)) has cell walls that are more uniformly thickened. It is said by Duchaigne to be the most frequent type. It is particularly common in the midrib of leaves, *Laurus nobilis* L. and *Ilex aquifolium* L. being cited as examples. This type also occurs in petioles e.g. in the Apiaceae, Araliaceae, and Magnoliaceae and in adult stems, e.g. in certain members of the Apiaceae and Lamiaceae.

Some authors also recognize a fourth type referred to as **lacunate collenchyma** or **Luckencollenchym** and the term is applied to collenchyma in which well-developed intercellular spaces are present. Duchaigne does not, however, accept this as a distinct type, but mentions that collenchyma that could be interpreted as lacunate can be seen in various Asteraceae and

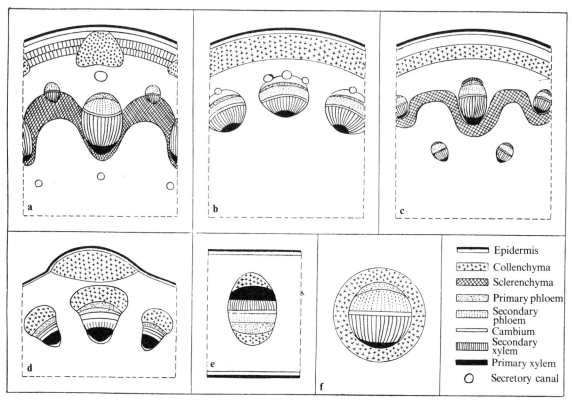

Fig. 6.1. Different examples of collenchyma distribution. (a) stem of *Foeniculum officinale* All., with peripheral strands; (b) petiole of *Hedera helix* L., with continuous peripheral cylinder; (c) young stem of *Piper carpunya* R., with deeply cortical cylinder; (d) young stem of *Medicago sativa*. L., with cortical strands next to the phloem; (e) lamina of *Viscum album* L., with strands impinging on the phloem and xylem; (f) circumvascular sheath in the flora axis of the Podostemonaceae. After Duchaigne (1955).

Malvaceae. The extent to which intercellular spaces are developed sometimes varies during the ontogenetic development of collenchyma.

Esau's (1936) and Majumdar's (1941) investigations both include important particulars concerning the ontogeny of collenchyma. The collenchyma strands in celery arise from the ground meristem in the young petiole by localized periclinal and anticlinal cell divisions. Repeated longitudinal divisions give rise to strands of elongated, thin-walled cells from which the adult collenchyma is differentiated. The ontogeny of the collenchyma in *Heracleum* is similar, although according to Majumdar the collenchyma is partly derived from prodesmogen cells which are also involved in producing the adjacent vascular bundles.

In his comprehensive study of the ontogeny of collenchyma Roland (1967) says the **colloblasts** become transformed first to procollocytes which develop

into constituent cells of the collenchyma (collocytes). Roland's investigation refers to *Clematis vitalba* L., *Daucus carota* L., *Glechoma hederacea* L., *Mercurialis annua* L. and *Sambucus nigra* L. In an earlier investigation Roland (1961) records the interesting observation that in the leaf of *Olea europea* L. the cells of the adaxial collenchyma keep pace with the increasing size of the leaf by becoming elongated. On the other hand the cells of the abaxial collenchyma, laid down at a very early stage in the development of the leaf, remain short. Roth (1958) has described the ontogeny of collenchyma in a species of *Beta*. For the ultrastructure of collenchyma see Beer and Setterfield (1958); Chafe (1970); Preston and Duckworth (1946); Roland (1964, 1965*a,b*, 1967); Wardrop (1969).

Walker (1960) has shown that mechanical shaking for 9 hours per day on 40 days increased the thickening of the walls of collenchymatous cells (as observed in transverse sections) of *Datura stramonium*. At the same time the lengths of the cells may be reduced. Growing the plants in darkness also reduced the thickness of the cell walls.

Venning (1949) compared celery seedlings grown in an air current to simulate wind with others grown under normal conditions and found that the number of collenchymatous strands was unchanged. On the other hand the air current increased the thickness of the cell walls, the diameters of the cells and the cross sectional areas of the strands.

Sclerenchyma

Sclerenchymatous cells, which are characterized by secondarily thickened walls, are an important component of the skeletal system of the plant. These elements, when mature, generally contain no obvious protoplasts, and they are commonly looked upon as dead cells. However Fahn and Leshem (1963) and other investigators mentioned by Fahn (1974) have found that fibres sometimes contain living protoplasts. Further on in this present discussion it is pointed out that sclereidal idioblasts, which, as we shall see, are one of the forms in which sclerenchyma occurs, enlarge by intrusive and symplastic growth during their ontogenetic development and often have persistent nuclei. This form of sclerenchyma must, therefore, also consist of living cells.

Sclerenchymatous cells occur very widely as a component of the tissues which give rigidity to stems and leaves. For example there are the **xylary fibres** which form the ground tissue of xylem (see also Vol. II). They also occur in the phloem of the leaf and stem in many plants. Others constitute the **'pericyclic' fibres** and stone cells of the stem and leaf. Sclerenchymatous cells are also a very common component of the supporting tissue of the veins of the leaf in the form of **bundle sheaths** whilst others are more superficial and occur in the peripheral parts of stems and

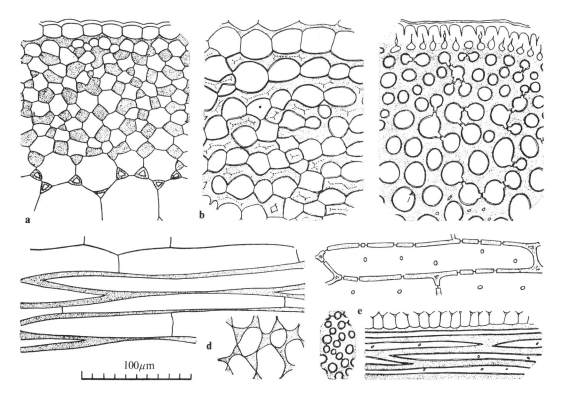

Fig. 6.2. Different types of collenchyma. (a) petiole of *Beta vulgaris* L., with angular collenchyma; (b) stem of *Sambucus nigra*, with tangential collenchyma; (c) annular collenchyma associated with median leaf vein of *Nerium oleander*; (d) angular collenchyma in the stem of *Cucurbita pepo* L.; (e) annular collenchyma in the stem of *Foeniculum officinale* All. After Duchaigne (1955).

leaves. As pointed out under leaf structure (p. 71–2) they also occur as terminal cells of the finer veins. They sometimes form continuous layers in fruits. They may also occur at the bases of trichomes, e.g. in certain Asteraceae, Begoniaceae, Euphorbiaceae, Fabaceae, and Melastomataceae.

Sclerenchyma is often well developed in plants exposed to dry conditions and it is undoubtedly because of the mechanical support which it provides that xerophytes in which it is present are prevented from collapsing. Sclerenchyma, particularly in the form of sclereidal idioblasts (see below), also occurs in certain aquatic plants. (See also under 'Ecological anatomy' in Vol. II.)

The component cells of sclerenchymatous tissue range from equidimensional to elongated and fibre-like. The more elongated sclerenchymatous elements sometimes become transversely partitioned by septa, the septa often being very thin and inconspicuous.

The concept of sclerenchyma as a distinct type of tissue is stated (Foster 1944) to have originated with Mettenius in 1865, who applied the term to both parenchymatous cells with secondarily thickened walls and to similar but narrower and more elongated cells that he classified as **prosenchyma**. Mettenius, however, did not apply the term sclerenchyma to thick-walled cells in the 'bast'. Later the term was used in different senses by individual authors. De Bary (1884) distinguished short sclerenchymatous elements, which he called **stone cells**, from more elongated elements which were termed fibres. A further distinction was drawn between bast fibres which occur in the bark and phloem and libriform fibres which form the ground tissue of wood. De Bary's ideas were taken up by subsequent workers and are still widely accepted at the present time. According to Foster (1944) Mettenius's (1865) original concept of sclerenchyma was perpetuated by Buch (1870), von Sachs (1882), and Wiesner (1890). The term **sclereid** had meanwhile been introduced for individual sclerenchymatous cells. This term is still widely used, but in the modern sense it is generally applied to relatively short sclerenchymatous cells which are contrasted with the longer fibres.

If the classification of sclerenchymatous cells had continued to be on a strictly morphological basis it might have remained comparatively simple. However, physiological anatomists, of whom one of the best known is Haberlandt, introduced an entirely new terminology for the various types of mechanical cells. In Haberlandt's (1914) vocabulary his collective term for all mechanical elements was **stereide**, and the stereides collectively made up the mechanical tissue which, following Schwendener, he termed **stereome**. Haberlandt refers to short stereides as **sclereides** or **sclerenchymatous cells** and these are contrasted with the longer sclereids to which he referred as **prosenchymatous cells**. The prosenchymatous stereides in Haberlandt's classification include **bast fibres** which occur in the bark and phloem. Bast fibres were contrasted with wood fibres (libriform cells) a term applied by Haberlandt, as by other authors, to mechanical elements in the wood. Finally Haberlandt included **collenchyma** amongst his stereides. In its modern sense collenchyma (see pp. 54–7) does not fit at all well into the concept of sclerenchyma.

Returning to the morphological classification of sclerenchymatous cells, our attention must now be given to mechanical cells, which are usually not sufficiently united to constitute a definite tissue, although they sometimes occur in groups. When not constituting a definite tissue they are distributed, often in an apparently haphazard manner, in the soft tissues of the organs for which they provide support. Because of their haphazard distribution, and because they are solitary or in relatively small clusters they are now generally classified as **idioblastic sclereids** or **sclereidal idioblasts**. In using these terms, however, it should be remembered that the cells were formerly called 'spicular cells' by Solereder (1908) and 'sclérites' by van Tieghem (1891). Other terms that were formerly used for idioblastic sclereids include '**internal hairs**', '**ground tissue hairs**' or '**trichoblasts**'.

Because sclereidal idioblasts are very variable in size and shape they were divided into the four following groups by Tschirch (1889, pp.301–2; see also Tschirch 1885):

(1) **Brachysclereids** or **stone cells**. These are more or less isodiametric in form. They usually occur in the pith, cortex and bark of stems and in the fleshy parts of many fruits e.g. those of *Pyrus* and *Cydonia*. They are often clustered.

(2) **Macro-sclereids** or **rod cells**. These are columnar cells, often forming an outer palisade layer in seed coats e.g. in the Fabaceae. They are sometimes referred to as '**Malpighian cells**'.

(3) **Osteosclereids** or **prop cells**. These are also columnar, but they differ from macro-sclereids in being dilated, lobed, or ramified at their ends. They usually occur in leaves.

(4) **Astrosclereids**. These are branched and often present a stellate appearance. Their branching is often very complex and they may assume very bizarre forms when mature.

Recent classifications of idioblastic sclereids

Variations of Tschirch's classification of idioblastic sclereids have been introduced more recently. For example T.A. Rao (1951*a*–*c*), in a preliminary survey, classifies idioblastic sclereids primarily on whether they originate in the epidermis, in the palisade tissue, or in the spongy mesophyll. He further subdivides them on their shapes, but the resulting groups do not precisely agree with those of Tschirch. Then again de Roon (1967), with special reference to the Marcgraviaceae, adheres to Tschirch's terms astrosclereids for branched sclereids with long or short arms. When the arms of an astrosclereid are short he classifies the cells as brachysclereids, and in this respect de Roon again falls closely into line with Tschirch. If the number of arms is reduced to between two and four the cell is termed a **librosclereid**. De Roon's term for an astrosclereid in which the central part of the cell is reduced and the arms long is **ophiurosclereid**. The same author classes astrosclereids that are exeptionally polymorphic as **idiosclereids**, but it seems to me that the difference between an idiosclereid and an astrosclereid is too small to be of much practical interest. De Roon, like Tschirch, uses the term osteosclereid for columnar sclereids that are lobed or ramified at the ends of their branches.

De Roon also varies the terminology for sclereids to correspond with their position in the leaf. Thus he refers to macrosclereids situated in the palisade parenchyma as **palosclereids**, and these are contrasted with **phellosclereids** which resemble cork cells and occur particularly in wound tissue. **Rhizosclereids** are so-called because they are situated in the palisade parenchyma but have root-like extensions into the adaxial part of the spongy parenchyma. A distinction between palosclereids and idiosclereids has been recognized by A.N. Rao (1971) in *Aegiceras corniculatum* (Myrsinaceae). Other modifications of Tschirch's classification of sclereids have been introduced by T.A. Rao (1959, 1951*a*,*b*).

In *Memecylon* (Melastomataceae) Subramanyam and Rao (1949) recognize ophiuroid and polymorphic types of sclereid. In this genus the ophiuroid sclereids are stated to have narrow, uniform lumina; the cell wall is homogeneous and not pitted. The same authors recognize various subdivisions of the ophiuroid and polymorphic types, but the lines of demarcation between them seem rather vague.

The idioblastic sclereids in the Melastomataceae have also received attention from Foster (1946) who examined these cells in 69 species of *Mouriri* (referred to in Foster's paper as *Mouriria*). The sclereids are very varied in form, ranging from small unbranched cells in the midrib and petiole to sclereids terminating the veinlets. Some of the sclereids terminating the veinlets have only rudimentary branches, whilst others are ramified, and there is a third type which extends obliquely or vertically across the mesophyll, becoming branched beneath each epidermis. There is also a fourth type of sclereid that is threadlike and acuminate, passing around and 'avoiding' the stomatal crypts. In a subsequent investigation Foster found that in *Mouriri huberi* Cogn. the terminal idioblasts arise singly or in pairs at the ends of the veinlets. They increase in length by vertical growth, extend vertically through the mesophyll and ramify beneath the epidermis.

It will be apparent to the reader, from the foregoing paragraphs, that the nomenclature for different types of idioblastic sclereid is somewhat confused and that some stability is greatly needed. My own recommendation is that, at least for the time being, Tschirch's classification should be followed and that Foster's subsequent very careful investigations should be taken into account. With cells that show such a range of form as that which is exhibited by sclereidal idioblasts it will be very difficult to devise a classification that will satisfy every investigator or cover every variation in cell form that is encountered. In such circumstances the needless multiplication of descriptive terms should be avoided unless and until a really comprehensive revision is undertaken.

Ontogeny of sclereidal idioblasts

The mother cells of sclereidal idioblasts are usually slightly larger than their neighbours and they often arise in the hypodermis or mesophyll and less frequently in the epidermis. In spite of the elaborate shapes which the idioblasts often assume when they are mature, most investigators believe that they remain uninucleate for so long as their nuclei persist. This applies e.g. in *Nymphaea* (Gaudet 1960) and in *Olea* (Arzee 1953*a*,*b*). A.N. Rao and Malaviya (1962) state, however, that they could find no nuclei in the mature idioblasts of *Dendrophthoë* (1 sp; Loranthaceae). Inamdar (1968*c*) likewise states that the nucleus ultimately degenerates in *Clerodendrum splendens* G. Don (Verbenaceae) but remarks that the cytoplasm is persistent. According to T.A. Rao (1951*a*,*b*,*c*) the nucleus of *Olea* degenerates after the cell becomes lignified.

After the sclereids have been initiated they increase

in size by a mixture of symplastic and intrusive growth. Growth of the sclereids has been described e.g. in *Olea* (Arzee 1953*b*) and *Osmanthus* (Griffith 1968), the sclereids in this last genus being initiated when the leaves have grown to half their mature length. According to Foster (1945*a,b*) the branches of the sclereids in the leaf of *Trochodendron* first arise as small protuberances which later become enlarged. At an advanced stage of branching the sclereids are stated to be 'amoeboid'. The exact course of development is apparently controlled by the nature of the environment provided by the surrounding cells. If intercellular spaces are numerous, the course of development of the sclereids is not the same as it is when they are surrounded by tissue that is more compact. The local environment of the sclereids thus appears to exercise some control over their ontogeny. Foster describes the idioblasts as endowed with 'rugged individualism'.

Foster (1944) notes that astrosclereids in the petiole of *Camellia* spp. arise from parenchymatous cells of the fundamental tissue system. The cell walls of the idioblast initials are of similar material to those of adjacent cell walls. As the enlarging branches of the developing idioblasts become elongated and grow between neighbouring cells that were previously in contact with one another, it is evident that the plasmodesmata must become ruptured. By dichotomous branching some sclereids become Y-shaped when mature, but others are stellate or irregular in shape. Details of the thickening of the cell wall are given. (For further information concerning sclereids at vein terminations see under Leaf topography on p. 00.)

Taxonomic distribution of sclereidal idioblasts

The table on pp. 207–14 is based on comparatively recent publications and includes just over 70 families of dicotyledons in which sclereidal idioblasts are known to occur. The idioblasts are not, however, equally prevalent in all of these families. For example they are noted as occurring only in the genus *Eryngium* in the large family Apiaceae, but they are much more widespread in such families as Annonaceae, Burseraceae, Capparaceae, Dilleniaceae, Ebenaceae, Euphorbiaceae, Gesneriaceae (*Cyrtandra*) (Bokhari and Burtt 1970), Hamamelidaceae, Menispermaceae, Oleaceae (Arzee 1953*a*), Proteaceae, Theaceae and Winteraceae. In the Marcgraviaceae they are present throughout the family. In other families they have been recorded only in certain genera. Furthermore it will be noted

that the families in which idioblastic sclereids have been recorded are by no means all closely related to one another. They are present in the supposedly primitive Winteraceae, as well as in one genus of the Apiaceae which is generally accepted as advanced. It must be remembered, however, that no comprehensive survey of the occurrence of sclereidal idioblasts has yet been made, and it may be that their distribution amongst the angiosperms as a whole will assume a more meaningful taxonomic pattern when more about them is known. At present the occurrence of idioblastic sclereids is mainly of diagnostic rather than taxonomic interest and this is because of their relatively restricted distribution. Morley (1953), however, says that the distribution of the different types of idioblastic sclereid in the genus *Mouriri* (Melastomataceae) to some extent follows a taxonomic pattern and this is true also in *Camellia* (Theaceae) according to Barua and Dutta (1959) and in the Gesneriaceae (*Cyrtandra*) according to Bokhari and Burtt (1970).

Publications amongst the older literature, to which Solereder (1908) refers, record the presence of 'spicular cells' in certain other families besides those included in the list on p. 00. The sclerenchyma is generally stated in these earlier publications to be present in the form of branched fibres in the mesophyll, or to form extensions from the bundle sheaths. It would seem, therefore, that the sclereidal cells mentioned in these older references are identical with those which are termed sclereidal idioblasts to-day. As the records are based on observations that were made some time ago, the data call for reinvestigation. Nevertheless for the sake of completeness, and because it may well be found that the records are substantially correct, these additional particulars are summarized below. Except where stated to the contrary the reader may assume that the 'spicular cells' are in the form of branched fibres in the mesophyll.

Asclepiadaceae (*Dischidia*, 1 sp); Bruniaceae (*Linconia*, 1 sp); Caesalpiniaceae (sclerenchyma of various kinds in certain species of *Cynometra*, *Heterostemon* (stone cells), *Macrolobium*, *Moldenhawera*, *Oxystigma*, *Saraca*, *Schotia*); Clusiaceae (reticulate cells in *Clusia rosea* L.); Combretaceae (sclerenchymatous extensions from around the vascular system common); Dipterocarpaceae (species of *Hopea* and *Vateria*); Epacridaceae (*Andersonia*); Ericaceae (*Diplycosia*, *Pernettya*, *Gaultheria*, *Rhododendron*); Ixonanthaceae (spiral tracheae in the mesophyll of *Ochthocosmos*) (c.f. Nepenthaceae); Lauraceae (species of *Actinodaphne* and *Ocotea*); Lecythidaceae (*Asteranthos*) Mimosaceae (sclerenchyma extending from the veins and spreading below the epidermis in *Inga* (14 spp.)); and *Pithecellobium*.

Fibres present but less well developed in *Affonsea* (1 sp),

Calliandra (1 sp), *Parkia* (1 sp), *Prosopis* (2 spp), *Schrankia* (1 sp), *Stryphnodendron* (1 sp); Monimiaceae (*Monimia*, 1 sp with lignified hypoderm); Myrtaceae (*Syzygium*, 1 sp); Nepenthaceae (spiral cells in mesophyll (c.f. Ixonanthaceae)); Sapindaceae (fibres in leaf tissue in species of *Cupaniopsis, Haplocoelum, Harpullia, Matayba, Paullinia, Serjania*); Symplocaceae (*Symplocos*, 1 sp in petiole); Tamaricaceae (*Reaumuria*, 1 sp); Thymelaeaceae (in a few species of *Daphne, Daphnopsis, Enkleia, Lasiosiphon, Lophostoma, Peddiea, Stephanodaphne*); Trigoniaceae (*Trigoniastrum*, 1 sp); Zygophyllaceae (*Nitraria*, 1 sp).

Special types of 'spicular cell' listed by Solereder include the following: (1) **Crystalliferous idioblasts** in Combretaceae (*Macropteranthes*), Rubiaceae and Schisandraceae; (2) **Elongated sclerenchymatous cells** in certain Combretaceae, Dilleniaceae, Loganiaceae; (3) **Multidirectional fibres** in the primary cortex of certain species of *Prunus* (Rosaceae); (4) **Spicular cells** in quinine bark (*Cinchona* spp Rubiaceae) and cells similar in shape in *Ancistrocladus* (Ancistrocladaceae) and *Styrax* (Styracaceae); (5) Long, thick-walled, **tuberculate fibres** in certain Apocynaceae; and (6) **Crystal-sheathed fibres** in Quebracho Bark (*Aspidosperma quebracho* Schlecht. (Apocynaceae)). This list could doubtless be extended.

Solereder draws a distinction between 'spicular cells' and more extensive groups of sclerosed, or at least thick-walled, tissue in the mesophyll. Although the distinction between these two categories of cells does not always appear to be very clearly defined, the families concerned are listed as follows.

Berberidaceae (*Berberis, Mahonia*); Capparaceae (*Capparis zeylanica*, abaxial hypoderm); Chloranthaceae (*Hedyosmum*, 2 spp); Clusiaceae (tendency for sclerosis of mesophyll in *Clusia*); Ericaceae (cells of the mesophyll uniformly thickened in species of *Vaccinium* (section *vitis idaea*); mesophyll composed of thickened cells except for the most adaxial of the palisade tissue in *Macleania*, 1 sp); Menispermaceae (lignified spongy tissue in a number of genera); Orobanchaceae (pitted cells in the mesophyll of *Epifagus*); Piperaceae (multi-layered hypodermis of thickened cells in *Piper*, 1 sp); Sapindaceae (lower layers of spongy mesophyll becoming sclerosed in species of *Matayba* and *Xerospermum*); Symplocaceae (spongy mesophyll composed of stellate, occasionally somewhat sclerosed cells, in *Symplocos*); Tamaricaceae (storage tracheids in mesophyll in some species of *Reaumuria* and *Tamarix*); Winteraceae (*Drimys, Zygogynum*).

Sclereidal hairs

In certain plants, structures occur which combine the characteristics of hairs and sclereids. An example was described by Guédès (1975) in the genus *Jovetia* (Rubiaceae). Here the adaxial epidermis consists of more than one layer of cells and many of the outer cells develop into sclereidal hairs. However it is interesting to note that the basal ends of the hairs penetrate more deeply into the leaf by intrusive growth, sometimes reaching the inner ends of the cells of the palisade sclerenchyma or even deeper.

Conducting elements of the phloem and xylem

In order to avoid repetition the conducting elements of the phloem (**sieve tissue**) are dealt with in Chapter 12. The conducting elements of the primary xylem (**tracheids** and **vessels**) are referred to under 'Stem' on pp. 175-6, whilst the corresponding elements of the secondary xylem are dealt with under Wood structure in Vol. II.

Literature cited

Parenchyma

Grew 1682; Jeffrey 1925

Collenchyma

Beer and Setterfield 1958; Chafe 1970; Duchaigne 1953, 1954a,b, 1955; Esau 1936; Foster 1949; Majumdar 1941; Müller 1890; Preston and Duckworth 1946; Roland 1961, 1964, 1965a,b, 1967; Roth 1958; Venning 1949; Walker 1960; Wardrop 1969.
Suggestion for further reading. Roelofsen 1959: general.

Sclerenchyma

1. Mentioned in text. Arzee 1953a,b; Barua and Dutta 1959; Bary, de 1884; Bokhari and Burtt 1970; Buch 1870; Fahn 1974; Fahn and Leshem 1963; Foster 1944, 1945a,b, 1946, 1947, 1949; Gaudet 1960; Griffith 1968; Guédès 1975; Haberlandt 1914; Inamdar 1968c; Mettenius 1865; Morley 1953; Rao, A.N. 1971; Rao, A.R. and Malaviya 1962; Rao, T.A. 1949, 1951a–c; Roon, de 1967; Sachs, von 1882; Solereder 1908; Subramanyam and T.A. Rao 1949; Tieghem, van 1891; Tschirch 1885, 1889; Wiesner 1890.

2. Literature cited in table on pp. 207–14. Arzee 1953a; Bailey and Nast 1944; Barua and Dutta 1959; Barua and Wight 1958; Bokhari 1970; Bokhari and Burtt 1970; Dickison 1969, 1970; Foster 1944, 1945a,b, 1946, 1947, 1955a,b, 1956; Griffith 1968; Inamdar 1968c; Katsumata 1971, 1972; Luhan 1954; Malaviya 1962; Morley 1953; Rajagopal and Ramayya 1968; Rao, A.N. 1971; Rao, A.R. and Malaviya 1962; Rao, T.A. 1949, 1950a,b, 1951a,b, 1957; Rao T.A. and Kulkarni 1952; Roon, de 1967.

Suggestions for further reading

Blyth 1958: primary extraxylary fibres
Dickison 1970: leaf of Dilleniaceae.
Rao, A.R. and Rao, C.K. 1972: root sclereids of *Syzygium*.
Rao T.A. 1951*c*: sclereids of *Diospyros*.
— and Bhupal 1973: typology of sclereids
— and Dakshni 1963: sclereids in *Memecylon*.
Spackman and Swamy 1949: septate fibres.
Tobler 1957: key work on sclerenchyma.
Wallis 1949: sclerenchyma in the diagnosis of vegetable powders.

Besides the references cited above there are important publications dealing with commercial fibres including many that are derived from plants. It is impossible to survey all of this literature here, but attention is drawn to the *Handbuch der Mikroskopie in der Technik*. This key work, written with collaboration from numerous listed specialists, is edited by H. Freund. (Freund 1951, 1970). These volumes also contain biographical data and portraits of many of the notable German workers in this field.

7

THE LEAF: GENERAL TOPOGRAPHY AND ONTOGENY OF THE TISSUES

C.R. METCALFE

Introduction

The ensuing account of the arrangement of the tissues and cells that go to make up the leaf follows on as a natural sequel to the section dealing with its external morphology (Chapters 3 and 4). The immediate discussion refers to the cellular organization of the leaf, especially the lamina, in general terms, while special aspects of leaf structure such as trichomes, stomata, and petioles are given separate treatment.

Epidermis

The epidermis, which normally consists of a single layer of cells, forms the outermost part of the leaf. The hairs (trichomes) (Chapter 5) and other superficial dermal appendages, such as scales, papillae, and the like, usually arise from the epidermis. The outer walls of the epidermal cells are impregnated with cutin, which often forms a special, superficial layer of the wall, known as the cuticle, discussed more fully on pp. 140-60. The subject of stomata is likewise elaborated on pp. 97-117. Immediately subjacent to the epidermis there is sometimes a morphologically distinct layer, the hypodermis, consisting either of transparent parenchymatous cells or, more rarely, of fibrous cells. This in turn is succeeded by the mesophyll, usually composed for the most part of photosynthetic, parenchymatous cells containing chloroplasts, but often including variously-arranged translucent cells as well. When there is no hypodermis the epidermis is in contact at its inner boundary with the cells of the mesophyll. The veins consist of vascular strands, often supported by mechanical cells, or provided with sheaths of translucent cells which are variously arranged in different taxa. However, the veins share the common characteristic of being divided into successively finer and finer strands which end in conducting elements that are in actual contact with the photosynthetic cells. The veins, especially the finer ones, are often embedded so deeply in the mesophyll that they do not project above the general level of the leaf surface. Alternatively they may form a raised network. We return to the subject of veins on p. 68.

Other features of the leaf which the systematic anatomist should note are the presence of mechanical cells, secretory structures (cells, cavities, and ducts), and cells containing ergastic substances such as crystals and opaline silica deposits. Laticifers are sometimes present but their occurrence is taxonomically restricted. Further details will be given in Vol. II.

Epidermis, hypodermis, dermal appendages, and papillae

Surface view preparations are particularly valuable for observing the types of hairs (trichomes), scales, and other dermal appendages. Sections of the leaf may also be needed to show how the bases of the dermal appendages are attached to the leaf surface. Trichomes show a wide range of structure and form, and, as has been indicated by Theobald, Krahulik, and Rollins (pp. 40-53) they can be of considerable diagnostic value at different taxonomic levels. Variations in hairiness also occur within a species, but these differences are often quantitative rather than qualitative. The taxonomic value of hairs has been recognized over and over again during the history of systematic anatomy. Solereder (1908) provided a very complete survey of the subject in so far as it was known in his time, but in his book the descriptive data are presented with a confusing wealth of detail which sometimes obscures the taxonomic significance of the characters under discussion.

The taxonomic value of the data concerning hairs and other dermal appendages recorded by Solereder (1908) was rather underestimated at the time when the first edition of this present work was written. It is to be hoped that the new classification of trichomes, presented by Theobald, Krahulik and Rollins, will serve a useful purpose by providing a new starting point for this subject which could be of great importance to systematic anatomists. The present failure to rely on data about trichomes depends very largely on the fact that there is no general agreement about the descriptive terms for the different types and the way in which they should be classified. Thanks to the

International Association of Wood Anatomists a body of informed opinion has been built up concerning the microscopical description of wood structure. A similar standardization of descriptive language now needs to be applied to trichomes and other dermal appendages such as scales. It is hoped that the important discussion by Theobald and his colleagues will help to fill the gap.

Returning to the epidermis itself, it is important for the systematic anatomist to be able to study the shapes of the cells as delimited by the anticlinal walls. This makes it necessary to study leaves in surface view preparations. Here it should be remembered that the epidermal cells vary considerably in size, shape, and orientation in different parts of the lamina of a single leaf. For example there are often marked differences between epidermal cells overlying the veins and those situated above the mesophyll between the veins. Epidermal cells from near the leaf margins are often quite different in appearance from those elsewhere. These local differences make it essential to ensure that comparisons are drawn only between cells from the corresponding part of the leaf surface, when cell shape and size are being used for diagnostic or taxonomic investigations. Other characters visible in surface view preparations include the thickness of the anticlinal walls, their waviness, the frequency and distribution of papillae (when present).

The anticlinal walls of epidermal cells vary so much in the extent to which they are straight, curved, or undulating that the use of the surface view appearance of cells in taxonomic studies is severely limited. For example, although the shape of the cells is seldom, if ever, reliable as a diagnostic character for a whole family, it is usually sufficiently fixed within a species to serve as a confirmatory identification character. At the same time there are other elementary facts with which experienced workers are usually familiar but which may come as a surprise to the novice. For example the anticlinal walls of the abaxial epidermal cells are very often more sinuous than the corresponding walls of the adaxial cells. Then again there are numerous taxa in which stomata are more numerous on the abaxial than on the adaxial surface. But the task of identification is often made more difficult because there are taxa to which these generalizations do not apply and with some taxa the converse is true. This lack of constancy in even the most elementary principles constitutes a real challenge to the investigator, who must always be on the lookout for the unexpected if he is to succeed. Those who expect identification or successful classification to be achieved by rule-of-thumb methods are not likely to get very far.

With these reservations, it may be noted that Solereder (1908, p. 1071) states that epidermal cells in which the anticlinal walls present a more-or-less zigzag course in surface view and show ridge-like projections at the apices of the angles occur in certain species belonging to many different families. Euphorbiaceae, Fabaceae, Lauraceae, and Ranunculaceae are cited as examples. In plants with narrow leaves the epidermal cells are often axially elongated. Examples occur in the Caryophyllaceae, Epacridaceae, Fabaceae, and Polemoniaceae, and there are doubtless many others. Sometimes the epidermal cells are almost prosenchymatous e.g. in species of *Candollea* and *Lathyrus*. Long cells of these kinds usually lie parallel to the median vein, although occasional exceptions have been recorded by Solereder in species of *Eutaxia*, *Silene*, and *Trifolium*. *Candollea* is also said to be remarkable for the occurrence of distinctly bordered pits, side by side with simple pits, in the anticlinal walls of the epidermal cells. Pitting of the outer walls of epidermal cells is much more common than is often realized. Solereder records it in Capparaceae, Celastraceae, Hippocrateaceae, and Sapindaceae. He also mentions that the pit canals in *Mortonia* (Celastraceae) are long and branched.

An epidermis consisting of specially small or specially large cells is noted by Solereder for a number of families. Where the cells are specially large in surface view they may also be palisade-like when viewed in transverse sections of the leaf. Sometimes the cell lumina are specially wide and serve to store water, or they may be filled with special contents. Water-storing cells in surface view preparations are most frequently in rows (files), but they may be reticulately arranged. The families in which Solereder noted epidermal cells of these distinctive types are listed in Vol. II.

Epidermal cells with thickened walls are common in xerophytes, and many botanists interpret this fact by assuming that the thick walls are of ecological rather than taxonomic significance. In the section on Ecological anatomy (Vol. II) I have emphasized that it is not always easy to distinguish clearly between taxonomic and ecological characters, so here again it is necessary to be on one's guard against making facile interpretations. Solereder (1908, p. 1072) interpreted the thickness of the walls of epidermal cells as having taxonomic significance at least in certain instances. For example he noted that epidermal cells with specially strong thickening of the inner periclinal walls are diagnostic for certain Epacridaceae and Euphorbiaceae and also recorded the occurrence of uniformly sclerosed cells in the families list on p. 200.

An epidermis of a rather special kind has recently been investigated by Guédès (1975). This is the multiseriate epidermis of *Jovetia* (Rubiaceae) which differentiates partly as hairs, whilst at the same time the cells of the adaxial epidermis show intrusive growth at their proximal ends by which they reach the inner part of the palisade parenchyma or penetrate even more deeply. The cells of the abaxial epidermis penetrate less deeply. Judging from the frequency with which hairs with deeply sunken bases are known to occur, it seems to me to be not improbable that intrusive growth of hair bases may also occur much more widely than has hitherto been realized.

Epidermal cells are sometimes papillose, and the papillae are often of diagnostic value either because of their localized distribution or distinctive appearance. Further reference to them will be found in Wilkinson's discussion in Chapter 10, pp. 148–52 and Theobald and his colleagues also mention them in their description of the different types of trichome (p. 45). Papillae are generally more common on the abaxial than on the adaxial surface of leaves, but they may occur on either or both surfaces.

Papillae in many taxa often remain thin-walled, but in other plants they become secondarily thickened e.g. in certain Ebenaceae, Geissolomataceae, Penaeaceae, Proteaceae, and Rosaceae. Another specialization is for the papillae to develop coronulate apices e.g. in certain Annonaceae, Araliaceae, Cornaceae, Ebenaceae, Fabaceae, Olacaceae, Sapindaceae, and Styracaceae (see also list on p. 198).

As already mentioned, epidermal cells are occasionally subdivided either horizontally or vertically by partitions which are much thinner than the other cell walls. The families showing this character are listed on p. 199. Families in which chlorophyll is present in the epidermis are also known.

While still on the subject of the epidermis, readers will be interested in the investigations of *Eucalyptus* leaves undertaken by Carr, Milkovits, and Carr (1971). These authors coined the term **phytoglyphs** for the collective microanatomical features of the epidermis in relation to taxonomy. The characters in question are stated to be genetically controlled. Phytoglyphs can be studied in surface view and in sections both with the electron and light microscopes.

Multiple epidermis and hypodermis

Apart from plants with subdivided epidermal cells, there are others with a translucent **multiple epidermis** or **hypodermis** (hypoderm). A genuine 'multiple epidermis', formed ontogenetically by periclinal divisions of the initially single-layered epidermis, is not always easy to distinguish from a true 'hypodermis', formed by the division of cells that are subjacent to the epidermis. For example, according to Solereder, the hypodermis of *Ilex aquifolium* is formed from the mesophyll, whereas in *Ficus elastica* the tissue also, but erroneously, called hypodermis is formed by the division of epidermal cells. Since it is often necessary to work with material that is too fragmentary to enable the development of the hypodermis or multiple epidermis to be seen in detail, it follows that in the literature of systematic anatomy the two types of tissue have frequently been confused.

The proportion of the total leaf surface beneath which a multiple epidermis or hypodermis is present also varies in different taxa and the hypodermis may be present beneath either or both surfaces. All stages are to found, ranging from a simple tendency for the epidermal cells to be locally divided parallel to the leaf surface, to a situation where the hypodermis is continuous and consists of a number of layers of inflated translucent cells. The cells of the hypodermis are often very much larger than those of the epidermis. When hypodermal cells are parenchymatous they may be polygonal or have an undulated outline in surface view whilst their walls vary in thickness. When the walls are thin they may contract on drying so as to produce a folding of the walls in much the same way as in a concertina. In other taxa the hypodermal cells are prismatic with an angle of the prism at right-angles to the leaf surface.

Special types of hypoderm include the following: (1) Collenchymatous hypoderm e.g. in the axis of certain Cactaceae; (2) Spongy hypoderm, e.g. in *Oedematopus obovatus* Tr. and Pl. (Hypericaceae); (3) Sclerotic hypoderm, composed of fibres and rod-shaped cells, occasionally connected with the sclerenchyma accompanying the veins. The genera concerned are listed on p. 0. All of these variants can be used for diagnostic purposes.

Gelatinization of the leaf epidermis

Epidermal cells that are partly or wholly mucilaginous are frequently said to be **gelatinized**. Solereder (1908) uses this term repeatedly and he regarded 'gelatinization' of the leaf epidermis as a reliable character for the diagnosis of species. According to Solereder, gelatinized epidermal cells are often larger than their neighbours and they are said to serve for water-storage. Gelatinization usually affects only the inner walls of cells. The cells concerned commonly have hemispherical or conical extensions into the sub-jacent mesophyll and at the same time the gelatinized portions of the inner walls are separated from the cell lumina by one or more cellulose lamellae. Gelatinized epidermal cells may be solitary or in groups. Alternatively the whole epidermis may be gelatinized, and gelatinization may extend to the hypodermis as well as the epidermis itself. Indeed Solereder warns his readers that statements in the literature do not always make it clear whether gelatinization refers to the epidermis or hypodermis. The families in certain members of which gelatinization of the epidermis and hypodermis are recorded by Solereder are listed on pp. 198-9.

The presence of crystals in epidermal cells is discussed in Vol. II.

Stomata

Surface view preparations are particularly valuable when studying the stomata and the cells which immediately surround them for diagnostic purposes. For further particulars see Wilkinson's account in Chapter 10 (pp. 99-102).

Mesophyll

The mesophyll as already noted usually consists, for the most part, of cells containing chloroplasts, and these photosynthetic cells constitute the chlorenchyma. A proportion of translucent cells may be present as well, but the number varies considerably in the leaves of different taxa.

(1) Chlorenchyma

The adaxial part of the chlorenchyma in most dicotyledons is composed of **palisade cells** which collectively form the **palisade tissue**. The cells of the abaxial chlorenchyma are usually much more irregular in shape and partly separated from one another by an extensive system of intercellular air-spaces. These cells constitute the **spongy mesophyll**. This arrangement of adaxial palisade and abaxial spongy tissue is sometimes reversed, especially in plants from specialized habitats. For example Solereder (1908) refers to species of *Frankenia* in which the adaxial surface of the leaf is adpressed to the axis whilst palisade tissue is developed beneath the abaxial surface. (See also under Ecological Anatomy in Vol. II). A lamina with one or more layers of adaxial palisade

cells accompanied by abaxial spongy tissue is said to be **dorsiventral** or **bifacial**. This may be contrasted with a homogeneous leaf in which palisade and spongy tissue are not distinguishable from one another. Families in which this type of structure is known are listed on p. 206. In an **isobilateral** (isolateral) **leaf** palisade tissue is developed towards both surfaces (Figs. 7.1(c), 7.2.(c)). Heinricher (1884) gives numerous, mostly European, examples of isobilateral leaves, with illustrations, and there is a list of families in which isobilateral structure has been recorded on p. 207 of this work. Another variant is for the adaxial surface to be very much reduced in area compared with the abaxial surface so that in transverse section the lamina is adaxially grooved and abaxially rounded, the true adaxial surface then being restricted to the adaxial groove. Sometimes the adaxial surface is eliminated altogether, producing the cylindrical or **centric leaf** (see p. 206).

In a centric leaf the chlorenchyma sometimes constitutes a continuous or interrupted peripheral zone of palisade cells radiately arranged around the circumference of the section (Fig. 7.1(b)). Alternatively the mesophyll may be more or less homogeneous, with little or no evidence of radiate structure. It is very common to find that in centric leaves of either type a considerable proportion of the mesophyll consists of translucent tissue, the number of chlorophyll granules in the cells gradually falling off towards the central axis of the leaf. Centric leaves occur in many xerophytes and they are further discussed under Ecological anatomy (Vol. II). Transverse sections also show that centric leaves vary considerably in different taxa, and these differences are often of considerable taxonomic interest.

Reverting to palisade tissue, it should be noted that the component cells may be regularly or irregularly stratified. Wherever an irregularity occurs it is caused by the unequal lengths of the cells. The number of layers of palisade cells also varies, not only from species to species, but sometimes in different specimens of a single species, or even in different individual leaves from a single plant. This quantitative variation restricts the taxonomic usefulness of the palisade tissue and points to the need for great caution in drawing conclusions based on mesophyll structure. On the other hand there are some specialized types of palisade cells which are often of greater taxonomic interest because of their restricted taxonomic occurrence. This applies, for example, to the so-called **Kranz structure** which is discussed more fully on p. 72. Another specialized form of palisade cell is the

funnel-cell. Funnel-cells are short, conical palisade cells narrowed at their outer ends but broadened at their abaxial ends. They are specially characteristic of plants from damp localities, and, according to Solereder, they occur for example in certain Gesneriaceae and Piperaceae. Similar cells have also been reported from the interior of the mesophyll adjacent to normal palisade parenchyma. Then there are the so-called **arm-palisade** cells in which each cell has from one to about eight inwardly directed flanges or invaginations. When the invaginations are at one or both of the narrow ends of the palisade cells, they appear to be H-shaped in transverse sections of the leaf. The structure of arm-palisade cells is reviewed in considerable detail by F.J. Meyer (1962). Solereder says that the systematic value of arm-cells may be either at the level of the species (e.g. in *Anemone* and *Phyllanthus*) or of the genus (e.g. in *Meliosma*.) There are numerous references to arm-palisade cells in the literature, but, even if all of the recorded occurrences of these cells could be quickly substantiated, a list of the families in which they occur would be of very limited taxonomic interest because they are so numerous. I am personally rather doubtful whether all of the records of arm-palisade cells are valid, and I suspect that some of them are due to faulty observation or misinterpretation. Further investigations seem to be needed before it can be decided whether their taxonomic distribution is really significant.

Palisade cells with rows of papillae on their longitudinal walls are termed conjugate palisade cells by Solereder. The papillae belonging to adjacent cells are in opposite pairs and in contact with one another. A tissue composed wholly of conjugate palisade cells is permeated by a well-developed intercellular system, the only contacts between the cells being through their papillae. Conjugate palisade cells show some resemblance to the peg-cells noted by Cutler (1969, pp. 119–20) in certain Restionaceae.

Ridged and **reticulate palisade cells** also occur. These have ridge-like or reticulate thickenings on their walls. Solereder records them in certain Candolleaceae (now Stylidiaceae), Clusiaceae (*Clusia*), Melastomataceae (seven genera) and Solanaceae (*Dyssochroma, Juanulloa, Markea*), Winteraceae (certain *Bubbia* spp. and *Belliolum*). Solereder also refers to narrow mesophyll **cells resembling fungal hyphae** in the Campanulaceae (*Lightfootia*), Moraceae (*Artocarpus* and *Ficus*), Rhamnaceae (*Reynosia* and *Zizyphus*). He also records mucilaginous swellings on the walls of both spongy and palisade cells, e.g. in certain Loganiaceae, many Melastomataceae and Menispermaceae and also in *Melananthus* in the Solanaceae. Lacunae filled with stellate tissue may also occur in the mesophyll. This relatively unusual character has been recorded e.g. in a few Caryophyllaceae and Lauraceae. **Green spots** are sometimes to be seen on the surface of petioles e.g. in the leaves of certain Apiaceae such as *Meryta*, where the spots are due to local interruptions of hypodermal collenchyma by cells containing a green pigment.

The spongy chlorenchyma is usually of less interest to systematic anatomists than palisade tissue. Nevertheless in spongy tissue useful diagnostic characters are provided by the arrangement and shapes of the cells and in the nature and sizes of the intercellular spaces.

(2) Translucent cells in the mesophyll

Besides the palisade and spongy chlorenchyma, the mesophyll, or indeed any part of a leaf, usually contains a proportion of **translucent cells** in which there is little or no chlorophyll. It must be remembered, however, that every grade between chlorenchymatous and non-chlorenchymatous cells is known to exist. The distribution patterns of translucent cells in the mesophyll vary considerably throughout the angiosperms and a large part of this variation shows little or no taxonomic relationship. It may be mentioned, however, that they are particularly abundant in succulent leaves. These receive further consideration under Ecological anatomy (Vol. II). Translucent cells are often well developed around the vascular bundles of the veins where they form bundle sheaths and in leaves that exhibit 'Kranz' structure.

(3) Palisade ratios

Before leaving the mesophyll, the use of **palisade ratios** in systematic anatomy must be considered. The concept of the palisade ratio can best be understood if it is first recalled that the epidermis of the leaf constitutes a multicellular translucent envelope surrounding the mesophyll. It follows that by examining cleared leaves in surface view it is usually possible to see the ends of the palisade cells lying beneath the cells of the epidermis. This does not apply to leaves in which there is a dense indumentum or an opaque adaxial hypodermis, both of which obscure the relationship of the palisade cells to those of the epidermis. However there are many species in which the palisade cells can be seen quite clearly through those of the epidermis, and this raises the question of whether the number of palisade cells lying beneath a single epidermal cell is sufficiently constant in any one species to be used for diagnostic purposes. According to Wallis and Dewar (1933) and Wallis and Forsdike (1938), the suggestion that the relationship between the cells of the epidermis and those of the subjacent mesophyll might be of taxonomic interest was first made by Zornig and Weiss (1925). Wallis and Dewar introduced the term palisade ratio for the average number of palisade cells beneath a single cell of the adaxial epidermis. The same authors, working with eight species of *Barosma*, described in detail their procedure for determining palisade ratios. The choice of a suitable clearing agent is important because the presence of water 'causes the mucilage to swell, thereby separating it from the palisade'. They therefore used chloral–phenol (equal parts by weight of chloral hydrate and phenol). Pieces of leaf 2–3 mm square were heated in this reagent in test tubes placed in a water bath for 15–20 minutes. Individual pieces were examined microscopically after being mounted in the same reagent. Powders were examined in the same way. With material which was resistant to the clearing treatment it was sometimes necessary to resort to boiling the pieces in lactic acid. A camera lucida was used to make drawings of four epidermal cells and the palisade cells beneath them. The number of palisade cells could then be counted and the figure thus obtained, when divided by four, (because the observations were made on groups of four cells), gave the palisade ratio for the leaf under examination. Palisade ratios in *Digitalis thapsi* L., *D. lutea* L., and *D. lanata* Ehrh. were subsequently established by Dewar (1933, 1934*a,b*), and by Wallis and Forsdike (1938), for *Atropa belladonna, Datura stramonium, Scopolia carniolica* and *Solanum nigrum*. In this group of species the last two authors also established that the · palisade ratio is constant in different parts of an individual leaf and showed the same to be true in leaves of a single species from a range of habitats, and finally in leaves of a single species collected over a sequence of years. (See also Vein termination numbers on pp. 70-1).

Veins

The arrangement of the veins in the lamina is an important component of the study of leaf architecture. This topic is dealt with in Chapter 4 by Hickey, whose studies have given a new impetus and precision to the subject. However, long before Hickey's recent

investigations, the veins were surveyed by numerous investigators, some of whose work is now to be considered. We may start with the extensive surveys of the vasculature of leaves made by Schuster (1908, 1910) and Glück (1919). There are also references in the literature to leaves in which the vasculature is of an unusual pattern. For example, Foster (1952) has published interesting data concerning the **feathery venation** that is characteristic of the Quiinaceae. The same author (Foster 1952, 1959*a,b*, 1961*b*, 1966, 1968, 1970, 1971) also made a very detailed survey of **leaves with dichotomous venation** which he interpreted as a primitive character, special attention being given to *Circaeaster*, the sole genus in the monotypic Circaeasteraceae. It may also be noted in passing that Wylie (1946) studied the leaves of 90 tropical and subtropical species from Florida and showed that there is a significant correlation between the organization of non-vascular tissues and the spacing of veins. This observation is said to support the view that the distribution of vascular strands in the blade is influenced in some degree by translocation through the living tissue between the veins.

Solereder (1908) considered certain characters of the veins and bundle sheaths to be of diagnostic value. He drew particular attention to the nature of the tissue immediately surrounding the vascular bundles in the small and medium-sized veins. In using these characters for diagnostic purposes there is need to ensure that the veins to be compared are situated in corresponding positions in the leaves under investigation. It must be remembered that the appearance of the cells that accompany an individual vascular bundle may vary considerably in sections taken at various levels throughout the length of the vein. Nevertheless with attention to these details reliable but restricted diagnostic information can be obtained. In the leaf veins of numerous angiosperms the vascular bundles are wholly or partly surrounded by a morphologically distinct or conspicuous **bundle sheath** consisting of one or more layers of cells. The presence or absence of a bundle sheath is often diagnostic at the species level, but the Begoniaceae is cited by Solereder as a family in which bundle sheaths are almost or completely lacking. When there is no bundle sheath the chlorenchymatous tissue of the mesophyll impinges directly on the vascular bundles. A plant with this type of structure was described by Solereder as having **embedded veins**. The same term was also used by Solereder for veins embedded in the mesophyll but sheathed by parenchymatous cells. Veins sheathed by large parenchymatous cells were noted in the first edition of this book in certain members of the families listed on p. 215, a few of which are illustrated in Figs. 7.1 and 7.2.

The sheath cells in *Gilia* (Polemoniaceae) have U-shaped thickenings and those in *Tribulus* (Zygophyllaceae) have thick, pitted cell walls. Veins in the next category to be considered have vascular bundles that are accompanied above and below, or are completely encircled, by sclerenchyma. Alternatively only a few sclerosed cells may accompany an individual vascular bundle. The families, genera, and species in which the vascular bundles of the veins are accompanied by sclerenchyma are so numerous and widely distributed throughout the dicotyledons that it would serve no useful purpose to attempt to survey them in general terms. This feature will be mentioned repeatedly in the family descriptions in future volumes of this work.

There are many taxa in which the vascular bundles of the veins are not only sheathed, but the sheaths themselves may extend towards, or come into contact with, the epidermis on either or both of the leaf surfaces. Vascular bundles in which the bundle sheath extensions reach both the adaxial and abaxial epidermis were described in the English translation of Solereder's work (1908) as **vertically transcurrent**. This term is not very familiar to-day, but it is a useful

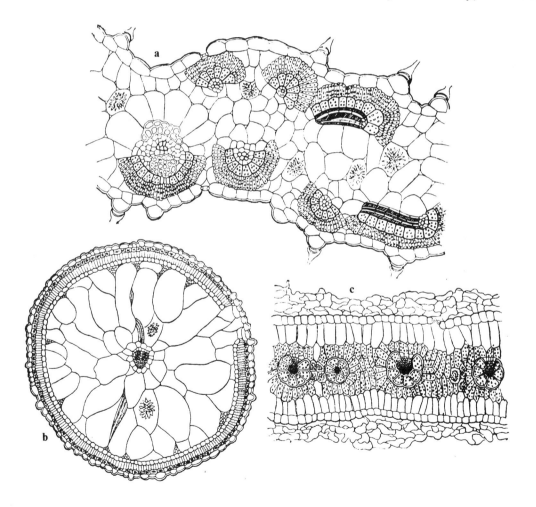

Fig. 7.1. **Transverse sections through leaves.** (a) *Bassia muricata* All., showing Kranz structure; (b) *Salsola longifolia* Forsk. Leaf centric in section; (c) *Atriplex halimenus* L. Leaf isobilateral. Note the large cells of bundle sheaths in (a) and (c). After Volkens, magnification not recorded.

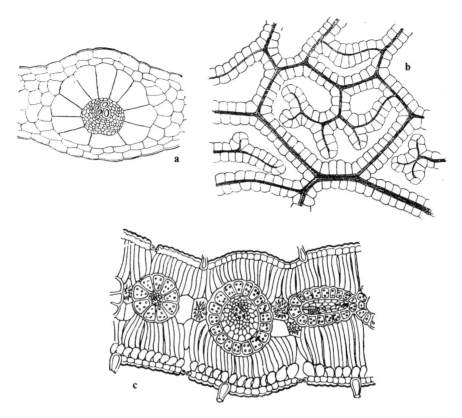

Fig. 7.2. **Bundle sheaths and chlorenchyma.** (a) Vascular bundle in the leaf of *Amaranthus blitum*, showing vascular strand and accompanying sheath cells; (b) *Amaranthus blitum* showing vascular strands and accompanying sheath cells in surface view; (c) *Tribulus alatus* Del., TS isobilateral lamina showing large cells of the bundle sheaths. (a) and (b) after Solereder, (c) after Volkens. Magnifications not recorded.

one for descriptive anatomy. Solereder says that small, vertically transcurrent veins are sometimes characteristic of a genus e.g. in some of the Epacridaceae. He also says that the small veins throughout a whole family, e.g. the Dipterocarpaceae, are sometimes all vertically transcurrent. The bundle sheath extensions of vertically transcurrent veins may consist of (1) translucent parenchyma or collenchyma, with cell walls that are usually either thin or slightly sclerosed; (2) sclerenchymatous girders, which may spread out beneath the epidermis to become T-shaped, or they may spread out still more widely so that the girders appear to be continuous forming a hypoderm of similar cells. (Families are listed on p. 215.)

The occurrence of bundle sheath extensions amongst dicotyledons was surveyed more recently by Wylie (1952), but this author does not use the term vertically transcurrent with reference to them.

Wylie points out that the bundle sheath extensions constitute vertical partitions which are laid down in such a way that they form a series of compartments in the leaf blade in which the chlorenchyma is situated. Out of 348 species of dicotyledons from Iowa, southern Florida, and the North Island of New Zealand that were examined by Wylie, 210 showed bundle sheath extensions. The extensions were found to be most frequent in thin leaves from northern deciduous forests and least common in the broad, evergreen leaves from New Zealand. In my experience, vertically transcurrent veins are of restricted taxonomic value and their most important use in systematic anatomy is for confirmatory diagnosis.

Vein-terminations and vein-islets

In many taxa the ultimate veins end blindly in the mesophyll in close association with chlorenchymatous cells. However, there are other taxa in which final terminations to the veins are lacking and instead the smallest veinlets form a closed network in which the meshes appear, in surface view preparations, to be occupied by 'islands' of chlorenchyma. These patches of chlorenchyma are called **vein-islets**, a term which is used especially by pharmacognoscists. Pharmacognoscists have also introduced the concepts of 'vein-islet' and **'vein-termination** (or veinlet-termination) numbers' as aids in microscopical identification.

Vein-islets vary in size and shape to only a limited extent within a species and their number per unit area of leaf surface is likewise relatively constant. This constancy enables the identity of fragmentary leaf material to be confirmed, and this is a well-established procedure amongst pharmacognoscists. Wallis (1960) had implicit faith in the vein-islet number for diagnostic purposes, believing their values to be independent of the size of the leaf and of the age of the plant. Whilst vein–islet numbers may often suffice for the identification of commercial samples of crude drugs, in my opinion it is questionable whether they are sufficiently precise for the identification of species in general taxonomic exercises. Even in pharmacognosy the method has its limitations. For example Levin (1929) in a study based on the genera *Barosma*, *Cassia*, and *Erythroxylum* found that the vein-islet number for a given species varies within narrow limits and indeed in a large genus the vein-islet numbers for closely related species tend to overlap. Nevertheless Levin found the vein-islet number to be sufficiently reliable to distinguish Alexandrian from Indian Senna (*Cassia* spp.), Bolivian from Peruvian Coca leaves (*Erythroxylon* spp.) as well as for the separation of *Digitalis lutea* from *D. purpurea*.

Hall and Melville (1951, 1954) introduced a new unit, the **veinlet-termination number** for diagnostic purposes. This value represents the number of veinlet terminations per square millimetre of leaf surface and the same two authors showed that this unit is statistically more reliable than the vein-islet number.

Vein-islet and veinlet-termination numbers have also been investigated by Gupta (1961). This author went a stage further than either Levin, or Hall and Melville, by introducing the concept of **absolute vein-islet** and **absolute vein-termination** numbers. These are obtained by multiplying the average vein-islet or vein-termination numbers by the areas of the leaves, measured in square millimetres, on which the counts were made. The absolute numbers were found to be constant. The technique for determining veinlet termination numbers is fully described in Gupta's

paper. They were all made from 4 mm^2 of leaf surface. An average of five readings was taken from the central region of the lamina avoiding the leaf margins, as well as the midrib and thicker veins. If necessary, trichomes were removed before the measurements were made. For this purpose leaves were dried and pressed flat. They were then rubbed over a rough surface such as filter paper or fine sand paper. The leaf was held flat with the finger tip and ground by a rotary movement, exerting as little pressure as possible.

Cells at the ultimate vein endings

The nature of the cells at the blind **terminations of the fine veins** must now be considered. We may start with the work of Strain (1933) who studied the vein endings in 118 species belonging to a miscellaneous collection of families and divided them into five major groups. In the following data presented by Strain the figures in brackets refer to the percentage of the species examined in which each type of ending was observed.

(1) Veins ending in one or two, long or short tracheids lying parallel to one another (40 per cent).
(2) Veins ending in single tracheids (3 per cent).
(3) Veins ending in three or more tracheids (17 per cent).
(4) Terminations multiple, composed of irregular or sometimes triangular tracheids (6 per cent).
(5) Terminations multiple, but each ending in a single tracheid (5 per cent).

The types of vein endings, according to Strain, were in no way correlated with the taxonomic affinities of the plants. Similarly in any one species no constant type of vein ending was noted in 'sun' and 'shade' leaves respectively.

Foster (1956) has drawn attention to the occurrence of **tracheoid idioblasts** at the vein terminations in *Pogonophora schomburgkiana* (Euphorbiaceae). The tracheoid idioblasts resemble typical tracheary elements in having spirally thickened, pitted walls, but they differ from them in form, size, and general topography. Terminal sclereids are also stated by Foster (1955a) to develop at the ends of the veinlets in certain Capparaceae, Hamamelidaceae, Melastomataceae, Polygalaceae, Rutaceae, Simaroubaceae (*Quassia*). Sclereids associated with the veinlets are also described by T.A. Rao (1950a) in *Linociera* and he adds that in three species they appeared to connect peltate glands on the leaf surface to the vascular bundles.

Tucker (1964) examined the veinlet terminations in cleared leaves of 152 species belonging to 10 genera with broadly magnoliaceous affinities. The leaf portions were cleared in 5 per cent sodium hydroxide and then in chloral hydrate before being stained in alcoholic safranin and mounted in synthetic resin.

Owing to ontogenetic changes that take place in the vein endings it was found necessary, in Tucker's investigations, to make sure that comparisons were restricted to fully mature leaves, but provided this precaution is taken the cellular organization of the vein endings was found to be reliably constant. The endings fell into the following categories:

Tracheoidal elements, with scalariform or scalariform-reticulate thickenings on at least one, and sometimes on two, walls. Tracheoidal elements were found to occur singly or in groups of two to three cells at each vein ending and they were noted in all but two of the species examined.

Undilated tracheids. Although cells of this kind are known to be common at the vein endings in representatives of many diverse, temperate families, *Liriodendron tulipifera* was the only species amongst those examined by Tucker in which undilated tracheids consistently terminate some of the veinlets in mature leaves.

Dilated tracheids, 'which may be clavate, gnarled, or ramified with short knobs or processes'. These were restricted to six of the species examined and these belonged to *Liriodendron*, *Magnolia* and *Michelia*.

Reticulate-walled tracheids. These were irregularly shaped, thin-walled tracheids with numerous reticulations. They were noted by Tucker in certain species of *Kmeria*, *Magnolia*, *Manglietia*, *Michelia* and *Talauma*.

Veins sometimes terminate in sclereids rather than tracheids. This applies, for example, in certain species of *Hibbertia* (Dilleniaceae), according to Dickison (1970). When this is so, the sclereids terminating the veins may show transitions to **idioblastic sclereids**, which are discussed more fully on pp. 58–61.

Lersten and Carvey (1974) have drawn attention to the vein endings in the leaves of the Ocotillo (*Fouquieria splendens* Engelm.) (Fouquieriaceae), a striking desert shrub from the south western states of North America. In this species a few leaves expand whenever the infrequent rain falls, and they become detached after being photosynthetically active for only a short time. The leaf is isobilateral and the spongy tissue poorly developed. In the distal part of the lamina, sclerified cells are present at the terminations of the minor leaf veins and become increasingly common towards the apex of the leaf where adjacent cells of the mesophyll also became sclerified. The authors were unable to decide whether the elements are tracheids or sclereids, so preferred to call them **veinlet elements.** The physiological significance of the veinlet elements is obscure, but it is interesting to note that the number of these cells varied in herbarium specimens from different sources.

Kranz structure

The term **Kranz structure**, clearly of German origin, strictly speaking means 'wreath structure'. It is applied to leaves which in transverse section show the medium-sized and small vascular bundles to be individually surrounded by circles of green tissue. Closer inspection shows that the green circles consist of conspicuous parenchymatous bundle sheath cells containing chloroplasts that are much larger than those in the assimilatory tissue of the mesophyll. Very often the assimilatory cells of the mesophyll, external to the bundle sheaths, are palisade-like in form and are arranged in a radiate manner around the vascular bundles, so that the mesophyll cells appear to enter into the composition of the green circles surrounding the vascular bundles. The palisade assimilatory cells are, furthermore, so situated that their inner ends next to the vascular bundles impinge on the sheath cells. At the same time the outer narrow ends of the most external layer of palisade cells are in contact with the epidermis, or touch the hypodermis if the leaf has one.

Kranz structure sometimes deviates from the most common type that has just been described. For example the sheath cells may be big enough to stand out in contrast to their neighbours, but the specially large chloroplasts may be lacking. Then again the sheath cells may be unequal in size or the sheath may be incomplete. Another situation is presented where the vascular bundles, or at least some of them, are incompletely surrounded by palisade chlorenchyma cells, or the mesophyll may be partly composed of translucent cells. The existence of these and other variations of Kranz structure can be overlooked, and it is sometimes difficult, from published descriptions, to be sure whether a particular leaf really shows Kranz structure or not.

The term Kranz structure has become familiar in recent years because of the discovery that photosynthesis in leaves with this type of structure often proceeds by the 'C_4 dicarboxylic acid pathway' (e.g. Laetsch 1968). According to Johnson and Brown (1973), in plants with this type of photosynthesis the first recognizable products, after CO_2 has been incorporated, are oxaloacetic, malic, and aspartic acids, all of which have 4-carbon molecules. This is in contradistinction to the more usual type of photosynthesis in which 3-phosphoglyceric acid is the first recoverable molecular product. From the data presented in some of the publications listed below it is evident that the relationship of the type of photosynthetic pathway to the presence or absence of Kranz structure is not absolute. It seems that the type of photosynthesis is bound up with the nature of the chloroplasts, the C_4 dicarboxylic acid pathway being followed only when the bundle sheaths contain chloroplasts of the specially large type mentioned above. Further research is still needed in which full attention is given both to the anatomical and physiological aspects of the subject. The investigations also need to be on as broad a taxonomic basis as possible so that the morphological variants of Kranz structure are fully covered.

When biochemical investigations are undertaken without accompanying anatomical studies, an over-enthusiastic biochemist might be tempted to claim that the leaf of a particular species must inevitably show Kranz structure because its photosynthesis proceeds by the C_4 dicarboxylic acid pathway, without confirming microscopically that the leaf does in fact show Kranz structure. I find that biochemists sometimes speak of the two alternative photosynthetic pathways as if the distinction between them is very clear cut. What I would like to know with more certainty is what happens about the photosynthetic pathway when the Kranz structure is incomplete. Perhaps by the time these words are printed this matter will have been fully investigated, but, if not, it does seem to be a subject that calls for further enquiry.

The following is a selection of publications concerning Kranz structure in which additional information is given. Moser (1934) noted Kranz structure in 79 species, and dorsiventral structure in 8 species, of *Atriplex*. Tregunna and Downton (1967) reported that certain Amaranthaceae, Chenopodiaceae, and Portulacaceae have 'low carbon dioxide compensations values'. El-Sharkawy, Loomis, and Williams (1967) investigated photosynthesis in *Amaranthus*. Frankton and Bassett (1970) refer to 4 species of *Atriplex*, belonging to the *A. patula* group, which do not show Kranz structure. Downton, Bisalputra, and Tregunna (1969) studied the ultrastructure of the chloroplasts of *Atriplex*. They found that the mesophyll chloroplasts began to degenerate when the bundle sheath chloroplasts reached a certain state of development. Crookston and Moss (1970) investigated the presence of large chloroplasts in the bundle sheaths of various dicotyledons in relation to the type of photosynthesis. Schöch and Kramer (1971) studied the ultrastructure of the large chloroplasts in the bundle sheaths of *Chamaesyce* (*Euphorbia*) *buxifolia* (Engelm, and Hitchcock) Small (Euphorbiaceae) and *Pectis leptocephala* (Cass.) Urb. (Asteraceae). Johnson and Brown (1973) also made ultrastructural studies on leaves with Kranz structure, but confined their attention to grass leaves, and this applies also to Downton and Tregunna (1968). Tregunna, Smith, Berry, and Downton (1970) investigated Kranz structure from the physiological standpoint. See also Metcalfe (1960, 1971) for the radiate structure of the mesophyll in grasses and sedges.

It is of special interest to readers of this work that leaves with Kranz structure, or in which the cellular organization tends to be of this type, have a restricted taxonomic distribution. It is perhaps more common amongst monocotyledons than in dicotyledons, particularly in the Gramineae and Cyperaceae. Amongst the dicotyledons the Amaranthaceae and Chenopodiaceae are the families most frequently quoted as showing Kranz structure. A more complete list of the families and genera in which it has been reported is given on pp. 214-15. In this list Kranz structure has been interpreted in a rather wide sense, and botanists who have studied the subject intensively may feel that some of the families on the list should have been omitted. It is hoped, however, that the information in the list will serve as a guide to taxonomic alliances amongst which Kranz structure may be expected.

Laetsch (1968) suggests that plants with Kranz structure probably first arose in tropical surroundings, and he goes on to say that the ability of the plants in question to maintain a maximum rate of photosynthesis under high light intensity could be related to their leaf anatomy and chloroplast specialization. It is interesting to note, as pointed out by Laetsch (quoting the work of other botanists cited in his paper) that the photosynthetic pathway in leaves with Kranz structure is very similar to 'the crassulacean acid metabolism (CAM) followed by many succulent plants'. Moser (1934) interpreted Kranz structure as a climax development from isobilateral structure.

Transfer cells
(see also under extra-floral nectaries on pp. 128 and 130)

In the elements known as **transfer cells** there are irregular ingrowths from the cell walls which provide an extensive interface between the wall and the protoplast into which the ingrowths penetrate. They are termed transfer cells because they are believed to be concerned with the local, mainly horizontal transfer of solutes. They may, for example, serve to connect the xylem to the phloem. They have been overlooked until comparatively recently and the current interest in them was initiated by Ziegler (1965) and by Gunning, Pate, and Briarty (1968). Since then they have been studied by Gunning and Pate (1969) and by Pate and Gunning (1969). They occur particularly in the phloem and xylem of minor leaf veins, as well as in various secretory structures and in haustoria. More recently they have been found to occur in association with nodes (Gunning, Pate, and Green 1970). Four types of transfer cell have been recognized by Pate and Gunning, two of which occur in the phloem and the other two kinds in the xylem parenchyma and bundle sheath respectively. Enlarging on their physiological functions, Gunning and Pate say that transfer cells serve the following purposes: (1) absorption of solutes from the external environment (e.g. in the epidermis of submerged leaves); (2) secretion of solutes to the external environment through nectaries and other glands; (3) absorption of solutes from an internal, extracytoplasmic compartment (e.g. in vascular parenchyma, haustorial connections, embryo-sacs, and embryos); and (4) secretion of solutes into an internal, extracytoplasmic compartment (e.g. tapetum of an anther; pericycle of a root nodule).

The taxonomic interest of transfer cells is, according to Pate and Gunning (1969), limited by their occurrence in a wide range of vascular plants. Writing about the occurrence of transfer cells in leaf veins, Pate and Gunning (1969) record that they were seen in 22 out of 200 families of dicotyledons of which representatives were examined. It is pointed out that they are much more common in dicotyledons than in monocotyledons and also that it is much more usual to see them in herbaceous than in woody plants. It seems rather curious to me that, if transfer cells are more prolific in herbs than in woody plants, there are not more of them amongst the many herbaceous representatives of the monocotyledons. This would appear to indicate that the taxonomic affinities of the plants in which these cells occur is more significant than the herbaceous habit in determining their distribution. Pate and Gunning say that transfer cells are particularly common amongst groups that are generally accepted as being phylogenetically advanced, such as the Asterales, Dipsacales, and Rubiales.

Turning to the occurrence of transfer cells in the vascular tissue at and near nodes, Gunning, Pate, and Green (1970) surveyed 190 species belonging to 32 families and found transfer cells in the form of modified vascular parenchyma in the nodes of floral leaves, bracts, bracteoles and cotyledons as well as of foliage leaves. Furthermore they were noted in more than half of the species examined and they were consistently present in certain families such as the Asteraceae, Fabaceae, and Lamiaceae. In the same article Gunning, Pate, and Green mention that, besides angiosperms, nodal transfer cells occur in horsetails, ferns and gymnosperms. The transfer cells at nodes are mostly associated with departing foliar traces. Phloem transfer cells also occur at the nodes, but they are much less conspicuous than those in the xylem.

Transfer cells in glands have been recorded by Gunning and Pate in quite a wide range of plants. The examples they give include the following: (1) in glands of insectivorous plants belonging to the Lentibulariaceae (*Pinguicula*) and Droseraceae (*Dionaea*, *Drosera*, and *Drosophyllum*); (2) in extrafloral nectaries (see also p. 124) of *Vicia* (Fabaceae); (3) in salt glands of Frankeniaceae (*Frankenia*), Plumbaginaceae (*Acantholimon*, *Armeria*, *Limoniastrum*, *Limonium*, *Statice*), and Tamaricaceae (*Myricaria* and *Tamarix*); (4) in glands of parasites and hemiparasites such as Orobanchaceae (*Lathraea*) and Scrophulariaceae (*Euphrasia*, *Odontites*, *Pedicularis*, *Rhinanthus*); (5) in haustorial structures e.g. in Cuscutaceae (*Cuscuta*).

Watson, Pate, and Gunning (1977) say that the distribution of transfer cells in the leaf veins of 118 species representing 85 genera and 35 tribes shows that their occurrence fits in with the taxonomic views of Dormer (1945a,b) rather than with the earlier views of Bentham and Hutchinson respectively. Transfer cells are generally absent from tropical 'pulvinate' tribes but present in temperate 'epulvinate' species.

It is clearly evident that further investigation of transfer cells will be needed before their physiological significance is fully understood and their taxonomic distribution fully recorded. Since they appear to play an important part in the lateral transfer of solutes, it would be interesting to know how this physiological function is achieved in plants in which transfer cells have not yet been detected.

Leaf ontogeny

Leaves are initiated from primordia which arise in a definite sequence just below the apical growing point of the stem. The sequence of events involved in these changes is predetermined and the details are usually fixed in and characteristic of the plants concerned. A combination of periclinal and anticlinal cell divisions produces the initial primordium, which, through apical growth, becomes peg-like. At first there is no evidence of flattening of the primordium to form the leaf blade, but lateral meristematic activity soon begins in two approximately opposite positions along the length of the hitherto peg-like primordium. The two lateral meristems are orientated in such a way that they face towards the parent axis from which the leaf primordium has arisen. The ultimate shape of the mature leaf is then determined by the relative rates of apical and lateral growth in the primordium.

The pioneer workers who studied this subject included von Ettingshausen (1845*a,b,* 1857), Prantl (1883), and Deinega (1898). The reader is also referred to later publications by Schuster (1908, 1910) and Foster (1949, pp. 185-9, 1952, 1961*a*). Hagemann (1970) has also given a modern elaboration and reappraisal of Prantl's results. Foster's (1952) article includes details concerning the unusual feather venation of *Quiina*. The general subject of leaf ontogeny has also been reviewed by Bünning (1956) and Hara (1957). Other publications concerning leaf ontogeny are those by Roth (1949: stipules and ligules, 1952: shield-shaped (scutiform) and peltate leaves); Troll and Meyer (1955: unifacial leaves). Kugler's (1928) discussion refers mostly to Coniferae and monocotyledons but a few dicotyledons are also mentioned. These comprehensive articles may be regarded as key works.

It is becoming increasingly clear that the changes that take place in the initiation and ontogeny of leaves are so intimately bound up with the development of the node and stem that they cannot usefully be considered in isolation. This has been emphasized in a recent article by Howard (1974), part of which is repeated in Chapter 8.

Literature cited

Bünning 1956; Carr, Milkovits, and Carr 1971; Crookston and Moss 1970; Cutler 1969; Deinega 1898; Dewar 1933, 1934*a,b*; Dickison 1970; Dormer 1945*a,b*; Downton, Bisalputra, and Tregunna 1969; Downton and Tregunna 1968; El-Sharkawy, Loomis, and Williams 1967; Ettingshausen, C. von 1845*a,b,* 1857; Foster 1949 pp. 185-9, 1952, 1955*a,* 1956, 1959*a,b,* 1961*a,b,* 1966, 1968, 1970, 1971; Frankton and Bassett 1970; Glück 1919; Guédès 1975; Gunning and Pate 1969; Gunning, Pate, and Briarty 1968; Gunning, Pate, and Green 1970; Gupta 1961; Hagemann 1970; Hall and C. Melville 1951, 1954; Hara 1957; Heinricher 1884; Howard 1974; Johnson, C. and Brown 1973; Kugler 1928; Laetsch 1968; Lersten and Carvey 1974; Levin 1929; Metcalfe 1960, 1971; Meyer, F.J. 1962; Moser 1934; Pate and Gunning 1969; Prantl 1883; Rao, T.A. 1950*a*; Roth 1949, 1952; Schöch and Kramer 1971; Schuster 1908, 1910; Solereder 1908; Strain 1933; Tregunna and Downton 1967; Tregunna, Smith, Berry, and Downton 1970; Troll and Meyer, 1955; Tucker 1964; Wallis 1960; Wallis and Dewar, 1933; Wallis and Forsdike 1938; Watson, Pate, and Gunning 1977; Wylie 1946, 1952; Ziegler 1965; Zörnig and Weiss 1925.

Suggestions for further reading

Björkman, Gauhl, and Nobs 1969: Kranz structure.
Björkman, Nobs, and Berry 1971: Kranz structure.
Dickison 1969: vasculature of the leaf.
Napp-Zinn (1973, 1974): a key work.
Pate and Gunning 1972: transfer cells.
Tumanyan 1963: general systematic anatomy of leaves.

8

THE STEM–NODE–LEAF CONTINUUM OF THE DICOTYLEDONEAE

RICHARD A. HOWARD

Introduction

Although the stem, the node, and the leaf are treated as individual structures in many textbooks of general botany or plant anatomy, an understanding of the internal anatomy of any one of these is impossible without information concerning the others. The stem–node–leaf is a continuum of cells and tissues. An excellent article by Wetmore and Steeves (1971) considers the continuum from the development point of view and was written for the physiologically oriented reader.

Leaf development (see also Leaf ontogeny, p. 74)

The development of the foliage leaf has been considered in detail for a relatively few dicotyledons, when one remembers the number of taxa recognized in the class. Yet morphologists suggest that there is a basic pattern of development which is common to leaves in general, and that the difference in final form of the mature leaf can be explained as variations on one morphogenetic theme (Wetmore and Steeves 1971).

Upon and within the apex of the stem, a primordium of a leaf is distinguishable in a definite relationship to other primordia, which is expressed eventually as the phyllotaxy of leaves on a stem. This arrangement may be established as alternate, opposite, whorled, or orixiate (Maekawa 1948). A distichous arrangement is considered to be superimposed.

Wetmore and Steeves (1971) reviewed the experimental work that has been done on leaf primordia, and Wetmore and Garrison (1966) are among the recent authors to state that 'incontrovertable evidence now exists that primordia when produced on the apex are uncommitted. It is their natural biochemical milieu which determines their developmental destiny whether leaf or bud.' Surgical isolation allows young leaf primordia to appear as buds.

The leaf primordium undergoes a period of growth, establishing a foliar axis, and perhaps, by basal elongation, the stem. The existence of recognizable areas within the primordium, variously termed *soubassement*, leaf buttress, leaf base, *Unterblatt*, and *Oberblatt*, has been suggested by some authors and refuted by others.

The axis of the foliar primordium develops an adaxial meristem responsible for the thickening of a midrib region. An apical meristematic zone of the primordium may continue temporarily the elongation of the primordium. Marginal meristems along the foliar axis develop either the lamina of a simple leaf or the leaflets of a pinnately compound leaf. Divisions at the apex of the primordium develop the leaflets of a palmately compound leaf. Multidirectional expansion of an apical meristematic area produces the ascidiate or peltate leaf blade. Subsequent intercalary growth may be responsible for the petiolar area or the regions of the rachis between blade lobes or pinnae of compound leaves. The timing and relative development of each of these meristematic potentials can explain the form and size of the ultimate leaf (Kaplan 1970*a*,*b*).

During its development, meristematic activity of cells within the leaf may be restricted and finally cease, the subsequent enlargement being caused only by increase in cell size. The dicotyledonous leaf is generally assumed to have limited terminal growth. Examples of continued apical growth have been reported in the Meliaceae (Skutch 1946), while variable development of laminar portions of the blade and the axis of simple leaves is evident in tendrils of *Mutis* spp, *Triphyophyllum*, and interrupted blade development in *Nepenthes* and *Codiaeum*. Residual meristematic activity of other cells is evident in the production of 'leaf-plantlet meristems' (Warden 1971–2), normally in *Bryophyllum* spp and in *Tolmiea menziesii*, or in the vegetative 'leaf cutting' reproduction of many Gesneriaceae, Begoniaceae, etc. Limitation of expansion of the leaf may be due, in part, to the maturation of tissues, that is, the production of xylem and the development of a cambium and secondary tissues, and even the maturation of the mesophyll cells. However, the pulvinal areas may never develop sufficient quantities of xylem tissue to be considered incapable of further elongation or meristematic development.

Vegetative propagation of dicotyledons from leaf

cuttings indicates a residual meristematic potential in some herbaceous plants. No woody plants are known to be reproduced from leaf or petiole cuttings alone. *Saintpaulia, Begonia, Peperomia,* are examples of plants which can be reproduced with ease from petiole plus blade cuttings. Reproduction from portions of the leaf blade are considered practical by the nursery-man only when they involve a portion of the midvein or a lateral vein. Regeneration is believed to come from dedifferentiation of some mature cells or from callus.

As the primordium develops into a recognizable leaf form, the vascular system is differentiated in stages, first as a procambial stage, and subsequently as xylem and phloem. The differentiation of the procambium is generally described as acropetal, taking place while the primordium is still undergoing cell division and even elongation. Differentiation of the phloem from the procambium is acropetal into the leaf primordium. Later, differentiation of xylem begins at the base of the leaf primordium and develops acropetally into the leaf and basipetally into the stem. The continuity of conducting tissue from the stem into the leaf is thus established and recognized as the primary vascular system.

In general, developmental studies have been concerned with primordia and very immature leaves. No literature seems to concern the development of the many varied and complex patterns of arrangement of mature vascular tissues in the petiole of the mature leaf, or how such complexities are established in an interpolated organ. The independence of bundle development is specified in a few ontogenetic studies. Although it appears from most studies that the median bundle extends into the developing leaf at a faster rate than lateral traces, Kaplan has shown (1970*a*) examples of the more rapid acropetal development of the primary lateral traces. The need for later-stage developmental studies represents a serious lack of information when one considers that the leaf primordium may: interpolate a petiole; separate the rachis with or without articulations in the development of pinnate compound leaves; develop one to many pulvini; and form accessory foliage organs or appendages such as stipules, stipels, thorns, or glands.

The node

Alternative definitions and classifications of nodes

The node is commonly defined as the position on the stem at which leaves occur — a superficial topographic orientation. Variations within this definition allow for the node to have one leaf, two or opposite leaves, and several or whorled leaves associated with a single node.

An alternative definition is based on the following considerations. When the leaf has abscised, a leaf-scar remains visible, commonly revealing in cross sections one or more vascular traces, which are the paths of conduction between the leaf and the vascular system of the stem. Where the traces enter (or depart from, depending on the point of view that is adopted) the vascular system of the stem in the higher plants, a gap or gaps in the vascular system can be found. This area of gaps has also been accepted as a 'node' permitting two contrasting considerations for descriptive purposes. The usual, widely accepted classification is generally attributed to Sinnott (1914), although it had been recognized much earlier. This classification established three nodal types, termed **unilacunar**, **trilacunar** or **multilacunar**, depending on the number of gaps. Pant and Mehra (1964) have suggested a third classification of nodes. The classes they recognized are termed **alacunar** when there is no gap, **unilacunar** when there is one gap and **multilacunar** when the number of gaps is more than one. In Europe, independently, emphasis has been placed on the number of traces instead of on the number of gaps. Pierre (1896) suggested '**Monoxylées** or **Monophalangoxylées**' and '**Trixylées** or **Triphalangoxylées**' depending on whether there were one or three traces entering the stem; while Hasselberg (1937) established the terminology '**Unifaszikular insertion**', '**Trifaszikular insertion**' and '**Multifaszikular insertion**'. Later, Marsden and Bailey (1955) combined the number of traces and gaps, recognizing a 'fourth type of nodal anatomy' when the traces were two at one gap. Takhtajan (1969) proposed a hypothetical but comparable 'fifth type' with two traces to a median and single traces to each of two lateral gaps.

Arnal (1962) discussed alternative characteristics for a node, suggesting that it has been defined either as a zone of insertion on the stem of a leaf-axillary-bud complex, or the zone of non-elongation of the stem. He acknowledged that the presence or absence of buds, including the formation of axillary and adventitious buds, was extremely variable. He also acknowledged that leaves may be present or absent, and that internodes might also be lacking, and concluded that the only criterion for a node was that of non-elongation.

Croizat (1960), by contrast, has called attention to the variety of structures that could be produced and be present superficially in an area which he called a 'nodal torus'. These structures include buds, shoots, flowers, inflorescences, scales, cataphylls, eophylls, leaves, stipules, trichimoids, spines, glands, and meristematic tissue. He recognized that the nodal torus could occupy a considerable portion of the stem, far exceeding the leaf-scar area and the gap region of the stele. He pointed out that these products of the nodal torus are found completely surrounding the axis, sometimes extending a distance above the leaf-scar or occurring below the leaf-scar.

Boke (1961), noting that areole meristems of the Cactaceae produce new spines seasonally for many years, considers these to be dwarf shoots rather than simple buds, thus constituting a single nodal torus.

Elongation of internodes

The internode is generally defined as the portion of stem between two successive nodes, again a topographical definition. The internode is the result of both cell division and cell elongation (Wetmore and Garrison 1966), and the amount of tissue between the topographic nodes is obviously variable. If no elongation occurs and the leaves are close together, the growth form is either that of a rosette or of a short shoot. The rosette growth form is often altered with the induction of flowering and the subsequent production of internodes, especially in biennials (Wetmore and Garrison 1961). Lateral short shoots may remain as closely associated nodes for many years in flowering trees, but short shoots of terminal growth have been shown to be under hormonal control for shorter periods of time (Gunckel and Wetmore 1946, Gunckel, Thimann, and Wetmore 1949, Titman and Wetmore 1955). A peculiar condition of double nodes has been reported in the genus *Anacharis* (Jacobs 1946), where occasionally an internodal area of the stem fails to elongate, and two leaf-bearing nodes are close together.

Surprisingly, the actual method of internodal elongation, with its initiation and cessation, has received little attention. Wetmore and Garrison (1961) studied the elongation of internodes in *Helianthus* and *Syringa*, and noted that in *Helianthus* elongation began at the base of the internodal area and proceeded acropetally into the supra-adjacent node, whereupon the next internode developed in a similar fashion. In *Syringa* several internodes did develop simultaneously though overlapping in timing. In *Helianthus*, cell enlargement is the dominant factor in internodal growth, whereas in *Syringa*, cell division is the more

prominent of the two processes, although in both taxa mitosis and elongation are involved. In many members of the Gramineae and Liliaceae, as well as in *Equisetum*, elongation and maturation occur in the opposite direction, with the basal area of the internode the last area to mature, often remaining meristematic for some time. Although uninvestigated to the present, a similar situation may prevail in the Chloranthaceae and other plants primarily with opposite leaves, in which dried specimens show a shrunken or collapsed zone just above a pair of leaves or at the base of the internodal area. Although the processes of cell division and cell elongation are presumably under biochemical control, there is no explanation why they cease within the internodal area, causing the internode to be considered mature.

The amount of elongation, i.e. the length of the internode, may vary considerably on a given shoot or on one plant. Within a given flush of growth or development of a shoot from a bud of a woody plant, the basal portion of the stem and the apical portion may have the nodes close together, and the middle portion have the nodes well spaced. An explanation is possible if the growth hormones causing the internodal elongation are related to the appendage of the node. Thus, the lower internodes associated with bud scales, cataphylls, or eophylls are shorter than those associated with larger foliage leaves. The comparative lesser length of the upper internodes still associated with full-size leaves immediately preceding the abrupt transition to bud scales of a terminal bud remains unexplained, as does the hypopodium development at the base of some branch systems where only cataphylls are present.

Anatomical description of the node

The continuity of the vascular tissue from the stem through the petiole to the apex of the leaf is evident, yet the nature of its path and the variations of pattern of bundle arrangement and position along its length have not received much consideration. A three-dimensional interpretation is desirable, yet extremely difficult to depict. Single sections in leaf-bearing regions, therefore, have generally been used to describe the node, and these remain useful. A single section also reveals the cortex as well as the cortical bundles if any are present. Sclerenchymatous tissue in the form of caps, rings, or isolated idioblasts can also be seen, together with laticiferous- and resin- cells when present. At the same time the single section shows the main vascular system of the node which consists of secondary and/or primary phloem, together with the secondary or primary xylem as well as the cambium. The vascular system may either consist of discrete bundles, or be continuous and its structure either collateral or bicollateral. The centre of the section is occupied by pith which varies in size and shape in different taxa. Embedded in the pith there are sometimes some complete or incomplete medullary bundles, together with occasional plates of sclerenchyma, idioblasts, and canals or cells containing latex or resin. It need hardly be said that any or all of the tissues and cells may have descriptive value when making comparisons.

In speaking of the traces, they are generally pictured as if they had originated in and 'departed' from the principal vascular system, leaving a gap in the vascular cylinder before entering the petiole. They may be seen passing through the cortical areas in cross section or in a horizontal longitudinal section. During their transit through the cortex they are seldom accompanied by sclerenchymatous cells.

Descriptions of three nodal types, known respectively as unilacunar, trilacunar, and multilacunar dominate the literature (Fig. 8.1(1–3), described in the underline). A fourth type was introduced by Marsden and Bailey (1955) when they found an example of a node where two traces were associated with a single gap. This was in *Clerodendrum trichotomum* Thunb. Nast and Bailey (1946) also discovered that, in *Euptelea*, the bud and leaf traces were intermingled, but they did not designate it by any name or number. Similarly, no numbering was applied to the 'split-lateral' or 'common gap' condition which I myself reviewed (Howard 1970*b*).

Cortical and medullary bundles

Added to the basic four types of nodal structure must be the stems where a cortical vascular system is present which may or may not contribute to the vascular supply of the leaf (Fig. 8.1(4)). This has been described for *Calycanthus, Chimonanthus,* and *Nyctanthes* (Lignier 1887, Fahn and Bailey 1957, Balfour and Philipson 1962, Kundu and De 1968) where the cortical system forms a girdling bundle at the node, in addition to having branches entering the leaf. Acqua (1887) illustrated but did not comment on the cortical bundles of *Buxus. Idiospermum australiensis* (now Calycanthaceae) (Blake 1972) also has a cortical system in the stem. Ogura (1937) noted that some species of *Blahdia* (*Ardisia*) of the Myrsinaceae have a cortical vascular system while others do not. A cortical system

was reported for *Rhynchopetalum* (*Lobelia*) (Bower 1884) without indication of its role in the leaf vascular supply. A special study is needed of cortical vascular systems, but as yet ample material has not been available. The general pattern of the relationships of the cortical vascular system to the leaf vascular supply seems to include the following variations: (1) a cortical system may run the length of the stem without association with the main vascular supply of the leaf; (2) the cortical system may run the length of the stem, giving rise to girdling branches at each node, while other branches enter the petiole; and (3) the cortical system may originate just above the node and enter the leaf at the next node.

The only developmental study of cortical vascular systems appears to be that of Balfour and Philipson (1962) who studied *Chimonanthus*. They reported that the cortical system developed independently and later than the principal vascular system in the shoot apex. The cortical sympodia are connected laterally and below the node, and each bundle gives off a branch which goes to the leaf as a lateral trace. Buds in *Chimonanthus* are vascularized both by the main system and the cortical system. One branch from each adjacent cortical bundle divides in the bud to form the two side cortical bundles of the bud vascular systems.

Medullary bundles have a varied role in their relationship to the principal vascular system of the stem and in their contributions to the vascular supply of the leaf. A list of families of dicotyledons in which cortical and/or medullary bundles are known is given on p. 00. Lignier (1887) described the path of medullary bundles in many taxa of the Melastomataceae. Col (1904), Wilson (1924), Maheshwari (1929, 1930), Davis (1961), and Pant and Mehra (1964) all indicate the complexities of the medullary bundles in the Amaranthaceae, Chenopodiaceae, Nyctaginaceae, and Polygonaceae.

Medullary bundles may be complete bundles with xylem and phloem present, or incomplete, in which case they generally consist of phloem tissue alone. Sclerenchyma has not been found in association with medullary bundles. The medullary bundles of the Piperaceae (De Candolle 1866) may divide and anastomose with each other, and they may enter the principal vascular system of the stem or depart from it. Medullary bundles which enter the principal vascular system before entering the leaf may be indistinguishable from bundles of the usual leaf traces. In the Melastomataceae, many medullary bundles often enter the petiole independently of the bundles of

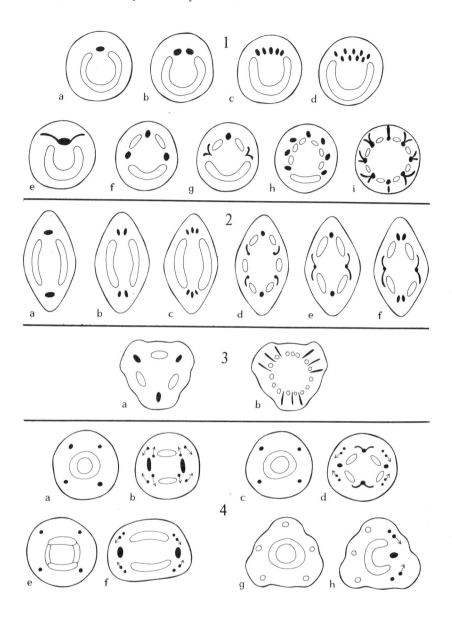

Fig. 8.1. Examples of nodal patterns.

1. Leaves alternate: (a) one trace from one gap; (b) two traces from one gap; (c) three or more traces from one gap; (d) alternating traces to the leaf and an axillary bud in *Euptelea*; (e) single trace showing vascular supply to stipules; (f) three traces from three gaps; (g) lateral traces of a trilacunar node with vascular supply to stipules; (h) multilacunar node without stipules; (i) multilacunar node with lateral traces showing vascular supply to the sheathing stipule.

2. Leaves opposite: (a) one trace from each gap; (b) two traces from each gap; (c) three (or more) traces from each gap; (d) opposite trilacunar nodes; (e) 'common gap' or 'split-lateral' condition where vascular supply of lateral traces to each of opposite leaves is supplies from a single lateral gap; (f) a 'split lateral' condition with two traces in each of the median gaps.

the principal vascular system and retain a medullary position in the petiole (Fig. 8.2(4)).

Departure of leaf trace bundles

In general, leaf trace bundles depart from the principal vascular system within the length of the stem that is represented superficially by the leaf base or leaf scar; i.e., the bundles pass out at an abrupt angle. Lateral traces depart from the vascular cylinder and run horizontally around the stem in the cortex before entering the leaf. On the other hand, the traces may depart from the vascular system a considerable distance below the leaf scar area of the stem and proceed upward in the cortex before entering the leaf. The terms **cathodic** and **anodic** have been applied to the traces to the left and right of the median trace, i.e., those away from or toward the direction of the spiral of the phyllotaxy. The idea of the consistent precocious departure of the cathodic trace is not substantiated. Attempts have also been made in the literature to characterize nodes of various plants on the basis of the number of bundles which are between the lateral traces or the lateral and the median trace (Record 1936). In general, the median trace is precocious in relation to the departure of the lateral traces. The fact that the traces may depart from a principal vascular system and be free in the cortex can often be determined by a superficial examination of the stem immediately below the leaf base. By far the most extreme example of early departure of a leaf trace is the one recorded by Johnston and Truscott (1956) for *Serjania*, where lateral traces may run in the cortex for 17 internodes before entering the leaf with the median trace.

In stems which have an opposite phyllotaxy the sets of traces associated with each leaf may depart at the same level, or one leaf may be higher than the other in terms of trace departure. Opposite leaves may be unilacunar as to their vascular supply, or trilacunar. Opposite or whorled multilacunar nodes have not been encountered. Whorled leaves may also be unilacunar or trilacunar. Carlquist (1955) has described a member of the Asteraceae with whorled leaves where one trace branches and a bundle enters the base of each of the adjacent leaves. The occurrence of a split lateral or a common gap has been reported for a number of families (Howard 1970*b*).

Vascular supply to the stipules

It is possible for stipules (see also p.18) to be vascularized by free bundles, by branches of a single lateral, or by laterals which run horizontally from their point of departure to the leaf base, fusing en route, and/or supplying branches which enter the stipules (Ozenda 1949).

Laticifers and resin systems at the node
(see also Vol. II for laticifers)

Nodal sections may also give some information on the continuity of latex and resin systems between the cortex of the stem and the leaf. The number of resin canals in the internodal area or in the petiole may be several times greater than the number of traces, but at the point of departure of the traces from the stem, only those resin canals associated with the traces persist, and the number of resin canals at the level of the abscission layer is usually equal to the number of traces (Artschwager 1943). The nature of the disappearance or the appearance of the intervening resin or latex canals has not been studied. In general, sclerenchyma caps to individual bundles will be less conspicuous in the nodal section than in the internodal or petiolar section.

Evidence of vascular supply to glands may also be present in the nodal section.

Variations of nodal patterns

Sinnott (1914) considered the node as a section through the area of the leaf trace gap, and gave a list of families of dicotyledons with unilacunar, trilacunar or multilacunar nodes. In most cases Sinnott reported only one nodal type per family. A few corrections of interpretation of Sinnott's much-cited work and a

---

3. Leaves whorled: (a) one trace from each gap; (b) trilacunar condition with three traces from three gaps to each of three whorled leaves.

4. Examples of cortical bundles: (a), (c), (e), (g) are internodal sections; (b) the cortical bundle enters the leaf and is replaced in the suprajacent internode by a bundle derived from the central vascular cylinder; (d) a 'split-lateral' with divided cortical bundles of which one portion enters the leaf and the other continues in the stem; (f) stem cortical bundles divide with portions of each entering the leaf and a portion continuing in the stem (a girdling nodal trace may also connect the cortical bundles); (h) alternate leaf arrangement with two of the cortical bundles dividing to supply vascular tissue to each leaf.

few reports of variations have since been published. Sinnott considered the trilacunar node to be primitive. Subsequently Bailey (1956), Ozenda (1949), Canright (1955), Takhtajan (1969), Pant and Mehra (1964), and Benzing (1967*a,b*) have arrived at differing conclusions. Trace number variation has been reported frequently in studies of seedlings where the mature nodal pattern is established three or four leaves after the cotyledonary node. Other authors have shown a variation in mature nodes of the trilacunar and multilacunar types as an inconsistency in the number of lateral traces on either side of the median trace. Kato (1966, 1967) reported such variation in *Citrus*, *Malus*, *Quercus*, and *Sorbus*, and Philipson and Philipson (1968) found a variation in *Rhododendron*.

Post (1958) studied nodes at successive levels in plants of *Swertia* and *Frasera* where traces varied from one to seven per node with a comparable number of gaps. This work may call for some further interpretation owing to the almost rosette-like structure of the stem base where the leaves are transitional to bracts of an inflorescence. My own studies on other Gentianaceae failed to demonstrate variations similar to those noted by Post.

Swamy and Bailey (1949) reported a difference in the node and trace number between leaves of *Cercidiphyllum* produced on long shoots (trilacunar) and those produced on short shoots (three-trace unilacunar). Their observations could not be repeated on material from the same tree where the present writer's recent studies showed all nodes on short shoots had regular trilacunar nodes.

Variations in unilacunar nodes may be encountered when sections are not examined from a sequence of levels. The double leaf trace found in many of the cultivated herbaceous Lamiaceae may show a fusion of the two traces above or below the level where a double trace is evident. A third trace may be present, weakly developed between two strong traces, and may fuse with one or the other of the double trace bundles or simply disappear (Swamy and Bailey 1950; Nakazawa 1956; Yamazaki 1965). In many herbaceous plants with a single gap, the interpretation of the number of traces is difficult when vascular development is weak. A single, broad, arc-like trace may be interpreted as two to five separate bundles.

Much nodal variation was reported by Howard (1970) in the case of the split-lateral traces of *Alloplectus ambiguus* (Gesneriaceae). In originally wild plants the pattern was altered from a clear-cut split-lateral trace, supplying vascular tissue to opposite leaves, to a situation of two trilacunar nodes when the plants were cultivated in a greenhouse. Such material was returned to native conditions in Puerto Rico where it has since been re-examined; in all stems studied the original split-lateral trace was again present.

Slade's (1952) study of cladode anatomy in New Zealand brooms showed a variation in the number of traces in successive leaves produced in a developing seedling. The earliest leaves may have three traces from three gaps. Later leaves may have pentalacunar or septalacunar nodes, produced either by the occurrence of flanking traces outside the lateral traces, or by the incidence of interpolating traces inserted between the median and the lateral traces of extremely flattened stems. A few species showed both flanking and interpolating traces. Unfortunately, this work does not show the continuity of traces through the internodes, nor were developmental studies reported.

Figs. 8.2–8.4. Selected sections of stems, petioles, and laminae.
The vascular pattern in different taxa is shown in camera lucida drawings of sections taken at successively higher levels from the node to the base of the lamina. The drawings for each taxon, excluding the leaf outlines, are at the same magnification. Sclerenchyma in heavy black.

Fig. 8.2.
1. (a) leaf of *Banksia serrata* (Proteaceae); (b) section of trilacunar node; (c) middle section of short petiole, median trace has produced two adaxial bundles while laterals have divided many times, first veins depart from centre of series of branches of the laterals; (d) base of blade, similar organization of bundles but each surrounded by ring of sclerenchyma.
2. *Hakea dactyloides* (Proteaceae): (a) leaf; (b) nodes; (c) petiole, lacking median derivatives of the median trace, but with four lateral traces derived from the two lateral traces shown in (b), all of the traces being embedded in heavily sclerosed tissue; (d) section of leaf in a position where the lateral traces have become the veins and the median trace has divided for the first time.

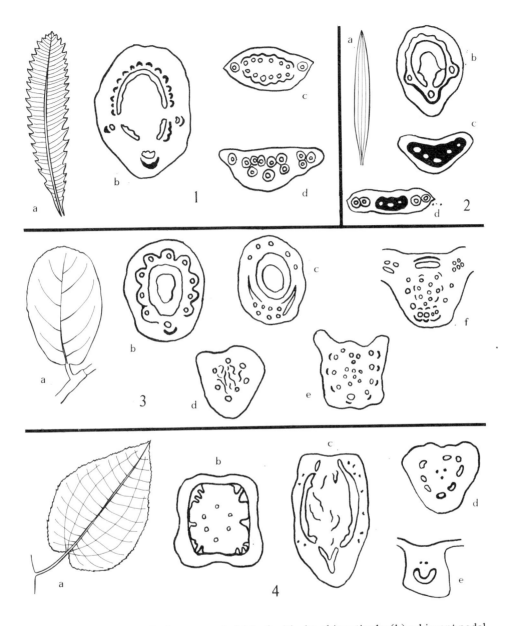

3. *Coccoloba rugosa* (Polygonaceae): (a) leaf with sheathing stipule; (b) subjacent nodal section showing precocious origin of traces; (c) nodal section, upper traces enter stipular sheath, median and laterals enter petiole; (d) basal section of petiole showing divergent trace branches in the 'cross zone,' note the absence of sclerenchyma in the lower pulvinal area; (e) midpetiole section showing the organization of vascular bundles in several concentric rings; (f) base of leaf blade with the organization of an adaxial plate derived from several lateral bundles.

4. *Amphilemma cymosum* (Melastomataceae): (a) leaf; (b) internodal section showing a circle of medullary bundles; (c) nodal section showing path of medullary bundles or their branches into the leaf base; (d) mid-petiole section with medullary bundles from the stem in a medullary position; (e) section through midrib with medullary bundles shown in solid black still present.

Fig. 8.3.

 1. *Symplocos glauca* (Symplocaceae): (a) leaf; (b) nodal section; (c) section at base of petiole; (d) section in middle of petiole, the rib-traces having arisen from the ends of the main vascular strand; (e) midrib section.

 2. *Peperomia hernandiifolia* (Piperaceae): (a) peltate leaf; (b) nodal section showing 'herbaceous' pattern of scattered bundles, three traces from three 'gaps'; (c) base of petiole, lines separate the products of the three traces; (d) upper end of petiole, where

In the course of my present study, many possible sources of variation were considered in leaves of mature stems. At an early stage in the study, 500 leaves and nodes were examined from single trees, involving crown and base leaves, sun leaves and shade leaves, those of vigorous shoots, and of short shoots or slow-growing shoots. Leaves were also obtained from single species, e.g., *Acer rubrum*, *Acer negundo*, from plants growing in states from Maine to Florida, and west to Colorado. No significant variation was found in any instance. Many commonly cultivated tropical species, e.g. *Hibiscus rosa-sinensis*, *Lantana camara*, *Nerium oleander*, were obtained from wild material or cultivated plants from temperate-area greenhouses or from botanical gardens around the world, and again vascular patterns proved to be consistent. Ecotypes were also considered, and no significant variation was encountered.

The patterns of traces and gaps accepted by Sinnott remain a good descriptive tool of anatomy and morphology. Supplemented with accessory data for cortical and medullary bundles when these are present, a section through the node of comparable materials will yield information of descriptive value.

Selected examples of stems, petioles and laminae are shown in Figs. 8.2–8.4 so as to demonstrate the changes in vascular structure at different levels. The names of the plants and structural particulars are given in the legends to the figures.

Literature cited

Acqua 1887; Arnal 1962; Artschwager 1943; Bailey 1956; Balfour and Philipson 1962; Benzing 1967*a,b*; Blake 1972; Boke 1961; Bower 1884; Candolle, De 1866; Canright 1955; Carlquist 1955; Col 1904; Croizat 1960; Davis 1961; Fahn and Bailey 1957; Gunckel, Thimann, and Wetmore 1949; Gunckel and Wetmore 1946; Hasselberg 1937; Howard 1970*b*; Jacobs 1946; Johnson and Truscott 1956; Kaplan 1970*a,b*; Kato 1966-7; Kundu and De 1968; Lignier 1887; Maekawa 1948; Maheshwari 1929; 1930; Marsden and Bailey 1955; Nakazawa 1956; Nast and Bailey 1946; Ogura 1937; Ozenda 1949; Pant and Mehra 1964; Philipson and Philipson 1968; Pierre 1896; Post 1958; Record 1936; Sinnott 1914; Skutch 1946; Slade 1952; Swamy and Bailey 1949, 1950; Takhtajan 1969; Titman and Wetmore 1955; Warden 1971–2; Wetmore and Garrison 1961, 1966; Wetmore and Steeves 1971; Wilson 1924; Yamazaki 1965.

Suggestions for further reading

Dickison 1973: nodes of *Xanthophyllum* (Polygalaceae).

Esau, 1943*a*: primary stem structure of *Linum*.

— 1943*b*: primary stem structure in general.

— 1945: primary stem structure of *Helianthus* and *Sambucus*.

Furuya 1953: organization of axillary branches in dicotyledons.

Gravis 1934, 1936: theory of foliar traces.

Kumar 1976: bilacunar two-traced nodes in *Geranium*.

Philipson and Balfour 1963: stelar evolution.

Slade 1971: stelar evolution.

Sugiyama 1972: nodal structure of *Magnolia*.

attached to the peltate blade; the single medullary bundle at the centre supplies the central veins descending from the midrib; the upper two bundles on either side of the section supply the basal veins of the leaf, the median trace and its two branches providing vascular tissue for the upper portion of the blade.

3. *Melicoccus bijugatus* (Sapindaceae): (a) leaf showing broadened rachis below upper leaflets; (b) nodal section; (c) base of petiole; (d) terete portion of petiole below first pair of leaflets; (e) section at the point of attachment of basal pair of leaflets; (f) section at point of attachment of upper pair of leaflets; (g) section of petiole of a leaflet.

4. *Aesculus hippocastanum* (Hippocastanaceae): (a) palmately compound leaf; (b) base of petiole, the bundles could have originated from either a tri- or pentalacunar node; (c) pulvinal section of petiole; (d) just above the pulvinus two branches from the median trace assume a medullary position; (e) middle of petiole; (f) apex of petiole at point of attachment of leaflets, medullary bundles of previous section have resumed association with the median trace and its branches; (g) section of petiole of leaflet.

Fig. 8.4.

1. *Sloanea dentata* (Elaeocarpaceae): (a) leaf showing upper and lower pulvini of petiole; (b) nodal section; (c) supranodal section showing dorsal association of lateral traces; (d) section of lower pulvinus; (e) section from middle of petiole where invagination, and foldings have produced a medullary ring shown with a surrounding layer of sclerenchymatous tissue and dorsal small bundles each with ring of sclerenchyma; (f) section of upper pulvinus taken where the vascular rings shown in (e) have broken up, or become dissociated, into numerous small strands; meanwhile the dorsal bundles have departed to form the lower veins; (g) midrib section showing the formation of the first strong secondary veins.

2. *Populus tremuloides*: (a) leaf, flattened petiole not well depicted; (b) nodal section showing patches of scattered sclerenchyma; (c) base of petiole; (d) lower portion of petiole; (e) middle of the petiole showing the organization of the vascular supply into a series of 'rings'; (f) apex of the petiole; (g) base of the blade, middle bundles of the flattened portion of the previous section supply the basal secondary veins.

3. *Aristolochia esperanzae* (Aristolochiaceae): (a) leaf; (b) nodal section showing the distinct bundles of the stem; (c) middle of the petiole showing the median bundle, two division products of each lateral trace, and the heavy sclerenchyma layer; (d) base of leaf blade with the departure of the basal veins.

9

THE PETIOLE

RICHARD A. HOWARD

Introduction

The petiole is interpolated in the development of the leaf. Studies of its elongation are relatively few. Masuda (1933) studied the petioles of woody and herbaceous plants in the botanical garden at Tokyo. He divided young petioles into equal portions with India-ink marks, and followed the elongation of each section. Three types of patterns of elongation were revealed in his studies. In type 'a' each zone elongated almost equally; in type 'b' the upper zone of the petiole showed a conspicuous elongation; and in type 'c' the lower zone showed the greatest amount of elongation. Most of the study was made in May or June, although one report is for a study conducted during October. Growth was completed in as few as 6 days, or over a period of 51 days. Type 'b' was regarded as the most common type, with elongation ceasing from the base upward in a time sequence. Tichoun (1923) has shown that the petiole in the majority of plants completes its elongation before the lamina reaches its maximum size.

My own studies, in the Boston, Massachusetts, area were made on woody plants in the Arnold Arboretum, and revealed additional growth patterns that cannot be brought into line with the data given by Masuda. Using plants growing out of doors, young petioles were marked as closely as possible with India ink from the axil of the leaf base to the base of the lamina. With the exception of *Liriodendron tulipifera*, all leaves completed the elongation of the petiole in a period of 12–15 days. In general, the petioles appeared to increase in length in such a way that the middle of the total length of the petiole increased fastest and stopped its growth earliest. The expansion and cessation proceeded in a wave from the middle of the petiole to each end. The apex of the petiole, especially if a pulvinus was present, and the base of the petiole increase in length the least or not at all. *Liriodendron tulipifera* in the Boston area continues to produce leaves until growth and development are stopped by frost. The upper quarter of the petiole seemed to continue elongation as in the type 'b' described by Masuda for *Liriodendron* and *Firmiana*.

The geographical middle of the petiole of leaves of most plants in the Boston area appears to be the first portion to increase in size and it is the earliest to mature. This is shown internally by the amount of lignified tissue revealed by a phloroglucinol stain. A cambium may develop first at the middle of the petiole, and sequentially toward each end, being absent in the pulvinus areas. Comparable developmental patterns of sclerenchyma were observed. The pulvinal areas show the least elongation and no development of secondary tissues or of sclerenchyma.

Further support for Masuda's type 'b' may be found in the statement of Funke (1929) that some petioles can, if necessary, become elongated even after the adult state appears to have been reached. Yin (1941) recorded growth in the length of the petioles of *Carica papaya* associated with diurnal movements of the leaves from a flexed to an upright position. The petiole was shown to grow on the abaxial side during the morning, and on the adaxial side during the evening, thus changing the orientation of the leaves.

Within a seasonal growth unit, for example in a temperate tree, the lowest leaves may have shorter petioles than those from the middle of the growth unit, and the upper leaves may also have petioles that are comparatively reduced in length. Anisophylly of leaves and of petioles of many plants with decussate leaves has been reported by Sinnott and Durham (1923), Heinricher (1910), Cook (1911), and Howard (1970b). Thus in most species of *Acer*, leaves borne horizontally have petioles of equal length, but in leaves that are members of a vertical pair the petiole of the lower member is longer than that of the upper member. Anisophylly of opposite leaves is characteristic of certain families such as the Gesneriaceae and Urticaceae and is revealed whenever the leaves and petioles of the two members of an opposite pair are unequal in size. No structural differences between the leaves or in the petioles of such plants have been reported. Plants in which leaves of two different morphological forms have been reported on any one individual plant are said to be heterophyllous. Transitions from juvenile leaves to mature leaves also occur and may be gradual or abrupt. There appear to be no differences in the anatomical structure of the petiole of such variants.

Superficial characters of the petiole can be of descriptive and taxonomic value. These may include length, colour, presence or absence of upper and lower pulvini, presence or absence of stipules, glands, thorns, colleters, pubescence etc. The adaxial surface of the petiole may be grooved for various distances, and the petiole may range from slightly ridged to strongly winged along the groove. Sections of an individual petiole reveal a characteristic outline which will vary from the base to the apex of the petiole.

Descriptive vascular patterns in the petiole

Grew (1675) was the first botanist to recognize that a cross section of a petiole can reveal different patterns of vascular bundles and he illustrated this in ten schematic cross sections.

In 1868 Casimir De Candolle proposed a theory of the leaf, likening the structure to that of the stem. The first comprehensive survey of petiole anatomy was that published by De Candolle in 1879. His survey of twenty 'families' led him to describe several fundamental concepts of the vascular structure he encountered. De Candolle proposed the terminology of an open system (*système ouvert*) versus a closed system (*système fermé*). In the open system the bundles, as seen in cross section, are arranged in an arc. In the closed system the bundles form a circle comparable to that in the stem. The bundles may be free or united. The system was invariably found to be open at the base of the petiole but it could become closed in the petiole and again open in the midrib or in the petiole of a leaflet. De Candolle also recognized a principal system (*système principale*, *système essentiel*), and an accessory system (*faisceaux détachés*), the latter composed of cortical bundles (*faisceaux intracorticaux*) and/or medullary bundles (*faisceaux intramédullaires*). The medullary bundles may either have the same orientation as those of the principal system or be inverted. It should be noted that the cortical and medullary systems recognized by De Candolle were recognized as belonging to the petiole alone and were not correlated with bundle systems in the stem. De Candolle classified the bundle arrangement in the families he studied.

He noted that groups of species in *Acer*, *Alnus*, *Aesculus*, *Mallotus* and other genera might have one system or the other. His division of species of *Fagus*, for example, would, today, represent the differences between *Fagus* and *Nothofagus*. Noting the variation in pattern encountered in a series of sections from the base of the petiole to the apex of the blade, De Candolle suggested that the most reliable comparative section could be obtained at the first meriphylle, that

is, the interval of the midrib between the departure of the first and second primary vein.

In the decade that followed, Vesque (1885), Petit (1886, 1887, 1889), Lignier (1888) and Acqua (1887) studied additional species, put forward new ideas, and introduced new technical terms relating to the vascular patterns of petioles.

Vesque (1881, 1885) appears to be the first worker who suggested that the middle of the petiole is the most reliable position from which a single section can be taken for comparative purposes. He studied a number of families and used the petiole anatomy to separate genera and families formerly united.

Petit (1886, 1887), in France, and Acqua (1887), in Italy, conducted broad studies of petiole vascular patterns and published their results competitively. Petit's study involved 500 species in 300 genera in 48 families. He recognized differences in patterns along the length of the petiole and stated 'it is in a terminal section that there will be presented for each plant the disposition most complicated and most regular which offers from one plant to another the greatest differences. It is, of consequence, the section most instructive. I have given this the name of the *characteristic*, for in many cases it will be sufficient, in order to recognize the family of the plants, and in some cases, its genus.' The names *coupe initiale* for the basal section and *coupe caractéristique* for the apical section are attributable to Petit. His work included, as a résumé, a key to the principal families of dicotyledons he studied on the basis of petiole vascular structure. Petit proposed a variation on the use of 'open' and 'closed' systems from that suggested by De Candolle. To Petit an open system was one in which the bundles were distinct or separated and a closed system was one in which the bundles were fused. Petit believed that herbaceous plants show individually distinct bundles (open structure) in the *coupe caractéristique* while in shrubby or woody plants the bundles were fused (closed) to form an arc or ring. He noted that perivascular or pericyclic sclerenchyma is generally lacking in herbaceous plants but present in woody plants and he attached taxonomic value to the presence or absence of this tissue.

Acqua's study was published in full before Petit's but following Petit's brief note indicating what he was doing. Acqua studied 19 families and proposed a classification of 13 patterns of vascular distribution in petioles. He correlated the position of the leaves with the vascular patterns. He recognized unilacunar, trilacunar, pentalacunar, and septalacunar nodes and others with a variable but greater number of bundles.

His unit number two was the double leaf trace of *Phlomis* and *Lamium*, later described as the fourth type of nodal anatomy by Marsden and Bailey (1955). Acqua also found in the Asteraceae examples of bundles from opposite leaves being associated with a single node, the 'split lateral' or 'common gap' (Howard 1970*b*).

Lignier (1887) published an extensive monograph describing the stem, petiole and leaf vasculature in the Calycanthaceae, Melastomataceae and Myrtaceae where cortical and medullary bundles of the stem are involved in the vascular supply of the leaf. Three-dimensional drawings are used and the complexities of the vascular system described in detail. This was followed in 1888 by an essay on the vascular system of the leaf and the stem of the phanerogams. Lignier stressed the fact that a single section of the petiole is inadequate for an understanding of the complex pattern developed in the length of the petiole. He felt that the most common pattern to be found in petioles is the arrangement of the bundles in a vascular arc. This arc could be a single broad bundle or consist of several bundles. The arc could also be expanded or divided, with extra bundles appearing at the ends of the arc. He designated the extra bundles as *surnuméraires* and called them *surnuméraires intérieurs* if they assumed a medullary position and *surnuméraires extérieurs* if they assumed a position in the cortex of the petiole. He noted that these corresponded only in some cases to comparable terms used by De Candolle.

Lignier proposed the idea that as the vascular arc increased in size within the petiole it would be forced into folds (*plis*). Folds could be either towards the interior (*pli interne*) or the exterior (*pli externe*) and if the folds are sufficiently deep they become separated from the vascular arc to form the medullary or cortical bundles of the petiole. Foldings with the separation of portions of vascular tissue could be recognized at the secondary and tertiary level. If in place of the vascular arc there is a complete circle and if a subsidiary arc is cut off from the circle by adaxial foldings, one or more vascular crowns can be recognized.

Morvillez (1919) accepted the idea of foldings as a descriptive approach in his study of petiole vascular patterns. He referred to the vascular bundle pattern seen in a cross-section of the petiole as a *chaine foliaire* which could be a continuous arc of tissue or separate bundles. If the arc curved inward at the adaxial ends a crosier could be formed and if portions of the crosier became isolated internal bundles or plates of tissue were established in a medullary position. Dehay (1935, 1942) and his students have used this descriptive vocabulary. Dehay chose for his illustrations the *section basilaire de la nervure mediane*.

An extensive survey of vascular structure of the petiole and the leaf has been published by Watari (1934, 1936, 1939) for the Fabaceae, *Acer*, and the Saxifragaceae. Watari considered the entire length of the vascular tissue from the node through the petiole and the lamina. His detailed work concerned the many branchings of the traces and their interconnections and are presented with sectional diagrams and complex three-dimensional reconstructions. For each group Watari proposed a classification based on the petiole structure.

Hare (1943) in a symposium on the taxonomic value of anatomical characteristics proposed a simple classification of the vascular structure of the petioles in transverse section as U-shaped, I-shaped or O-shaped, the latter being the hollow cylinder of the petiolar stele. Hare postulated that the structure is related to the mechanical stresses set up by the weight of the lamina and the lateral movements of the leaf. He also felt that the distinctive features of vascular patterns should be regarded as mainly adaptive and functional and of little phylogenetic significance. He concluded that 'characters derived from the petiole therefore can be used with confidence, but their value for purposes of classification varies widely at different taxonomic levels'. The proposals of Hare were incorporated in the first edition of this present work (Metcalfe and Chalk 1950) as twelve diagrammatic outlines.

I myself (Howard 1962) proposed the following classification relating the nodal structure at the level of the leaf gaps to the vascular patterns occurring in sections through the lower pulvinus, and at successively higher levels in the petiole or into the midrib of the lamina.

Node 1–1, simple trace, flat, slightly curved or U-shaped

1. Trace continuous:
   a. Without rib traces – *Allamanda* (Fig. 9.1(a)).
   b. With rib traces – *Graptophyllum* (Fig. 9.1(b)).
   c. Trace split longitudinally, later fusing – *Eugenia* (Fig. 9.1(c)).
2. Trace forming open or closed siphonostele:
   a. Vascular system open, with terminal rib traces – *Ilex* (Fig. 9.1(d)).
   b. Vascular system open with lateral rib traces – *Actinidia* (Fig. 9.1(e)).
   c. Vascular system open, without rib traces – *Celastrus* (Fig. 9.1(f)).
   d. Vascular system closed, without rib traces – *Terminalia* (Fig. 9.1(g)).
3. Trace invaginating at ends:
   a. Ends inroll – *Lyonia* (Fig. 9.1(h)).

Figs. 9.1–9.5. Selected petiole vascular patterns.
Successive sections of any one petiole were taken at its base, in the middle of the length of the petiole, at the apex of the petiole or the base of the blade below the departure of the secondary veins. The illustrations are presented in groups of three, which show the vascular pattern at each of the three levels chosen for investigation, the lowest being on the left and the highest on the right of each group. The leaves in Figs. 9.1–9.4 may be alternate, opposite or whorled, simple or compound, and with or without stipules. Leaves in Fig. 9.5 are always alternate. The stems to which the petioles are attached may have, or lack, cortical and/or medullary vascular systems.

Fig. 9.1 (a)–(j). Nodes with one trace from one gap.

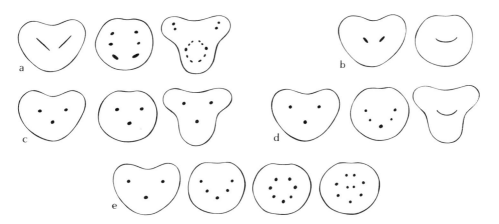

Fig. 9.2 (a)–(e). Nodes with two or three traces from one gap.

b. Forming siphonostele with medullary plate – *Cordia* (Fig. 9.1 (i)).

c. Inverted plate or simple arc – *Capparis* (Fig. 9.1 (j)).

Node 2 or more traces from one gap

1. Trace bipartite:
   a. In petiole – *Clerodendron* (Fig. 9.2 (a)).
   b. Below petiole – *Calycanthus* (Fig. 9.2 (b)).
2. Trace tripartite or more:
   a. Bundles free, forming an arc – *Bougainvillea* (Fig. 9.2 (c)).
   b. Bundles fused in an arc – *Solandra* (Fig. 9.2 (d)).
   c. Bundles free, in ring – *Hernandia* (Fig. 9.2 (e)).

Node 3–3, bundles free

1. Three bundles throughout the petiole – *Pedilanthus* (Fig. 9.3 (a)).
2. Lateral traces divide:
   a. Petiole with five traces – *Pittosporum* (Fig. 9.3 (b)).
   b. Petiole with many traces in U-shaped pattern – *Miconia* (Fig. 9.3 (c)).
   c. Free traces in a ring with medullary bundles – *Aesculus* (Fig. 9.3 (d)).
   d. Free traces in a ring without medullary bundles – *Sambucus* (Fig. 9.3 (e)).
3. Median trace divides and the division products assume a dorsal position – *Hibiscus* (Fig. 9.3 (f)).

Node 3–3, bundles fuse to form an arc

1. Bundles fuse and form an arc, flat or variously curved – *Lonicera, Betula* (Fig. 9.3 (g)).
2. Bundles fuse and form a flat arc with dorsal free traces – *Cornus* (Fig. 9.3 (h)).
3. Bundles fuse and invaginate at ends – *Congea* (Fig. 9.3 (i)).

Node 3–3, bundles fuse to form a siphonostele

1. Vascular system:
   a. formed by the simple fusion of traces – *Cotinus* (Fig. 9.3 (j)).
   b. formed subsequent to division of the median, the branches of which form dorsal bundles – *Acer* (Fig. 9.3 (k)).
2. Vascular system with accessory bundles:
   a. one accessory large bundle situated dorsally – *Hamamelis* (Fig. 9.3 (l)).
   b. small multiple accessory bundles dorsal in position – *Carya* (Fig. 9.3 (m)).
   c. accessory bundles in medullary position – *Bauhinia* (Fig. 9.3 (n)).

Node 3–3, bundles fuse to form more complex patterns

1. By invagination forming one or many medullary bundles or plates – *Quercus, Tilia* (Fig. 9.4 (a)).
2. Siphonostele invaginating to form included or dorsal accessory bundles – *Fagus* (Fig. 9.3 (b)).
3. Siphonostele formed, then lateral invaginations giving rise to a dorsal, smaller siphonostele or plate over U-shaped arc – *Carpinus* (Fig. 9.4 (c)).
4. Polystelic types:
   a. axillary bud included, petiole not compressed – *Platanus* type.

b. axillary bud not included, petiole compressed – *Populus* type (Fig. 9.4 (d)).

Nodes multilacunar, many traces from equal number of gaps

1. Traces remain free:
   a. bundles form ring or 'U' – *Ricinus* (Fig. 9.5 (a)).
   b. bundle pattern invaginates – *Rhizophora* (Fig. 9.5 (b)).
   c. by anastomosis forming concentric rings – *Coccoloba* (Fig. 9.5 (c)).
2. Traces fuse:
   a. ring simple – *Dendropanax* (Fig. 9.5 (d)).
   b. ring with included bundles – *Macaranga* (Fig. 9.5 (e)).

My subsequent work has shown this classification can be applied successfully to a majority of the vascular system patterns found in the dicotyledons as a whole. It does not, however, cover all of the possible combinations of node–petiole vascular structure with other useful diagnostic characters listed below.

A simple system of nomenclature for the node-petiole vascular types would be desirable. The generic names given here are those in which I first observed a particular pattern. However, variations of these patterns have been found within genera or families, suggesting that the use of family or generic names for the various patterns is not desirable. The alternative use of serial numbers would soon become unwieldy. Even the terminology which has been used, historically or in this contribution, presents difficulties. The individual vascular bundles of one section may be organized into a broken ring (eustele) for a short distance along the petiole, and the bundles may become more closely associated later to form a complete ring surrounding the pith (siphonostele). Medullary bundles may be present in the middle of the petiole, but not near the base nor at the apex of the petiole. The terms 'invagination' and 'inrolling' are visually descriptive terms of motion relating to successively different patterns of vascular bundles; but the terms are incorrect when considered from a developmental point of view.

Although previous workers have differed in their opinions on the taxonomic value of the anatomy of the node and the petiole in the recognition of taxa, I have shown that it is possible to use a combination of characteristics to create a key to sterile material of a local flora. The same success could probably be achieved with families and genera of dicotyledons. Such a dichotomous key would involve a great many characteristics and some of these would be repeated in several categories. For example, a simple flat arc in the petiole might arise from a unilacunar node or a trilacunar node while the relatively simple U-shaped pattern of vascular tissue in the middle of a petiole

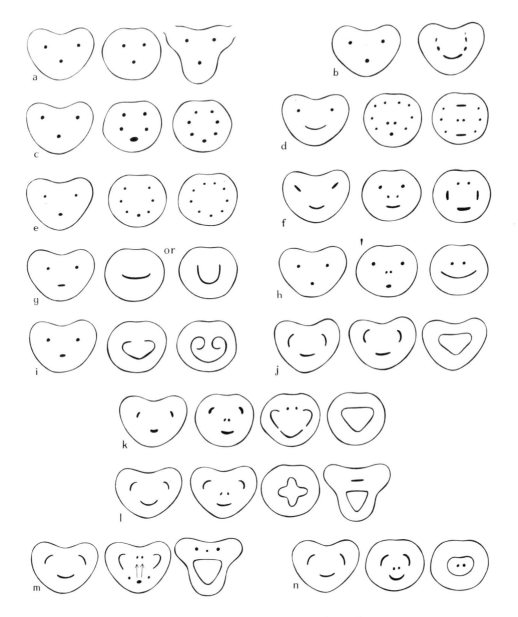

Fig. 9.3 (a)–(n). Nodes with three traces from three gaps.

could be associated with a unilacunar, trilacunar, or multilacunar node. The same characteristics would be involved in compiling a description of the vascular system extending from the internode to the leaf tip.

The following additional diagnostic characters are a few of many that would be useful.

Internodal area: cortical or medullary bundles present or absent.

Nodal area per leaf: leaf gaps 1 (unilacunar), 3 (trilacunar), or 5 or more (multilacunar). In addition the question of how many of the traces in each group are represented either by protoxylem strands or metaxylem would have to be considered.

Leaf position: opposite, alternate, whorled; distichous or decussate.

Leaf form: simple, compound, including sub-units for pinnate or palmate with bi- or tri-compound as well as unifoliolate

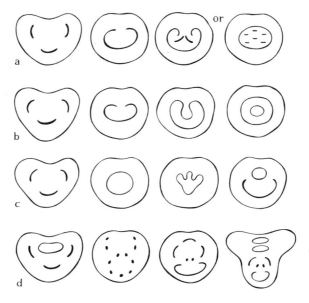

Fig. 9.4 (a)–(d). Nodes with three traces from three gaps.

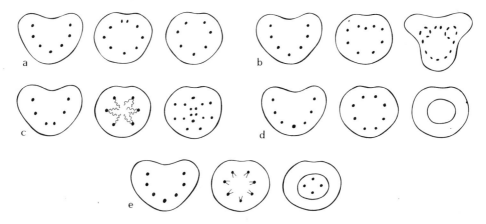

Fig. 9.5 (a)–(e).Multilacunar nodal types with five or more traces from an equal number of gaps; leaves alternate only, but stipules may be present or absent.

types. Leaves entire or lobed. Venations equally pinnate or pli-nerved or palmately veined.

Stipules: present or absent, varying in type and position.

Pulvinus: present at the base or apex of the petiole or at both or neither.

Petiole: length, characteristics such as terete, flattened, grooved, winged or decurrent.

Petiole structure: basal or apical vascular networks present or absent. Traces free or united; branched or unbranched, reticulate equally or unequally. Bundles collateral or bicollateral, complete or incomplete. Traces in an arc or ring, with accessory bundles to those of the stele in the form of rib bundles, an adaxial plate or supplementary stele, or as medullary bundles.

A great many individual papers over the years have illustrated the vascular pattern of the petiole as diagrams, camera-lucida drawings or photographs. In interpreting any one of these publications it is necessary to determine where the section was taken, i.e., as a *coupe initiale*, in the middle of the petiole, as a *coupe caractéristique*, or as a section in the meriphylle.

When diagrams are given to show the relationship

of the vascular bundles to one another, the method of indicating the various traces has become more or less standardized, the only variations being in the use of different languages or abbreviations. Thus traces seen in cross-sections are usually distinguished as median, lateral and ventral and they are commonly lettered and numbered as M for the median and M^1 for branches of the median. L, Lg or Ld are used for the laterals to the left (*gauche*) or right (*droit*). Branches of the laterals are shown as L^1, L^2, L^3 etc. Bundles in an adaxial position relative to the median and laterals may be designated as A, Ad or V (ventral). This is done without specific reference to their origin, which may be from independent gaps as distant laterals, as branches of the median or of the laterals.

The role of the traces

In most herbaceous dicotyledons and in some families which are considered to be woody (Euphorbiaceae, Saxifragaceae, Ranunculaceae) the principal traces which enter the base of the petiole remain distinct until they reach the blade. Interconnecting branching may be present but is minimal. Even when the individual traces are broadened by the development of a cambium and the addition of secondary tissue, careful observation of the points of primary xylem will permit the identification of the traces.

In general the median trace is usually unbranched in the petiole but it may produce branches which move to medullary or adaxial positions. The branches either remain free or fuse to form a single adaxial trace; alternatively they may become incorporated in the vascular ring of the petiole. The median trace of a trilacunar node, or a node with a higher number of gaps, does not generally contribute to the vascularization of the veins of the lower part of the blade. Often the median trace and its adaxial branches may extend and be recognizable even at the apex of the lamina (Sugiyama 1972).

In a trilacunar node the lateral traces may remain independent of the median trace or become intimately associated with it. If the median trace forms branches which are seen in an adaxial position the lateral traces will be interpolated into the vascular pattern. In a trilacunar node of a plant having stipules the lateral traces may either themselves enter the stipules or they may only supply branches which enter the stipules. Within the lamina the lateral traces tend to supply the vascular tissue of the basal veins, the basal lobes, or the vascular system from one-third to two-thirds of the

basal portion of the lamina, while the median trace supplies all of the vascular tissue of the upper portion of the lamina. In pentalacunar or septalacunar nodes the outer lateral traces supply the lower veins of the lower lobes of the lamina successively.

Although the median trace is usually the dominant trace in the petiole, it may also be the first to be eliminated. Swamy and Bailey (1950), Nakazawa (1956), and Yamazaki (1965) describe the median trace in *Sarcandra* and *Chloranthus* as an unbranched trace which becomes progressively more indistinct in the costa and ultimately disappears before reaching the apex. Yamazaki (1965) pointed out that in *Liriodendron* the median trace had little relationship to the lateral venation of the leaf. Bailey and Swamy (1949) described a double trace condition in the petiole and blade of *Austrobaileya*. The traces originated from different parts of the 'eustele' and remained independent to the apex of the blade. Each bundle supplied lateral branches as veins for the leaf. *Trimenia* (Money, Bailey, and Swamy, 1950) also showed a double trace in the petiole, while *Piptocalyx* (Trimeniaceae) exhibited two traces at one gap and four at the other of a pair of opposite leaves. The two middle traces of the four fused so that the leaf received three traces. In *Ascarina* of the Chloranthaceae, Swamy (1953) noted two traces at the base of the petiole which remain distinct for most of the length of the leaf blade. The two traces present in the gap area of *Clerodendrum* (Marsden and Bailey 1955) divide in the petiole several times, and the middle ones fuse to form a single strand within the blade.

Some attention has been given by various authors to the orientation of the bundles within the petiole and the leaf blade. In general, the position of the phloem relative to the xylem, i.e., peripheral or toward the centre, is of little significance in single sections of the petiole. In a series of sections the changes in orientation are successive, so that a bundle with normal phloem orientation can be reversed in position a few sections later. The bundles obviously do not twist in development, but the position of the xylem and the phloem results from special paths of differentiation of these tissues in the procambial state.

In multilacunar nodes with sheathing stipules (or a stipular sheath), the most distant trace or traces from the median may enter only the stipule and contribute nothing to the petiole. Alternatively, the distant traces may move on a horizontal path, associating with the nearer lateral, and enter the petiole. Branches from these traces may supply vascular tissue to the sheathing stipules.

Suggested method of study

In the information to be reported for individual families, in succeeding volumes of this work, nodal and petiole patterns will be described and illustrated. Most of the patterns will have been obtained from fresh material or material collected and preserved in alcohol. To obtain the vascular pattern in three dimensions, a mature leaf and a portion of the stem to which it is attached are selected and sections are cut progressively from the internode and then through the node, the petiole and the blade. The first series of sections taken at random intervals ranging from the stem to the lamina will determine where the critical areas of pattern change occur. The changes can, however, be expected (1) at the node, (2) immediately above the basal pulvinus, (3) at the upper pulvinus or the base of the blade, (4) at the points where the leaflets are attached, or (5) where the veins depart. Satisfactory sections can be cut freehand with a razor or a safety razor blade and stained in phloroglucinol and hydrochloric acid. This temporary stain does not work well on material stored in a liquid preservative containing formalin and this applies to FAA. No permanent slides can be obtained by this method.

Phloroglucin is a generally accepted stain for lignified tissue but the coloration is not always precisely the same because the term lignification is used by plant anatomists in a rather loose sense. Srivastava (1966) for example has indicated the nature of the variation of this staining. Lignified tissue in some plants such as members of the Thymelaeaceae assumes a weak coloration with phloroglucin, or none at all.

In many of my own investigations no fresh or preserved material was available and herbarium specimens had to be used. This obviously restricted the amount of material available for study and called for critical sectioning. The herbarium material was softened in boiling water or by soaking in sodium hydroxide. Woody portions often required softening in commercial hydrofluoric acid. The standard technique of embedding in paraffin, tissue mat, or celluloidon etc. was also used for material that was difficult to obtain or when satisfactory sections could not be cut by hand. The embedding technique was also followed when there were problems in interpreting the structure. Longitudinal sections are occasionally necessary for one or several angles of approach. Clearing and selective staining methods were also used.

The vascular system in the node, the petiole and lamina, with the associated structures of stipules, buds and branches, still offers a broad field for further investigation. Descriptive data can be given for only a few taxa in any family. There is no study of an individual family in which all of the genera or all of the species have been investigated fully or even sufficiently for purposes of description. Developmental studies of the several vascular systems are relatively few. Hopefully, individuals with access to abundant material having unusual vascular patterns, and with sufficient time, will undertake the clarification of these aspects of plant anatomy. For all who attempt to use this information I can only stress again the need to visualize in three dimensions types of structure that most frequently have been illustrated, until now, in two dimensions.

Literature cited

Acqua 1887; Bailey and Swamy 1949; Candolle, C. De 1868, 1879; Cook 1911; Dehay 1935, 1942; Funke 1929; Grew 1675; Hare 1943; Heinricher 1910; Howard 1962, 1970b; Lignier 1887, 1888; Marsden and Bailey 1955; Masuda 1933; Metcalfe and Chalk 1950; Money, Bailey and Swamy 1950; Morvillez 1919; Nakazawa 1956; Petit 1886, 1887, 1889; Sinnott and Durham 1923; Srivastava 1966; Sugiyama 1972; Swamy 1953; Swamy and Bailey 1950; Tichoun 1923; Vesque 1881, 1885; Watari 1934, 1936, 1939; Yamazaki 1965; Yin 1941.

10

THE PLANT SURFACE (MAINLY LEAF)

HAZEL P. WILKINSON

PART I : STOMATA

Introduction

The stoma is here pictured as normally consisting of an elliptical **pore** in the epidermis of leaves, herbaceous stems and floral parts, surrounded by two specialized, kidney-shaped epidermal cells, the **guard cells** (Figs. 10.1 and 10.2). Frequently, elevated extensions of the cuticular membrane, known as outer **stomatal ledges** or rims, rise from the guard cell surface like an incompletely roofed dome (Plates 2 and 3). In median transverse sections the guard cells are seen as two rounded or triangular thick-walled cells between which there is usually a narrow pore, although this may rather infrequently be completely occluded. The outer stomatal ledges (rims) appear as archways extending

in a protective manner over the stomatal pore and outlining an outer cavity (vestibule) (Fig. 10.1). Sometimes similar but reduced inner ledges project towards one another, thus forming an inner cavity, while on the other side of these ledges there is a substomatal cavity within the mesophyll of the leaf.

Taking dicotyledons as a whole, the guard cells in some species are immediately surrounded by epidermal cells that are indistinguishable from their neighbours, either in their morphological form or in the nature of their cell contents, such as crystals or other ergastic substances. In other species the cells next to the guard cells are morphologically distinctive because of their shape, size and orientation in relation

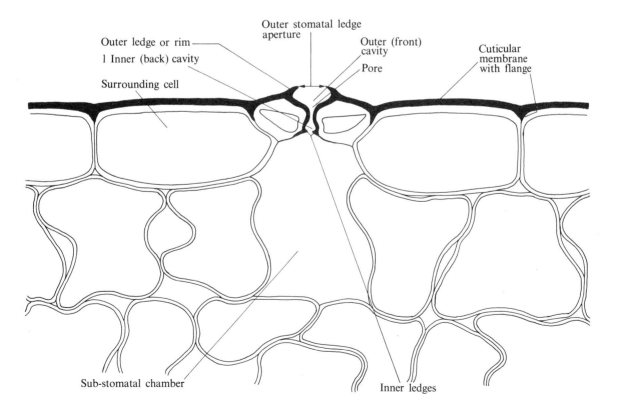

Fig. 10.1. Diagrammatic representation of the stomatal apparatus: transverse section.

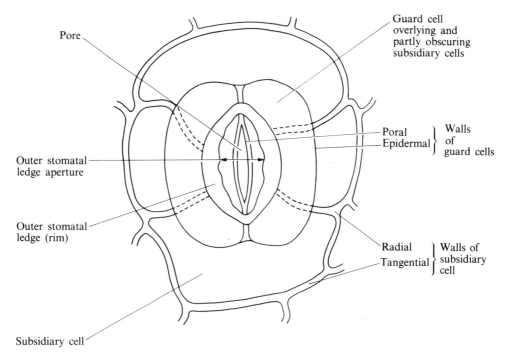

Pore

Guard cell overlying and partly obscuring subsidiary cells

$\left.\begin{array}{l}\text{Poral} \\ \text{Epidermal}\end{array}\right\}$ Walls of guard cells

Outer stomatal ledge aperture

Outer stomatal ledge (rim)

$\left.\begin{array}{l}\text{Radial} \\ \text{Tangential}\end{array}\right\}$ Walls of subsidiary cell

Subsidiary cell

Fig. 10.2. Diagrammatic representation of the stomatal apparatus: surface view.

to the guard cells. Whenever this distinction exists, the cells surrounding the guard cells are known as **subsidiary cells**. The term subsidiary cell is not used, however, when the epidermal cells surrounding the guard cells cannot be readily distinguished from their neighbours.

Examination of the epidermis in surface view shows that there are wide variations in the sizes of stomata as well as in their frequency and distribution. Naturally these variations in the morphology and distribution of the adjacent epidermal cells raises the question of how far they can be reliably employed by the systematic anatomist, and the main purpose of this present section is to discuss this question and to give an account of the various types of stomata.

Stomatal apparatus in transverse section

A general plan of an unspecialized stoma in transverse section is given in Fig. 10.1. Sections of the stomatal apparatus, however, show but few characters of diagnostic importance. Transverse sections nevertheless clarify some of the features that are to be seen in surface view. For example, sections reveal the extent to which the stomata are sunken in, or raised above, the leaf surface. It is also possible to see whether

the outer chamber is entire or partially divided into compartments. This happens, for example in some members of the Rhizophoraceae where the outer stomatal ledges face towards each other from the outer ends of the opposed guard cells (Fig. 10.5(h), p. 108). Sections also make it easier to see whether inner stomatal ledges are present or absent, and whether the cuticle which lines the pore is smooth or crenate (Fig. 10.5(j)). The shape of the guard cells and the thickness of their walls are also visible. Furthermore, sections show that the extent to which sunken stomata are depressed below the leaf surface may be due to the thickness of the cuticle, the height of the epidermal cells, or the presence of a hypodermis which may be up to several layers thick. With raised stomata, sections show whether they are only slightly elevated or supported on columns of cells. Many excellent diagrams of stomata in transverse section are given by Hryniewiecki (1912a,b; see also Napp-Zinn 1974).

The guard cell outline varies slightly in different species from roundish, often with thin walls, to somewhat triangular, usually with especially thick walls. Care has to be taken when considering this feature that the sections to be compared are from corresponding regions across the guard cells.

Subsidiary cells as a basis for stomatal classification

Before the first edition of this work was published it was customary to classify stomata, for purposes of diagnostic recognition, according to Vesque's (1889) time-honoured scheme. This scheme was widely accepted, especially in the field of pharmacognosy, because it was found by prolonged practical experience to provide reliable characters for the recognition of crude drugs. Vesque's scheme was based on the orientation of the subsidiary cells in relation to the guard cells and it will be recalled that his four principal classes were termed ranunculaceous, cruciferous, caryophyllaceous and rubiaceous. The classes were evidently so named because they were first noted in, or were known to be highly characteristic of, members of the families after which they were named. However, as factual knowledge about stomata increased it became apparent that Vesque's names for the classes were not appropriate, since it was gradually revealed that each kind occurs in many other families besides those after which they were named. Another difficulty is that Vesque's terms have not always been used in exactly the same sense. They were originally applied to distinguish between the modes of ontogeny of the stomata, but, in the course of time, they were much more frequently used to classify the stomata on the arrangement of the cells surrounding them, as seen in the mature leaf, and the ontogenetic aspects of the terminology were dropped. The reason for this change of emphasis is the very practical one that when his task is to identify fragmentary material or very numerous samples, a busy taxonomist seldom has enough time, material or opportunity for complete ontogenetic investigation. It is a great pity that this practical problem exists because the same or very similar groupings of subsidiary cells around a mature stoma may sometimes be reached along more than one ontogenetic pathway (see also p. 103). It follows, consequently, that groupings of subsidiary cells around a stoma may or may not have a strictly equivalent status. When it is not equivalent, the stomata evidently do not afford any evidence of taxonomic affinity between the plants in which they are exemplified. On the other hand, the taxonomic significance of the stomatal type will obviously be much greater wherever it can be shown that the plants being compared are alike in their stomatal ontogeny as well as in the appearance of the stomata of the mature leaf. Nevertheless, similarity of the stomatal apparatus in the mature leaf often provides a reliable diagnostic character, especially when taken in combination with other characters, even when the ontogeny of the stomata is unknown or different. The moral of all this is that, although it is not always possible to do so, whenever an opportunity exists we ought to determine the ontogenetic pathway by which the subsidiary cells are developed as well as the arrangement of the subsidiary cells in the mature leaf. For further discussion of this topic, see Tomlinson in *Anatomy of the Monocotyledons* (1969, Vol. III, p. 390) and Metcalfe, ibid. (p. 392); Tomlinson (1974, pp. 117–21); Fryns-Claessens and van Cotthem (1973), Neischlova and Kaplan (1975). See also the examples of stomatal ontogeny on p. 103 of this work.

In the first edition of this work (Metcalfe and Chalk 1950), following discussions with Mr. H.K. Airy Shaw of the Kew Herbarium, it was tentatively suggested that Vesque's term ranunculaceous should be replaced by anomocytic, cruciferous by anisocytic, caryophyllaceous by diacytic and rubiaceous by paracytic. Later on the term actinocytic was added for a type of stoma that Vesque did not use in his classification and tetracytic for a type of stoma that was first recognised at Kew in certain monocotyledons. Airy Shaw's terms for stomata are now widely used in descriptive anatomy and they have also found their way into official publications such as the *British Pharmacopoeia* and *British Pharmaceutical Codex*. There is therefore no longer any need to regard these terms as tentative and they have been adopted throughout this present work. Terms introduced by other workers which are also now generally accepted have been added to Airy Shaw's terms and all of these are defined below.

Classification of stomata based on shapes and arrangement of subsidiary cells

Actinocytic: a term originally used in the first edition of this work for stomata encircled by radiating subsidiary cells; now more precisely applied to stomata surrounded by subsidiary cells that are somewhat radially elongated (modified from Stace 1965*a*; see also Fig. 10.3(b)).

Actinocytic stomata are rather uncommon, but they have been found e.g. in *Euclea pseudebenus* E. Mey. (Ebenaceae) and some members of the Ancistrocladaceae. Giant or hydathodic stomata are also frequently actinocytic.

Allelocytic: This term is applied to stomata with an alternating complex of three or more C-shaped subsidiary cells. (See also diacytic and parallelocytic (Payne 1970).)

Amphi-: The Greek prefix amphi-, meaning around, double, or on both sides is sometimes applied to leaves and stomata. For example, leaves are said to be amphistomatic when the stomata are present on both surfaces. Dilcher (1974) uses the same prefix in amphianisocytic, amphibrachyparacytic etc. See also amphicyclic below.

Amphicyclic: should mean with a double ring of subsidiary cells. According to van Cotthem (1970), it has been used to indicate stomata with two or more rings of subsidiary cells. He suggested the use of dicyclic, tricyclic, . . ., polycyclic.

Anisocytic: Airy Shaw's term for stomata surrounded by three cells, one of which is usually smaller than the other two. (Fig. 10.3(c–e)). (Families listed on pp. 202–3.)

Anomocytic: epidermal cells around the guard cells not distinguishable from other epidermal cells (Fig. 10.3(a)). (Families listed on pp. 201–2.)

Bicyclic: subsidiary cells in two rings (Baas 1975); see dicyclic.

Brachyparacytic: two cells flanking the sides of the guard cells but not completely enclosing them; the subsidiary cells may or may not be elongated parallel to the long axis of the guard cells (Dilcher 1974).

Cyclocytic: subsidiary cells forming one or two narrow rings around the guard cells; number of cells four or more (Stace 1963, 1965a), but modified to two or more by van Cotthem (1971) (Fig. 10.3(f–h)). The term is also used in a more restricted sense by Dilcher (1974). The somewhat similar term 'encyclocytic' was proposed by Stromberg (1956), before the term cyclocytic was introduced. Cyclocytic stomata are frequent in certain genera of the families listed on p. 203.

Diacytic: stomata enclosed by one or more pairs of subsidiary cells whose common walls are at right angles to the guard cells (Fig. 10.3(j,k)). The term as now used is a modification of Airy Shaw's term in the first edition. Diacytic stomata include those that are **diallelocytic**. These have an alternating complex of three or more C-shaped cells of graded sizes at right angles to the guard cells (Payne 1970). Development of the stomata is mesogenous (see p. 00) (Pant 1965, Payne 1970). The term is equivalent to **amphidiacytic** of Dilcher (1974). (Families listed on p. 203.)

Diacytic stomata are characteristic of the Caryophyllaceae. The diallelocytic type is found e.g. in *Plectranthus australis* R.Br. (Lamiaceae) and '*Hemigraphis exotica*' (Acanthaceae) (Fig. 10.3(n)). (See list of families on p. 203.)

Diallelocytic: see diacytic.

Dicyclic: subsidiary cells in two rings (van Cotthem 1970b).

Encyclocytic: see cyclocytic.

Epiamphistomatic: greater percentage of stomata on upper epidermis.

Epistomatic: stomata on upper surface only.

Helicocytic: stomata surrounded by a helix of four or more cells (modified from Payne 1970) (Fig. 10.3(i)). Equivalent to amphianisocytic of Dilcher (1974). Helicocytic stomata are common in families in which the anisocytic type also occurs, e.g. Brassicaceae, Crassulaceae and Begoniaceae (see list on p. 203).

Hemiparacytic: similar to paracytic but guard cells accompanied by only one subsidiary cell lying parallel to the stomatal pore; subsidiary cell longer or shorter than guard cells. (Found occasionally among paracytic stomata e.g. in *Glinus latioides* L. and '*Trianthema lancastrum*' (both Aizoaceae)).

Hexacytic: a modified tetracytic type with an additional pair of lateral subsidiary cells (van Cotthem, 1970). The polar cells may be larger or smaller that the lateral cells. See also parahexacytic–monopolar, parahexacytic-dipolar, brachyparahexacytic–monopolar, and brachyparahexacytic–dipolar of Dilcher (1974, pp. 98–100) (Fig. 10.3(p–r)).

Hypoamphistomatic: greater percentage of stomata on lower epidermis.

Hypostomatic: stomata on the lower surface only.

Laterocyclic: paracytic stomata in which the two lateral subsidiary cells surround the guard cells completely (van Cotthem 1970b); e.g. *Calystegia sepium* L. (Convolvulceae), Fig. 10.3 (x, y). See **paracytic**.

Laterocytic: with more than two subsidiary cells on at least one lateral side of each guard cell pair. Anticlinal walls between adjacent subsidiary cells radiating from the guard cell pairs are equal in thickness to other cell walls. Laterocytic stomata should not be confused with paracytic stomata with subdivided subsidiary cells or with parallelocytic stomata. They are characteristic of many Celastraceae (den Hartog, van ter Tholen and Baas, 1978).

Paracytic: stomata accompaned on either side by one or more subsidiary cells parallel to the long axis of the pore and guard cells. The subsidiary cells may or may not meet over the poles and may or may not be laterally elongated. (Modified from Airy Shaw's original definition in the first edition of this work; Fig. 10.3(v–y)). Sometimes the subsidiary cells are subdivided and, when this is so, the later-formed anticlinal walls are thinner.

Paracytic stomata are characteristic of the Magnoliales and Rubiaceae, but they also occur in some genera of the families listed on p. 202. Subsidiary cells in the Rubiaceae usually meet over the poles, while those of several genera in other families e.g. *Drimys* (Winteraceae) and *Linum* (Linaceae) usually fall short of the poles, one or both of which may be flanked by perigenous (see p. 202) neighbouring cells. Several variations similar to those in Fig. 10.3(v–y) may occur in a single family, e.g. Aizoaceae (Dupont 1962). Several subsidiary cells parallel to the pore are common in some families, e.g. Pittosporaceae.

Parallelocytic: stomata with an alternating complex of three or more C-shaped subsidiary cells of graded sizes parallel to the guard cells (Payne 1970), (Fig. 10.3(o)). (Families listed p. 203.)

It should be noted that the plants listed by Payne as having this type of arrangement were recorded in the first edition of this work as being of the paracytic type. The subsidiary cells of parallelocytic stomata are mesogenous (see p. 102) (Paliwal 1967) or perigenous (see p. 102) (Stebbins and Jain 1960).

Polycyclic: stomata with several rings of subsidiary cells.

Polycytic: stomata with five or more cells enclosing the guard cells (Dilcher 1974).

Quadricytic: Field (1967) used this term as equivalent to tetracytic.

Staurocytic: stomata surrounded by three to five similar subsidiary cells with anticlinal walls arranged crosswise to the guard cells. The subsidiary cells are more or less radially elongated, but the cell walls at right angles to the long axis

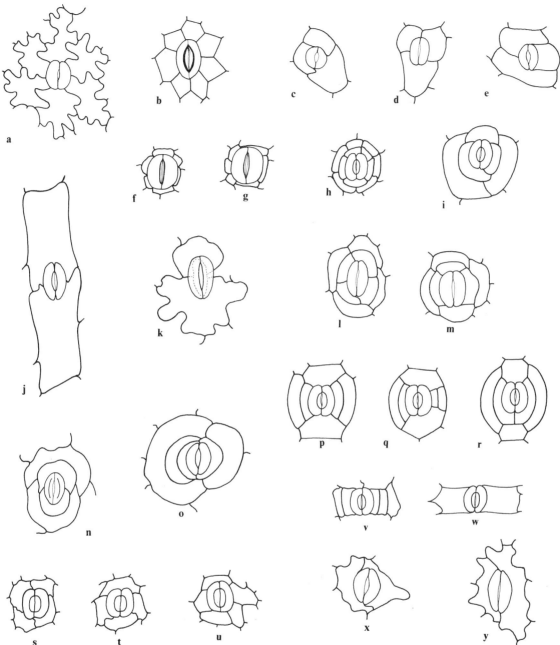

Fig. 10.3. Stomatal types. (a) anomocytic, *Caltha palustris* L. (*Ranunculaceae*); (b) actinocytic, large stomata of *Lannea stulmannii* (Engl.) Engl. (Anacardiaceae); (c–e) anisocytic, *Brassica oleracea* L. (Brassicaceae); (f–h) cyclocytic, (f–g) *Parishia maingayi* Hook f., (h) *Schinopsis marginata* Engl. (Anacardiaceae); (i) helicocytic, *Begonia argenteo-guttata* (Begoniaceae); (j, k) diacytic, (j) *Dianthus* sp (cultivated carnation) (Caryophyllaceae), (k) *Stachys officinalis* (L.) Trev. (Lamiaceae); (l) cyclocytic and staurocytic and (m) cyclocytic and tetracytic, *Piper betle* L. (Piperaceae); (n) diallelocytic, *Plectranthus australis* (Lamiaceae); (o) parallelocytic, *Portulaca oleracea* L. (Portulacaceae); (p–r) hexacytic, Stapelieae (Asclepiadaceae); (s–u) staurocytic, *Norantea guianensis* (Marcgraviaceae); (v–y) paracytic, (v, w) *Dracontomelon* spp (Anacardiaceae); (x, y) laterocyclic, *Calystegia sepium* L. (Convolvulaceae). ((p–r) redrawn from D. V. Field 1967.)

of the pore are variously orientated. The term staurocytic was first introduced by van Cotthem (1968, 1970) who used it with reference to fern stomata. Later on, Fryns-Claessens and van Cotthem (1973) said that staurocytic stomata are characteristic of the Marcgraviaceae (Fig. 10.3(s–u)). Two types of staurocytic stomata are distinguished by Dilcher (1974). In one type there are anticlinal walls at right-angles to the guard cells. In the second or **anomotetracytic** type the subsidiary cells are arranged in an irregular and variable pattern.

Subsidiary cells: These are the epidermal cells next to the guard cells, but in order to be designated as subsidiary cells they must differ structurally from other epidermal cells, e.g. in size, shape, in possession of papillae or in their contents. Subsidiary cells are so called no matter whether they are derived ontogenetically from the same or different mother cells as the guard cells.

Tetracyclic: with four rings of subsidiary cells.

Tetracytic: stomata surrounded by four subsidiary cells, two of them parallel to the guard cells, the remaining pair being polar and often smaller. One polar cell is, or both are sometimes replaced by a single or a pair of ordinary epidermal cells, and this may happen at either pole or at both poles of the stoma. The term as here defined is modified from the definition given by Metcalfe (1961). See also Dilcher (1974, pp. 96–97). Tetracytic stomata are equivalent to quadricytic in the sense used by Field (1967). (Families listed on p.203.)

Tetracytic stomata are highly characteristic of numerous monocotyledonous families (see Metcalfe and others in *Anatomy of the Monocotyledons*, Vols. I–VI); amongst dicotyledons, recorded e.g. in *Tilia* and in some Asclepiadaceae.

Tricyclic: stomata with three rings of subsidiary cells.

Tricytic: stomata with three subsidiary cells around the guard cells; often cyclocytic.

Intermediate types of subsidiary cell arrangement

Subsidiary cells may not always fit exactly into the types listed above and some authors have experienced difficulty in deciding how to describe such arrangements. As an example, anisocytic, staurocytic, and tetracytic stomata may simultaneously be cyclocytic (Plate 5, A,C). In such situations, the subsidiary cells could be designated as stauro-cyclocytic, tetra-cyclocytic, etc. (Fig. 10.3(l,m)). Sometimes the dividing line between tetracytic, staurocytic, and actinocytic may be somewhat blurred. Similarly, subsidiary cells intermediate between actinocytic and cyclocytic are quite common. Jansen and Baas (1973), Baas (1975) and Hartog, van ter Tholen and Baas (1978) have used the term 'complex' for several intermediate types, e.g. complex anisocytic and complex laterocytic. Some authors have also found difficulty in deciding whether surrounding cells are truly anomocytic or somewhat actinocytic. Obviously the only solution here is to state the fact.

Whilst this present account was being prepared,

Dilcher (1974) published an article in the *Botanical Review* in which the descriptive terms which were then in vogue were subdivided and more precisely defined. Unfortunately dicotyledonous cuticles are not always sufficiently constant in structure to conform with Dilcher's suggestions. This is made clear by reference to Fig. 10.3(p–u). The confusion which has arisen makes it evident that changes of terminology should, in future, be introduced only after most careful consideration. It serves no useful purpose to coin a fresh term for every minor or occasional variation that is encountered.

Ontogeny of stomata

Many of the basic facts about stomatal ontogeny have been well established since the latter part of the nineteenth century, as can be seen by the writings of de Bary (1884), Strasburger (1866), Vesque (1889) and Solereder (1908). Interest in the subject was revived when Florin (1933) investigated the development of leaves in gymnosperms. His initiative has led to the publication of the results of numerous investigations concerning stomatal ontogeny in a wide range of plants. Florin's work revealed two types of development. In the first or perigenous type the two guard cells originate by a single division of the stomatal initial whilst some of the neighbouring cells become modified independently as subsidiary cells. Florin termed the adult appearance of this type as haplocheilic (simple-lipped). In the second or mesogenous type the subsidiary cells and guard cells are produced from the same initial. The term used by Florin for the mature condition was syndetocheilic (compound-lipped). The ontogenetic terms are still used with reference to the development of dicotyledonous stomata. From about 1965 onwards, Pant, followed later by other workers, began the examination of stomatal development in a number of dicotyledonous families. Pant revised ontogenetic terminology in 1965 and the whole subject has been reviewed very thoroughly with all the additional terminology fully explained by Fryns-Claessens and van Cotthem (1973). In Pant's classification the perigenous group is not subdivided, the mesoperigenous group is subdivided into three types and the mesogenous into four main subgroups. Fryns-Claessens and van Cotthem's revised classification gives many subdivisions of the basic types of ontogeny. The perigenous type is split into six subtypes, each giving rise to a different appearance in the adult leaf. The subtypes are; (1) aperigenous (anomocytic); (2) monoperigenous (hemiparacytic); (3) diperigenous (paracytic); (4) tetraperigenous (tetracytic); (5) hexaperigenous (hexacytic); (6) polyperigenous or cycloperigenous (cyclocytic). The mesoperigenous group is shown to have nine developmental pathways which lead to anomocytic, diacytic, hemiparacytic, anisocytic, staurocytic, cyclocytic, and paracytic dicotyledonous types of adult stomata as well as additional types for other plant groups. Mesogenous stomata have eleven different developmental pathways giving rise to diacytic, paracytic, cyclocytic, allelocytic, anisocytic, helicocytic, and tetracytic adult dicotyledonous stomata, in addition to others found in other plant groups.

From the above it is apparent (as has already been mentioned on p. 99) that adult stomata which appear alike may have had different developmental pathways. For example, the ontogeny of the diacytic stoma is of two types according to Inamdar (1969b): mesoperigenous in the Caryophyllaceae and mesogenous in the Acanthaceae and Verbenaceae. Similarly, in 'Clematis integrifolia' (Ranunculaceae), mesoperigenous anisocytic ontogeny appears as the anomocytic arrangement in the adult leaf (Décamps 1974). Published data from various other sources also indicate that paracytic stomata may be derived from perigenous, mesoperigenous or mesogenous origins. Less frequent examples also occur where stomata that are alike in their initial stages assume a different appearance when adult. An example of this occurs in the Brassicaceae for which the ontogeny is usually mesogenous and the adult stomata are anisocytic. However, according to Paliwal (1967), the adult stomata of *Cheiranthus cheiri* L. and *Iberis amara* L. have three subsidiary cells in their development, but at maturity they do not remain distinct and the stomata often appear anomocytic. Similarly, Zubkova (1975) studied the ontogenesis of anisocytic and anomocytic types of stomata in leaves of 50 species of Brassicaceae and drew attention to the diversity of ways in which the mature pattern is attained. The status of genuinely anomocytic stomata is difficult to assess because their ontogeny may be perigenous or mesoperigenous.

In view of the above mentioned facts, the arrangement of the adult subsidiary cells should not be used to assess the phylogenetic position of the species without due caution.

Phylogeny of stomatal types

Although there has been much speculation and some reasoned argument concerning the phylogenetic relationships of the various stomatal types, there is at present no generally accepted consensus of opinion concerning the course of stomatal evolution. This is clearly shown from the following excerpts of recorded data.

Florin (1933, 1958) believed that the perigenous (haplocheilic) type of stoma ante-dated the mesogenous (syndetocheilic) type in geological time. The Devonian fossils *Asteroxylon* and *Rhynia* (Psilophytopsida) have anomocytic stomata, while the earliest paracytic stomata so far discovered occur in Mesozoic Bennettitales and, according to Mersky (1973), also in Lower Cretaceous angiosperm cuticles.

In the dicotyledons much interest has centred on the stomata in the supposedly primitive Magnoliales in which both perigenous and mesogenous types occur. Paliwal and Bhandari (1962) and Pant and Gupta (1966) found that the stomatal development of all members of the Magnoliaceae studied by them is mesogenous. Payne (1970) found either mesoperigenous or sometimes mesogenous stomata in *Liriodendron*. Stomata of *Trochodendron* and *Tetracentron* (Bondeson 1952) and likewise those of *Schisandra grandiflora* Hook.f. & Thoms. (Jalan, 1962) are mesoperigenous. Paracytic stomata occur in all genera of the Magnoliaceae, although *Liriodendron* has a few that are anomocytic intermingled among them (Baranova, 1972). *Trochodendron* has peculiar anomocytic stomata with the guard cells partly overlying the subsidiary cells, so that in surface view the adjoining cells appear very narrow, sometimes appearing almost as if the stomatal apparatus was paracytic. *Bubbia perrieri* Cap. (Winteraceae), a geographically isolated species from Madagascar, has anomocytic–perigenous stomata. The Himantandraceae, Eupomatiaceae, Annonaceae, Canellaceae, and Illiciaceae all have paracytic stomata. The guard cells of various (mainly tropical) members of the Magnoliales show interesting features such as lamellae (chordal bars), pores or lobe-like thickenings which appear somewhat similar to structures found in the Cycadales and Bennettitales. Details of these features will be given under the various families of the Magnoliales.

Stebbins and Khush (1961) believed that the primitive stoma had several subsidiary cells and that this primitive type gave rise independently to more specialised stomata with two subsidiary cells or none. Although Cronquist (1968) suggested that anomocytic stomata are primitive, he now accepts the viewpoint presented by Baranova in 1972. Cronquist (1976, private communication) has restated his opinion very clearly as follows: 'Although anomocytic stomates are probably primitive for vascular plants as a whole, paracytic stomates are probably primitive *within* the angiosperms.'

Takhtajan (1969), basing his conclusions on the work of Baranova, considers that the primitive angiosperm stoma was mesoperigenous-paracytic. However, Paliwal (1969) points out that the stomatal ontogeny in the supposedly primitive Ranales and Magnoliales resembles that of supposedly advanced groups such as the Rubiales and Asterales. Reference to the tables on pp. 201-3, showing the distribution throughout the dicotyledons of the various types of stomata, clearly demonstrates that there is no general correlation between the stomatal type and the phylogenetic level of the families in which the different types are exemplified. All of these facts taken together clearly demonstrate that it would be wiser, at least for the time being, to use stomatal characters for diagnostic purposes only, except perhaps when considering different stomatal types within a restricted taxonomic group, of which a good example is given by van Staveren and Baas (1973) and Baas (1974). Baas considers that in the Icacinaceae, paracytic and anomocytic stomata are primitive for the family and that the other types represented are derived. These other types are cyclocytic, anisocytic and intermediates between them. Baas's conclusions are based on comparisons with the specialization levels in the anatomy of the wood and nodes (Bailey and Howard 1941) and in pollen morphology (Lobreau-Callen 1973). Baas says that there is a strong tendency for cyclocytic stomata to be restricted to the most specialized genera and for anisocytic as well as two intermediate types to occur in genera at an intermediate level of specialization. On the other hand, paracytic and anomocytic stomata, and intermediates between them, are wholly restricted to genera belonging to the most primitive group.

Some workers (Guyot 1966; Chappet 1969; Gorenflot 1971; Patel and Inamdar 1971) have demonstrated that stomata of more than one type are developed during the life history of an individual species. Guyot and Gorenflot worked on members of the Apiaceae and Saxifragaceae respectively and Chappet on *Vicia faba* L. (Fabaceae). Guyot (1966) found that the first type of stoma to develop on the cotyledons was anomocytic and that anomocytic or anisocytic stomata appeared on the first foliage leaf. Successive leaves showed the introduction of small percentages of other types. In the adult leaves anomocytic stomata were still the most frequent. Chappet observed that the sequence of stomatal types

in *Vicia faba* was as follows: anomocytic–perigenous, anomocytic–mesoperigenous, anisocytic–mesoperigenous, and anisocytic–mesogenous. This succession is the result of an increasing number of unequal divisions undergone by the initial mother cell. Guyot made a set of diagrams showing possible routes of development of tetracytic, paracytic and diacytic types from the anomocytic in the Apiaceae.

Gorenflot suggests that too much attention has been paid to the stomatal type that shows the highest percentage in the adult leaf and not enough to the type showing the smallest percentage. The latter, he thinks, may reflect the current evolutionary status of the plant in question, particularly if this corresponds with the majority of the other plant characters (karyological, biochemical, palynological, etc.). This somewhat new approach is certainly worth investigating in other families.

Cotyledons do not always have anomocytic stomata, as some authors seem to think; for example, they are anisocytic in *Kostermansia malayana* Soegeng (Baas 1972*b*, p. 339 and his Fig. 9). When only adult leaves are available for investigation, it might be helpful to record the percentages of stomatal types present. How unfortunate it is that, comparatively speaking, very few fossil leaf cuticles are preserved. However, the best use can be made of these by inspection for developing stomata and arrested stomatal development. While the fossil record for dicotyledonous plants and their ancestors remains so incomplete, views on whether a particular type of stoma is primitive or advanced is largely a matter of speculation.

Further notes on subsidiary cells

It has already been mentioned (p. 102) that subsidiary cells can sometimes be readily distinguished from adjacent living epidermal cells by the nature of their contents, whether or not their shape is distinctive. Examples are the gum-like contents in the subsidiary cells of the Trochodendraceae and the homogeneous, yellow-brown pigmentation absent from adjacent epidermal cells (in addition to being paracytic) in *Clusia* (Hypericaceae) according to Howard (1969). Translucent subsidiary cells, contrasting with the mottled appearance of the remaining epidermal cells, have been noted by the same author in *Tabebuia rigida* Urb. (Bignoniaceae). A granular substance in the subsidiary cells of *Microsemma salicifolia* Labill. (Thymelaeaceae) was reported in the first edition of this book.

Sometimes subsidiary cells are conspicuous because they have a thicker cuticle than the surrounding cells (Plate 5(A)). I have myself noted several examples in the Anacardiaceae, e.g. *Lithraea molleoides* Engl. and *Loxostylis alta* Spreng. De Paula (1974) has also reported this feature in *Chrysochlamys, Kielmeyera corymbosa. Rheedia* and *Symphonia* (Hypericaceae) Other subsidiary cells are made obvious by their specially thin cuticle. Examples are *Lophopetalum torricellense* Loes. (Celastraceae) observed by Ding Hou (personal communication), *Laguncularia* (Combretaceae, Plate 4(B) of Stace 1965a), and I have made a similar observation on *Antrocaryon klaineanum* Pierre (Anacardiaceae, Plate 5(C)).

Papillose subsidiary cells occur, for example, in *Melanochyla auriculata* Hook.f. (Plate 5(G,H), Fig. 10.4(1)) and although the unspecialized cells are also papillose in *Semecarpus* (Anacardiaceae), I have myself noted that the papillae of the subsidiary cells have more elaborate lobes or frills which at least partly obscure the stomata. By way of contrast, subsidiary cells without papillae surrounded by papillose cells are a feature shown by several species of *Cratoxylum* (Hypericaceae; Baas 1970) (Fig. 10.4(r)).

Lobe-like extensions from the guard cells into the lumina of the adjacent cells are a feature found in *Penaea myrtoides* Linn., *Saltera sarcocolla* (L.) Bull. (Penaeaceae; Fig. 10.4(a)) as well as in *Strephonema* spp (Combretaceae; Stace 1965a, p. 47) and a species of *Boea* (Gesneriaceae; Sahasrabudhe and Stace 1974, their Figs. 3,5, pp. 57,58). Straight-walled surrounding cells are in conspicuous contrast to the other epidermal cells which have sinuous walls in *Micropholis* (Sapotaceae; Howard 1969).

Various chemical reagents used to test the distinctiveness of subsidiary cells have been listed by Patel (1978).

Variation in the appearance of the guard cells and their outer stomatal ledges[1] (rims)

Solereder (1908, pp. 1078-9) strongly emphasized the diagnostic importance of the morphology of the guard cells and of their cuticular ledges. The outline of the pair of guard cells as seen in surface view is usually constant at the species level and is sometimes characteristic of a genus. The range of possible variation is limited, since the shape can only be (1) broader than long, (2) roundish, (3) broadly elliptical, (4) narrowly elliptical, and (5) angular. Length-to-breadth ratios can make the types (1) to (5) more precise. The first four categories are common throughout the

[1] For stomatal ledges see also p. 97.

flowering plant families. The fifth is of more restricted occurrence, but is, for example, characteristic of the Papaveraceae (including Fumariaceae). The stomatal poles where the guard cells meet may be obtuse, truncate, rounded or retuse. According to Bongers, Jansen, and van Staveren (1973), there is some evidence which suggests that stomatal shape may be associated with plant habitat. These authors found that plants from open vegetation usually have almost circular guard cells, whereas those of forest plants are elongate.

Also located at the stomatal poles are the cuticular thickenings of the common wall between the guard cells known as 'T' pieces (sometimes reduced to a rod or bar) (Fig. 10.4(t,u), Plate 5(D,F)). These have been recorded by many authors in diverse families. Bailey and Nast (1948) and Baranova (1972) reported T pieces in certain species of the Illiciaceae, whilst in the Combretaceae, Stace (1965b) noted them in some species of *Pteleopsis, Ramatuela, Strephonema,* and *Terminalia,* but found them absent from the species he examined in *Anogeissus, Bucida, Calopyxis eriantha* Tul., *Conocarpus, Guiera,* and *Thiloa.* Some authors have found this character to be very variable at the species level, e.g. Singh and Kundu (1962) who examined 11 species of *Digitalis* (Scrophulariaceae).

Ovoid-shaped, thin-walled areas at the ends of the guard cells, so common in the Gramineae, are occasionally seen in dicotyledonous stomata, as in *Eriandra fragrans* van Royen and van Steenis (Polygalaceae) (Fig. 10.4(k)).

The guard cells of stomata may stand out in contrast to the remaining cells of the epidermis because they lack ergastic substances characteristic of the other epidermal cells. Such cells were referred to as 'negative idioblasts' by Weber, H. (1959), who gave *Scutellaria altissima* L. (Lamiaceae) as an example. Another example is shown by *Stylosanthes* (Fabaceae) in which crystals are present in all the epidermal cells except the guard cells and subsidiary cells. Conversely, when the guard cells contained secreted materials from which the surrounding cells were free, they were termed 'positive idioblasts' by Weber, H. (1955).

Outer stomatal rim or ledge

The outer stomatal rim or ledge appears under the s.e.m. as a single undivided structure which is perforated down some three-quarters of its length during development. Under the light microscope it may appear as two structures, one from each guard cell. The perforated margin may be smooth or somewhat ragged.

The highly cutinized outer rims of the guard cells

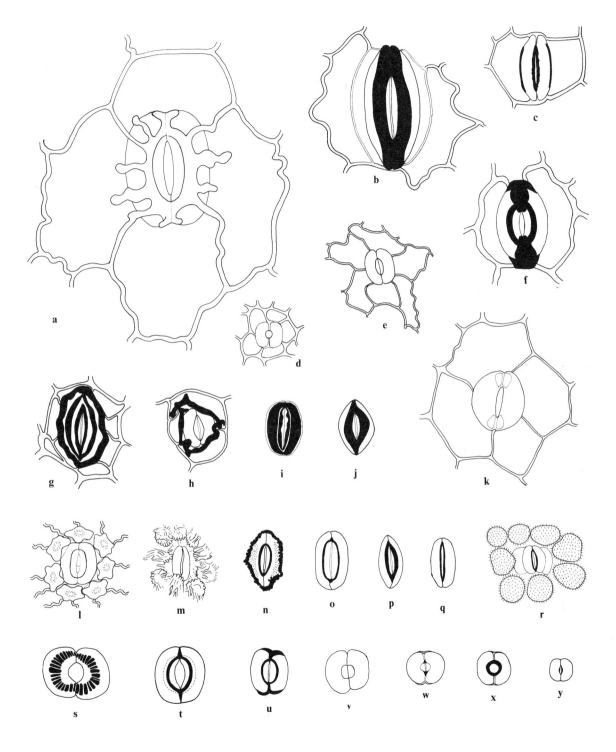

50μm

are often conspicuous and easy to measure (Plates 4(E), 5(A, C–E)). Their outlines usually follow those of the guard cells. Their diagnostic value has been demonstrated, for example, n the mangroves by Stace (1966). There is a single outer stomatal rim in *Conocarpus, Laguncularia* and *Lumnitzera* whereas the rims of *Bruguiera, Ceriops, Kandelia* and *Rhizophora* are conspicuously two lipped (Figs. 10.4(g,i), 10.5(h), Plate 2(I)). Some fine examples of two-lipped ridges have also been recorded by Jähnichen (1969, his Figs. 3,4, 5a,b) in fossilized species of *Cyrilla* (Cyrillaceae) from the Lusatian and Nether Rhine Tertiary browncoal deposits, and also described and figured by Grambast (1954) for various living species of *Ficus*. Baranova (1972 p.455) found that in the Magnoliaceae the thickness and form of the outer stomatal rim are of diagnostic value at the species level or, in *Magnolia* itself, at the sectional level. Strikingly tall, upright outer stomatal rims often occur on the stem stomata of xerophytic plants, as for example in some species of *Fabiana* (Solanaceae; Böcher and Lyshede 1972, their Figs. 26,28a, plate X,a,b,c). Examples showing how the outline of the outer stomatal rim may vary in different genera are illustrated in Figs. 10.4, 10.5(e), and also in Plates 2(A–G,I,) 3(A,C–E), 4(D,E) and 9(A,C). I have myself noted that the particularly large, round rims of the leaf stomata of *Hyaenanche globosa* (Gaertn.) Lamb. and Vahl (Euphorbiaceae) (Plate 2(C)) provide a remarkable contrast to the long and narrow ones of *Coriaria nepalensis* Wall. (Coriariaceae, Plate 3(E)).

Peristomatal rims, cuticular cups (ramparts), and sunken stomata

Occasionally the subsidiary cells bear ridges of cuticle which surround or overlap the guard cells, partly obscuring them (Fig. 10.4(g,h,m,n), Plates 2 (D–G,I) and 4(E,H)). I propose the term peristomatal rim for low ridges of cuticle and I have myself observed that *Schinopis marginata* Engl. (Anacardiaceae) provides a fine example of peristomatal rims in which the margin appears scalloped and the surface striate (Fig. 10.4(m)). Another member of the Anacardiaceae that shows narrower but nevertheless thick and distinctive ridges is *Rhus thyrsiflora* Balf.f., and there are wider ones in *Buchanania obovata* Engl. (Plate 2(D)). The rims in *Heeria argentea* (E. Mey.) O. Kuntze are even more solid and have the appearance of thick, wide structures. Similar rims occur in *Ceriops decandra* (Griff.) Hou (Rhizophoraceae) (Plate 2(I) and in Figs. 10.4(g) and 10.5(h)). On the subsidiary cells of *Schefflera venulosa* var. *venulosa* Harms there are crescentic ridges of cuticle (Plate 2(E)) which produce a very distinctive pattern on abaxial leaf surfaces of, this species. A considerably deeper 'cup' or 'flask' may sometimes occur as has been noted in the leaves of *Franklandia fucifolia* R.Br. (Proteaceae), and on the xerophytic stems of *Monttea aphylla* (Miers) Benth. and Hook. (Fig. 10.5(g)). The cup or flask may be formed from cuticule only, cuticle and epidermis, or may even include deeper tissues, particularly on green stems. There is no clear distinction between rim, cup,

Fig. 10.4. Various stomata in surface view, illustrating variation in size, outline, and morphology. (a) *Saltera sarcocolla* Bull. (Penaeaceae), viewed from below to show more clearly the nodular wall underneath the guard cells; (b) *Illicium floridanum* Ellis, with thick and wide outer stomatal ledge; (c) *Coriaria nepalensis* Wall.; (d) *Sorindeia juglandifolia* Planch. ex Oliv. (Anacardiaceae); (e) *Acer Flabellatum* (Rehd.) Fang.; (f) *Illicium philippinensis.* Merr. showing in particular the characteristic cuticular thickening of the guard cell polar regions; striae over guard cells and subsidiary cells not shown; (g) *Ceriops decandra* (Griff.) Hou (Rhizophoraceae) with double outer stomatal ledge and peristomatal rim; (h) *Schefflera venulosa* Harms. var. *erythrostachys* (Araliaceae), with conspicuous cuticular ramparts on the subsidiary cells; (i) *Bruguiera gymnorrhiza* (L.) Lamk. (Rhizophoraceae) with double outer stomatal ledge nearly obscuring the guard cells; (j) *Laguncularia racemosa* Gaertn. f. (Combretaceae), with wide outer stomatal ledge; (k) *Eriandra fragrans* V.R. and Steenis (Polygalaceae), showing thin areas at poles of guard cells; (l) *Melanochyla auriculata* Hook. f. with papillose subsidiary cells; (m) *Schinopsis marginata* Engl. with scalloped and ridged subsidiary cells overlapping the guard cells; (n) *Rhus thyrsiflora* Balf. f. with thick cuticular peristomatal rim; (o) *Lannea stuhlmannii* (Engl.) Engl. with apiculate poles to the outer stomatal ledge; (p) *Rhus succedanea* L.; (q) *Rhus semialata* Merr.; (l–q) all Anacardiaceae; (r) *Cratoxylum glaucum* Korth., showing paracytic, non-papillose subsidiary cells and papillose surrounding cells (Hypericaceae); (s) *Cyrilla racemosa* Loud. with ridged outer stomatal ledge, viewed from below; (t) *Lannea gossweileri* var. *gossweiler* Exell and Mendonça, showing thick polar rods to the guard cells; (u) *Buchanania siamensis* Pierre with polar T-pieces; (v) *Mangifera caesia* Bl.; (w) *Mangifera indica* L.; (x) *Sorindeia mildbraedii* Engl. and Brehmer; (y) *Astronium graveolens* Jacq. (t–y) all Anacardiaceae. Scale as shown.

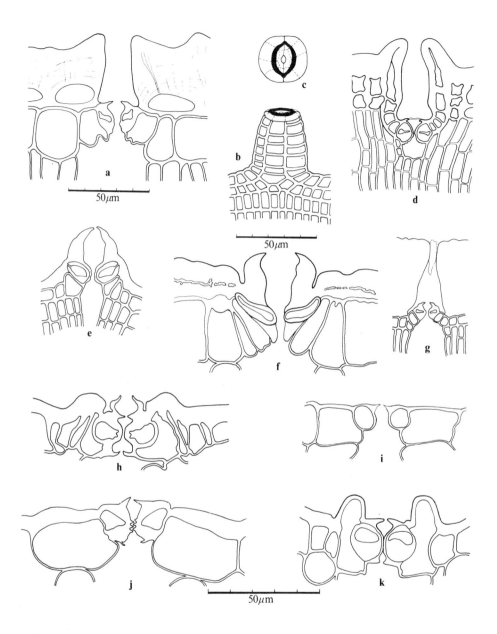

Fig. 10.5. **Stomata in transverse section and side view** (stem and leaf) (a) Stoma sunken in pit formed by epidermis with thick cuticular wall in leaf of *Franklandia fucifolia* R. Br. (Proteaceae), (b) side view of stoma raised on a column of cells on the abaxial surface of the leaf midrib in various *Dracontomelon* spp (Anacardiaceae); (c) surface view of stoma at top of column of cells; (d) stoma at bottom of 'pitcher' formed by multiple epidermis and thick cuticular wall in stem of *Bredemeyera colletioides* (Phil.) Chod. (Polygalaceae), note diaphragm between outer stomatal ledges; (e) raised stoma with enormous upright and curving outer stomatal ledges in a stem of *Fabiana denudata* complex (*Fabiana viscosa* Hook. and Arn., *F. denudata* Miers and their hybrids) (Solanaceae), note large front cavity; (f) guard cells sunken, but stoma with wide, projecting, outer ledges in a leaf of *Hyaenanche globosa* (Gaertn.) Lamb. and Vahl

and flask, the difference being merely one of the height of the cuticle and sometimes the height of other structures as well.

A somewhat similar effect is produced by stomata accompanied by papillose subsidiary or surrounding cells, as in *Garrya elliptica* Douglas (Garryaceae Fig. 10.5(k)) and in *Melanochyla auriculata* Hook. f. (Anacardiaceae; Plate 5(G)). Certain types of papillae, discussed more fully on p. 148, have a lobed or frilled 'corona' and, in these, the lobes or frills overlapping the guard cells are larger than those over other epidermal cells. This can be seen in many species of *Melanochyla* (Fig. 10.12(c)) and *Semecarpus* (both Anacardiaceae). Trichomes may also overarch stomata, as in some tropical and subtropical *Quercus* spp. (see Uphof, Hummel and Staesche 1962, p.128, their Fig. 52).

Stomatal crypts (Fig. 10.6, p. 115)

These are sunken areas of leaf surface (usually abaxial) each perforated by several stomata and often densely hairy as in *Banksia* and *Dryandra* (Proteaceae) and *Mouriri* (Melastomataceae). According to Morley (1953), 21 of the 74 species of *Mouriri* he studied have only surface stomata, while the remaining 53 have stomata in crypts as well as on the surface. Some of these crypts were found to be arranged in a specific manner while others can be grouped into various types according to their shape as seen in transverse section (Fig. 10.6). Stomatal crypts on the lower surface of the leaf are well known in *Nerium oleander* Linn. (and *N. odorum* Ait., Apocynaceae). Deep stomatal pits may also occur in the stems of plants, sometimes produced by the development of a multiple layered epidermis, as for example in the 'stomatal pitchers' on the xerophytic stems of *Bredemeyera colletioides* (R.Phil.) Chod. (Polygalaceae; Fig. 10.5 (d); Böcher and Lyshede 1968) (see also p. 107).

Lignified stomata

Lignified stomata are known to occur for example in *Camellia japonica* L. (Heilbronn, 1916). Kaufmann (1927), working mainly on Gramineae and Cyperaceae, but also on a few dicotyledons, discovered that in the stomatal ledges of *Laurus noblis* Cav. and *Quercus ilex* L. almost the whole guard cell membrane is lignified. *Hedera helix* L. (Araliaceae) and *Mahonia aquifolium* (Pursh.) Nutt. (Berberidaceae) also have indistinctly lignified stomata that are best seen in transverse section.

Plugged stomatal pores

Stomata of Coniferae in which the pores are plugged with resinous material have been known for many years. Their presence in Monocotyledons and Dicotyledons was demonstrated by Wulff at the end of the nineteenth century (1898). Mühldorf (1922) found that the stomatal pores of *Helleborus niger* L. are filled with granular material. Plugged stomata in the Winteraceae in general have been described in some detail by Bailey and Nast (1944; see especially their plate II, figs. 7–11), where the granular deposit filling the stomatal aperture is referred to as 'alveolar'. The same feature was reported by Bondeson in 1952 for *Drimys winteri* Forst., (see also plate 2(H) for *Drimys lanceolata* Baill.) and by Baranova (1972) who examined 19 species, covering all genera in the Winteraceae (except *Exospermum*) and found alveolar plugs in all of them. However, according to Vink (1970) and Bongers (1973) several representatives of the Winteraceae do not have alveolar plugs. There is also disagreement among authors as to the exact nature of the plugs. Wulff (1898) and Ziegenspeck (1941) considered them to be waxy, whereas Bailey and Nast (1944) suggested a cutinaceous composition. Vink and Bongers both found that some were cutin-

---

(Euphorbiaceae ?); (g) stoma sunken in deep pit formed by extremely thick cuticular layer in a stem of *Monttea aphylla* (Miers) Benth. and Hook. (Scrophulariaceae); (h) double outer stomatal ledges and raised cuticular layer over subsidiary and adjacent cells in a leaf of *Ceriops decandra* (Griff.) Hou (Rhizophoraceae); (i) simple, round, unornamented guard cells, level with the adaxial leaf surface of *Nymphaea gigantea* var. *alba* (Nymphaeaceae); (j) guard cells with characteristic ridges on inner, poral walls in leaf of *Helleborus niger* L. (Ranunculaceae); (k) papillose subsidiary cells and stoma of leaf of *Garrya elliptica* Douglas (Garryaceae).

(d), (e) and (g) drawn from photographs in Böcher and Lyshede (1968, 1972). (d) × 194, (e) × 330, (g) × 134; all the others, scale as shown.

aceous and others waxy. Likewise both of these authors discovered that the distribution of the alveolar material was variable, but Bongers in particular found that the distribution and qualitative features of the alveolar material could be classified and used in generic descriptions. Certain members of the Cactaceae have their stomata closed by the development of thylloid cells, with or without thickenings, which according to Bukvic (1912) arise from the subsidiary cells and/or those of the mesophyll. I have myself also noted that the stomatal pores of *Schinopsis quebracho-colorado* (Schlecht.) Barkley and Meyer are frequently blocked with a lacquer-like substance.

Stomata absent, reduced, or of low frequency

Stomata are rarely absent amongst dicotyledons but this situation does occur for example in the submerged leaves of waterplants and in the reduced leaves of saprophytes and parasites. Solereder (1908) recorded complete absence of stomata in the submerged leaves of the Ceratophyllaceae, Droseraceae (*Aldrovanda*), Nymphaeaceae, Podostemonaceae and Ranunculaceae (*Ranunculus* and *Batrachium*). This applied also in some saprophytic Monotropaceae and some parasitic Balanophoraceae, Rafflesiaceae, and Orobanchaceae. Grüss (1927*a*) found stomata to be absent from, or infrequent in, the aerial leaves of *Nuphar luteum* Sibth. (Nymphaeaceae). In the first edition of this book other examples are also recorded. Occasionally the leaves of marsh plants lack stomata, especially if they are at least partly submerged as are the foliage leaves of *Genlisea* (Lentibulariaceae), although some stomata are present on their peculiar ascidiform leaves. (Families listed on p. 204.)

Plants with a particularly low frequency of stomata are often hydrophytes or plants especially adapted to storing water. This, according to Howard (1969), applies to the two epiphytes *Peperomia emarginella* (Sw.) DC. with 22 large stomata per mm^2 and *P. hernandiifolia* (Vahl) A. Dietr. In *Pilea krugii* Urb. (Urticaceae) there are 65 rather small stomata per mm^2 (Howard 1969).

Control of stomatal size

Since the early years of botanical investigation, workers in plant anatomy have noticed that some plants have small stomata and others large ones. Furthermore, size is often correlated with density. Some authors consider that stomatal size is too vari-

able to be of diagnostic value, while others have found the contrary to be true. My own experience has indicated that stomatal size shows a much wider range in some taxa than in others, so it can sometimes be a useful diagnostic character when dealing with taxa in which the size ranges are restricted.

It has been known for many years that diploid plants usually have smaller stomata than their polyploid relatives (Aalders and Hall 1962; Sax and Sax 1937; Stebbins 1950; Carolin 1954; Stone 1961; etc.). Related species often have stomata of similar size. Carolin found that, in *Dianthus*, the closer the stomatal sizes of the parents of a polyploid (race or species) are to one another the greater the likelihood that the stomatal size of the polyploid will significantly exceed that of both its parents. The closer the parents are phyletically, the greater the probability that the stomata of the progeny will be larger than those of the parents. Although this is often true for other examples of polyploids as well, other authors have not found it to be a principle which holds true universally.

Shade, a humid atmosphere, and moist soil conditions are all known to be coincidental with smaller stomata, while full sunlight and drier conditions seem to produce larger stomata. On the other hand, stomata are less easily influenced by these factors than are ordinary epidermal cells, according to Tarnavschi and Paucă-Comănescu (1972). These authors worked on several herbaceous plants and found that stomatal dimensions decreased slightly at higher altitudes, although they found this character to be less variable than stomatal numbers.

Pataky (1969) examined the leaf epidermis of *Salix* in different regions of the leafy crown. Although the length and width of the guard cells changed slightly with their position in the crown, the length-to-width ratio did not.

Most authors agree that stomatal size is usually sufficiently stable to be used as a diagnostic character. This, however, is true only if the full size range is given. It is better still to record the most frequently occurring size, or the length-to-width ratio, of the stomata.

Many authors do not give actual measurements of stomatal size, contenting themselves with designations such as 'small' or 'large', so that an actual comparison cannot be made. However, from experience, it appears that the term 'small' is generally applied to stomata of which the guard cells are less than c. 15 μm long and 'large' to stomata of more than c. 38 μm long. Small stomata often have a high density (see p. 110) and large stomata a low density (see pp. 110, 115).

Particularly small stomata often occur in plants with microphyllous leaves, and these leaves often have a thick cuticle and/or are densely hairy. Examples are provided by the deeply furrowed, abaxial surface of the heath-like leaves of the Empetraceae, Epacridaceae and Brunoniaceae, as well as by the leaves of certain Verbenaceae, such as *Casselia* (from Brazil) and *Coelocarpum* (from Socotra and Madagascar). The Malayan trees belonging to the (*Gonystylus* Thymelaeaceae) and the alpine and arctic members of the Saxifragaceae also have small stomata. Among the smallest whose size has been quoted are those of *Miconia pycnoneura* Urban (Melastomataceae), a forest shrub in which they are 10 μm long (Howard, 1969). Stomata of *Acer campbellii* Hook. f. et Thoms. are 10 μm (Lorougnon, 1966) and those of *Rhus copallina* L. (Anacardiaceae) 6 μm (Carpenter and Smith, 1975). (Families listed p. 204.)

There appear to be only a few families in which particularly large stomata have been reported, i.e. *c.* 40 μm or more long. Some examples are the especially large elliptical guard cells which occur in species of *Austrobaileya* (Austrobaileyaceae) as well as in *Kadsura* and *Schisandra* (Schisandraceae). Large stomata have also been recorded in certain genera of the Gesneriaceae and in *Stegnosperma* (Phytolaccaceae). Stomata up to 46 μm long have been recorded in *Hedyosmum arborescens* Sw. (Chloranthaceae) and *Peperomia emarginella* (Sw.) DC. (Piperaceae) according to Howard (1969). Even larger stomata, 45–88 μm long, occur in several species of *Phelline* (Phellineaceae) (Baas, 1975).

Especially large stomata have often been noted in the literature as occurring over veins or adjacent to them. Alternatively they may be rather isolated in the centre of areolae. Bünning (1956) has described clear areas around stomata (and hairs) as 'fields of inhibition', the extent of which depends on the size of the initial (see also Korn, 1972). These large stomata have been called water- or hydathode–stomata by many authors and are thought by some (including Stace, 1965*a*), to secrete drops of water. Dunn, Sharma, and Campbell (1965) used the term 'primary stomata' for markedly large stomata, because they appear to begin their development earlier than the surrounding smaller stomata. These authors also mentioned that well-developed radiating striae ('wrinkles of stress'), were generally present around primary stomata, thus making them stand out distinctly. This is a common phenomenon, familiar to all who examine leaf cuticles (Plate 3(C)). Sitholey

and Pandey (1971) proposed the term 'giant' stomata for abnormally large stomata to distinguish them from true water-stomata on the one hand and normal stomata on the other. They reported the presence of such giant stomata in the leaves of *Mangifera indica* L. (Anacardiaceae) and *Limonium acidissima* L. (Rutaceae). I have myself found abnormally large stomata to be quite common in several other genera of the Anacardiaceae, e.g. *Lannea*, *Melanochyla*, *Pistacia* and *Pseudosmodingium*. These stomata all have relatively well-developed outer rims (ledges), which is in contrast to the water porès (hydathodes) of *Tropaeolum majus* L., described and figured by Haberlandt (1914, his fig. 186D). Van Cotthem (1971) mentions relatively large stomata, resembling hydathodes, lying over veins in *Brosimum* and *Ficus* species (both Moraceae).

Recently Korn and Frederick (1973) have discovered that, in *Ilex crenata* var. *connexa*, specially large stomata, with a mean length of 41.5 μm, begin to appear when the leaves are 6–10 mm long. The large stomata, referred to as D-stomata, are distributed in an orderly pattern. Korn and Frederick have devised a method whereby the positions of the stomata can be plotted in BASIC language and the data fed to a Honeywell G430 computer.

It is not clear whether these large stomata are in fact permanently open or whether they close in the same way as normal stomata. There may not be a distinct dividing line between these large stomata and some types of hydathodes (see p. 117).

Solereder (1908) noted the occurrence of stomata of two distinct sizes in certain members of the Juglandaceae. Subsequent work has shown that the differences in stomatal size in an individual leaf may be due to the fact that the stomata are not all initiated at the same moment during the ontogeny of the leaf. This was found, for example, by Pant and Gupta (1966) in their work on the Magnoliaceae. Dunn, Sharma, and Campbell (1965), after studying the imprints of stomata from the mature leaves of 226 species of miscellaneous dicotyledons, concluded that stomatal size is an unreliable character, and supported this contention by pointing out that in *Maclura pomiferum* Schneid. the size (length ?) of the stomata ranged from 11.6–42.2 μm which covered the figures which the same authors obtained from a wide range of species. It is indeed true that some species have a wide size-range, sometimes in three to four size categories as they correctly reported, but this in itself can be a useful character to contrast with other species which have a narrow size-range. Some examples which

I have noted myself and which stand out in contrast to *Maclura* are *Mangifera indica* L. with stomata **18–21** μm long and 18–21–24 μm wide (most frequent in bold type) or *Sorindeia madagascariensis* Thouars, 15–**18**–21 μm long and 15–**18**–21 μm wide. However, this also shows very clearly that it would be useless to use size to distinguish species with small roundish stomata. Evidently stomatal size can be used as a diagnostic character, provided due precautions are taken.

Gorenflot (1971) refers to two types of epidermis that occur in the Saxifragaceae. In the first or 'regular' type the stomata are normally arranged and appear to be developed synchronously. In the second or 'mixed' type new stomata are formed between adult stomata.

From the above remarks it is evident that more investigation into the different ages at which stomata are developed in young leaves is required and that their relationship to size classes in the adult leaf should be more clearly established. The particular question of stomatal size shows how necessary it is for authors to make clear whether or not they are including the large hydathode stomata in their measurements.

Stomata raised on columns

Solereder (1908) observed that projecting stomata may be found in species from damp localities and gives several examples, e.g. *Santiria* (Burseraceae; his fig. 43,A,B), '*Cineraria cruenta*' (Asteraceae), and some species of *Cordia* (Boraginaceae) as well as in *Fabiana* (Solanaceae) in which the leaves are said to be coated with varnish. Solereder also noted projecting stomata on the stems of various Cucurbitaceae. Haberlandt (1914) recorded the same feature on the fruit stalks and leaves of *Adenopus*, *Cucurbita*, *Luffa*, *Physedra* and *Sphaerosicyos* (Cucurbitaceae; see his fig. 188). Similar structures occur on the abaxial surfaces of leaf midribs of the tropical-rain-forest trees, *Dracontomelon* (Anacardiaceae, Fig. 10.5(b,c); Wilkinson 1971). In *Dracontomelon* the raised stomata have either a round or very short, elliptical pore which contrasts somewhat with the more usual elongate elliptical pore and suggests that the round pore may be permanently open. The guard cells have thick outer stomatal rims. These peculiar stomata are undoubtedly a modification of the especially large stomata mentioned above.

Stomata in groups

The occurrence of stomata in groups is a useful generic diagnostic character in some families. Furthermore, the number of stomata per group, or the characteristic position of the stomata, may help to distinguish one species from another. The following examples are taken from Solereder (1908) or the first edition of this book, unless otherwise stated. *Grindelia squarrosa* (Pursh.) Duval (Asteraceae) has small groups of stomata on both surfaces of the leaf overlying areas of assimilatory tissue, according to Holm (1910). Certain genera in the Caryophylleae have groups of stomata overlying patches of pale coloured parenchyma situated at the ends of veins in the teeth of young leaves (hydathodes). Stomata are recorded as being in groups in *Macrococulus pomiferus* Becc., *Eleutharrhena macrocarpa* Forman (Cutler, 1975) and in *Antizoma* spp (Menispermaceae) as well as in *Castela* and *Soulamea* (Simaroubaceae). Among the Ochnaceae the stomata are in pairs or groups of three in *Sauvagesia* and in *Godoya* they are in crowded groups between the network of veins. Some species of the Begoniaceae have stomata in groups of two to five, whereas in a large number of other species they occur singly, according to van Cotthem (1971). *Napeanthus* (Gesneriaceae) has stomata in groups of two to eighteen (Solereder 1908; van Cotthem, 1971). Recently, Wiehler (1975) found that one of the characters he could use to separate *Gasteranthus* from *Besleria* (Gesneriaceae) is that the stomata are grouped into islands in the former, but scattered over the lamina in the latter. In *Irvingia* the stomata tend to be in parallel groups of about five and to a lesser extent in *Klainedoxa* (both Ixonanthaceae). Members of the Himantandraceae have their abaxial stomata clustered around the bases of peltate scales, as recorded in the first edition of this work (p.36) and by Baranova (1972, her fig. 12), and an exactly similar arrangement is found in *Kostermansia malayana* Soegeng (Baas, 1972b, his fig. 4, Bombacaceae). (Families listed p. 203.)

Stomata restricted to particular areas of leaf surface

It is common knowledge that stomata in leaves with a dorsiventral lamina (see p. 99) either occur on both surfaces (amphistomatic), or, more often, they are wholly or almost exclusively abaxial (hypostomatic). In isobilateral leaves they are to be found in all parts of the leaf surface. Certain deviations from these basic patterns are also generally familiar, a well known example being the exclusively adaxial (epistomatic) stomata in floating leaves.

Numerous examples of restricted or specialized stomatal distribution also occur and these can serve as useful diagnostic characters because of their restricted occurrence. For example, in *Saxifraga*, the various species can be divided into four groups based on stomatal distribution. In the first group they are confined to the adaxial surface; in the second group they occur on both surfaces at the tip of the leaves; in the third group they are restricted to the leaf margins; and in the fourth group they are confined to the middle on the adaxial and the margins of the abaxial surface. Stomata in *Chrysosplenium* are all abaxial in the Alternifolia group, but occur on both surfaces in the section Dialysplenium. They are occasionally in groups in both genera. A.W. Hill (1931) found that in *Daphne* (Thymelaeaceae) the stomata of *D. cneorum* L. are dispersed over the whole abaxial surface except the midrib, while those of *D. petraea* Leybold are confined to two bands on either side of the midrib leaving the median area and margin free of stomata. The hybrid between these two species has only the margin free from stomata.

In *Mimosa cruenta* Benth. stomata occur throughout the adaxial surface, but on the abaxial side they are confined to one of the two longitudinal halves of each leaflet. They are absent from the abaxial side of the basal leaflet. These features correspond with the 'sleep-position' of the leaves of the species in question, the parts of the epidermis which bear stomata being covered, while the parts free from stomata are not covered during 'sleep'. Other species of *Mimosa* and *Piptadenia* (Fabaceae) have different patterns of stomatal distribution.

High altitudes appear to induce an increase in the proportion of species with amphistomatic leaves. Spinner (1936) examined some 404 species of mountain plants from the Andes and the Alps, of which 63 per cent were amphistomatic on the Andes and 83 per cent on the Alps. Furthermore, of the total species having stomata predominantly on the adaxial surface, there were 12 per cent on the plains, 25 per cent on the Alps and 41 per cent on the Andes. Conversely, there were 75 per cent of the species with hypostomatic leaves on the plains, 41 per cent on the Alps, and 23 per cent on the Andes. (See also Wagner 1892; Lohr 1919; Leick 1927; and Espinosa 1932.)

Orientation of stomata[1]

Particularly narrow leaves often have stomata arranged parallel to their main veins. Examples of families in which certain species show this feature are Campanulaceae, Epacridaceae, Fabaceae, and Proteaceae (Solereder, 1908). *Littorella uniflora* (L.) Ashers (Plantaginaceae) in its terrestrial forms has stomata arranged in longitudinal rows on *both* surfaces of the leaf (first edition of this book). Some plants with parallel veins, nevertheless, have random orientation of stomata, as in *Needhamia* (Epacridaceae) according to Watson (1962), and transversely to them in *Lysinema* (Solereder, 1908).

In dicotyledons with broad leaves the stomata are usually orientated in a random manner. However, alternative patterns sometimes occur. For example, stomata are at right-angles to the veins in *Hololachna*, *Myricaria*, *Reaumuria* (with some exceptions), and in *Tamarix* (all belonging to the Tamaricaceae; Solereder, 1908). A similar arrangement occurs in some species of Santalaceae, in *Nuytsia floribunda* R. Br. (Loranthaceae), according to Solereder, and in *Laguncularia racemosa* (L.) Gaertn.f. (Combretaceae; Stace,1965b). Many species of the Epacridaceae with reticulate venation have stomata parallel to the long axis of the leaf, except *Wittsteinia* in which they are randomly orientated (Watson, 1962).

Stomatal number, frequency, absolute number, and Stomatal Index

Pharmacognoscists in the 1920s unsuccessfully attempted to use the number of stomata per unit area of leaf surface as a diagnostic character for fragmentary material. Timmerman (1927), working with *Datura* (Solanaceae), was one of those who found the stomatal frequency, and also the ratio of the number of stomata per unit area in the adaxial and abaxial surfaces, to be too variable to have any practical diagnostic value. Nevertheless it was thought to be useful in descriptions to give the range of stomatal size and also to quote the most commonly occurring figures for stomatal frequency. Loftfield (1921) likewise found the stomatal frequency in *Malva rotundifolia* L. to be diagnostically unreliable, but noted that the ratio of stomata to other epidermal cells was more uniform. In addition, he found that the difference in stomatal frequency between sun and

[1] See also lists on p. 204.

shade leaves is accounted for by the extent to which the epidermal cells become enlarged rather than by a difference in the proportion of stomata produced.

Salisbury (1927) likewise emphasized that the frequency of stomata is high when the size of the epidermal cells is low and that the frequency is low when the epidermal cells are large. To express the stomatal frequency independently of the size of the intervening epidermal cells, he introduced the

Stomatal Index $\dfrac{S}{E + S} \times 100$, where S denotes the number of stomata per unit area and E the number of epidermal cells of the same area. By means of this index it can be shown that the number of stomata formed in the epidermis is no greater for sun-leaves than for shade-leaves. He furthermore maintained that 'the increased stomatal frequencies in plants grown on dry soil as compared with those grown on wet soil, of small leaves as compared with large leaves ... are all shown to be due chiefly to differences in the growth of the epidermal cells, that is, to differences in the spacing of the stomata and not to differences in the proportion of stomata developed.' Those variations of the Stomatal Index which do occur are due to internal factors, mainly humidity and nutritional conditions, according to Salisbury.

Very careful tests were made of the Stomatal Index by Rowson (1943a,b, 1946) checked by the use of the standard deviation and other statistical procedures. Working on *Cassia* (Fabaceae), he made over 140 counts of *C. acutifolia* Delile which had a Stomatal Index value ranging from 11.4-13, and over 170 counts on *C. angustifolia* Vahl which revealed a Stomatal Index ranging from 17.1-20. He also demonstrated that these differences of Stomatal Index value were statistically significant. On the other hand, he noted that the stomatal number varied considerably for each species and that this was of little value as a differential character. Rowson also investigated the possibility that the Stomatal Index might vary significantly in different parts of the same leaflet. Accordingly he made counts in ten different positions of both upper and lower surfaces in *C. angustifolia* which showed that the Stomatal Index for various positions on the lower surface varied from 16.2-19.7, while that for the upper surface ranged from 16.3-18.5, almost within his original estimation of 17.1-20. After further research Rowson (1946) came to the following conclusions concerning the Stomatal Index: (1) it does not vary significantly at different positions upon the leaf surface, (2) it is independent of leaf size and plant habitat, (3) it is the *same* for different

varieties within a species, (4) co-generic species may be differentiated by means of the Stomatal Index, and (5) the Stomatal Index value is more uniform upon the lower than the upper surface, except in isobilateral leaves. However, van Staveren and Baas (1973) found that the Stomatal Index of *Apodytes dimidiata* E. Meyer ex Arn. (Icacinaceae) was too variable (7-21) to be of any use as a taxonomic character.

The use of the Stomatal Index has been criticized by Lück (1966) who says that it is nothing more than a relative frequency and that the variation of the Stomatal Index depends upon the value of the frequency. He comments upon the extremely small size of the samples of leaf material upon which the Stomatal Index is often used. He recommends the use of a formula to correct the possibility of an abnormal distribution curve, which often results from stomata of small samples and from the use of the Stomatal Index which deals only with percentage results. Most statistical textbooks recommend using a simpler formula, the **arcsin transformation**. Where the difference between species is not clear cut and the Stomatal Index is used, or if a statistical examination is required, then the arcsin transformation must be used. It is suggested that reference should be made to the statistical tables of Fisher and Yates (1963), in particular to the transformation table, Table X, p. 74. If the difference between species is clear cut and the Stomatal Index is used as supporting evidence for other characters, it is not necessary to take the extra trouble of calculating the arcsin transformation. It would be helpful to readers if authors would state the methods they use.

Stober (1917), by comparing winter and summer leaves of a number of herbaceous species, demonstrated that stomata are more abundant in stem leaves than in rosette leaves. In 80 per cent of the experimental plants they were found to be more numerous in the abaxial than in the adaxial surface. Stober also noted that the number of stomata is almost equal in both surfaces of mesophytic leaves in which both sides are almost equally exposed to light and air. Stober and subsequent workers also found high stomatal frequencies in xerophytic plants, whilst in broad rosette leaves the stomata are larger but less frequent. The sizes of the stomata were also correlated with the number per unit-area in over 60 per cent of Stober's specimens.

Many workers, e.g. Yapp (1912), Timmerman (1927), Salisbury (1927), and others have subsequently found that the stomatal frequency is highest near the leaf apex, lowest towards the base and also greater at the leaf margins than near the midrib. In contrast, a few other workers, Eckerson (1908), Skene (1924), and Odell (1932), found stomata to be most numerous at the base of leaves and adjacent to the midrib. It has been found by various workers that the leaves on herbaceous and some shrubby plants situated at progressively higher levels on the plant, have correspondingly smaller cells and hence greater density of stomata and epidermal cells. This is sometimes referred to as Zalensky's Rule, after the researches of that botanist published in 1904. Yapp (1912) working on *Filipendula ulmaria* (L.) Maxim found a range from about 300–1300 stomata per mm^2 for the abaxial surface of leaves from lower and upper parts of the same shoot.

Similarly, many authors report that both the stomatal frequency and the Stomatal Index are less variable on the lower than on the upper surface.

Gupta (1961) investigated the diagnostic reliability of the Absolute Stomatal Number, i.e. the Stomatal Number multiplied by the area of lamina. He found this value to be constant in fully differentiated leaves, and noted that there was a marked correlation between the number of stomata and the area of lamina on leaves of a single branch. Individuals of the same species grown in different environments showed the same type of correlation. Sharma and Dunn (1968, 1969) have criticised the Absolute Stomatal Number severely since they found (1) it varies with leaf size; (2) that even if the values for both sides of the leaf are added together it only becomes more disproportionate; (3) it is very easily modified.

In spite of all the difficulties enumerated above, many workers consider that the stomatal density is a useful additional character for distinguishing species when comparable areas of leaf are used. Researchers are agreed that the mid-lamina region of the abaxial surface in the centre of the leaf is the least variable and therefore most suitable.

Plants characterized by an especially high or low density of stomata

Solereder (1908) noted that there are plants with a large number of stomata, which together with the neighbouring or subsidiary cells form almost the entire epidermal surface, and that this feature might be employed for the diagnosis of species and occasionally even of more extensive taxonomic groups. Some examples of the feature given by him are *Macropanax*, *Schefflera*, and *Tubidanthus* (Araliaceae), while I myself have noted that *Ozoroa* and *Heeria* are good examples in the Anacardiaceae.

Two of the highest recorded frequencies are those of *Veronica cookiana* Colenso (Scrophulariaceae) with 2200 per mm^2, according to Adamson (1912), and *Miconia pycnoneura* Urban (Melastomataceae) at 2230 per mm^2 (Howard, 1969). Many authors have found that xerophytic plants have high densities of stomata and most of the above examples support this view. Nevertheless not all xerophytic plants have such high values as those given above. For further details on stomatal density, see the comprehensive works by Salisbury (1927, 1932), Maximov (1929, 1931), Kropfitsch (1951a,b), Larsen (1961), Rangaswamy (1962), Gindel (1969), Pazourek (1970), Knecht and Orton (1970), and Knecht and O'Leary (1972).

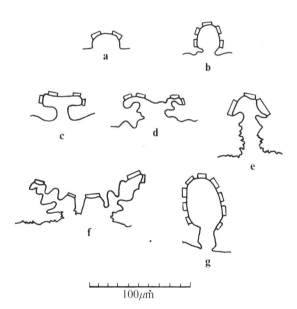

100 µm

Fig. 10.6. Cross-sections of stomatal crypts of *Mouriri* spp. The positions of the stomata are indicated diagramatically. (a) Type I crypt; (b) Type II crypt; (c) Type III crypt; (d) *M. eugeniaefolia*; (e) *M. pusa*; (f) *M. cearensis*; (g) *M. crassifolia*. (Redrawn from Morley 1953.)

Stomata on floral parts, fruits, and seeds

Stomata are known to occur in floral organs and fruits as well as in leaves and stems. The epicarp of fruit walls possesses stomata at least when young; these later give rise to lenticels and often mark the site where periderm is initiated. Internal stomata occur in the endocarps of many fruits and seeds. For further particulars, see Netolitzky (1926), Gassner (1955), Barton (1967), and Vaughan (1970); stomata on anthers, Kenda (1952); on perianth, of a large number of families, Maercker (1965b).

Abnormal stomata

Abnormal stomata occur sporadically but widely, but they are especially characteristic of certain families such as the Solanaceae (Ahmad, 1964a and b), and Rubiaceae (Pant and Mehra, 1965). Dutta and Mukerji (1952) found the abnormalities to be of great diagnostic value in some medicinally used species of *Datura*. Abnormalities in numerous guises include the following: (1) stomatal pore permanently closed; (2) cuticular ledges from the guard cells much thicker than in normal stomata; (3) stomata with no guard cells, but subsidiary cells present; (4) stoma reduced to only one guard cell accompanied by subsidiary or ordinary epidermal cells; (5) guard cells unusually narrow; (6) adjacent stomata with common subsidiary cells (Pant and Mehra, 1965).

Control of guard cell movement

The attention of readers who are especially interested in this subject is directed to the comprehensive account by Meidner and Mansfield (1968), Aylor, Parlange and Krikorian (1973), Levitt (1974), Louguet (1974), and Allaway and Milthorpe (1976).

For a consideration of the relative merits of the use of the s.e.m. and light microscope for the examination of stomata and cuticles see pp. 155-6.

Literature cited

Aalders and Hall 1962; Adamson 1912; Ahmad 1964a,b; Allaway and Milthorpe 1976; Aylor, Parlange, and Krikorian 1973; Baas 1970, 1972b, 1974, 1975; Bailey and Howard 1941; Bailey and Nast 1944, 1948; Baranova 1972; Barton 1967; Bary, de 1884; Böcher and Lyshede 1968, 1972; Bondeson 1952; Bongers 1973; Bongers, Jansen, and van Staveren 1973; Bukvic 1912; Bünning 1956; Carolin 1954; Carpenter and Smith 1975; Chappet 1969; Cotthem, van 1968, 1970a,b, 1971, 1973; Cronquist 1968; Cutler 1975; Décamps 1974; Dilcher 1974; Dunn, Sharma, and Campbell 1965; Dupont 1962; Dutta and Mukerji 1952; Eckerson 1908; Espinosa 1932; Field 1967; Fisher and Yates 1963; Florin 1933, 1958; Fryns-

Claessens and Cotthem, van 1973; Gassner 1955; Gindel 1969; Gorenflot 1971; Grambast 1954; Grüss 1927a; Gupta 1961; Guyot 1966; Haberlandt 1914; Heilbronn 1916; Hill, A.W. 1931; Holm 1910; Howard 1969; Hryniewiecki 1912a,b; Inamdar 1969b; Jähnichen, von 1969; Jalan 1962; Jensen and Baas 1973; Kaufmann 1927; Kenda 1952; Knecht and O'Leary 1972; Knecht and Orton 1970; Korn 1972; Korn and Frederick 1973; Kropfitsch 1951a,b; Larsen 1961; Leick 1927; Levitt 1974; Lobreau-Callen 1973; Loftfield 1921; Lohr 1919; Lorougnon 1966; Louguet 1974; Lück 1966; Maercker 1965b; Maximov 1929, 1931; Meidner and Mansfield 1968; Mersky 1973; Metcalfe 1961; Metcalfe and Chalk 1950; Morley 1953; Mühldorf 1922; Napp-Zinn 1974; Neischlova and Kaplan 1975; Netolitzky 1926; Odell 1932; Paliwal 1967, 1969; Paliwal and Bhandari 1962; Pant 1965; Pant and Gupta 1966; Pant and Mehra 1965; Pataky 1969; Patel 1978; Patel and Inamdar 1971; Paula, de 1974; Payne 1970; Pazourek 1970; Rangaswamy 1962; Rowson 1943a,b, 1946; Sahasrabudhe and Stace 1974; Salisbury 1927; 1932; Sax and Sax 1937; Schoch 1972; Sharma and Dunn 1968, 1969; Singh and Kundu 1962; Sitholey and Pandey 1971; Skene 1924; Solereder 1908; Spinner 1936; Stace 1963, 1965a,b, 1966; Staveren, van, and Baas 1973; Stebbins 1950; Stebbins and Jain 1960; Stebbins and Khush 1961; Stober 1917; Stone 1961; Strasburger 1866; Stromberg 1956; Takhtajan 1969; Tarnavschi and Paucă-Comănescu 1972; Timmerman 1927; Tomlinson 1969, 1974; Uphof, Hummel, and Staesche 1962; Vardar and Bütün 1974; Vaughan 1970; Vesque 1889; Vink 1970; Wagner 1892; Watson 1962; Weber, H. 1955, 1959; Wiehler 1975; Wilkinson 1971; Wulff 1898; Yapp 1912; Zalensky 1904; Ziegenspeck 1941; Zubkova 1975.

Suggestions for further reading

Arraes 1970: statistical results show stomatal index is not of value for distinguishing *Cassia* spp.
Baranova 1968: stomatography and taxonomy.
Bobisut 1910: function of stomata in *Nepenthes*.
Bremekamp 1947: peculiar stomata of shade plants of *Streblosa* (Rubiaceae).
Bünning and Sagromsky 1948: structure of leaf stomatal types.
Chandra, Kapoor, Sharma and Kapoor, 1969: epidermal and venation studies in Apocynaceae – I.
Davies and Kozlowski 1974: stomatal responses of

five woody angiosperms to light intensity and humidity.

Fahn and Dembo 1964: walls of guard cells in desert plants continue to thicken after maturity.

Gadkari 1964: stomatal frequencies of cotyledonary leaves of different varieties of Indian cotton.

Gangadhara and Inamdar 1975: action of growth regulators on cotyledonary stomata.

Gay and Hurd 1975: the influence of light on stomatal density in the tomato.

Guttenberg 1959: general physiology and anatomy of stomata; 1971, pp. 203-219: stomatal anatomy.

Inamdar, Bhatt, Patel, and Dave 1973: structure and development of stomata in vegetative and floral organs of some Passifloraceae.

Kapoor, Sharma, Chandra, and Kapoor 1969: epidermal studies in Apocynaceae — II.

Kaul 1970: stomatal size, frequency and index of ecotypes and ecads of *Mecardonia dianthera* (Scrophulariaceae).

Ketellapper 1963: mechanism and control of stomata.

Kralj and Sušnik 1967: polyploids of *Humulus lupulus* L. distinguished by stomatal size and chromosome number.

Kritikos and Steinegger 1948, 1949a,b: stomatal index of diploids and tetraploids of *Lobelia syphilitica* and other *Lobelia* spp.

Kutík 1973: relationships between quantitative characteristics of some stomata and epidermal cells of leaves.

Mankevich 1971: influence of ploidy on anatomical characters.

Michele, de, and Sharpe 1974: a parametric analysis of the anatomy and physiology of stomata.

Moncontié 1969: stomata of the Plantaginaceae.

Paliwal 1972: classification, and development of abnormal types of stomata.

Parkin 1924: discussion concerning arborescent families with paracytic stomata and herbaceous ones with anomocytic stomata and their relation to phylogeny.

Pisek, Knapp, and Ditterstorfer 1970: structure and amount of opening of stomata.

Porsch 1903: stomata of submerged plants.

Raschke 1975: stomatal action (review of literature).

Ryder 1954: general morphology and anatomy of leaves.

Sagromsky 1949: formation of stomatal patterns including the formation of groups.

Schittengruber 1953a,b: stomata on white flecks on leaves of *Pulmonaria* and *Cerinthe major*.

Schürmann 1959: influence of hydration and light on the structure of stomatal initials.

Sen, 1958: stomatal types in the Centrospermae including intermediate types.

Smith 1937: number of stomata in *Phaseolus vulgaris* studied with the analysis of variance technique.

Vesque 1885: leaf anatomy of the principal gamopetalous families.

Walker and Dunn 1967: environmental modification of the leaf cuticle of Alaska pea plants.

Weber, H. 1949a,b: on the structure of stomatal groups on shoots.

Weiss 1865: on the number and size ratio of stomata.

PART II : HYDATHODES [1]

Introduction

The term **hydathode** was introduced by Haberlandt (1894, 1914), for structures secreting water. This included hydathodes with epithem and glandular trichomes. In this section we are concerned mainly with the former (for trichomes, see p. 40). Hydathodes in their simplest form are apertures in plant surfaces, especially leaves, through which liquid, as opposed to water vapour, is secreted. They occur predominantly on the toothed margins of herbs, deciduous shrubs, and trees. The apertures frequently have guard cells, as in stomata. Others of a more complex type, with epithem, are usually located near vein terminations at the apex of leaves. Their structure follows a simple basic pattern, Fig. 10.7(b). The tracheids from about one to three vein endings fan out at the base of the epithem, which is a cushion of colourless, often loosely arranged, parenchyma cells. This tissue is frequently separated from the leaf mesophyll by a distinctive sheath of cells which may contain tannin-like substances (Perrin, 1972b). Alternatively the sheath cells may be suberized, or have casparian strips (Sperlich, 1939). The epithem is usually towards the upper surface of the leaf, so that it is covered by the adaxial epidermis, in which there are from one to several stomata, which may be incompletely differentiated, or larger or smaller than ordinary stomata. They are usually considered to be permanently open (Haberlandt 1914; Reams 1953; Stevens 1956; and others). Some authors such as Esau (1965a), Belin-Depoux (1969), and Perrin (1972b) note that hydathodes sometimes resemble extrafloral nectaries (see

[1] Families listed p. 205.

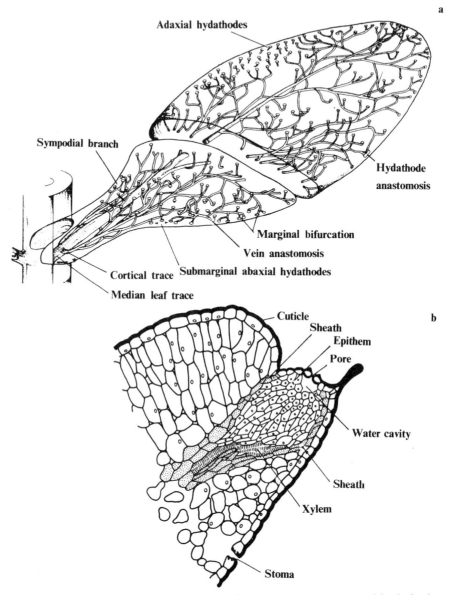

Fig. 10.7. Hydathodes. (a) Stereodiagram of the venation pattern and hydathodes of *Crassula argentea* Thb. Some veins have been excluded for clarity and vein thickness has been exaggerated for emphasis; × 2.3; from Rost (1969). (b) Hydathode of leaf of *Saxifraga lingulata* Bell. in longitudinal section. Tannin-containing sheath cells are stippled; after Häusermann and Frey-Wyssling (1963), modified by Esau (1965).

p. 127). This applies for example to the hydathodes in certain species of *Saxifraga*, where the punctiform mitochondria near the nuclei recall those which Eymé (1963) noted, in the nectar-secreting cells of the extra-floral nectaries of three species of Ranunculaceae. The water secreted from hydathodes may contain minerals, sugars, and other substances in solution. For example, the secreted liquid frequently contains small amounts of calcium, but when this is present in high concentrations, the hydathodes are more in the nature of chalk glands or cystoliths (see Vol. II). It thus follows that there is no clear distinction between hydathodes and other secretory structures such as extrafloral nectaries, glands and cystoliths which may be variously categorized according to individual viewpoints. These refer to such criteria as gross morphology, ultra-structure, or concentration of solutes in the liquid secreted.

Physiology and classification of hydathodes

The process by which liquid water is secreted from hydathodes is known as **guttation**, a term introduced by Burgestein (1887). Munting (1672, 1696), a Dutch botanist, was apparently the first to suggest that liquid water might issue from special apertures in the leaf surface (see Ivanoff 1963). The classical work on hydathodes was, however, produced during the nine-teenth century, as shown by the writings of Habernicht (1823, *Calla aethiopica* L.), A.P. De Candolle (1825, quoting Fr. Léandro on guttation in *Caesalpinia*), and Schmidt (1831, guttation in *Arum colocasia*). The reader is also referred to the descriptions of hydathodes (water pores) in the more general writings by Mettenius (1856), Unger (1858), Duchartre (1859, 1868), Borodin (1870), de Bary (1877, 1884), Volkens (1884), Haberlandt (1894-1924) and Decrock (1901). De Bary's observations are indeed so accurate that even today they provide a good basic description for hydathodes. For example, citing Mettenius (1856), he correctly pointed out that the number of apertures per individual hydathode varies in different taxa. His examples include one very large, wide, open pore at the apex of each leaf tooth in *Primula sinensis* Sabine, or on the adaxial side of each leaf tooth and at the leaf apex in *Saxifraga orientalis* Jacq. One to two similarly situated hydathodes occur, for example, in *Sambucus nigra* L. and *Prunus padus* L. The corre-sponding numbers are three in *Cyclamen*; three to six in *Ulmus procera* (*U. campestris* auct. angl.) and *Helleborus niger* L.; six to eight in *Crepis sibirica* L. and *Corylus avellana* L.; and much larger numbers per group, in for example *Tommasinia verticillaris* Bert. (Apiaceae) and *Alchemilla vulgaris* L. De Bary also pointed out that the hydathodes are often situated above enlarged endings to the veins. In addition, he drew a distinction between hydathodes that are supplied by one or by two veins, but the validity of this distinction has been questioned by Belin-Depoux (1969) who found only one vein ending beneath the hydathodes which he examined, whereas Perrin (1972*b*) found one or three veins.

Rost (1969) has studied the vascularization of the hydathodes of *Crassula argentea* Thunb. In this species the hydathodes occur over the whole adaxial surface, some of them being arranged in an irregular row near the leaf margin (Fig. 10.7(a)). There are about 300 hydathodes in a typical leaf, each consisting of the enlarged end of a vein in contact with an epithem. Each hydathode is sheathed by tanniniferous cells and there are eight to ten pores per hydathode, the pores being slightly smaller than those of the normal air-stomata.

Another point made by de Bary, which is still recognized as valid, is that the size of the hydathode stoma frequently varies according to their number, the smaller pores being more numerous than those that are larger. De Bary also drew a distinction be-tween hydathodes with pores of small diameter, e.g. in certain species of *Crassula*, *Ficus*, and *Saxifraga*, and hydathodes in which the apertures are much larger and permanently open, as in the Papaveraceae and Tropaeolaceae. He also noted that the guard cells are sometimes immovable and that they may die at an early stage, e.g. in *Aconitum*, or disappear, as in *Hippuris* and *Callitriche*. The term epithem (or epi-thema) was coined by de Bary for small masses of translucent cells in the mesophyll which serve as internal portions of the hydathodes. Once again, de Bary's fine observations on the hydathode struc-ture can be used. The cells of the epithem are small and colourless (without chloroplasts); they have dense cytoplasm and large nuclei. Each mass of epithematous tissue may communicate with the exterior through from one to several stomata which are usually larger than normal air-stomata, and there may be a large subjacent air-chamber. Bundle endings leading to the epithem either terminate below them, or the tracheids may be insinuated between the epithematous cells. The epithemata vary in shape in different taxa; for example, they are small and narrow in *Fuchsia* and *Primula sinensis* Sabine, broad in the leaf teeth of *Papaver* and *Brassica*, and elongated or oval in different species of *Crassula*. To elaborate de Bary's classification of hydathodes, it may be noted that he recognized the following types.

I. *Active hydathodes.* These are entirely composed of epider-mal cells which are not directly connected to the vascular system and from which water secretion takes place solely from these epidermal cells. Active hydathodes may be further subdivided into the following groups.

A. **Unicellular hydathodes.** These occur e.g. in *Gonocary-um pyriforme* Scheff (Icacinaceae), *Anamirta cocculus* (L.) Wight and Arne and *Arcangelisia* (Menispermaceae) (but see also Haberlandt 1894, 1898; Spanjer 1898; Krafft 1907; van Staveren and Baas 1973; and Wilkinson 1978 for diverse opinions on these structures).

B. **Multicellular trichomatous hydathodes.** These include: of *Plumbago lapathifolia* Willd., scutate glands from the scale leaves of *Lathraea squamaria* L. (see also Groom, 1897); (2) glands divided into foot and body regions, as in *Phaseolus multiflorus* Willd. and *Piper nigrum* L.; (3) long, stiff, tapering hairs in *Machaerium oblongifolium* Vog. (Fabaceae). These exude water through the lateral walls of the dead terminal cell.

II. *Passive hydathodes.* These are multicellular structures opening to the exterior via water-pores directly connected to the vascular system. They are of two kinds.

(A) **With epithem**, e.g. in *Primula sinensis*, *Papaver* and *Geranium* spp. and in certain representatives of the Urticaceae and Moraceae.

(B) **Without epithem**, e.g. in *Vicia sepium* L., the calyces of various Bignoniaceae e.g. *Spathodea campanulata* Beauv., as well as in certain members of the Scrophulariaceae, Solanaceae, and other families.

The 'trichome-hydathodes' of *Monarda fistulosa* L. var. *fistulosa* (Lamiaceae) which exude 450 μm^3 of an aqueous solution per minute (Heinrich, 1973) and those of *Cicer arietinum* L. (Fabaceae) which exude 1000 μm^3 per minute (Schnepf, 1965), have the morphological appearance of glands. Similarly, the curved secretory glands of *Phaseolus multiflorus* Willd., classified by Haberlandt as trichome hydathodes, have been shown by Perrin (1972b) to resemble glands in their characteristics. Earlier, Stocking (1956) had suggested that such structures should be considered as glands that secrete dilute nectar.

The systematic anatomist will naturally ask himself whether the different kinds of hydathode can be used as reliable taxonomic characters. It might alternatively be expected that, because hydathodes secrete water, their distribution would be related to habitat conditions rather than to the taxonomic affinities of the plants in which they are exemplified. There is no simple answer to this question because the recorded data are too sporadic. Such evidence as there is goes to show, however, that the presence of hydathodes and their histological characters can be valuable in taxonomy for confirmatory diagnosis. This may perhaps be clarified by the following references to more recent work. Frey-Wyssling (1941) recorded guttation in 340 genera belonging to 115 families which included both monocotyledons and dicotyledons. According to Esau (1958), however, Frey-Wyssling did not make it clear whether the guttation took place through water-pores or ordinary stomata. Ziegenspeck (1949) demonstrated the occurrence of hydathodes of one sort or another of the kinds that we have been discussing in representatives of 270 families of dicotyledons. His investigation covered 1–72 species from each of the families that he studied, but the Asteraceae

received special attention and 158 species were examined.

Perrin (1972b) examined hydathodes in more than 150 species belonging to diverse families, but mainly from the Caryophyllaceae, Ranunculaceae, Brassicaceae, Saxifragaceae and *Papaver somniferum* L. He draws attention to the following variations in hydathodes:

(1) variations in the number of hydathodes per leaf e.g. in *Stellaria indica* L. and *Ranunculus ficaria* L.; (2) slight variations in the bulk and shape of the epithem e.g. in *Geranium robertianum* L.; (3) variations in the number, size, and shape of apertures in hydathodes e.g. in *Ranunculus repens* L., *Saxifraga aizoides* L., *Stellaria media* (L.) Vill. and *Veronica officinalis* L.; (4) the occurrence of abaxial apertures in species where they are normally adaxial; (5) variations in the extent to which hydathodes are sunken; (6) the epithemial cells vary in size and shape. These may be isodiametric, or slightly elongate and without air spaces, or moruliform (lobed) and with large air-spaces. (7) The precise relationship of the tracheids to the epithem cells may vary. (8) Other variable characters include sizes of substomatal air-cavities and (9) the presence, absence, or degree of development of the sheath surrounding the epithem.

Chakravarty (1937), in a more restricted study, made a survey of the hydathodes in selected representatives of the single family Cucurbitaceae. In some species he found epithem hydathodes and in others trichome hydathodes. Hydathodes of this last type might more appropriately be regarded as secretory glands. He also found extrafloral nectaries resembling hydathodes.

Belin-Depoux (1969) noted differences in the structure of hydathodes within a single genus. This applies for example in *Saxifraga*. The hydathodes of *S. oppositifolia* L. and *S. aizoon* Jacq. are in pits and each hydathode is surrounded by a well-defined sheath of tanniniferous cells. Those of *S. hypnoides* L. and *S. decipiens* Shr., on the other hand, are not in pits and the surrounding sheaths are translucent. The first of these types of hydathode is to be found in simple leaves and the second in lobed leaves. A relationship between hydathodes and the external morphology of leaves also occurs in *Campanula rotundifolia* L. (Rea, 1921), where the hydathodes are on the adaxial surface near the ends of the principal veins. However, since the leaves at successively higher levels on the plant are correspondingly narrower, it follows that there are fewer principal veins and in consequence fewer hydathodes (Rea's Fig. 1). Rea also records that the apertures of the hydathodes from shade leaves of the same species were more numerous and larger than those of the hydathodes from comparable leaves from either normal or sun shoots.

The presence, absence, and shape of hydathodes all have diagnostic value. For example in *Pyrenacantha* (family Icacinaceae) they may be pyriform or globular and they may or may not be projecting, and these characters have been used in a diagnostic key (Villiers 1973, Lucas 1968). Villiers also uses presence or absence of hydathodes in a key to three species of *Chlamydocarya* Baillon, another member of the Icacinaceae. Stern (1974) states that hydathodes are an almost constant feature of the leaves of *Escallonia* (Escalloniaceae). These hydathodes are associated with the tip of the primary vein and the marginal serrations. In different species of *Pyrenacantha* the hydathodes may have a clump of unicellular hairs and/or a loosely organized parenchymatous 'callosity' (epithem?). A loosely-organized multicellular trichome may also be present. The callosities and multicellular trichomes are frequently filled with dark-staining tanninoid substances. Most usually these hydathodes are supplied by a single flared veinlet. Some hydathodes are without detectable 'water-pores', but most species have one or a few on the adaxial surface.

The hydathodes of various species of *Ficus* were noted by such early workers as de Bary (1884), Haberlandt (1894, 1895), Möbius (1897), Renner (1907), and Molisch (1916). Haberlandt (1894) considered that epithem hydathodes were restricted to the subfamilies Ficeae and Conocephaleae and that capitate trichomes situated over vascular bundles occurred in the Artocarpoideae. In 1895 he illustrated a typical hydathode of *Ficus elastica* Roxb. Molisch reported the orange appearance of the hydathodes of *F. javanica* Reinw. ex Streud. More recently Lersten and Peterson (1974) have investigated in detail the anatomy of *F. diversifolia* Blume. They report that the hydathodes are scattered over the adaxial surface of the leaf and appear as white spots in the young leaves, but are pink in the older ones. Each hydathode is seen as a slightly depressed epidermal pad studded with twenty to thirty water pores (small stomata). The anatomical structure is that of a typical hydathode, with an epithem of small irregular cells and conspicuous intercellular spaces, the whole being surrounded by a close-fitting sheath of unpigmented cells. The base of each hydathode is associated with a conspicuous vascular junction of three or more veins, from which strands of tracheary elements (not phloem) extend upwards for a short distance into the epithem tissue. Guttation was not observed in plants kept in a greenhouse.

According to Tucker and Hoefert (1968) the only known hydathodes which develop from a shoot apex are those occurring at the tip of *Vitis vinifera* L. tendrils. Another particularly unusual type of hydathode has recently been discovered by Kaplan (1970*b*). He has described the leaves of *Lilaeopsis occidentalis* Coult. and Rose and of *Oxypolis greenmannii* Math. and Const. (Apiaceae) as entirely lacking in obvious signs of pinnae, except for a series of reduced nodal appendages along the adaxial surface. He concludes that these small appendages are equivalent to reduced, transformed pinnae which serve as hydathodes at maturity.

Since hydathodes serve for the secretion of liquid water it is not surprising to find them in aquatic and marsh plants. For example very large stomata, with guard cells slightly sunken below the surrounding radially-arranged papillate cells, occur in the so-called 'light-spot' which overlies the vein anastomoses at the centre of the peltate leaves of *Nelumbo nucifera* Gaertn. (see Vouk and Njegovan, 1949). Although the apertures were found to be 10 μm in diameter when 'closed' and 40 μm when open, it is noteworthy that there was no evidence of guttation from the 'light-spots' which Vouk and Njegovan examined. The same authors nevertheless interpreted them as pneumatophores serving for aeration and transpiration. In young submerged leaves of *Ranunculus fluitans* Lam., Mortlock (1952) found a hydathode at the tip of each of the approximately 90 leaf segments, with three to five apertures in every hydathode. In the functionless hydathodes of older leaves of the same species there was but one aperture and this was to be seen between the markedly lobed epithem cells which appeared to secrete a gummy substance. In relatively young leaves of the marsh-inhabiting *Caltha palustris* L., Stevens (1956) found a hydathode with an average of ten water-pores on the adaxial surface of each lobe of the crenate leaf margin. By the time that the leaf is mature there is only one apical opening to each hydathode and, when the leaf stops growing, brown material which has meanwhile appeared between the epithem cells soon blocks the pores.

Kurt (1929) attempted to throw some light on the possible relationship between the organisation of hydathodes and the ecology of various Saxifragaceae in relation to their structure. He was able to arrange the species he examined in two groups. In the first he found that in plants from a moist habitat which have a 'mossy' habit or cushion form, there are only one or two hydathodes and the stomata are small. The hydathodes are not in pits; papillae are absent; there is little guttation and only a small amount or no excretion of calcium carbonate. In the second, in rosette plants from a dry habitat, the hydathodes usually have one to four larger stomata often in deep pits; papillae are sometimes present; the amount of guttation is greater and abundant calcium carbonate is excreted.

From the above observations of Kurt and other authors, guttation is considerably affected by humidity, soil moisture and temperature. Munting (1696) indicated that guttation may not be apparent during the heat of the day but is very active in the evening and at night when the temperature is sufficiently low. For a general review of the subject of guttation see Ivanoff (1963).

In spite of the fact that hydathodes are usually considered to be water-secreting organs, there is some evidence that marginal epithem hydathodes may, in some conditions, absorb water. That the leaf as a whole, and certain hairs in particular, are capable of absorbing water and chemical sprays is of course well known. Wood (1970) made some elementary tests on a number of plants. He took strips of leaf 1 cm wide from the margin of thin-textured leaves and immersed them in Indian ink. He reported that 'in some cases a mass flow of ink particles into the hydathode can be seen under the microscope'. Curtis (1943) has also shown that Neutral Red will penetrate through the hydathodes into the vascular network of both maize and squash leaves. Wood examined 73 British deciduous trees and found that 48 (66 per cent) have hydathodes, whereas of the 21 evergreens he looked at, only 3 (14 per cent) have hydathodes. He pointed out that the ability to absorb water from light rain, dew or mist, might be an advantage to plants growing in an adequate soil during typical British summers, when water is deficient and there is no frost. He contrasted these conditions with those of the tall, tropical-rain-forest trees, which are exposed to a high rainfall and have poor soil. He examined the leaves of 35 broad-leaved tropical dicotyledonous trees, 34 of which belonged to 17 different families, and discovered that only one had marginal hydathodes.

Secretion of chemical substances by hydathodes

It is beyond the scope of this work to deal with the chemical nature of the material secreted by hydathodes in detail, but the following further particulars may be added to those which have already been mentioned. The secretion of calcareous material by certain Saxifrages has been noted by numerous authors

such as de Bary (1877), Waldner (1877), Gardiner (1881), Volkens (1883), Haberlandt (1894), and Kurt (1929). More recently Curtis (1944) found calcium to be the most abundant element secreted from *Brassica oleracea* L., *Cucurbita pepo* L., *Cucumis sativus* L., and *Solanum lycopersicum* L. Perrin (1972*b*), by analysing the guttation liquid from *Caltha palustris* L., *Papaver rhoeas* L., and *Stellaria media*, demonstrated that phosphorus is the most abundant element to be secreted under natural conditions, and it is six times more plentiful than calcium, while other common elements are present in smaller amounts. Goatley and Lewis (1966), Sheldrake and Northcote (1968), and Perrin (1972*b*) found glucose, galactose, fructose, and arabinose in the guttation liquid from various plants. The amount of these sugars decreased in the same sequence. According to some authors, vitamins are present in very small amounts and some evidence of enzyme and auxin activity has been recorded. For further data concerning the occurrence of acid phosphatase in various secretory tissues, see the publications by Ziegler (1955), Vis (1958), Frey-Wyssling and Häusermann (1960), Cotti (1962), and Figier (1968).

Ultrastructure of epithem cells

Finally it should be noted that the ultrastructure of typical epithem cells from various dicotyledons, and also the glandular cells of the 'trichome-hydathodes' of *Phaseolus multiflorus* in relation to secretion, have recently been investigated by Perrin (1972*b*) as well as the material secreted by these trichome-hydathodes. He found that the epithem cells show various unusual features. These are to be seen in the plasmalemma, in the peripheral zone of the hyaloplasm, golgi apparatus, chondriosome plastids, nucleus, endoplasmic reticulum, and paraplasmic cell wall. Perrin also noted that plasmodesmata are nearly always absent. With regard to the glandular cells of *P. multiflorus*, Perrin discovered that there is a preliminary stage of preparation for secretion and a second stage during which secretion usually takes place. The same author (1971) also reported the presence of transfer cells (see also p. 128) in the epithem of hydathodes in *Taraxacum officinale* Weber, *Cichorium intybus* L., and *Papaver rhoeas* L. Pate and Gunning (1972) similarly record wall ingrowths in epithem cells of a number of plants. They noted that the wall ingrowths are particularly well developed where the epithem is in contact with xylem elements (p. 73).

Similarity of hydathodes to floral nectaries

Some floral nectaries are similar to hydathodes in that they have in their epidermis several to numerous large stomata which seem to be permanently open, as for example in the various members of the Centrospermales (Zandonella, 1967) and certain Phaseoleae (Waddle and Lersten, 1973). The main difference between these two structures seems to lie in the nature of the vascular supply, i.e. mainly or exclusively phloem in floral nectaries and xylem in hydathodes, which in turn reflects upon the composition of the liquid secreted (see also Frey-Wyssling 1935, pp. 330-2 and pp. 336-8). Consideration of similarity to extrafloral nectaries, glandular trichomes, squamellae and colleters is given on p. 127.

Literature cited

Bary, de 1877, 1884; Belin-Depoux 1969; Borodin 1870; Burgestein, 1887; Candolle, A.P. De 1825; Chakravarty 1937; Cotti 1962; Curtis 1943, 1944; Decrock 1901; Duchartre 1859, 1868; Esau 1958, 1965a; Eymé 1963; Figier 1968; Frey-Wyssling 1935, 1941; Frey-Wyssling and Häusermann 1960; Gardiner 1881; Goatley and Lewis 1966; Groom 1897; Haberlandt 1894, 1895, 1897, 1898, 1914; 1924; Habernicht 1823; Häusermann and Frey-Wyssling 1963; Heinrich 1973; Ivanoff 1963; Kaplan 1970b; Krafft 1907; Kurt 1929; Lersten and Peterson 1974; Lucas 1968; Mettenius 1865; Möbius 1897; Molisch 1916; Mortlock 1952; Munting 1672, 1696; Pate and Gunning 1972; Perrin 1971, 1972b; Rea 1921; Reams 1953; Renner 1907; Rost 1969; Schmidt, L. 1831; Schnepf 1965; Sheldrake and Northcote 1968; Spanjer 1898; Sperlich 1939; Staveren, van and Baas 1973; Stern 1974; Stevens 1956; Stocking 1956; Tucker and Hoefert 1968; Unger 1858; Villiers 1973; Vis 1958; Volkens 1883, 1884; Vouk and Njegovan 1949; Waddle and Lersten 1973; Waldner 1877; Wilkinson 1978; Wood 1970; Zandonella 1967; Ziegenspeck 1949; Ziegler 1955.

Suggestions for further reading

Auer 1962: cotyledons of *Pulsatilla vulgaris*.

Betts 1920: New Zealand rosette plants: Plantaginaceae.

Böcher and Lyshede 1972: xerophytic apophyllous plants from the South American shrub steppes.

Burck 1910: water secretion in plants.

Burström 1961: development of structures resembling hydathodes on submerged pea leaves.

Chemin 1920: *Lathraea*.

Cumming 1925: *Atriplex babingtonia* Woods.

Davidson 1973: *Tetrameles* and *Octomeles* (Datiscaceae).

Edelstein 1902: hydathodes on the leaves of woody plants.

Gardiner 1883: the significance of water glands and nectaries.

Goebel 1897: *Tozzia* and *Lathraea*.

Govier, Brown, and Pate 1968: guttation of leaf glands of *Odontites*.

Heimann 1950: influence of periodic illumination on the guttation rhythm of *Kalanchoë blossfeldiana*.

Hohn 1950: hydathodes and their function.

Hülsbruch 1932: *Dysophylla* (Lamiaceae).

Kristen 1969: light and electron microscope investigation on the Hydropoten of *Nuphar lutea* and *Nymphoides peltata*.

— 1971: also of *Nelumbo nucifera*.

Lavier-George 1937: *Hottonia palustris*.

Lepeschkin 1906: mechanism of active water secretion of plants.

Lippmann 1925: guttation.

Minden, von 1899: anatomy and physiology of water-secreting organs.

Napp-Zinn 1973: survey of subject.

Nestler 1896: guttation.

Neumann-Reichart 1917: anatomy and physiology of hydathodes.

Perrin 1970: ultrastructure of glandular hairs of *Phaseolus multiflorus*.

— 1972a: crystalline inclusions in the epithem cells of the hydathodes of *Cichorium intybus* L. and *Taraxacum officinale* Weber.

— and Zandonella 1971: nuclear invaginations in hydathode cells.

Philipson 1948: hydathodes at apex of leaves, bracts and sepals of *Lobelia dortmanna* L.

Renaudin 1966: scale leaves of *Lathraea clandestina* L.

— and Garrigues 1967: ultrastructure of hydathode glands of *Lathraea clandestina* L.

Ross and Suessenguth 1925-6: *Lafoënsia*.

Roth and Clausnitzer 1972: *Sedum argenteum*.

Sastre and Guédès 1974: *Sauvagesia erecta* L.

Scherffel 1928: *Lathraea squamaria* L.

Schmidt, H. 1930: *Saxifraga*.

Slatyer 1967: guttation.

Slavik 1974: guttation.

Stern, Sweitzer, and Phipps 1970: *Ribes*.

Taub 1910: water secretion of the Urticaceae.

Treub 1889: *Spathodea*.

Tswett 1907: members of the Lobeliaceae.

Unger 1861: *Saxifraga crustata*.

Uphof, Hummel, and Staesche 1962: water-secreting hairs.

Vardar 1950: hydathodes in submerged plants.

Vouk 1916: guttation in *Oxalis*.

Weingart 1923: *Crassula schmidtii* Regel.

Wilson; J.K. 1923: the nature and reaction of water from hydathodes.

Yuncker and Gray 1934: *Peperomia*.

PART III : EXTRAFLORAL NECTARIES

Introduction

Extrafloral nectaries are glands on the vegetative organs of plants from which nectar is secreted, the term nectar being used rather loosely by plant anatomists mainly for sugary, but sometimes for resinous and other types of secretion. For particulars of their first recognition by Hall (1762), a student of Linnaeus, and of their subsequent early investigation, the reader is referred to publications by Odhelius (1774), Kurr (1833), Sainte-Hilaire, de (1847), Caspary (1848), Martinet (1872), Poulson (1875), Delpino (1886), Bonnier (1879), Davis (1883), Morini (1886), Nieuwenhuis von Uexküll-Güldenband (1907), Haupt (1902), Schwendt (1907), and Böhmker (1917).

Types of extrafloral nectaries

Extrafloral nectaries are large enough to be seen with the naked eye or simple hand lens, for which reason they are frequently mentioned in the general literature of taxonomy, as well as in histological descriptions. They fall into two basic types which, for convenience, may be referred to as hair nectaries and epidermal nectaries. The hair nectaries consist of local concentrations of glandular trichomes, the individual hairs usually being similar to, or identical with, those that are distributed in a more isolated manner elsewhere on the plant surface (Fig. 10.8 (m)). Hair nectaries occur: (1) level with the surface of the organ on which they are present; (2) in shallow depressions; (3) in deep pits of diverse structure situated near the plant surface, the pits often being provided with an ostiole; or (4) in hollow bodies. Hair-gland nectaries may be situated close to the terminal branches of the normal vascular system of the plant, but they are stated in the literature to have no special vascular supply of their own.

Epidermal glands, as the term implies, share the common character that they are lined with a palisade epidermis from which secretion takes place (Fig. 10.8(b)). Sometimes occasional epidermal cells may be transversely divided, or the whole epidermis may consist of more than one layer. Unlike the hair glands, the epidermal glands usually, but not invariably, overlie a specially vascularized parenchymatous zone. Apart from sharing the common character of a palisade epidermis, glands of this second type vary considerably in their morphological form. Zimmermann's (1932) broad survey shows that they may be: (1) in pits; (2) level with the surface of the organs on which they occur; (3) cup-or basin-shaped; or (4) raised or stalked. Epidermal nectaries are occasionally multiple structures.

It has been recorded by Solereder (1908) and other writers that extrafloral nectaries of both kinds have a strictly localized distribution on the plants which bear them and are rarely found on the upper surface of the leaf (except in (6) below). Some of the principal sites are: (1) over the veins, especially at the proximal end of the lamina; (2) in the angles of the principal veins; (3) in the angles of the secondary veins; (4) on the surface of the lamina but without being in direct contact with the veins; (5) on the leaf

---

Fig. 10.8 Extrafloral nectaries. (a) base of abaxial surface of leaf of *Aleurites moluccana* Willd. (Euphorbiaceae), with two large oval nectaries at the distal end of the petiole; (b) LS part of flat gland of *Spathodea campanulata* Beauv. (Bignoniaceae); (c) TS midrib and one axillary gland; (d) TS petiole and nectaries of *Aleurites moluccana*; (e, f) horn-like projections with palisade-like epidermal cells in *Croton amabilis* Muell. Arg. (Euphorbiaceae); (g) thorn-like projections at the distal end of the petiole of *Adenopus breviflorus* Benth. (× 2.5 Cucurbitaceae); (h) enlargement of apex of one of them (× 14); (i–k) petiolar nectaries of *Acacia* spp. (Fabaceae); (i) *A. cornigera* Willd. (× 2), (j) *A. ruddiae* Janzen (× 2); (k) *A. collinsii* Safford (× 2); (1) TS midrib of *Ayenia*

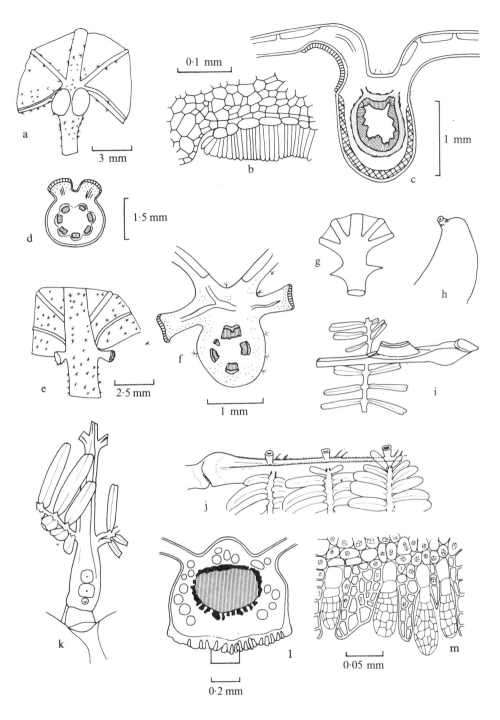

cuatrecassae Cristób. (Sterculiaceae) with multiapertured hair nectaries; (m) high-powered detail of part of nectary in (1). (a–h) slightly modified from Schnell, Cusset and Quenum, (1963); (i–k) drawn from photographs by Janzen (1974); (l, m) from Cristóbal and Arbo (1971). Vertical lines in (c) represent stele; dots in (d) and (f) represent crystals of calcium oxalate.

margins, especially where there are teeth or sinuses; (6) on the leaf rachis; and (7) on the petiole, especially at the proximal and distal ends. This has been confirmed by Schnell, Cusset, and Quenum (1963) who surveyed the extrafloral nectaries in 12 tropical dicotyledonous families, whilst Cusset (1965) studied the extrafloral nectaries of the Passifloraceae more intensively.

The following are selected examples of the types mentioned above. Depressed, elliptical, epidermal nectaries (diameter *c.* 1 mm) occur in *Acridocarpus longifolius* (G. Don) Hook.f. (Malpighiaceae). Hair nectaries in pits, crypts, and cavities are common amongst the Malvaceae notably in *Kydia calcycina* Roxb. and several species of *Hibiscus*. Cristóbal and Arbo (1971) examined extrafloral nectaries on the midrib of *Ayenia* (Sterculiaceae) Fig. 10.8 (l,m). Arbo (1972, 1973) has described depressed uni- and multi-apertured nectaries in *Byttneria* and *Megatritheca* respectively. The multi-apertured nectaries were found to be subdivided by an intruding epidermis.

Flat hair nectaries occur e.g. in species of *Hibiscus* and *Thespesia populnea* (L.) Soland. (Malvaceae). The nectaries in *Spathodea campanulata* Beauv. (Bignoniaceae) (Fig. 10.8(b,c)) are epidermal and there is one on either side of the midrib at its proximal end. They are more uncommon on the adaxial surfaces of leaves, but they have been seen, e.g. in *Monotes kerstingii* Gilg at the bases of the veins, and in various *Shorea* species (both Dipterocarpaceae) where disc glands occur on the adaxial surfaces of the leaves and their stipules. Glands in the form of groups of peltate scales also occur, for example in the abaxial vein axils of *Paulownia* (Scrophulariaceae), *Catalpa* (Bignoniaceae) and in *Anisoptera*, *Doona*, *Hopea*, and other genera of Dipterocarpaceae.

Various forms of paired, raised structures ranging from slightly elevated swellings to distinctly projecting rods, horns, or auriculate nectaries are frequently located at the distal end of the leaf petiole or just on the base of the lamina. Less commonly they occur a little lower on the petiole and sometimes also on the subtending axis at the proximal end of the petiole. In compound leaves, extrafloral nectaries may also occasionally be found on the leaf rachis. Short but rather large, oval swellings covered with tall, palisade-like secretory epidermal cells occur in *Aleurites moluccana* Willd. (Euphorbiaceae) at the distal end of the petiole (Fig. 10.8(a,d)). Horn-like structures, also covered with palisade epidermal cells, occur in various species of *Croton* (Euphorbiaceae), e.g. *C. amabilis* Muell. Arg. (Fig. 10.8(e,f)) with two, and

C. macrostachys Hochst. with four, located at the distal end of the petiole. Thorn-like projections are found towards the distal end of the petiole of *Adenopus breviflorus* Benth. (Cucurbitaceae), (Fig. 10.8(g,h)). Distinctly stipitate, slender glands characterise the leaves of *Croton ciliato-glanduliferus* Ortega. They occur in groups at the base of the petiole, at the base and on the sides of the midrib abaxial surface, and also around the margin of the leaf. Each of these structures has a rounded apex covered with palisade-like epidermal cells. Very similar structures occur on the highly specialized, modified leaves of *Triphyophyllum peltatum* (Hutch. & Dalz.) Airy Shaw (Dioncophyllaceae; Metcalfe, 1951) in which the glandular structures are red in colour.

Marginal glands, such as those on leaf teeth and in sinuses, are very common in some families such as Euphorbiaceae, Passifloraceae, Rosaceae, and Salicaceae. Schnell, Cusset, and Quenum (1963) have pointed out the similarity of the structure of many marginal glands to that of the stipulate/horned type, even to the presence of vascularization (although very slender).

From the examples cited in the literature it appears that herbaceous and ligneous plants show a homologous range of morphological types of extrafloral nectary. Many authors have drawn attention to the frequent concentrations of calcium oxalate crystals in the neighbourhood of extrafloral nectaries in diverse taxonomic groups of plants.

Taxonomic significance of extrafloral nectaries

The list on p. 205 of dicotyledonous families in which extrafloral nectaries have been recorded shows that they occur in a number of miscellaneous families. Solereder states that they are sufficiently widespread in certain families to be diagnostic for the families concerned. In other families they may occur in all of the species of a genus, e.g. in *Qualea* (Vochysiaceae), or they may be confined to certain species only. It is evident, however, that the term extrafloral nectary, whilst descriptively convenient, is applied to a range of structures that may well be of mixed origin. In the circumstances it would be surprising if they had any broad taxonomic significance. Nevertheless Solereder's opinion that they are of diagnostic value appears to be fully justified. For example it is interesting to note that hair nectaries of various kinds are to be found amongst certain Caryophyllaceae (Solereder's Fig. 25, p. 108) as well as in members of the Convolvulaceae,

Fabaceae, Olacaceae, Oleaceae, Malvaceae, Menispermaceae, and Polygonaceae. Epidermal nectaries, on the other hand, occur in the Loganiaceae (*Fagraea*), Marcgraviaceae (*Marcgravia*), Nepenthaceae (Solereder's Fig. 164, p. 678), and Rubiaceae (*Coprosma*).

A survey of the literature also shows that the presence or absence of extrafloral nectaries, as well as their morphology, density and position on the plant, and even their colour, have all been used for the separation of taxa. Dorsey and Weiss (1920) made a statistical analysis of the petiolar nectaries of over 3000 leaves belonging to 15 species and interspecific hybrids of the plum. They concluded that, after allowing for the full range of variability within each taxon, it is still possible to distinguish species and cultivars by differences in the disposition of the nectaries. They note, however, that the number and position of the nectaries vary with the age of the tree, the position of the leaves on the plant, and conditions that are unfavourable for growth. In consequence these same authors justly comment that when observations are unavoidably restricted to a few herbarium specimens some caution should be exercised before drawing conclusions.

Van Royen and van Steenis (1952) report a distribution for extrafloral nectaries within the Polygalaceae which supports the separation of *Eriandra* from three other genera: absent in *Eriandra*; present in *Barnhartia*, *Diclidanthera*, and *Moutabea*.

In another investigation Léonard (1957) separated *Gilbertiodendron* and *Pellegriniodendron* from *Macrolobium* (*sensu lato*) (Fabaceae) because they possess marginal glands (concurrently with other morphological characters) which are lacking from *Macrolobium* (*sensu stricto*). Working with other members of the same family, Bhattacharyya and Maheshwari (1971*b*) utilized extrafloral nectaries in subgeneric, specific, and even varietal delimitations within the genus *Cassia* L. This showed, for example, that the nectary-bearing species of the subgenus *Senna* Benth. can be divided into two groups. One of these has petiolar nectaries, e.g. in *C. occidentalis* L. A second group has nectaries between the leaflets on the rachis, e.g. in *C. tora* L. and *C. surattensis* Burm.f. The second of these groups can be subdivided still further as follows: (1) With one nectary between the lowermost pair of leaflets, e.g. in *C. obtusifolia* L. and *C. absus* L.; (2) with one nectary between the two lowermost pairs of leaflets, e.g. in *C. tora* L. and *C. sturtii* R.Br.; (3) with one nectary between the first three to four pairs of leaflets, e.g. in *C. surattensis*; and (4) with one nectary between all pairs of leaflets,

e.g. in *C. auriculata* L. Reverting to *C. tora* and *C. obtusifolia*, which, as has just been pointed out, fall into sections (2) and (1) respectively, it may be noted that *C. tora* has straight nectaries with tapering ends, pale yellow in colour, but with darker apices. In *C. obtusifolia* the nectaries are curved and orange-coloured with linear beak-like structures at their apices. These differences fall into line with phytochemical investigations which have shown that the seeds of these same two species contain different (although related) classes of chemical substances.

Janzen (1974*a*, pp. 28,29) has referred to the morphology and number of nectaries in keys to the species of Swollen-Thorn Acacias from Central America. The basic types of extrafloral nectaries developed on these various *Acacia* species are illustrated in Fig. 10.8(i-k).

Relationship of extrafloral nectaries to glandular trichomes, squamellae (nectarthodes), colleters, hydathodes, floral nectaries, and domatia

Where extrafloral nectaries consist of aggregated glandular trichomes, resembling those that occur singly elsewhere on the plant surface, it is clearly evident that the nectaries and the hairs have a common origin. This applies in *Osmanthus* (Oleaceae; Metcalfe 1938*b*), the Malvaceae (Ragonese 1960; Schnell, Cusset, and Quenum, 1963) and *Byttneria* (Sterculiaceae; Arbo, 1972). Rao, L.N. (1926) and Schnell, Cusset, and Quenum (1963, their Fig. 20) have shown how in *Spathodea stipulata* Wall. (Bigno-niaceae) the glands are represented by a continuous transitional series of structures ranging from unicellular trichomes to multicellular sacs lined with an epidermis of palisade secretory cells.

The term **colleters** was used by Schumann (1891) for certain secretory structures on the interpetiolar stipules of *Pentas lanceolata* (Forsk.) K. Sch. (Rubiaceae), but Solereder (1908) described the same structures as glandular shaggy hairs. Nevertheless, in Solereder's glossary, the term 'colleters' refers to 'large hairs secreting mucilage and found on buds'. Basically similar structures have recently been described in detail and termed **squamellae** (Ramayya and Bahadur 1968), **nectarthode** (Lewis, 1968), and, correctly, **colleters** (Lersten 1974*a,b*). These are all multicellular, glandular structures, usually composed of an elongated axis bearing oval to very elongate glandular epidermal cells, occasionally with a reduced axis but very elongated, separated epidermal cells. They secrete various mucilages, gums, or resins and are to be found on the leaves, stipules, and bud scales in many families of dicotyledons. The orange-coloured secretory glands described by Lewis for *Malus pumila* Mill. var. Jonathan apple leaves are located in pairs along the adaxial surface of the midrib, singly at the tips and notches of the serrated margins, on secondary veins and fine reticulations, as well as at the apices of the stipules. Those described by Ramayya and Bahadur are arranged in rows at the base of the adaxial surface of the leaves of *Allemanda cathartica* L. and *Tabernaemontana divaricata* (L.) R.Br. (both Apocynaceae) and secrete a yellowish brown resinous substance. According to Lersten (1974*b*) 'rubiaceous colleters have now been shown to vary sufficiently in certain taxa to be considered as additional useful taxonomic characters'. He recognizes several types of colleter (see his Fig. 1): (1) a standard type with regular columnar epidermal cells, (2) standard with irregular epidermal cells, (3) standard with reduced (flat) epidermal cells, (4) intermediate, (5) dendroid, and finally, (6) brushlike, in which the axis is reduced and the elongated epidermal cells are separated from one another for their entire lengths. He also discovered a morphological correlation between the bacterial leaf nodule symbiosis and dendroid colleters in *Pavetta* and *Psychotria*.

Structures that are transitional between extrafloral nectaries and hydathodes (see p. 117) also exist. This applies for example when trichomes serve to secrete water instead of nectar. Lepeschkin (1921) and Janda (1937) thought that extrafloral nectaries may be derived from hydathodes. Janda, citing the Malvaceae as an example, also pointed out that this family is particularly abundant in tropical regions where the humid atmosphere inhibits transpiration. Extrafloral nectaries with a palisade epidermis are structurally similar to hydathodes with epithem, and also to floral nectaries. However, in hydathodes, the ultimate branches of the vascular supply consist of tracheids only, in contrast to those of extrafloral nectaries where phloem elements are also present and floral nectaries which tend to consist of phloem only. Esau (1965*a*) notes a relationship between the proportion of phloem cells in the vascular tissue and the concentration of the secretion. Frey-Wyssling (1955) likewise records a sugar concentration of up to 50 per cent where phloem predominates, and as low as 8 per cent where xylem is the main component of the vascular tissue. However, in spite of these similarities to both epithem-hydathodes and floral nectaries, unlike either of them, extrafloral nectaries are not usually recorded as possessing stomata (but see reference to *Populus*

deltoides below). On the other hand, extrafloral nectaries show another similarity to floral nectaries in that they do not often possess an endodermal-like band of cells below the secretory tissue, as is frequently the case in epithem-hydathodes. For exceptions see Zimmermann (1932; *Pithecoctenium muricatum* Mac.) and Metcalfe (1951; *Triphyophyllum peltatum* Airy Shaw).

Many botanists from the mid-nineteenth century onwards have searched for resin-secreting structures on the bud scales and leaves of many dicotyledonous plants. Curtis and Lersten (1974) have reviewed the literature on this subject for *Populus* spp. They investigated the stipular bud scales, emerging leaves, and stipules of *Populus deltoides* Marsh, which has glands at the tips of marginal teeth and at the tips of small stalks arising at the junction of the lamina and petiole. The first leaves to emerge lack resin-secreting structures. However, there are several large stomata occurring abaxially near the margin. Nectar was seen to ooze from some of the non-glandular teeth of such leaves emerging from buds forced in the laboratory. On subsequent leaves all teeth are glandular, with more secretory cells and larger vascular bundle endings. Both resin and nectar are secreted. Under laboratory conditions early and late leaves occasionally exuded clear, sweet nectar that covered the tooth, accumulating until it formed drops or ran down the leaf surface. The authors think it likely that guttation occurs through the large stomata proximal to the secretory gland. Although this portion of the tooth is an extrafloral nectary by definition, it lacks the specialized internal structure usually associated with nectaries or hydathodes. Nevertheless, they consider that *Populus* has a rather unspecialized type of hydathode that can also act as a nectary under certain conditions.

There is, in some instances, further evidence of similarity between extrafloral nectaries and hydathodes, in that both sometimes possess transfer cells. In their article on transfer cells, Pate and Gunning (1972) note that 'There is a structural and functional gradation between hydathodes and nectaries ... and wall ingrowths are as much a feature of the one as they are of the other.'

Extrafloral nectaries in crypts show an overlap in structure with domatia which have a glandular lining. According to Stace (1965*a*, p. 35), gland-like structures are to be found in the domatia of certain members of the Combretaceae. Lundstroem (1887), reported similar structures in *Anacardium occidentale* L., and I can confirm their presence in this and 7 other species of *Anacardium*. I have also seen elongated cells lining the domatia of *Swintonia schwenkii* T. and B. (both Anacardiaceae). In contrast to these statements, Jacobs (1966, p. 5) states that all authors are agreed that domatia are non-glandular (see also p.136 of this work). Since extrafloral nectaries sometimes occur in the same type of situation as domatia, i.e. in the vein axils on the abaxial surfaces of leaves, the difference of presence or absence of glands is probably of doubtful significance. The relationship seems at least to be very close, and domatia and extrafloral nectaries are sometimes indistinguishable.

Theories concerning the origin of extrafloral nectaries

In certain plants such as *Vicia faba* L. the stipules bear nectaries on part of their surface. Other plants such as *Jatropha gossypifolium* L. (Euphorbiaceae) and *Smeathmannia laevigata* Soland. ex R.Br. (Passifloraceae) have nectaries at the base of the petiole, which is also a position in which stipules commonly occur in other plants which have them. Nectaries sometimes occur at the junctions between the lamina and the distal end of the petiole. Various authors have thought that nectaries in this last position may represent modified parts of the lamina itself or even stipules. *Sambucus nigra* L. bears leaf-like stipules at the bases of the leaflets of some of its imparipinnate leaves, but in other leaves of the same species there are stipitate glands in the corresponding position (Schnell, Cusset, and Quenum 1963, their Fig. 24). These authors visualized the developing lamina as if it were composed of a number of coalesced segments to which they referred as *articles foliaires*. Cusset (1965) subsequently made detailed observations on the ontogeny of the extrafloral nectaries of the Passifloraceae. He suggested that the term *articles foliaires* should be replaced by a new term *métamères*. These are areas of meristematic tissue arranged in a series along the margins of the lamina in very young, developing leaves. When mature they may appear in one or other of several alternative forms. These include (1) *pseudo-poils glanduleux* (complex, vascularized organs); (2) marginal glands and their immediately subtending tissue (*adénophores*); (3) submarginal laminar glands and their *adénophores*; and (4) parts of the innervated lobes. Some examples of these metameric origins are shown by *Passiflora molissima* Bailey in which each glandular tooth corresponds to a single *métamère*, whereas those of *P. quadrangularis*

L. correspond to several *métamères* whose innervation is fused with that of a secondary nerve. The petiolar glands of *P. alba* Link and Otto and *P. quadrangularis* are formed from several *métamères*. Cusset also considers the *métamère* theory in the light of phylogeny, with reference to 'the primitive leaf'. Detailed research in many families into the ontogeny of marginal and petiolar nectaries may help to elucidate this idea.

Chemical nature of and conditions favouring nectar secretion

The chemical nature of the material secreted from extrafloral nectaries varies in different taxa. For example in *Impatiens* and *Ricinus* the glistening secretion is transparent, sugary, and viscous. According to Figier (1972c) there is a considerable amount of sucrose in the nectar from the petiolar glands of *Impatiens holstii* Engl. and Warb. He considers this to be linked with the presence of sieve tubes and transfer cells with wall ingrowths, which appear among the glandular cells. (See also p. 128, and under Hydathodes, p. 122.) Similarly Kalmán and Gulyás (1974) report that the terminal branches of phloem run directly into glandular tissue below the columnar epidermal cells of *Ricinus communis* L. from whence the fluid is transported to the glandular tissue by cells of a 'transfusion nature'. The secretion of a sugary fluid has also been reported by Delpino (1886) in '*Olea fragrans*' (now *Osmanthus fragrans* Lour.). Glucose and fructose in different proportions are stated by Percival (1961) to be secreted from the stipular nectaries of *Vicia sativa* L. and *V. sepium* L. (see also Lüttge, 1961, 1962a,b). Petiolar glands of certain *Prunus* species are stated by Gregory (1915) and Knapheisowna (1927) to contain sugar and tannin, and those of *Turnera ulmifolia* L. to contain about equal amounts of sucrose, glucose and fructose (Elias, Rozich, and Newcombe, 1975). Secretion is, at least in some instances, most active before the leaves are mature. Francis Darwin (1876) noted that from the petiolar nectaries of *Acacia sphaerocephala* Cham. and Schlechtd. 'the secretion was so abundant as to drip out onto the floor of the glasshouse'. The secretion from *Osmanthus fragrans* is most active in August according to Delpino. Metcalfe (1938b) refers to maximum secretion during the summer months from *Osmanthus ilicifolius* Mouillef and records that the proportion of leaves on which the glands are secreting varies during the day as well as from day to day. In *Clerodendrum splendens* G. Don, growing in

Gujurat State, India, secretion is stated by Inamdar (1968c) to be more active on cloudy days and during winter.

The route by which secreted material reaches the surface of the glands has been the subject of numerous investigations. Readers who are interested in the parts played by the cell wall, the cuticle, canals in the cell wall or ectodesmata (sometimes mistakenly referred to in the literature as plasmodesmata), should consult the following publications: Nieuwenhuis von Uexküll-Güldenband 1914; Schumacher and Halbsguth 1939; Lloyd 1942; Schnepf 1959, 1968; Franke, W. 1960a,b; Mercer and Rathgeber 1962; Uphof 1942; Untawale and Mukherjee 1969; Martin and Juniper 1970; Figier 1972a,b,c; Kalmán and Gulyás 1974. The subject of transfer cells is treated very thoroughly by Gunning and Pate 1969 and Pate and Gunning 1972. (see also pp. 73, 122, 128 of this work.)

Association of extrafloral nectaries with ants and other insects

The physiological function of extrafloral nectaries is still rather obscure, and many of the statements in the literature are speculative. Although some detailed observations and experiments (see Janzen *et al.*, below) have been made since the classical writings of Wettstein (1888), Schimper (1903), and Haberlandt (1914), the position is relatively unchanged today. Wettstein, Schimper, and Haberlandt, like other previous and subsequent writers, refer to the association between extrafloral nectaries and sundry insects, particularly ants. The important relationship between ants and the leaflets of the Bull's Horn Acacia (mainly *A. cornigera* L. and *A. sphaerocephala* Schlecht. and Cham. was shown when Belt (1874) demonstrated that the extrafloral nectaries secrete a honey-like liquid. Ants are attracted by this secretion and also consume the yellow protein bodies at the tips of the leaflets (now termed Beltian bodies). In doing so they protect the trees from the ravages of leaf-cutting insects and other predators, especially herbivores. The attraction of ants by extrafloral nectaries is also mentioned by Francis Darwin (1876) and Nieuwenhuis von Uexküll-Güldenband (1907). The relationship of obligate-acacia ants to extrafloral nectaries has been dealt with in great detail by Janzen (1966, 1967a,b, and 1974a) and also by Carrol and Janzen (1973). Hetschko (1908) observed 58 species of insect visiting the stipular nectaries of *Vicia sativa* L., while Bequaert (1922) commented that ants are seldom absent from

the large stipular glands of *Vicia faba* and other species of *Vicia*. Springensguth (1935) lists 509 different species of insect he found on the extrafloral nectaries of several species of plant. Observations were made on *Osmanthus ilicifolius* by Metcalfe (1938b) who noted that wasps were attracted by the active leaf nectaries during the summer. Authors disagree over the extent to which nectaries are important in attracting insect pollinators.

Literature cited

Arbo 1972, 1973; Belt 1874; Bequaert 1922; Bhattacharyya and Maheshwari 1971b; Böhmker 1917; Bonnier 1879; Carroll and Janzen 1973; Caspary 1848; Cristobal and Arbo 1971; Curtis and Lersten 1974; Cusset 1965; Darwin, F. 1876; Davis, J.J. 1883; Delpino 1886; Dorsey and Weiss 1920; Elias, Rozich, and Newcombe 1975; Esau 1965a; Figier 1972a,b, and c; Franke 1960a,b; Frey-Wyssling 1935, 1955; Gregory 1915; Gunning and Pate 1969; Haberlandt 1914; Hall, B.N. 1762; Haupt 1902; Hetschko 1908; Inamdar 1968c; Jacobs 1966; Janda 1937; Janzen 1966, 1967a,b, 1974a; Kalmán and Gulyás 1974; Knapheisowna 1927; Kurr 1833; Léonard 1957; Lepeschkin 1921; Lersten 1974a,b; Lewis 1968; Lloyd 1942; Lundstroem 1887; Lüttge 1961, 1962a, b; Martin and Juniper 1970; Martinet 1872; Mercer and Rathgeber 1962; Metcalfe 1938b, 1951; Morini 1886; Nieuwenhuis von Uexküll-Güldenband 1907, 1914; Odhelius 1774; Pate and Gunning 1972; Percival 1961; Poulsen 1875; Ragonese 1960; Ramayya and Bahadur 1968; Rao, L.N. 1926; Royen, van and Steenis, van 1952; Sainte-Hilaire, A. de 1847; Schimper 1903; Schnell, Cusset and Quenum 1963; Schnepf 1959, 1968; Schumacher and Halbsguth 1939; Schumann 1891; Schwendt 1907; Solereder 1908; Springensguth 1935; Stace 1965a; Untawale and Mukherjee 1969; Uphof 1942; Uphof, Hummel and Staesche 1962; Wergin, Elmore, Hanny and Ingber 1975; Wettstein, von 1888; Zimmermann 1932.

Suggestions for further reading

Agthe 1951: sources of plant nectar.
Anderson 1972: colleters on *Crusea* (Rubiaceae).
Aufrecht 1891: thesis on extrafloral nectaries.
Bazavaluk 1936: petiolar glands of *Prunus*.
Belin-Depoux and Clair-Maczulajtys 1974, 1975: leaf glands of *Aleurites moluccana* Willd.

Bellini 1909: *Paulownia imperialis* Sieb. and Zucc.

Bernhard 1964, 1966: Crotonoideae.

Bhattacharyya and Maheshwari 1971*a*, 1973: Leguminales.

Böcher and Lyshede 1972: on xerophytic plants from South American shrub steppes.

Bonnier 1878: general.

Brocheriou and Belin-Depoux 1974: ontogeny of secretory pockets in various Myrtaceae.

Brown 1960: ants, acacias, and browsing mammals.

Cammerloher 1929: structure and function of extrafloral nectaries.

Carlquist 1959*b*: on leaves and bracts of *Holocarpha virgata*.

Chakravarty 1937: Cucurbitaceae.

Chavan and Bhatt 1962: *Blastania fimbristipula* Kotschy and Peyr.

Chavan and Deshmukh 1960: ontogeny of extrafloral-nectaries in *Gmelina*.

Cristóbal 1971: *Byttneria*.

Curtis and Lersten 1973: *Populus deltoides*.

Daumann 1967: *Impatiens*.

Dave and Patel 1975: developmental study of extrafloral nectaries of *Pedilanthus tithymaloides*.

Dop 1927: Bignoniaceae.

Elias 1972: *Pithecellobium macradenium*.

Elsler 1907: *Diospyros discolor* Willd.

Fahn 1974: general plant anatomy.

Fahn and Rachmilevitz 1970: ultrastructure and nectar secretion in *Lonicera japonica*.

Figier 1968, 1971*a*: ultrastructure of the stipular nectaries of *Vicia faba* L.

Frei 1955: presence or absence and type of vascular supply to floral nectaries.

Frey-Wyssling 1933: *Hevea brasiliensis*.

Frey-Wyssling and Häusermann 1960: interpretation of shapeless nectaries.

Groom 1894: *Aleurites*.

Hardy 1912: on some species of *Acacia*.

Horner and Lersten 1968: secretory trichomes (colleters) of *Psychotria bacteriophila*.

Hove, van and Kagoyre 1974: stipular glands in various Rubiaceae.

Inamdar 1969*a*: foliar nectaries in *Bignonia chamberlaynii*.

Janzen 1969: *Müllerian bodies of Cecropia*.

Keuchenius 1916: *Hevea brasiliensis*.

Kirchmayr 1908: *Melampyrum*.

Körnicke 1918: *Hibiscus*.

Kuntze 1891: Malvaceae.

Lersten 1972: stipular glands and trichomes in relation to bacterial leaf nodule symbiosis in *Psychotria*.

Lersten and Curtis 1974: colleter anatomy in *Rhizophora mangle*.

Lüttge 1971: structure and function of plant glands.

Maheshwari, J.K. 1954: Verbenaceae in general and *Duranta plumieri* Jacq. in particular.

—— and Chakrabarty 1966: *Clerodendrum japonicum* (Thunb.) Sweet.

Morini 1886: general.

Nesmeyanova 1968: *Hibiscus cannabinus*, L.

Nestler 1899: *Phaseolus multiflorus* Willd.

Ojehomon 1968: *Vigna unguiculata* (L.) Walp.

Parija and Samal 1936: in *Tecoma capensis* Lindl.

Parkin 1904: *Hevea brasiliensis* Muell.- Arg.

Patel and Inamdar 1974: some Gentianales.

Petaj 1916: *Ailanthus glandulosa*.

Pijl, van der 1951: *Honkenya filicifolia* Willd.

—— 1955: concerning myrmecophytes; useful definitions of terminology.

Poulsen 1877: *Batatas edulis*.

—— 1918: *Carapa guianensis* Aubl.

Ragonese 1973: *Dimorphandra* and *Mora*.

Rathay 1889: general.

Reed 1917: *Gossypium*.

—— 1923: *Ricinus communis*.

Reinke 1875: general.

Rettig 1904: ant-plants.

Rickson 1969, 1971: Beltian bodies in *Acacia cornigera* and Müllerian bodies in *Cecropia peltata*.

Roth 1974*b*: *Passiflora*.

Salisbury 1909: *Polygonum*.

Schnell 1963*a*: marginal and laminar surface glands.

—— and Cusset 1963: leaf glands in general.

Schremmer 1969: *Salix eleagnos* Scop.

Simon-Moinet 1965: homologies between the foliar and floral nectaries of *Impatiens balsamina*.

Terraciano 1897–9: Bombacaceae.

Trelease 1881: *Populus*.

Vis 1958: histochemical demonstration of acid phosphatase in nectaries.

Weber, F. 1951: *Impatiens*.

Weingart 1920: *Phyllocactus*.

PART IV : DOMATIA

Introduction

The term **domatium** (from the Greek: δωμάτιον; a little house) was introduced by Lundstroem in 1887. A rough translation of his definition is 'those formations or transformations on plants adapted to the habitation of guests, whether animal or vegetable, which are of service to the host, in contrast to cecidia, where such habitation is injurious to the plant'. He was the first to make a methodical study of domatia and discussed their possible cause and role. He used the term 'acarodomatia' because he saw mites frequently within the leaf domatia. Today, the term domatia is usually applied to depressions, pockets, sacs, or even to tufts of hairs in the principal vein axils where they occur exclusively on the abaxial surfaces of leaves. Some authors also apply the term to the revolute, basal margins of the leaves of certain plants. Lundstroem's definition of domatia is sufficiently broad to include cavities of various forms, which are found in almost any vegetative part of the plant. They might, for example, take the form of swollen hypocotyls, stems, or stipular thorns, perforated by one or several holes (see pp. 137-8). In the present work we are mainly concerned with leaf domatia. These are less frequently associated with insects than the other types in which animal occupants often find shelter. Domatia are found predominantly on woody plants of humid tropical or subtropical regions and although they are also present on plants from colder regions, they have not yet been discovered in plants from permanently dry sources. One of the few recorded examples of a herb having leaf domatia which I have been able to trace is *Eupatorium riparium* Regel (Asteraceae), (Hamilton 1896). This species has recently been re-examined by Jacobs (1966), who reported that it is ligneous towards the base, and also that he considers the 'domatia' to be nothing more than hairless cavities formed by the adjoining spaces under the projecting nerves. This view is supported by Y.M. Brouwer (University of Queensland, private correspondence) who has frequently examined *Eupatorium riparium* in the field.

De Barros (1961*b*), in a species of *Oxalis*, interpreted as domatia the hairs which he observed where the leaflets are attached to the petiole. Most authors would disagree with this opinion since the hairs are not in the axils of veins, and they neither surround nor are they situated in a cavity.

Many further examples could be quoted from the early literature to show that domatia and glands have often been confused, but these are not relevant to the present context and we must now pass on to see how leaf domatia have been classified.

Classification of leaf domatia

Classifications of leaf domatia have been proposed by Lundstroem (1887), Hamilton (1896), Penzig and Chiabrera (1903), Roechetti (1905), Briquet (1920), de Wildeman (1938), Chevalier and Chesnais (1941*a,b*), Stace (1965*a*), and Jacobs (1966). The main classifications are as follows:

1. Tufts of hairs in vein axils

Equivalent to the *Haarschöpfe,* type 1 of Lundstroem; 'group 4' of Hamilton; *ciuffi densi di pelli* or 'type H' of Penzig and Chiabrera; *lophique* of Briquet; *domaties en pinceau* of Chevalier and Chesnais; *domatium fasciculatum* of Stace; 'type 3' of Jacobs. Examples are numerous: *Strychnos gardneri* A. DC., (Loganiaceae); *Dracontomelon edule* Skeels (Anacardiaceae); *Juglans sinensis* Dode (Juglandaceae); *Alnus glutinosa* (L.) Gaertn. (Betulaceae), Fig. 10.9(a,b.).

2. Pockets

Often shaped like flattened funnels with wide distal openings. Equivalent to *Taschen* or *Düten*, 'type 4' of Lundstroem; 'group 2' of Hamilton; *tasche o borsette*, 'type T without hairs', and 'type Th with hairs' of Penzig and Chiabrera; type *ascique* of Briquet; *domaties en pochette* of Chevalier and Chesnais; *domatium marsupiforme* of Stace; 'type 2' of Jacobs. Examples: *Parrotia persica* C.A. Mey. (Hamamelidaceae); *Macaranga domatiosa* Airy Shaw (Euphorbiaceae) Fig. 10.9 (c,d).

3. Sacs: i.e. projecting, extended pockets

Equivalent to *Beutel* of Lundstroem; group 2 of Hamilton; 'T' and 'Th' of Penzig and Chiabrera; *thylacique* of Briquet;

Fig. 10.9 Leaf domatia. (a, b) hair tuft domatia of *Alnus glutinosa* (L.) Gaertn. (Betulaceae; (a) × $\frac{1}{3}$, (b) × 2.3); (c, d) pocket domatia of *Macaranga domatiosa* Airy Shaw (Euphorbiaceae; (c) × 2.3; (d) × $\frac{1}{3}$); (e) revolute margins at the base of a leaf of *Duroia saccifera* Hook f. (Rubiaceae; after Hall 1967, hairs omitted; × $\frac{1}{3}$); (f, g) pit domatia of *Pennantia cunninghamii* Miers (Icacinaceae, (f) × $\frac{1}{3}$, (g) × 2.3); (h, i) sac domatia of *Dysoxylum fraseranum* Benth. (Meliaceae; (h) × 1.6, (i) × $\frac{1}{3}$); (j, k) sac domatia towards the base of a leaf of *Cola marsupium* K. Schum.; (j) abaxial; (k) adaxial (Rubiaceae; after Schnell and de Beaufort 1966; both × $\frac{2}{3}$). Petioles omitted or incomplete.

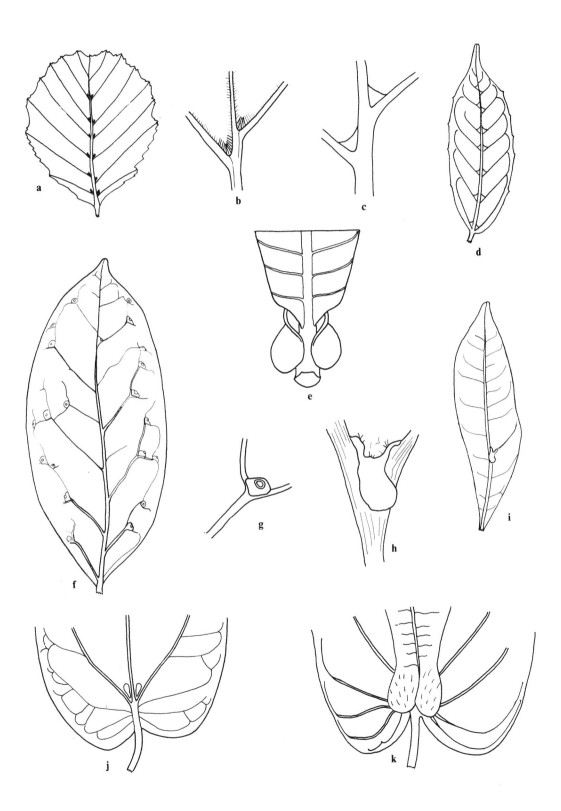

domaties en pochette of Chevalier and Chesnais; *domatium marsupiforme* of Stace. Not mentioned by Jacobs. Examples: *Maieta guianensis* Aublet (Melastromataceae); *Cola marsupium* K. Schum. (Sterculiaceae) Fig. 10.9(j,k).

4. Pits or crypts; the roof sometimes appearing as a raised dome.

Equivalent to *Grübchen* of Lundstroem; 'Group 1' of Hamilton; *fossette*, 'type G without hairs', and 'type Gh with hairs', of Penzig and Chiabrera; *bothrique* of Briquet; *domaties en pertuis* of Chevalier and Chesnais; *domatium lebetiforme* of Stace; 'type 1' of Jacobs. Examples: *Conocarpus lancifolius* Engl. (Combretaceae); *Pennantia cunninghamii* Miers (mainly in secondary and tertiary axils (Icacinaceae)), Fig. 10.9(f,g).

5. Revolute margin at base of leaf

Equivalent to *Zurückbiegungen* of Lundstroem; not recognized by Hamilton; 'type F' of Penzig and Chiabrera; *ptychique* of Briquet; *domaties en ourlet* of Chevalier and Chesnais; *domatium revolutum* of Stace; not recognized by Jacobs. Examples: *Oreomunnea pterocarpa* Oerst. (Juglandaceae); *Duroia saccifera* Benth. and Hook. f. (Rubiaceae), Fig. 10.9 (e).

It must be clearly understood that there are intermediates between the five classes of domatia that have just been mentioned. Furthermore the domatia may include hairs, either within themselves or at their rims. Sometimes intermediates between all of the types 2–4 are found together in a single species or even on one plant. An example of the latter is *Dysoxylum fraseranum* Benth. (Meliaceae) which appears, at least in herbarium specimens, to have both triangular pocket domatia and also sac-like structures which protrude and overhang the subtending secondary veins (Fig. 10.9(h,i)). However, it should be borne in mind that domatia in herbarium material may be distorted.

Most domatia occur in the primary vein axils, i.e. between lateral veins and the midrib, or between the more-or-less equal veins at the base of a palmate-peltate type of leaf. In addition, they are quite common in secondary and tertiary vein axils. They are usually solitary, although more than one may rarely be found in the same axil. The number of domatia per leaf may vary between one and a very large number, for example, up to 370 have been recorded for *Anacardium occidentale* L. by Lundstroem. Another example is *Terminalia catappa* Gaertn. in which there may sometimes be 'pit' and at other times 'pocket' domatia. This Malayan species shows a certain amount of geographical variation, according to Jacobs. Specimens from the western part of its range more frequently have two to three collateral domatia (some-

times fused) in the primary axils, than do specimens from the eastern part of its range. Schnell (1966*b*) points out the distinct polarity shown by the leaves of *Cola marsupium* K. Schum. (Sterculiaceae) and some species of *Tococa* and *Maieta* (Melastomataceae) with regard to the size of their domatia, which are large at the base of the leaf, but gradually become smaller towards the apex.

Uses in taxonomy

When considering domatia as a taxonomic character, one must bear in mind the possible confusion with galls, particularly in species whose relatives have genuine domatia (see p. 135).

Jacobs (1966) made a very valuable and detailed study of leaf domatia. According to him, 'Domatia may be of limited taxonomic significance as a supporting character'. However Professor van Steenis (1968) replied to Jacobs by calling his statement a 'truism, because hardly any systematic character has absolute value, and though even having restricted value in one group, genus, species etc., it may break down in another allied group or occur erratically in that group'. Furthermore, he says, 'This should, however, not lead to the conclusion that the constant occurrence of domatia is to be regarded only as a 'supporting character', which implies that it should have less value than the other vegetative or generative characters'. Jacobs also contends that, 'the regularity of their occurrence is also variable in single leaves, individuals, and taxa, and therefore is to be tested purposely in every case anew.'

Van Steenis also called Jacobs's second statement 'a truism, as all systematic characters must be tested purposely in every case anew, including domatiae'. As a named example where domatia are useful in distinguishing species it is perhaps rather appropriate to mention certain *Nothofagus* spp, about which van Steenis has said, 'To me the presence or absence of domatia is of vital importance for the distinction of *N. fusca* and *N. truncata*'. However, Jacobs himself gives other examples where domatia are of taxonomic use. These include *Jasminum domatigerum* Lingelsh. (Oleaceae) and *Shorea domatiosa* P.S. Ashton (Dipterocarpaceae), which have been so named because of their domatia. Recently *Macaranga domatiosa* (Euphorbiaceae) has been similarly designated by Airy Shaw (1971). According to Dr. H.O. Sleumer (in conversation with Jacobs), the closely knit group consisting of *Clethra barbinervis* Sieb. and

Zucc., *C. fargesii* Franch. and *C. sleumeriana* Hao (Clethraceae), all from south-east Asia, is distinguished at a glance from all congeners because of their domatia, which are constantly present in these species.

True leaf domatia are characteristic of the woody members of certain families, as for example, the Rubiaceae (Howard, J.E. 1862; Penzig and Chiabrera 1903; Wheeler 1942; Chevalier 1942; Beille 1947; Schnell and Beaufort 1966; Hallé 1966) and Melastomataceae (Schnell 1966*a,b*; Whiffin 1972), although they do not exist in all genera and species of these families. Furthermore, considerable variation in the morphology of these structures often occurs in a single family, i.e. tufts, crypts, domes, and pockets all occur in both the Juglandaceae (Chevalier and Chesnais 1941*a,b*) and Rubiaceae. On the other hand, morphologically similar domatia occur in very diverse families such as Combretaceae, Dipterocarpaceae (Guérin, 1906) and Hamamelidaceae (Lundstroem 1887, 1888).

A further example (Jacobs 1966) where domatia are useful in distinguishing species is that of *Rinorea* (Violaceae) which has domatia in only two of the *circa* twelve Indo-Malesian species, as in *R. longiracemosa* Craib (range Burma to Borneo and Java), which helps to distinguish it from its allies, *R. javanica* and *R. pachycarpa* Craib.

Are domatia normal or pathological structures?

Are domatia naturally occurring, genetically determined structures, or are they initiated by insects, fungi or some other outside agent? Various early authors believed that there is a close association between animal organisms, particularly mites, ants, and domatia. One of the earliest was Spruce (1896, but not published until 1908) who, as a result of his journey on the Amazon, formulated the Lamarckian hypothesis that these myrmecophytic structures, whose formation was initiated by ants, ultimately become hereditary. Lundstroem (1887) suggested that domatia may be interpreted in more than one way, viz.: (1) they may, like galls, be pathological; (2) they may serve to catch insects; (3) they may have only an indirect connection with their tenants; and (4) they may be of use to the plant as dwellings for commensals. Lundstroem adopted the last of these ideas, but also drew an interesting parallel between galls and domatia. He was also inclined to agree with the Lamarckian hypothesis. In 1884, Beccari published his beautifully illustrated work

on the ant plants of Malesia and Papua. The plants mainly belong to the Rubiaceae but also the Nepenthaceae, Melastomataceae, and Asclepiadaceae. At the time when his article was published he thought that domatia, including the swollen hypocotyls and stems, were an hereditary fixation induced by the ants. However, by 1904, he was convinced that the swellings appeared independently of action by the ants. This view was formed not only as a result of his *Wanderings in the great forests of Borneo*, but also because of the results of work published by Forbes (1880, 1885) and Treub (1888). These authors raised young *Myrmecodia* (Rubiaceae) from seed and found that the tuber is a normal production of the plant and that the galleried inner structure arises in the absence of ants. Among later authors, Ross and Hedicke (1927), de Wildeman (1938), Mani (1964), and Schnell and Beaufort (1966) also agree that domatia are not galls, but preformed structures made use of by ants.

Good evidence for the spontaneous or genetically imposed origin of leaf pocket domatia has been presented by Schnell and Beaufort (1966) and by Schnell (1966*b*). They have reminded us of *Scaphopetalum thonneri* de Wildem. et Durand (Sterculiaceae), which has alternate leaves and a corresponding alternate disposition of the unpaired pocket domatia. This strongly indicates that they originate during the apical development of the branches and that their formation is not due to external causes. The same authors also compare the tropical African genus *Scaphopetalum* with the American *Maieta* (Melastomataceae) which is heterophyllous, and has a large and a small leaf opposite each other at a node. Leaf pockets are found only at the bases of alternately large leaves. Schnell and Beaufort likewise compare the African *Vitex* (Verbenaceae) and *Canthium* (Rubiaceae) with the American *Cecropia* spp. (Urticaceae) all of which have stem myrmecodomatia whose orifices are disposed at 90° from one node to the next, i.e. alternating with the leaves and reflecting the internal weak point of the vascular tissue. Bailey (1922), commenting on *Vitex*, notes that the disposition of these pores is related to the leaf phyllotaxy (see also p. 137 of this work and Fig. 10.10(d)).

The swollen stipular thorns of *Acacia* spp have been interpreted as galls by some authors, notably Bequaert (1922, p. 373), Houard (1922, 1933), and Mani (1964), but without giving proof or naming the responsible organism. For example, Mani says they are 'caused by gall mites', Glover-Allen[1] reported a

[1] See Wheeler (1913), footnote on his p. 130.

'single small larva' inside stipular thorns of *Acacia fistula* Schweinf., and Winkler (Winkler and Zimmer, 1912) discovered a beetle larva in intact swollen thorns of an East African *Acacia*. These vague references may be discounted in view of later evidence. Most authors are convinced that the stipular swellings are hereditary, basing this conclusion on the study of germinated seedlings which developed them without apparent stimulation from outside agencies. Field observations also show that stipular thorns develop without being pierced by ants (see Ascherson 1878, Paoli 1929, Eggeling 1952, and Monod and Schmitt 1968). Monod and Schmitt have also given a good review of the literature on this topic, and they themselves were successful in obtaining swollen thorns on young plants of *A. drepanolobium* Harms under cultivation. They were unsuccessful in causing the thorns to swell on plants grown in a sterile chamber in a green-house. However, they cite Dr. M.D. Gwynne and Professor Brian Hocking as having successfully obtained them on plants under sterile cultivation after only two years. Monod and Schmitt also looked at transverse sections of these structures and found no evidence of the cellular proliferation or enlargement which are characteristic of galls. These authors refer to the swellings as 'myrmecodomatia' and 'pseudo-galls'. For further details concerning the anatomy of the stipular thorns of *Acacia* see p. 137.

Jacobs (1966) was the first investigator who attempted to demonstrate experimentally whether leaf domatia are inherited or caused by mites and he was able to conclude that the absence of mites does not prevent domatia from developing.

There are some galls which are difficult to distinguish from domatia because both the galls and the domatia occur in leaf axils of the same species (for examples see Jacobs 1966 and Darlington 1968). Galls can, however, be distinguished from domatia either because they consist of a very large number of cells or because they are made up of abnormally large cells. In order to find out whether tufts of axillary hairs are to be interpreted as galls or domatia it would be necessary to follow their development under experimental conditions.

Anatomy of domatia

Few authors have examined the anatomy of domatia. Hamilton (1896), working mainly on Australian plants, found the domatia to be 'distinguished by peculiarities in the minute structure of the part of the leaf overlying them'. The principal differences of leaf structure he mentions are that the 'spongy' parenchyma is compact (particularly at the summit of the domatium) and that the cells contain large chloroplasts. In the few species he examined, the palisade tissue above a domatium is usually composed of two compact rows of cells. In *Pennantia cunninghamii* Miers, the palisade cells are short oblong cells whose long axis is horizontal instead of vertical, bottle-shaped in *Canthium oleifolium* Hook. and of normal palisade shape but very narrow in *Dysoxylum fraseranum* Benth. Neither the external nor the internal epidermis is particularly specialized, although in some species the epidermis lining the cavity bears unicellular, rarely septate hairs without living contents. Stomata are absent from the domatia in some species but present in others. According to Lundstroem, a gall-forming infection always affects the colour and structure of the chlorophyll, but the area overlying a domatium appears to be richer in chlorophyll than other parts of the leaf. This has been confirmed by Schnell (1960) and Jacobs (1966). Greensill (1902) noted that there were no stomata lining the domatia on various species of *Coprosma* (Rubiaceae), and that chloroplasts were absent from the two hypodermal layers. Likewise, Penzig and Chiabrera (1903), in 426 species from 44 families noted that stomata were absent from or diminished in number in domatia.

Sampson and McLean (1965) reported that *Elaeocarpus dentatus* Vahl has two to four layers of sclerenchyma immediately beneath the epidermis or beneath a one-layered hypodermis on the abaxial side of the domatium and part of the roof of the domatium. I myself have noted the following interesting examples of domatia. *Arcangelisia flava* (Lour.) Merr. and *A. tympanopoda* (Laut. and K. Schum.) Diels (Menispermaceae) have bowl- or pouch-shaped domatia in the major vein axils. The epidermal cells lining the domatia are of normal size, but stomata are absent. Above the epidermis surrounding the cavities of the domatia there are several rows of thick-walled sclereids (Wilkinson 1978, her Fig. 5A). Eight species of *Anacardium* possess domatia. In each species there are a few rows of collenchyma above the abaxial epidermis around the base of the domatium, while the main body of the sac is surrounded by several rows of thick-walled parenchyma. Each of these species has a crowded mass of glands within the domatium. Similarly, *Swintonia schwenkii* T. and B. (Anacardiaceae) has domatia which are surrounded by a sac of thick-walled parenchyma cells and the abaxial epidermis lining the domatium is composed of thickly cutinized, elongated, palisade-like cells which nearly occlude the lumen.

Anatomy of Myrmecodomatia (ant domatia)

Certain plants habitually have hollow, swollen stem nodes and internodes (termed cauline hollows by some authors). The walls of these cavities are pierced by one or several apertures cut by ants, at, or a little above, or below the nodes (see Bequaert 1922; Schnell 1966; Schnell and Beaufort 1966; Hallé, 1967). These structures are common in the Fabaceae, Passifloraceae, Rubiaceae, and Verbenaceae. *Schotia africana* Baillon (Fabaceae) is an example of a species with internodal myrmecodomatia and their cavities have ant apertures towards the distal end of the internodes, where they are always on the side away from a leaf petiole. Swollen node domatia are also common in species of *Vitex* (Verbenaceae), which has quadrangular stems and opposite and decussate leaves. The ant holes occur a little below the nodes as two symmetrical pores on the two faces not carrying leaves, rotating through 90° from one node to the next. In all instances, ants are able to detect the point of least resistance (Fig. 10.10(d)).

Various myrmecophytes of the Belgian Congo, with swollen, hollow stems (occupied by ants), were studied anatomically by Bailey (1922). He examined in detail the anatomy of *Cuviera angolensis* Hiern (Rubiaceae), *Plectronia laurentii* Wildem. (Rubiaceae) (Fig. 10.10(a–c), *Barteria fistulosa* Masters and *B. dewevrei* Wildem. and Durand (Passifloraceae), *Vitex staudtii* Guerke, and *V. littoralis* Decne (Verbenaceae). Quoting directly from Bailey, the salient features he discovered were that,

In all, there is apparently an inherent tendency towards the formation of a heterogenous pith, the central succulent portion of which collapses and dries up leaving an internal chamber or cavity. They are all characterized by similar peculiarities in the differentiation of their fibrovascular cylinders, which are more or less closely correlated with phyllotaxy. Certain sides or radii of the stele tend to be thinner, to contain fewer vessels, and to differentiate later than others.

The means by which the ants are able to detect such points has been discussed by Wheeler (1913). These stem domatia also have peculiar 'callus heteroplasia', situated in the thinner sides of the myrmecodomatia. The callus heteroplasia are formed by the young cambium and cortex when these tissues are exposed by the removal of the underlying cells of the medulla and xylem. Their formation may be caused by the activities of ants or coccids and are certainly used as food by the latter. In addition, *Cuviera* sp. and *Plectronia laurentii* Wildem. have internal 'fungus gardens' (see Bailey, 1922, p. 596).

The anatomy of the stipular thorns of *Acacia*[1]

A study of the stipular thorns of *A. drepanolobium* Harms and *A. seyal fistula* Schweinf. by means of transverse sections and whole cleared preparations has been made by Monod and Schmitt (1968). In cleared preparations, they noted that in the petiolar region of young intumescences, a central vein is separated into two; each branch forms a meridian arc around the swelling, one above and one below. In turn, each of these fans out into finer veins at the periphery of the domatium wall, uniting again at the bases of the spines and passing into them (Fig. 10.10(f)). From their examination of transverse sections, these authors discovered that the large-celled medullary parenchyma begins to disappear very early, when the domatium is only 5-6 mm in diameter. By the time it has expanded to 16 mm, it is already hollowed out and the parenchyma is reduced to a simple, thin layer adhering to the domatium wall. The latter consists of an epidermis, below which are several rows of cortical collenchyma, within which is a band of sclerenchyma in contact with the numerous small vascular bundles already noted in the whole, cleared preparations mentioned above (Fig. 10.10(e)).

Note on *Myrmecodia* and *Hydnophytum* (Rubiaceae) and *Dischidia* (Asclepiadaceae)

The grossly swollen, tuberous hypocotyls of *Myrmecodia* and *Hydnophytum* are permeated by a network of chambers and passages inhabited by ants, Fig. 10.10(g). These cavities are formed by the activity of phellogen layers which arise in the parenchyma. The cells enclosed by the layers of cork subsequently die, and so cavities are formed. There are many apertures to the outside. For further particulars, see Forbes (1880, 1885); Treub (1883, 1888); Groom (1893); Scott and Sargant (1893); Beccari (1884, 1904); Rettig (1904); Miehe (1911a,b); Kerr (1912); and Janzen (1974b).

Leaves of some species of *Dischidia* are shaped like hollow bladders, which usually contain debris, largely carried in by nesting ants. Into the bladder grows an adventitious root which appears to ramify in proportion to the amount of debris present. Interesting

[1] See p. 135 for the morphology of these thorns.

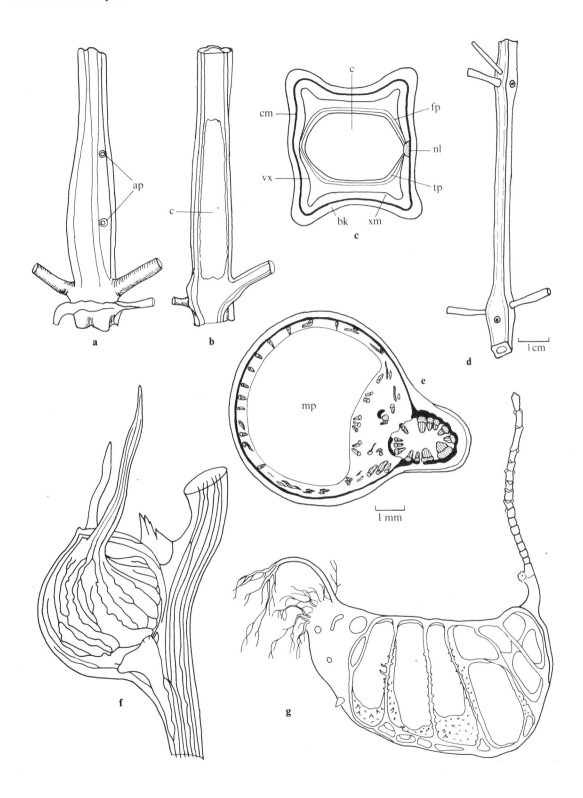

descriptions of these structures are given by Teuscher (1967) and Janzen (1974*a,b*).

Domatia versus extrafloral nectaries

According to Jacobs (1966, p. 277), 'All authors have agreed that non-glandular structures in the nerve axils at the underside of the leaves on woody dicotyledons are to be called domatia'. However, later in his own work (p. 300), when discussing *Viburnum* he says 'In *V. platyphyllum* Merrill also from the Philippines, there seemed to be a homology between the (largely) glandular formations sometimes found at the basal nerve axils and the domatia higher along the midrib. And in *V. vernicosum* from Borneo there seemed to exist a series of intergrades between the basal glands of the common orbicular type which varnish the young leaves, and the domatia of the ordinary kind in the higher nerve axils.'

Jacobs was however, unable to verify this 'suspected but seldom apparent homology between domatia and glands'.

Guérin (1906) also recorded domatia containing glands in *Shorea aptera* Burck. (now = *S. palembanica* Miq.) and *S. lepidota* Bl. I myself examined *Shorea aptera* at Kew and found that it had glands in the domatia, similar to those distributed elsewhere on the leaf.

Lundstroem (1887) reported that the domatia on the leaves of *Anacardium occidentale* L. contained glands, although Jacobs has not commented on this fact. I have personally examined this species by means of cuticular preparations and leaf sections. Distinctly pocket-shaped, marsupiform domatia were found to be carpeted with dense aggregations of glands, slightly modified from those seen scattered over the abaxial surface of the leaf. Similarly, *A. giganteum* Hancock, *A. microcarpum* Ducke, *A. microsepalum* Loes, *A. othonianum* Rizz., *A. parvifolium* Ducke, *A. pumilum* St. Hil., and *A. spruceanum* Benth. have domatia of variable morphology, but all lined, at least in part, with glands. It is interesting to note however, that *A.*

humile St. Hil. has only dense aggregations of glands in the abaxial vein axils and no domatia. *Tinospora glabra* (Burm. f.) Merr. (Menispermaceae) shows both conditions within the one species, i.e. domatia carpeted with glands in some leaves and absence of domatia but with patches of glands in nerve axils in other leaves (L.L. Forman, Kew Herbarium, personal communication).

Barros (1961*a,b*) has referred to certain structures on the leaves of *Norantea brasiliensis* Choisy as domatia. She describes them as slit-like and bordered with hairs. Nevertheless, they are not situated in the axils of veins. Similar pits on the abaxial surfaces of the leaves of *Norantea guianensis* Aubl. (Weber, 1956), *Marcgravia sintenisii* Urban and *M. rectiflora* Triana and Planch. (Howard, R.A. 1970*a*) have been described as extrafloral nectaries. Their anatomical investigations reveal an apparently glandular area of cells at and around the base of these pits, although according to Weber, nothing is known about their function. Nevertheless, since they are not in the axils of veins and are thought to have a glandular function, they should perhaps be regarded as extrafloral nectaries.

Ontogeny of axillary leaf domatia

Tô Ngoc Anh (1966 and also in Schnell, Cusset, Tchinaye, and Tô Ngoc Anh, 1968) examined the ontogeny of the leaf domatia of various species of *Coffea* and several species of temperate and other tropical plants. Her work shows the appearance in vein axils of islands of young cells whose development is delayed. In the very young stages there is nothing to distinguish them from the neighbouring tissues. She emphasizes (as have a number of other authors) the constant and remarkable localization of domatia in the axils of veins of very diverse families without taxonomic relationship. She suggests that this indicates their origin from ancient structures, like those of extrafloral nectaries. She thinks that the lack of

---

Fig. 10.10. Myrmecodomatia. (a–c) *Plectronia laurentii* de Wildeman; (a) part of a branch with swelling above a node and showing two apertures; (b) LS of the same myrmecodomatium ($\times \frac{1}{3}$, from Beauquaert 1921–2); (c) TS quadrangular, hypertrophied stem ($\times$ 2, from Bailey 1921–2). (d) Shoot of *Vitex myrmecophila* Mildbr. showing two swollen nodes with apertures disposed at 90° from one node to the next (from Schnell and Beaufort, de 1966); (e) TS stem at a young stage of development of the stipular thorn of *Acacia drepanolobium* Harms (large-celled medullary parenchyma (mp) absent from older specimens); (f) swollen stipular thorn of the same species showing vascular supply in a cleared preparation ((e) and (f) from Monod and Schmitt, 1968); (g) LS swollen hypocotyl of *Hydnophytum guppyanum* Becc. ($\times \frac{1}{3}$), Ap, aperture; bk, bark; c, cavity; cm, cambium; fp, layer of medullary tissues which consists of flattened, thick-walled cells; mp, medullary parenchyma; nl. nutritice layer or callus; tp, remains of thin-walled pith tissue; vx, vesselless xylem; xm, xylem containing numerous vessels.

secretory epidermis in domatia is no reason for weakening the comparison with extrafloral nectaries. Furthermore, as she points out, a number of extrafloral nectaries are also disposed in a hollow, and in both cases it appears that they arise in a zone of reduced development so that the peripheral tissues around this zone (which have normal development) overarch the depression. Tô Ngoc Anh also indicates that a comparison of the ultrastructure of epidermal cells of leaf domatia and extrafloral nectaries would prove valuable. In addition, she draws a comparison between the development of domatia and the epiphyllous buds of *Cardamine* and certain ferns, the retarded development of buds, and the 'prickles' of *Mourera fluviatilis* Aubl. (Podostemonaceae). Finally, Tô Ngoc Anh concludes from her investigation that domatia arise spontaneously, i.e. are genetically inherited.

Bequaert 1921-2; Briquet 1920; Chevalier 1942; Chevalier and Chesnais 1941*a,b*; Darlington 1968; Eggeling 1952; Forbes 1880, 1885; Greensill 1902; Groom 1893; Guérin 1906; Hallé 1967; Hamilton 1896; Houard 1922, 1933; Howard, J.E. 1862; Howard, R.A. 1970*a*; Jacobs 1966; Janzen 1966, 1967*a,b*, 1974*a,b*; Kerr 1912; Lundstroem 1887, 1888; Mani 1964; Miehe 1911*a,b*; Monod and Schmitt 1968; Paoli 1929, 1930; Penzig and Chiabrera 1903; Rettig 1904; Rocchetti 1905; Ross and Hedicke 1927; Sampson and McLean 1965; Schnell 1960, 1966*a,b*; Schnell and Beaufort, de 1966; Schnell, Cusset, Tchinaye, and Tô Ngoc Anh 1968; Scott and Sargant 1893; Spruce 1908; Stace 1965*a*; Steenis, van 1968; Teuscher 1967; Tô Ngoc Anh 1966; Treub 1883, 1888; Weber, H. 1956; Wheeler 1913, 1942; Whiffin 1972; Wildeman, de 1938; Wilkinson 1978; Winkler and Zimmer 1912.

Literature cited

Airy Shaw 1971; Ascherson 1878; Bailey 1922; Barros, de 1961*a,b*; Beccari 1884, 1904; Beille 1947;

PART V : THE CUTICLE

Introduction

The **cuticle** (cuticular membrane) is basically a two-layered sheet of non-cellular material which covers all mature parts of the shoot, floral parts, and trichomes. A cuticle is believed to be present on the roots of various angiosperms according to Scott, Hammer, Baker, and Bowler (1958) and Scott (1963, 1966), but other authors such as Bonnett and Newcomb (1966), have not found this to be true for other plants. The major chemical components of the cuticle are lipids, waxes, and cutin, the last being a bipolymer comprising fatty and hydroxy-fatty acids. The chief structural component is cutin with waxy substances or wax precursors embedded within the membrane and exuded over its surface. Small amounts of other substances, such as phenolic compounds are also occasionally found in the membrane.

For a complete history of the study of cuticle, readers should consult works such as those by Edwards (1935), Roelofsen (1952), Stace (1965*a*), Martin and Juniper (1970), Napp-Zinn (1973), and Sargent (1976*c*). This last publication contains a useful set of diagrams illustrating the evolution of ideas on the structure of the cuticle and also a glossary of terms applied to the cuticle by various authors. Only a very brief history is given here.

It is significant that Brongniart (1834), who was primarily a palaeobotanist, was the first to report that he had separated the cuticle from the epidermis of leaves. In 1832, he 'macerated' a large number of leaves of both dicotyledons and monocotyledons by leaving them to rot in water. From these rotting leaves the cuticle separated intact. He described this non-cellular 'pellicule' (the cuticle), as transparent, colourless or pale grey, sometimes showing a network of lines indicating the junction of epidermal cells underneath, sometimes granular and with elongate openings corresponding to the position of many stomata. He realized that the cuticle covered most parts of plants. He was also aware of the work of Henslow (1832) who isolated cuticle from the corolla, stamens, and style of *Digitalis purpurea* L. with the aid of dilute nitric acid. Further progress was made by von Mohl (1842-52), who first drew attention to the anatomical difference between the cuticle proper and the cuticular membrane. This was the result of investigating the cuticle in transverse sections (of which he produced some fine diagrams), as well as in surface view.

In spite of the work of these authors, most botanists at that time dismissed the cuticle merely as a secretion from the epidermal cell walls (Treviranus 1835; Schleiden 1861) or as a thickening of the outer wall (Meyen 1837 and Wigand 1854).

The knowledge of this separable leaf cuticle was of considerable value to palaeobotanists, and it was in this field that further progress was next made. The following selected examples are taken largely from Edwards (1935) and Stace (1965a). One of the earliest references to fossil cuticles is by Brodie (1842) who remarked 'When the sandstone is freshly broken, the epidermis of the fossil frequently peels off'. Weber (in Wessel and Weber 1855) examined some leaves from browncoal deposits and observed in the transparent epidermis the form and arrangement of stomata and epidermal cells. He first named the leaves from their external appearance and then proceeded to compare the cuticles with those of living species, and found that the structure confirmed the identifications. An early systematic treatment of cuticles was that of Bornemann (1856) on fossil cycad cuticles. From the early twentieth century up to the present time, a succession of classic works appeared, which included the numerous studies of fossil and recent plant cuticles, namely, those by Nathorst (1907-12) on cycads and lycopods, by Florin (c. 1920-58) on gymnosperms, and finally by Harris (c. 1926–present, particularly 1956, 1961, 1964, 1969, and Harris, Millington, and Miller, 1974) on the Jurassic flora of Yorkshire. The first really significant work on fossil and recent dicotyledonous cuticles was that of Bandulska (1923-31). She investigated the cuticular anatomy of coniferous and dicotyledonous leaves from the Eocene (Lutetian) flora of Bournemouth, England. Before Bandulska could attempt to identify these fossil cuticles, she had to make an intensive cuticular study of certain present-day dicotyledonous families whose presence in England during the Eocene was indicated on various grounds. She was able to assign some of the fossil cuticles to clearly recognizable genera in the Lauraceae: *Aniba*, *Lindera*, *Litsea*, and *Neolitsea* (1926) and *Cinnamomum* (1928) and others in *Fagus* and *Nothofagus* to the Fagaceae (1928). Many of the other cuticles remain unidentified, or of uncertain identity, even today, largely due to the lack of sufficient information about present day dicotyledonous leaf cuticles.

By way of complete contrast Odell (1932) in a lengthy paper attempted to prove that epidermal and cuticular features of angiosperms are unsatisfactory for diagnostic work. The mistake that Odell made was to take each cuticular feature separately, and then, on grounds of variation, or inapplicability in individual cases, to reject it as useless in taxonomy. In addition, she failed to distinguish characters which might be of family, generic, or specific rank.

It would not be proper to leave this brief history of the anatomical study of leaf cuticles without mentioning the work of Solereder (1908). He laid down some sound guiding principles; for example, he speaks of the varying systematic value shown by individual cuticular characteristics and says that one should take all the features into consideration and test their systematic value in each case (see also p. 134). Over the last 25 years, there have been numerous papers on cuticular studies. Unfortunately many of them are descriptions of only a few species, and often without reference to classical taxonomic relationships, or to other disciplines such as palynology or cytology.

The ultrastructure of cuticle has also received the concentrated attention of many workers. Reverting to the concept that the cuticle is two-layered (see p. 140), this view was first based on studies with the polarizing microscope by numerous workers including Frey (1926), Anderson (1928, 1935), Meyer (1938), Roelofsen (1952), and Sitte and Rennier (1963). These investigators suggested that the cuticle is attached to the outer part of the walls of the epidermal cells by an intermediate zone of pectinaceous material. This has since been confirmed by means of transmission electron microscopy (Roelofsen 1959; Sitte and Rennier 1963; and Juniper and Cox 1973).

Priestley (1943) considered that the outer epidermal wall is relatively complex in nature and that it shows gradation, usually from a pure cellulose wall, facing the epidermal protoplast, through layers containing varying quantities of pectic and fatty substances to an outermost sheet of cuticle free from cellulose and usually free from pectic compounds. He refers to the outer layer as 'cuticle' and the inner zone as the 'cuticular layer'. The term 'cuticular membrane' was used by Roelofsen (1952, 1959) for the two-layered membrane lying above the pectic layer. According to him, this is composed of an inner cuticular layer consisting essentially of a cellulosic framework incrusted with cutin between its cellulose microfibrils, and an outer, usually thinner part of the cuticle proper, which is made up chiefly of cutin adcrusted (Sitte 1955) onto the cuticular layer. Esau (1958) substituted the term cutinization for incrustation and cuticularization for adcrustation. The terminology of Roelofsen and Esau was adopted by Stace (1965a), whose diagram illustrating the gross zonation is shown in Fig. 10.11(a) and by Martin and Juniper (1970) (their Fig. 11C). However, Sitte and Rennier (1963) object to the cuticle and cuticular layer, preferring cuticle proper for the first and cuticular layers for the second. They object to the term cuticular layer, which they consider unreliable, since this term has been used to denote layers which contain cellulose (in addition to cutin and wax) and this has a diffuse distribution of uncertain limits. They suggest that their term, 'cuticular layers', would indicate cutin-containing layers, regardless of whether they contain cellulose or not (see Fig. 10.11).

Those who are concerned with the ultrastructure of the cuticle, particularly in association with its wax content and formation, tend to prefer developmental terminology. Schieferstein and Loomis (1959) comment that the primary cuticle (first formed) appears to form rapidly on the walls of any living cell which is exposed to air. They state that 'The formation of the primary cuticle is normally followed by secondary developments involving the deposition of waxes, and perhaps cutin, within and beneath the primary cuticle. The outer portions of the epidermal wall, or the entire wall, may be infiltrated with waxes.' Sargent and Gay

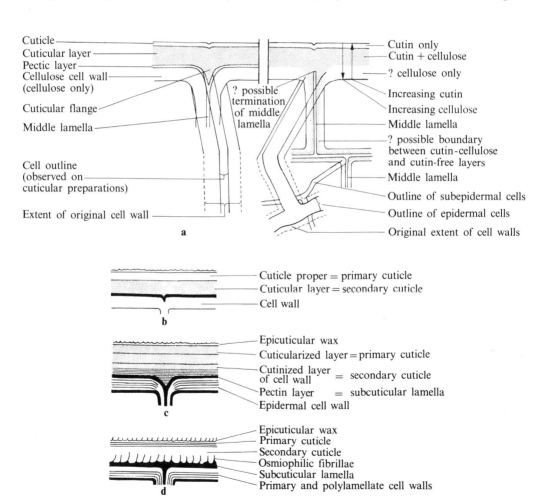

Fig. 10.11. Various interpretations of the cuticular and outer periclinal cell wall layers. (a) from Stace (1965a); (b) from Sitte and Rennier (1963) modified by Sargent (1967c); (c) from Martin and Juniper (1970) modified by Sargent (1967c); (d) from Sargent (1967c). Further diagrams giving other interpretations can also be found in Sargent (1976c).

(1977) point out that the primary cuticle as defined by Sargent (1976*a,b*), formed during epidermal differentiation, should not be confused with the more rigid secondary cuticle which is intussuscepted between the primary wall and cuticle later in differentiation. Sargent (1976*a,b,c*, and personal communication), states that the secondary cuticle is 'cutinous material differentiated external to the primary cell wall and traversed by osmiophilic fibrillae' (Fig. 10.11(d)). At the same time, she considers the secondary cuticle to be quite distinct from the cutinized cell wall which may be present below it. According to these definitions, the cuticular membrane familiar to anatomists is composed of the outer primary cuticle (cuticle proper), the secondary cuticle and, in addition, the cutinized part of the cell wall below it, which may also include the subcuticular lamella. In most present-day anatomical descriptions, the cuticular membrane is usually referred to simply as 'the cuticle'. At present authors do not agree as to whether the cutin and wax precursors reach the leaf outer surface by migrating throughout the outer epidermal cell walls, or by means of special channels, such as **teichodes** (formerly called **ectodesmata**; see Franke 1971; Esau 1977).

Recently, Carr, Milkovits, and Carr (1971) coined the term **phytoglyph** for the microanatomical features of the surface of leaves, including the microanatomy of the cuticle. Phytoglyphic analysis relates to the combination of three methods; light microscopy of stained cuticles, scanning electron microscopy of leaf surfaces, and light microscopy of thin sections of the cuticle and associated structures. The microanatomy of leaf cuticles is considered in the next section under Cuticular ornamentation.

Cuticular ornamentation

Although various early botanists and palaeo-botanists appreciated the diagnostic use of certain cuticular characters such as stomata, trichomes, and cell-wall outlines (see review of literature by Stace 1965*a* and Sinclair and Sharma 1971), very few considered cuticular ornamentation.

Ornamentation of the outer surface of the cuticle largely consists of striae, which are ridges or folds of cuticle. These may occur on the flat surface of the cuticle, or on trichomes and papillae. Papillae themselves are another form of ornamentation which is considered separately on p. 148. The morphology of the inner surface of the cuticle (p. 151) and cuticular thickness (p. 160) are also treated separately.

External cuticular surface

Illustrations were made of the striae on the adaxial epidermal cells of *Helleborus foetidus* L. by von Mohl as early as 1842 (his Figs. 5-7). In 1884, de Bary reported that the free outer walls of *Rochea coccinea* DC. (Crassulaceae) were warty, that those of *Helleborus niger* L., *H. foetidus* L., and *Dianthus caryophyllus* L. had broad, blunt bands (striae), while those of *Rumex patientia* L. and *R. obtusifolius* L. had thin, sharp bands. He furthermore remarked that, 'The bands often run nearly straight and parallel, and are then usually longitudinal relative to the whole body, ... not uncommonly they are wavy and branched (e.g. *Helleborus, Pyrus communis* L.), and in the majority of cases they are continuous from one cell to the next.'

Investigators may be surprised at the appreciation of the value of cuticular ornamentation already shown by Solereder in 1908, as shown in the following extract.

The cuticle varies in thickness and, as seen in surface view, is either smooth or provided with granular or verrucose thickenings, or striated. The thickness of the cuticle as well as the degree of marking on its surface may, in extreme cases, be utilized for systematic purposes; in other cases a certain amount of discretion is necessary, i.e. these features should not be employed until abundant material of the species in question has been examined; the two features, moreover, are not always developed in the same way on the two surfaces of the leaf. A more important character from the systematic point of view is the kind of marking on the cuticle.

In other parts of his book, Solereder makes it quite clear that he considered that markings on the cuticle 'are usually solely of value in specific diagnosis'. Certain of the 'extreme cases' which he mentions are good examples of the qualitative nature of some cuticular markings. One of these is that of deep folds, 'which resemble mountain ranges in a surface view of the leaf, and in transverse section appear as prongs or rods, which vary in size and are often branched. They are especially well developed on the lower side of the leaf in all species of the genus *Oxythece*.' The s.e.m. representation of the abaxial surface of *Oxythece leptocarpa* Miq. (Plate 6(H)) bears out Solereder's description. In the Capparaceae, Solereder noted that 'the varied marking of the cuticle' is one of the characters useful for specific diagnosis. Elsewhere he remarks, 'It is a noteworthy fact that the cuticle is not striated in any of the Aristolochiaceae, but it frequently has a granular structure.' Likewise in the Hamamelidaceae he comments that it shows a 'Cuticle strongly developed and granular in *Bucklandia, Corylopsis, Diocoryphe, Rhodoleia* and *Trichocladus.*'

Again, according to Solereder, Brändlein (1907) described some 17 genera of the Flacourtiaceae (Samydaceae) as having a cuticle which 'may frequently be compared with the markings on an etched glass plate, so that in these cases we may describe it as "etched"; but striation and granulation also occur'. Obviously this is an interesting example which needs reinvestigating. Hüller (1907) noticed that the cuticle of '*Polemonium humile*', and unnamed species of *Collomia*, *Gilia*, and *Phlox*, may be finely or coarsely striated. He also recorded that *Gilia intertexta* Steud., *G. leucocephala* A. Gray, *G. minima* A. Gray, and *G. navarretia* Steud. have a granular or verrucose cuticle. Such a description certainly indicates that the cuticular ornamentation is of no diagnostic value at the family level, but the degree of use at the generic and specific levels was not considered further.

In the late nineteenth and early twentieth centuries, the first atlases for use in pharmacognosy were published (e.g. Berg 1861; Vogl 1887; Moeller 1889; Tschirch and Oesterle 1893–1900; Greenish and Collin 1904). A little later, numerous papers and some theses were published on the anatomy of leaves used as drugs (e.g. Guérin and Guillaume 1908; Kurer 1917; Levin 1929; Dewar 1933, 1934a,b; Wallis and Saber 1933; Dewar and Wallis 1935; Rowson 1943a,b, 1946; Forsdike 1946; Wallis 1946, 1952; see also p.154). However, only a few articles actually make use of striations and granulations of the cuticular surface. The thesis of Kurer (1917) was upon the diagnostic use of cuticular striae and protuberances on hairs and epidermal cells, to distinguish adulterants from official leaves of certain compounds. He was particularly interested, for example, in distinguishing the adulterants in china tea (14 adulterants), tobacco, senna, *Digitalis* (15 adulterants), *Mentha*, and *Atropa belladonna* L. He was able to distinguish all of the 14 adulterants of *Digitalis purpurea* L. from the official drug, on the presence, absence, or position of striae and granules on the leaf and hair surfaces and on trichome morphology alone. Apparently the single adulterant of *Atropa belladonna* was *Solanum nigrum* L. According to Kurer, the cuticle on both sides of the leaf of the first of these two species is striate, while that of *Solanum nigrum* is smooth. In addition, the hairs of the *Atropa* have a smooth cuticle, while the uniseriate hairs of the *Solanum* have a tuberculate cuticle. Similarly, Dewar and Wallis (1935) found that the cuticle is striate in *Digitalis thapsi* L. but smooth in *D. purpurea* L., *D. lanata* Ehrh. and *D. lutea* L. It is interesting to note that Paganelli Cappelletti (1975) reinvestigated the leaf

cuticular characters of *Atropa belladonna* under the s.e.m. and his final remark is that observations of cuticular striations are of great usefulness for diagnostic purposes, and enable one to distinguish even small leaf samples of this species from adulterants. He appears to have reached this conclusion without having seen the thesis of Kurer, or considered trichome ornamentation.

Martens (1931a,b, 1933a,b; 1934a,b,c, 1935) studied in detail the cuticles of the petals and staminate hairs of *Tradescantia* in the hope of discovering why they become striated. He supposed that there might be a relationship between protoplasmic streaming and cuticular marking. Moreover he thought that the form of the folds was determined by the rate of their formation, as well as by the direction of cell stretch during the formative period. For example, parallel striation would result from a one-sided direction of cell-wall growth. Van Iterson (1937) repeated Martens's experiments and added some interesting ideas of his own. He noted that the time and place of development of the cell helped to determine the exact pattern of cuticular striations. He also considered that structural peculiarities of the cuticular layer could be a cause of the pattern of folds.

Returning to the possible taxonomic use of cuticular ornamentation, we find that Lavier-George (1936) used cuticular features as one of the characters of diagnostic value to separate species of *Philippia* (Ericaceae). She compared homologous leaves of *Philippia* from various sites (her p.176). She also appreciated the requirements of a 'good' taxonomic character (her p.197) for she says 'Les caractères qualitatifs des épidermes, seuls utilisables en systématique, et précisant les constantes spécifiques de ce groupe, peuvent se résumer dans le tableau ci-dessous'. In addition, she wisely combined a study of these characters with those of the stamens and stigma which are accepted as being taxonomically stable. In section D of her key there are two examples where presence/absence, or type of striae were used to separate species, i.e. cuticle striate in *P. danguana* H. Perrier de la Bâthie, cuticle smooth in *P. ibytiensis* H. Perrier de la Bâthie; striae branched in *P. ciliata* Benth., and striae parallel in *P. jumellei* H. Perrier de la Bâthie. In addition to these examples, she was able to distinguish other features of cuticular striae which I mention here because, from my experience of studying leaf cuticles, I consider they could be qualitative rather than quantitative. For example, the cuticle of the adaxial cells of *P. myriadenia* Baker has many short striae confined within each cell boundary, whereas in *P. capitata*

Baker there are long, continuous striae which pass over several cells. In *P. andringitensis* H. Perrier de la Bâthie the striae are fine and warty. (compare (B), (D), and (E) on Plate 6). In addition she also seems to have been able to distinguish (1) fine concentric striae confined to each cell, as in *P. lecomtei* H. Perrier de la Bâthie; (2) random orientation of striae confined to cells as in *P. ciliata* Benth. I have not myself been able to obtain material to verify these last 'types'.

An important work on cuticular studies was that of Stace (1965*a*; see also p. 151 of this work). His opinions, based on the investigation of the cuticles of mangrove families, concur very closely with my own which are based on a similar investigation of the Anacardiaceae (Wilkinson 1971). Some of Stace's remarks concerning striations are particularly pertinent. For example he considered striations to be of great taxonomic value in some groups, e.g. they are highly characteristic of three or four species of *Macropteranthes*. There is, however, considerable variation in the degree of development of the cuticular striations in some species. Stace also warns us about the variability of striations. He examined a single specimen of *Combretum molle* R.Br. ex G. Don, which was the only known gathering of the species from South West Africa. The specimen

Table 10.1 *Principal types of cuticular folding*

Type of folding	Examples of species where noted
(1) Striae (These are very common)	
a. Striae long and parallel	*Philippia capitata* Baker (Ericaceae); Plate 6(D)
b. Orientation random	*Rhus typhina* L. (Anacardiaceae); Plate 6(A)
c. Striae short; not overlapping the cell	*Philippia myriadena* Baker (Ericaceae); Plate 6(B)
d. Striae fine and warty	*Philippia andringitensis* H. P.de B. (Ericaceae); Plate 6(E)
e. Striae fine but not warty; undulate etc.	
(2) Filigree	
a. Coarse and dense	*Choerospondias axillaris* (Roxb.) Burtt & Hill. (Anacardiaceae); Plates 4(H) and 6(C)
b. Fine	*Lannea alata* (Engl.) Engl. (Anacardiaceae)
(3) Reticulate: with crests and buttresses	
a. Coarse	*Hevea brasiliensis* Muell-Arg. (Euphorbiaceae); Plates 4(F) and 6(C)
b. Fine and open	*Oncotheca balansae* Baill. (Oncothecaceae)
(4) Ridges (wider than ordinary striae)	
a. Roundish in circumference	*Lewisia cotyledon* B. L. Robinson (Portulacaceae); Plate 6(F)
b. With tall, deep sulci at sides	*Oxythece* (= *Neoxythece*) *leptocarpa* Miq. (Sapotaceae), Plate 6(H)
(5) Wrinkles	
a. Widely spaced and fine.	*Pseudosmodingium perniciosum* Engl. (Anacardiaceae); Plate 4(E)

is particularly distinctive in its very conspicuously striated cuticle, the striations being so strong as to obscure the cell outlines. However, he later examined several specimens of the same species from tropical south Africa, where this species is abundant. The degrees of striation ranged from the pronounced type just described to the presence of only a few striations placed radially to each trichome base. This last situation was noted in many or even most species of the genus. The two extremes in this series have an entirely different aspect, since the cell outlines are obscured on the one hand and clearly represented on the other. Stace comments that the cause of the differences is not known, but 'it would seem to be environmental as the climate of the South West African habitat is relatively very dry'. In my own research on the leaf anatomy of the Anacardiaceae, I found a parallel example in *Pistacia terebinthus* L. This species shows a wide range of cuticular ornamentation from coarsely striate to smooth. From the scanty information gleaned from herbarium labels, it appeared that there might be some association with habitat conditions such as the amount of rainfall and the nature of the soil. However, Bergen (1904) noticed that the sun leaves of *Pistacia* were striated while those of shade leaves were not. In contrast to this, Ahmad (1962) said that he had examined herbarium specimens of *Cestrum diurnum* L., *C. parqui* L' Hérit. and *C. purpurescens* (? *C. purpureum* Standley) collected from different parts of India and from abroad and that his study of their striations 'clearly brings out that different climatic and adaptive factors do not affect their development and form'. It is possible that the amount of striation may therefore be a stable, reliable character in one taxon but not in another.

After studying 226 species of dicotyledons, Dunn, Sharma, and Campbell (1965), reported that the most xerophytic species generally had the most wrinkled cuticle, while the most mesophytic or water-loving species had the smoothest surface.

Mueller (1966a,b) studied cuticular patterns in *Vaccinium* (Ericaceae) and found that adult foliage, either from the same plant in different years or from greenhouse cuttings, is recognizably the same. The differences appeared to be in the density and prominence of the striations. He also found that suspected hybrids had variable cuticular patterns which always differed from that of the parents, and furthermore, that hybrid patterns were not alike from one population to another, nor were they always intermediate. He suggests that the cuticular pattern for a species is under rather strong genetic control.

Rather conflicting results with regard to striation (and granulation) have recently been obtained by Baas and his associates. Following a detailed examination of 25 genera in the Icacinaceae, van Staveren and Baas (1973) reported that some genera show a rather constant kind of striation (or granulation) of the cuticle, whereas in other genera, such characters may be restricted to certain species only. Jansen and Baas (1973) examined *Kokoona* and *Lophopetalum* (Celastraceae) and in their report refer to two collections of *Lophopetalum pallidum* Laws., one (KL 156 1566) with a coarsely granular cuticle, and the other, (b.b. 18997) with a striated cuticle. It is not unknown for a species to have both features on a single specimen. This is hardly surprising since striations are a feature on the outside of the cuticle and granulation is usually on the inner surface (see p. 154). Similarly, in the Aquifoliaceae, Baas (1975) found variation in the cuticles of *Phelline* and *Sphenostemon*, although the monospecific *Oncotheca* had a smooth cuticle in the three specimens examined. On the other hand, he discovered a distinctive cuticular type in the monospecific *Nemopanthus*, which has conspicuous, rather widely spaced wrinkles on the abaxial surface (his Plate 6, Fig. 3). A very similar cuticular striation was found on the leaves of *Pseudosmodingium perniciosum* Engl. (Anacardiaceae; Plate 4(E)) by Wilkinson (1971).

It seems desirable that types of striation should be classified for ease of description and to make for greater clarity. This has been done in Table 10.1, which should be considered together with Plates 4 and 6. No doubt there are many other variants in each of the above groups and likewise, in time, other groups will be discovered. The cuticular ornamentation of *Hevea brasiliensis* was described by Rao A.N. (1963) as reticulate and has been included in the above types.

Since the early days in which leaf cuticles have been examined, most anatomists have noticed that striae radiate from, or are concentric around, stomata. It is now known that the earlier formed stomata are the larger ones with the longer, more conspicuous striae (Dunn, Sharma, and Campbell 1965, p. 189). Striae which radiate from all points of the stomata are the most common, particularly when the stomata are anomocytic. Sometimes striae occur characteristically as two wings (alae) on either side of a stoma (Plate 3(E)), often when the subsidiary cells are paracytic. Less common examples of concentric (Plate 3(A)), or of both concentric and radiating striae (Plate 3(D)). These last two examples can be of diagnostic value.

In spite of the foregoing emphasis on cuticular striations, it should be recognized that the outer surface of the cuticle may be entirely smooth (Plate 4(A)).

Trichome ornamentation

Concerning the ornamentation of hairs, de Bary (1884 p. 73) commented 'What has been said of the walls of the epidermal cells holds in the main for those of hair-structures. Projections of the outer surface, in [the] form of ridges, warts, or even of those sharp prickles represented in Fig. 21 B, appear in hairs more commonly than in the epidermal cells.' This subject is treated briefly by Uphof, Hummel and Staesche (1962, pp. 39–41).

Trichome ornamentation is rarely used in leaf anatomical diagnoses, but is well worth consideration for diagnostic value between families or genera. It might be possible to classify hair ornamentation in some manner, such as that suggested below.

A. Smooth
e.g. *Anamirta cocculus* (L.) Wight and Arn., Menispermaceae (Plate 3(F)).
B. Striate
(i) Fine, not spiral, e.g. *Achillea ageratifolia* Benth. and Hook. var. *serbica* (Asteraceae).
(ii) Coarse spiral striae with knobs, e.g. *Garrya elliptica* Dougl. ex Lindl. (Garryaceae; Plate 3(G)).
C. Warty
(i) Long warts arranged in longitudinal rows, e.g. *Heeria argentea* (E. Mey.) Kuntze (Anacardiaceae, Plate 3(B)).
(ii) Roundish warts, e.g. *Hydrophyllum canadense* L. (Hydrophyllaceae; Plate 3(H)).
D. Wax particles
(i) Fine flakes, appearing as fine striae; spiral, e.g. *Gunnera tinctoria* Mirb. (Plate 10(B)) and less distinctly in *G. manicata* Linden.
(ii) Other types.

This classification will undoubtedly need to be augmented after further investigations have been made.

Van Staveren and Baas (1973, p. 353) use the character of presence/absence of cuticular markings on hairs in their synoptical key for the Icacinaceae. In this key, 6 genera are recorded as having hairs with cuticular markings and 13 genera have hairs with a smooth cuticle. However, one awkward genus fell into both categories: *Apodytes brachystylis* F. Muell. has hairs with verrucose markings, while *A. dimidiata* E. Meyer ex Arn. has hairs with or without such markings.

Other external markings

Although much attention has been paid to striae, there are other cuticular markings which can be of diagnostic value.

The line of contact between the anticlinal flange and the periclinal wall may be indicated by a groove in the outer surface of the periclinal wall. On the other hand the anticlinal wall may be represented by a raised ridge above the cuticular surface, which in turn may be notched in the midline, or raised into an angular projection, or alternatively just a gently sloping bank (see 'Cuticulae' der C.I.M.P., 1964, Plate 1, Figs. A,B). Von Mohl (1842) observed that anticlinal walls could be raised as ridges above the general level of the cuticle. His examples are all monocotyledons. His illustration of the cuticles of *Phormium tenax* Forst. (his Fig. 31) and of *Ruscus aculeatus* L. (his Fig. 32) in transverse section also clearly shows anticlinal walls raised above the general cuticular level.

Lange (1969) made a brief summary of types of external marking, which he divided into (A) Gross morphology and (B) Microrelief. He also drew attention to certain less common features such as grooves, concavities, convexities, and sacs. The following examples are all taken from his work.

(1) Grooves may occur in the midline of the area where the anticlinal cuticular wall (flange) joins the periclinal wall, e.g. in *Hakea salicifolia* (Vent.) B.L. Burtt (Proteaceae).
(2) Concavities, i.e. depressions in at least the central region of the cell, e.g. *Dysoxylum fraseranum* Benth. (Meliaceae).
(3) Convexities, i.e. roundish cells which form a low dome, e.g. *Acacia melanoxylon* R.Br. (Fabaceae).
(4) Sacs i.e. oblong cells slightly raised.

With regard to microrelief, he has observed a range from smooth through degrees and varieties of granulation, e.g. fine in *Dissiliaria tricomis* Benth. (Euphorbiaceae), coarse in *Pittosporum undulatum* Vent. (Pittosporaceae) and warty (a monocotyledonous example). In addition to this he discussed internal ornamentation (see p. 155).

Alveolar material

Bailey and Nast (1944) reported that, in the Winteraceae, an alveolar layer may be present, overlying the cuticle proper; the composition is apparently cutinaceous. (For its relation to stomatal plugs, see p. 109.) This cuticular feature was studied in much greater detail by Bongers (1973) and the following information has been obtained from his work.

Alveolar material occurs in some or all species of six genera of the Winteraceae, although absent from some species of *Drimys* and *Bubbia perrieri* Capuron. In cuticular macerations examined by light microscopy it is impossible to establish whether the granular appearance is due to the internal structure of the cuticular layer or to alveolar material on the outer surface. Under the light microscope, alveolar material appears as fine to coarse granules, and therefore should be checked in transverse section and, if possible, by s.e.m. examination (Plate 2(H)). Furthermore, in cuticular preparations, the features of the underlying cuticle proper and the cuticular layer are often obscured by this material. With the use of the s.e.m., this material appears as a three-dimensional reticulum.

Bongers was able to classify the distribution and qualitative nature of alveolar material in the Winteraceae as follows.

A. Distribution
    (1) Absent.
    (2) On rims of stomata only.
    (3) Present as stomatal plugs.
    (4) Extending over neighbouring cells.
    (5) Covering the whole epidermis.
B. Qualitative nature
    (1) Homogeneous, i.e. alveolar material without protruding, solid pieces of cutin.
    (2) Heterogeneous, i.e. with protruding, solid pieces of cutin.
    (3) Alveolar material with tectum-like structures.

These categories were used by Bongers in his generic descriptions.

Domes and papillae (see also pp. 45, 105)

These are projections of the outer periclinal cell wall, including the cuticular membrane and cuticle proper. They often also include the cell lumen. Less commonly they are solid, being largely composed of cuticle and cellulose. Papillae may range in prominence from mere domes (sub-papillose of some authors), through a continuous series of heights until they are more appropriately classified with trichomes (Fig. 10.13). There is no clear distinction between unicellular trichomes and papillae, but the latter are not known to be composed of more than one cell. Barthlott and Ehler (1977) have suggested a series of terms for the curvature of the outer periclinal wall (i.e. convex, half-sphere, dome, cone, papilla, hair-papilla, and hair), based on the length-to-breadth ratio of the projection. Papillae may be smooth-walled or striate and in the more elaborate forms, 'coronulate', lobed, or branched, and with or without a waxy covering. Occasionally papillae may arise from only the central region of the outer cell wall but often the entire width of the outer wall is involved. In a few examples, there are several papillae per cell. Papillae are particularly common on the abaxial surfaces of leaves, sometimes imparting a greyish, white, or dull appearance to the surface. Papillae adjacent to stomata may be more elaborate than those over ordinary epidermal cells. They may even be interlocking. Sometimes veins or leaf margins are papillose and areolar cells are not, while at other times the converse is true (Plate 5(B,H)).

A good review of the literature on this subject is given by Napp-Zinn (1973, 1974). In the present article, a few examples of the different types of papillae which may have diagnostic value are given.

Distribution of papillae among dicotyledonous families

Papillae seem to be distributed among 'primitive' and 'advanced' dicotyledonous families alike. In addition, they appear to be present in some water plants, herbs, shrubs, lianes and trees, and also rather more frequently in tropical species. With regard to supposedly primitive examples, they are present on *Nelumbium speciosum* Willd., *Helleborus niger* L., *H. macranthus* Guerke and *Drimys brasiliensis* Miers (*D. winteri* Forst.). Certain members of the Menispermaceae such as *Menispermum canadense* L., *Stephania japonica* Miers, and *Diploclisia glaucescens* (Bl.) Diels are all lianes, the last two tropical. On the other hand, they occur in *Oxalis carnosa* Molina, *Reseda lutea* L. (at leaf margins), *Thymus vulgaris* L. and *Salvia* spp., all of which are herbs. They also occur in various *Philippia* and *Rhododendron* species which are shrubs and small trees. Many members of the Euphorbiaceae, e.g. *Macaranga indica* Wight (a tropical tree) and *Manihot coerulescens* Pohl (a tropical shrub) also have papillose leaves. Many members of the Ebenaceae, Simaroubaceae, Lauraceae, and woody Fabaceae such as *Dalbergia, Sophora,* and *Swartzia,* as well as *Melanochyla* and *Semecarpus* (Anacardiaceae), all of which are tropical and subtropical trees, likewise have papillose leaves.

Function and diagnostic value of papillae

Haberlandt (1914) thought that papillose epidermal cells might have the function of concentrating the limited light below the canopy of the tropical rain forests by acting as lenses. In *Aquilegia vulgaris* L. and *Vinca major* L. and also certain monocotyledons, the epidermal cells have a central protuberance which resembles a concavo-convex lens. He points out that one would expect shade leaves to have an adaxial epidermis with well-developed power of light perception and quotes Gaulhofer (1908) as having found this expectation to be realised in a great many examples. He also says that if papillae are present in both sun and shade leaves, as in *Cercis siliquastrum* L., *Prunus padus* L., *Fagus sylvatica* L. etc., their light-condensing action is more powerful in the case of shade leaves. Both Haberlandt and Stahl (1896) refer to the leaves of some tropical-rain-forest plants as being 'velvety' (e.g. Melastomataceae, *Ficus barbata* Wall., *Cissus discolor* Blume, *Begonia* spp) due to the presence of conical papillae. Stahl not only supported the 'absorption of light' theory, but also suggested that the papillae facilitated the removal of water from the leaf surface, obviously 'useful' to a tropical-rain-forest plant. Brenner (1900) reported that *Crassula portulacacea* Lam. has non-papillate cells under normal conditions but develops papillae when grown in damp situations, which would also suggest that, in this example, the presence or absence of papillae is an unreliable diagnostic character.

With regard to the Begoniaceae, Solereder (1908) comments that a papillose epidermis is very widely distributed, but that it is not of systematic value for large groups of species, and may only be employed with caution even for specific diagnosis, since papillae may sometimes be present and sometimes absent in the same species. He also says that in this family, the presence of papillae on the lower epidermis is relatively rare (e.g. *Begonia heracleifolia* Cham. et Schlecht. and other species).

However, Gogelein (1968), after revising the genus *Cratoxylum* (Hypericaceae), concluded that the distribution patterns of papillae in different species growing under more or less similar conditions suggest that the presence or absence of papillae is not determined by the environment. Baas (1970) found papillae to be an unreliable diagnostic character in *Cratoxylum*, but thought that in some species a kind of geographical pattern exists for the occurrence of papillae. He gives *C. formosum* (Jack) Dyer ssp. *formosum* as an example, since a high proportion of specimens from Java and Sumatra have papillae, a lower proportion in Borneo and the Philippines have papillae, while they are entirely lacking in plants from Celebes and Southeast Asia.

Three of the seven species of *Swintonia* examined by Goris (1910) were papillose, and one of these, *Swintonia schwenkii* Teijsm. and Binnend. had well developed, striate papillae in one specimen, but only domes in another. Goris concluded from his investigations that the presence, absence, or prominence of papillae may depend on the climate and he pointed to the need to examine numerous specimens to clear up the problem. I myself also investigated the leaf anatomy of *Swintonia* (Wilkinson 1971) and am able to support the above information. In addition I have examined some 13 species of *Semecarpus*, 7 of *Melanochyla*, and one of *Holigarna*, all papillose members of the Anacardiaceae. From this investigation there appear to be two distinct types of papillae in *Semecarpus*, and *Melanochyla*; (i) with a frilly-striate crown (Figs. 10.12(a,e) and Plate 7(E,I)) and (ii) with a smooth, lobed crown (Figs. 10.12(b–d and f) and Plates 5(I) and 7(F,I)).

From my own observations on papillose members of various families, it would seem that the presence and prominence of papillae are diagnostically unreliable because they vary with the climate or distribution of the species. On the other hand there are various morphologically distinct types of papillae which can be used for diagnostic purposes (see list below and Plate 7).

An interesting series of types of papillae within one family has been described and illustrated by Reule (1937). After examining the epidermis of 81 species of Aizoaceae (*Mesembryanthemum sensu lato*), he recognized seven more-or-less distinct types of epidermis connected by transitional forms. Type I were non-papillose and in Type II only extreme types were domed. His Type III (*Lithops* type) all had domed epidermal cells, e.g. *Lapidaria margaretae* Schwant., with smooth domes; *Corpuscularia lehmanni* Schwant. with a 'molar'-like crown to the dome, and *Conophytum perpusillum* N.E. Brown, each cell with a small, apically rounded papilla at its centre (Fig. 10.13(a–c)). This last example is transitional to Type IV (*Kegelzellentypus* or skittle-cell type). Type IV cells are provided with more elongated, cone-like papillae, as for example in *Cheiridopsis candidissima* N.E. Brown (Fig. 10.13(e)). In Type V, typified in *Cheiridopsis pilansii* L. the cuticle is extended into hair-like projections as shown for *Muiria hortenseae* N.E. Brown in Fig. 10.13(g). In Type VI (*Riesenzellentypus* or giant-celled type) some of the epidermal cells are enlarged into balloon-like structures as in *Trichodiadema densum* Schwant (Fig. 10.13(h)), and *T. bulbosum* Schwant. In Type VII (*Eigener Typus, Zellkuppen mit Haaren*), a proportion of the epidermal cells are in some way modified, e.g. the hair-like groups of cells of *Odontophorus marlothii* N.E. Brown (Fig. 10.13(f)), which in this example are a modification of Type IV. All these types have been included in the wider classification given below.

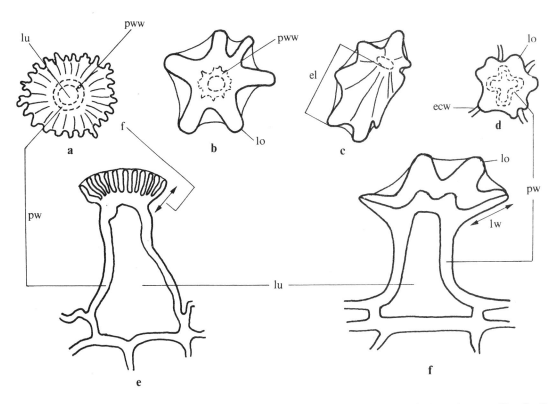

Fig. 10.12. Frilly-striate and smooth-lobed papillae. (a) Surface view of a frilly-striate papilla; (b–d) smooth-lobed papillae, (b) on an unspecialized epidermal cell, (c) on a subsidiary cell showing longer lobes on the guard cell side, and (d) papilla with rudimentary lobes on a minor vein cell; (e) longitudinal section of a frilly-striate papilla; (f) longitudinal section of a smooth-lobed papilla on an unspecialized cell. ecw, epidermal cell wall; el, extension of lobes; f, frill; lo, lobe; lu, lumen; lw lobe width; pw, papilla wall; pww, papilla wall width.

Classification of types of Papillae (Figs. 10.12, 10.13, and Plate 7)

A. Domes
1. One per cell
   (a) involving only a central circumscribed area of the outer wall, e.g. *Alysicarpus* (Fabaceae), *Conophytum perpusillum* N.E. Brown (Aizoaceae) Fig. 10.13(c).
   (b) Involving entire outer wall, e.g. 'Romneya hybrida' (Papaveraceae), *Lithops* spp. and *Lapidaria margaretae*. Schwant.
2. Several per cell, e.g. *Dillwynia hispida* Lindl. (Fabaceae; Plate 7(A))

B. *Papillae (sensu stricto)*
1. Globular or spherical
   (a) striate, e.g. *Idesia polycarpa* Maxim. (Flacourti-aceae; Plate 7(H)).
2. Cones
   (a) Containing silica, e.g. *Pachylobus macrophyllus* (Oliv.) Engl. (= *Dacryodes*, Burseraceae), see Napp-Zinn, 1973, his Fig. 120, IV, V, and *Petrea volubilis* L. (Verbenaceae); *Cheiridopsis candidissima* N.E. Brown (Aizoaceae), Fig. 10.13(e).

(b) Cones, solid, e.g. *Thymus vulgaris* L. (Lamiaceae), see Napp-Zinn, 1973, his Fig. 120, III, (after Martinet, 1872); *Geissoloma* sp. (Geissolomataceae).
(c) Cones with lumina, i.e. trichome-like, e.g. *Orphium frutescens* E. Meyer and *Swertia* (Gentianaceae) which show a series from single to two- and three-celled hairs.
(d) Filiform, solid cuticle, e.g. *Muiria hortenseae* N.E. Brown (Aizoaceae, Fig. 10.13(g)).
3. Cylindrical (finger-like)
   (a) Smooth, e.g. *Rhododendron parryae* Hutch. (Eric-aceae, Plate 7(B)), in this example, covered with wax.
   (b) Striate, e.g. *Rhus typhina* L. (Anacardiaceae, Plate 7(C)).
   (c) Echinate cuticle, e.g. *Combretum zenkeri* Engl. & Diels. (Stace 1965a).
4. Coronulate
   (a) Molar-like crown, e.g. *Conophytum perpusillum* N.E. Brown (Aizoaceae, Fig. 10.13(c)).
   (b) Frilly–striate, e.g. *Semecarpus densiflorus* (Merr.) Steenis (Fig. 10.12(a,e), Plate 7(E); *Holigarna kurtzii* King (both Anacardiaceae).
   (c) Smooth-lobed, e.g. *Semecarpus vitiensis* (A. Gray) Engl. (Plate 7(F and I); *Melanochyla bracteata* King,

Plate 5(I) Fig. 10.12(b–f), (Anacardiaceae).
5. Branched, e.g. *Willughbeia grandiflora* Dyer, ex Hook.
  f. (Apocynaceae, Plate 7(D,G) and *Gibbaeum album*
  N.E. Brown (Aizoaceae).

Doubtless other types exist, and with the variations of smooth, striate, non-waxy, and waxy, many useful distinctions will be found. A different type of projection from the periclinal cuticle is occasionally seen, as in *Coriaria nepalensis* (Plate 4(D) and just visible in Plate 3(E)). In this type, the central region of epidermal cells is elevated into flat-topped, table-like structures.

Internal cuticular surface

Certain basic facts concerning the internal surface of the cuticle, such as the occurrence of granulation and flanges, were noted by von Mohl (1842, 1847), Haberlandt (1910, 1914), and Linsbauer (1930). These features were rather neglected until their importance was realized by palaeobotanists. The working group of palaeobotanists of the International Commission of the Palaeozoic Microflora (Arbeitsgruppe 'Cuticulae' der C.I.M.P.) of 1964, attempted to achieve uniform description by evolving a new terminology for it.

Palaeobotanists are often confronted by the inner surfaces of leaf cuticles in their fossil material (see Dilcher 1974). The extensions of cuticle into cellulose anticlinal walls of the epidermal cells have been referred to as 'lamellae' or 'pegs', by Solereder (1908), and as 'cuticular flanges' by Haberlandt (1910, 1914), Stace (1965a, p. 19) and many other authors. They are termed 'ribs' by Lange (1969) and 'spandrels' by Frey-Wyssling (1976). Among the earliest illustrations of flanges are those of von Mohl (1842). His are all monocotyledonous examples viz. V-shaped in *Aloë margaritifera* Burm. (= *Haworthia margaritifera* Haw.) and in *Lomatophyllum borbonicum* Willd. (his Figs. 25 and 30), short in *Ruscus aculeatus* L. (his Fig. 32), and massive and wedge-shaped in *Aloë* sp (his Fig. 12). Haberlandt (1914) noted that 'In transverse section [of the cuticle] each flange presents a wedge-shaped, or more rarely a lanceolate, outline'. According to Lange, this is surprisingly rare, although Baker (1970) disagrees with this because he found that the V-shaped type occurs in at least half of the membranes examined by him. Lange records a hemi-cylindrical type of 'rib' in *Dysoxylon fraseranum* Benth. and a U-sectioned type, for which he gives a monocotyledon example (*Aloë succotrina* Lam.). Wedge-, top-, rod-, or spindle-shaped flanges have

been reported and illustrated by Öztig (1940) in various Aizoaceae, especially *Lithops* spp.

Cuticular flanges are sometimes absent as for example in water plants and herbs with membranous leaves. When present they may be as long as the anticlinal wall itself as in many xerophytic plants. Sometimes, in addition, the inner periclinal wall and even, in extreme xerophytes, the walls of subepidermal layers may also be cutinized. However, although the above is generally true, de Lamarlière (1906) found a 'cuticulae interne' lining the aerenchyma of water plants. The cuticle usually extends over the guard cell outer chamber, as well as the pore and inner chamber, sometimes forming inner stomatal ledges, and finally ends in a stomatal flap on the side of the adjacent subsidiary cell or spongy mesophyll.

The number of inner layers of cuticle has seldom been used for diagnostic purposes. However, Lavier-George (1936) used the number of inner layers of cuticle to separate *Philippia trichoclada* Baker, which has three layers, from *P. myriadenia* Baker which has two layers.

Undulation of the anticlinal walls in surface view

The cuticular flanges, like the cellulose walls to which they belong, may be straight, curved or undulate. It is convenient, when describing anticlinal walls, as seen in surface view, to omit the term 'anticlinal' and refer only to the 'walls'. It is also highly desirable to use a uniform terminology and I suggest that the basic patterns depicted by Stace (1965a) and reproduced here in Fig. 10.14 might serve as a basis. For ease and clarity of description, however, some authors may prefer to use a number in lieu of a lengthy description.

Only a few of the numerous arguments and discussions in the literature concerning the value or imperfections of wall undulation as a stable, diagnostic character are mentioned here. Watson (1942) made a good survey of the literature prior to that date, and the following information has been taken from his work. Differences in waviness between sun and shade leaves were first reported by Areschoug (1897) and later confirmed by Anheisser (1900) and others. These authors found that undulations are consistently more pronounced in shade leaves. Numerous investigators also agree that there is a greater tendency towards waviness on the abaxial surface, although a few exceptions to this are known. Yapp (1912), Neese (1916), and Rippel (1919) reported progressive decreases in waviness from the base to the tip of the plant. Haberlandt (1926, 1930, 1934, 1935) evolved

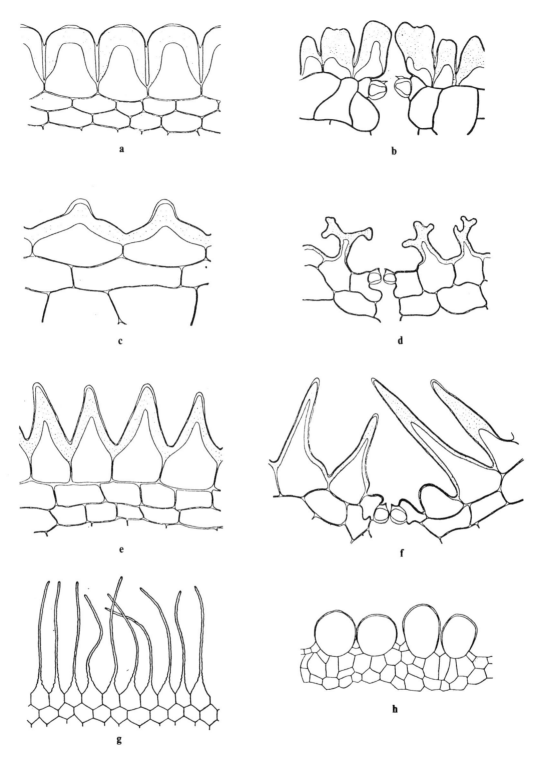

Fig. 10.13. Various types of papillae and structures intermediate between papillae and hairs. (a) *Lapidaria margaretae* Schwant. (b) *Corpuscularia lehmanni* Schwant. (c) *Conophytum perpusillum* N.E. Brown. (d) *Gibbaeum album* N.E. Brown. (e) *Cheiridopsis candidissima* N.E. Brown. (f) *Odontophorus marlothii* N.E. Brown. (g) *Muiria hortenseae* N.E. Brown. (h) *Trichodiadema densum* Schwant. (c)–(e), (g), and (h) × 77.5; (a), (b), and (f), × 280. (From Reule 1937).

Type		Frequency	Wavelength	Amplitude
1		0	x	0
2		0·5	$2x$	y
3		1·0	x	y
4		1·0	x	$2y$
5		2·0	$x/2$	y
6		2·0	$x/2$	$2y$
7		2·5	$2x/5$	$2y$
8		2·5	$2x/5$	$2y$

Fig. 10.14. Basic patterns of anticlinal walls (flanges) as seen in surface view (adapted from Stace 1965a). Type 1, straight; type 2, curved; type 3, loose, wide U-shaped curves of shallow amplitude; type 4, loose, V-shaped curves of deep amplitude; type 5, tight, frequent, U-shaped curves of shallow amplitude; type 6, tight, acutely angled, U-shaped curves of deep amplitude; type 7, tight, sharply angled, V-shaped curves of deep amplitude; type 8, tight, deep convolutions of omega (Ω) shape. Reproduced by kind permission of the Trustees of the British Museum (Natural History).

the hypothesis that the straightness of the anticlinal walls of epidermal cells in sun leaves is due to the inhibiting effect of sunlight on the genes for waviness. Watson's (1942) own investigations were on plants of *Hedera helix* L., all of equal age and produced by vegetative propagation from a single stock. His results show that in sun leaves the number of undulations is least, at 6·3 crests per cell, whereas in both slight and deep shade, the number of undulations rises to 8·7 crests per cell. In addition he found that in sun leaves both the cuticle and cell walls rapidly thickened and hardened thereby making the cell walls immutable before they could become very undulate. By way of contrast in leaves from deep shade the cuticle hardens much less rapidly and the radial and inner tangential walls remain very delicate and plastic much longer so that the side walls are finally equally wavy at the inner and outer ends. In slight shade, the cell walls harden more rapidly, so that they are less wavy at their inner than at their outer ends. Dilcher and Zeck (1968) found the same effect on wall undulation in sun and shade leaves of *Fagus grandifolia* Ehrh., but in *Quercus alba* L. and *Q. rubra* L., the adaxial epidermis was the same for both sun and shade leaves. The total

results of their investigation led them to state that neither the shape of the leaf nor the position from which the cuticular sample was taken hindered the identification of the leaf cuticle. On the other hand, Sharma and Dunn (1968), who grew *Kalanchoë fedschenkoi* Hamet and Perrier under five different grades of environment, found that the undulations were always markedly reduced in the more xeric habitats. Wall undulation (as well as the number of stomata on leaf surfaces) was used by Ramayya and Rajagopal (1968) to separate the genera *Talinum* and *Portulaca*. The first of these two genera had sinuate walls and mostly V-shaped, together with a few U-shaped, sinuses. This configuration contrasted with the straight or sinuate walls, and mostly U-shaped sinuses, of *Portulaca*. Some variation in wall undulation between different leaves of the same species is always to be expected. Variation between types 1–3 or 2–5 (Fig. 10.14) are common, but it would be unusual to find types 7 and 8 or 2 and 7 within the same species.

When investigating the leaf cuticles of any family which has a wide, well-documented distribution, the following remarks of Baas (1975) should be remembered:

High mountain species in the tropics always have thick leaves, and mostly also thick cuticles. Their anticlinal epidermal cell walls are usually straight to curved, and rarely undulated. The percentage of species with undulated walls increases with decreasing altitude in the tropics: c. 17% for high mountain habitats; c. 45% for mountain habitats between 1000 and 2500 m altitude; and c. 60% for the tropical lowland (cf. Bongers, 1973, for comparable results in *Drimys piperita* Hook. f.). In the temperate and subtropical regions c. 60% of the species have undulated anticlinal epidermal cell walls.

However, one would also agree with another statement in the same article: 'Latitude and altitude are of course very poor and rough indicators of the ecological conditions of the species involved. Data from the literature and on herbarium labels were unfortunately insufficient to overcome this lack of ecological information for most species studied.'

Internal microrelief of the anticlinal wall

When using the light microscope, the cuticular flanges of a leaf (or fruit) often appear to be composed of granules or 'beads' of cuticle. This is because the walls are interrupted by areas of non-cutinized material which appear as minute pores or pits (Plates 4(B,C) and 8(C,D)). Such walls are often said to be 'beaded'. The beading may be of various grades from fine to coarse and the cuticular flanges completely or only partly perforated. In many other plants, however, the anticlinal walls show no sign of pitting (Plates 4(A) and 8(B,H)). Haberlandt (1910; his plate 12 and Fig. 8) described and illustrated a fine example of coarse beading in the adaxial epidermis of *Prunus laurocerasus* L. The distinctly beaded walls of *Digitalis lanata* Ehrh. enabled Dewar and Wallis (1935) to distinguish this species from *D. purpurea* L., *D. lutea* L., and *D. thapsi* L. which have non-pitted walls. Several types of wall pitting were found to be of some diagnostic value in the Anacardiaceae (Wilkinson, 1971). Ferguson (1974) illustrated in his plate 18(B), the adaxial anticlinal walls of *Cocculus hirsutus* (L.) Diels, which appear to be made up entirely of granules of cuticle. Whether the anticlinal cuticular flanges are pitted or not was found to vary in 5 out of 15 species of *Ilex*, when tested for infraspecific variability by Baas (1975). Many more similar tests on other plants are clearly needed.

The presence or absence of beading was studied by van Staveren and Baas (1973) in Malesian Icacinaceae. For example, of the nine species of *Phytocrene* Wall., *P. borneensis* Becc., *P. bracteata* Wall., *P. interrupta* Sleum., and *P. macrophylla* (Bl.) Bl. all have pitted cuticular flanges, while the other five species do not. Similarly six of the eight species of *Rhyticaryum* have pitted cuticular flanges and the other two species do not, whereas they were present in only two of the ten species of *Stemonurus* examined. However, a subtle difference is recorded in two of the four species of *Citronella* D. Don and in most of the 28 species of *Gomphandra* Wall. ex Lindl. that were examined. Here, the anticlinal walls are said to be 'seemingly interrupted', which is either due to the granular structure of the flange cuticle itself or to the cuticle overlying these walls.

Lange (1969) made some s.e.m. examinations of the inner surfaces of the cuticles of an unspecified number of species. He noted that the 'ribs' (i.e. flanges) may be smooth, e.g. *Stenocarpus sinuatus* Endl. (Proteaceae), 'broken' e.g. *Persoonia falcata* R.Br. (also Proteaceae), or 'extensively interrupted', which correspond to the various types of beading seen with the light microscope.

A number of other features of internal sculpturing of the cuticle sometimes occur and these may also be of diagnostic value. The following three characters have been recognized.

(1) Specially thin or thick areas in the loops of the undulations. Thin areas in the loops of sinuous walls of the adaxial cuticle have been reported by van Staveren and Baas (1973) in 4 of the 28 species of *Gomphandra*; in one of the

two species of *Mappianthus* (*M. iodides* Hand. Mazz.) and in one species of *Pyrenacantha* (*P. repanda* (Merr.) Merr.) studied by them (all belong to the Icacinaceae).

(2) Small groups of cells with long, straight, anticlinal and thick periclinal walls. Small groups of cells of this kind sometimes stand out in contrast to the surrounding cells which have short, undulating, anticlinal walls and thin periclinal walls. This occurs e.g. in *Parishia maingayi* Hook. f. (Plate 4(G)).

(3) The cuticular flanges at the corners of cells may project beyond the lengths of the cell walls (Plate 8(F)), sometimes ending in long, whip-like extensions. These may appear, in surface view, as seen with the light microscope, as thickenings at the cell corners.

Internal microrelief of the periclinal wall

For many years anatomists have recorded the granular appearance of some cuticles; some have noted the irregular appearance of the cuticle-cellulose wall interface as seen in transverse section. The coarsely granular inner surface of the cuticle of *Aloë margaritifera* Burm. (= *Haworthia margaritifera* Haw.) was clearly illustrated by von Mohl as early as 1842 (his Fig. 26). Solereder (1908) likewise refers to the penetration of the cellulose membrane into the cuticularized portion of the outer wall in the form of lamellae or pegs. He notes this last situation chiefly in certain Fabaceae belonging to the tribe Podalyrieae and Genisteae, as well as in certain Lythrarieae and Proteaceae. This penetration of the cuticle into the cellulose walls commonly causes what is termed 'false pitting' or 'internal striation' of the outer walls of the epidermal cells when seen from the surface. Haberlandt (1914) also noted these features, for, with regard to the cuticle-cellulose wall interface he remarks, 'When present, these cutinised layers are as a rule sharply marked off from the underlying cellulose layers; the surface of contact between the two may be smooth, but is sometimes uneven, owing to the fact that the cutinised layers project at a number of points in the form of minute teeth or ridges of various shapes' (see his Plate 12, Figs. 1–6). Nevertheless many anatomists have failed to note whether the granulation was upon the inner or outer surface.

Kurer (1917), see also p.144, distinguished various species of *Digitalis* by differences in the distribution of granular epidermal cells. For example, only the epidermal cells overlying the nerves were granular in *D. lutea* and *D. purpurea*, whilst the adaxial epidermal cells were uniformly granular in *D. ambigua*. All epidermal cells, on both surfaces, were granular in *D. laevigata*, none in *D. parviflora*. Similarly in *Solanum tuberosum* the adaxial epidermis was granular but non-granular in *Solanum nigrum* L. In *Vaccinium vitis-idaea* L. the adaxial epidermis is granular, whilst the surface is smooth in *V. uliginosum* L.

During the course of my own investigations into the leaf anatomy of the Anacardiaceae (Wilkinson, 1971), the following six categories have been observed with the light and s.e. microscopes.

(1) Smooth, as in *Schinus longifolia* Speg. (Plate 4(A)) and *Lithraea caustica* (Molina) Hooker and Arnott (Plate 8(B)).

(2) Finely granular as in *Sorindeia* spp.

(3) With granules of moderate size, e.g. in *Lannea stuhlmannii* (Engl.) Engl. (Plate 8(D)).

(4) Coarsely granular, as in *Schinus molle* L.

(5) Flocculant granular (i.e. having a fluffy or curdled appearance) as in *Buchanania arborescens* Bl. (Plate 8(F)) and *Schinus terebinthifolius* Raddi (Plate 4(C)).

(6) Crustose granular as in *Buchanania obouata* Engl. Plate 8(C)).

As a result of these and other observations, it appears that coarser granulation occurs in leaves that are more xerophytic. In addition, since granulation is usually on the inner cuticular surface, it follows that a cuticle may at the same time be striate, since the striae are commonly located on the outer surface. However, internal granulation may occur in rows, giving the appearance of granular striae (Plate 4(C)).

The literature clearly shows that not much is known about the reliability of granulation as a stable character. However, Baas and his associates have made some attempt to test it, only to find that granulation varies within a species. According to Jansen and Baas (1973), *Lophopetalum pallidum* Laws. has a coarsely granular cuticle in specimen KL. 1566, but is striated in bb. 18997. Van Staveren and Baas (1973, p. 354) in their detailed examination of the leaf characters of the Icacinaceae commented that 'Some genera, however, exhibit a rather constant kind of striation and/or granulation in their cuticle. In other genera such characters may be restricted to part of the species only.' In their paper they have a good micrograph of a transverse section through the upper epidermis (their Plate II, Fig. 11) of *Hartleya inopinata* Sleum. showing the granular cuticular layer lying between the cuticle proper and the outer epidermal wall.

The leaf anatomy of 95 species of *Ilex* has been described by Baas (1975). He was usually able to study only one specimen of each species, but the infraspecific anatomical variation was studied in a restricted number of specimens belonging to 15 species. In two of these species, which were not named, the texture of the cuticle was found to vary.

A recent trend in research is the investigation of the inner surface of cuticle with the s.e.m. (Lange 1969; Baker 1970; Dilcher 1974; Alvin and Boulter 1974). Although most of the work has been carried out on fossilized and recent gymnosperms there are several indications of its probable usefulness in dicotyledonous studies. Alvin and Boulter have shown the importance of adequate maceration, so that the cellulose matter is completely removed from the microchannels to show the characteristic surface sculpturing. Furthermore, a method of describing the granulation of the internal surface under the s.e.m. is being developed in which the types of sculpturing may be denoted by letters. The size, shape, arrangement of the protrusions of cuticle and depressions (indicating the former positions of 'cellulose' pegs) may be given.

More detailed analyses are required to test the value of the morphology of microrelief of the inner periclinal wall, particularly with regard to granulation.

Relative importance of s.e.m. and light microscopy in cuticular studies

A correct picture of the morphology of the cuticle can most reliably be obtained when observations are made both with the s.e.m. and the light microscopes. Each type of instrument has its advantages and disadvantages. The light microscope shows the anticlinal wall (flanges) particularly well, and this applies also to the number and type of the subsidiary and neighbouring cells which abut on the guard cells of the stomata. The light microscope is also an effective instrument for studying glands, the bases of hairs, and the body cells of hairs. However, the type of anticlinal wall undulation and the arrangement of the subsidiary cells can sometimes be seen more clearly with the s.e.m. This occurs when the cell walls are slightly raised or depressed. The absence, presence, and degree of 'pitting' are also visible with the light microscope, as can be seen in Plate 4(A–C). However, it is clearly evident that the same characters can be seen more easily when the internal surface of the cuticle is examined with the s.e.m. This is evident if the photographs shown in Plate 4(A–C and G) are compared with those shown in Plate 8(B–D,F, and H). Differences in the thickness of the cuticle (Plate 4(G) and 5(A,C)), including polar thickenings, can be seen with the light microscope. On the other hand, poral and peristomatal rims, striae, papillae, and hair ornamentation are better studied with the s.e.m. (Plates 2,3,6, and 7), while wax morphology can be properly seen only with the s.e.m. (Plates 9 and 10).

While the inner and outer vestibules (cavities) and

pores of stomata can be seen quite clearly with the light microscope, in transverse sections, more complete information can be obtained under the s.e.m. and also from silicone-rubber replica casts likewise under the s.e.m. (Idle 1969).

Literature cited

Ahmad 1962; Alvin and Boulter 1974; Anderson 1928, 1935; Anheisser 1900; Arbeitsgruppe 'Cuticulae' der C.I.M.P. 1964; Areschoug 1897; Baas 1970, 1975; Bailey and Nast 1944; Baker 1970; Bandulska 1923–31; Barthlott and Ehler 1977; Bary, de 1884; Berg 1861; Bergen 1904; Bongers 1973; Bonnett and Newcomb 1966; Bornemann 1856; Brändlein 1907; Brenner 1900; Brodie 1842; Brongniart 1834; Carr, Milkovits and Carr 1971; C.I.M.P. (Commission Internationale de Microflore du Paléozoique) 1964; Dewar 1933, 1934a,b; Dewar and Wallis 1935; Dilcher 1974; Dilcher and Zeck 1968; Dunn, Sharma and Campbell 1965; Edwards 1935; Esau 1958, 1977; Ferguson 1974; Florin 1921–58; Forsdike 1946; Franke 1971; Frey 1926; Frey-Wyssling 1976; Gaulhofer 1908; Gogelein 1968; Goris 1910; Greenish and Collin 1904; Guérin and Guillaume 1908; Haberlandt 1910, 1914, 1926, 1934a,b, 1935; Harris 1926, 1956, 1961, 1964, 1969; Harris, Millington and Miller 1974; Henslow 1832; Hüller 1907; Idle 1969; Iterson, van 1937; Jansen and Baas 1973; Juniper and Cox 1973; Kurer 1917; Lamarlière, de 1906; Lange 1969; Lavier-George 1936; Levin 1929; Linsbauer 1930; Martens 1931a,b, 1933a,b, 1934a,b,c, 1935; Martin and Juniper 1970; Meyen 1837; Meyer, M. 1938; Moeller 1889; Mohl, von 1842–52; Mueller 1966a,b; Napp-Zinn 1973, 1974; Nathorst 1907–12; Neese 1916; Odell 1932; Öztig 1940; Paganelli Cappelletti 1975; Priestley 1943; Ramayya and Rajagopal 1968; Rao A.N. 1963; Reule 1937; Rippel 1919; Roelofsen 1952, 1959; Rowson 1943a and b, 1946; Sargent 1976a,b,c; Sargent and Gay 1977; Schieferstein and Loomis 1959; Schleiden 1861; Scott, F.M. 1963, 1966; Scott, F.M., Bystrom, and Bowler 1963; Scott, F.M., Hamner, Baker and Bowler 1958; Sharma and Dunn 1968; Sinclair and Sharma 1971; Sitte 1955; Sitte and Rennier 1963; Solereder 1908; Stace 1965a; Stahl 1896; Staveren, van and Baas 1973; Treviranus 1835–1848; Tschirch and Osterle 1893–1900; Uphof, Hummel, and Staesche 1962; Vogl 1887; Wallis 1946, 1952, 1966, 1967; Wallis and Saber 1933; Watson 1942; Wessel and Weber 1855; Wigand 1854; Wilkinson 1971; Yapp 1912.

Suggestions for further reading

Andersson 1977: papillae.
Arzt 1933: presence of internal cuticle.
Baker 1971: chemical and physical characteristics of cuticular membranes.
Barthlott and Ehler 1977: electron microscopy of the surface of the epidermis.
Bollinger 1959: absorption of UV light by cuticle.
Candolle, de, A.P. 1827, 1841: nature of plant epidermis.
Chafe and Wardrop 1973: ultrastructure.
Cohn 1850: discussion of the nature of the cuticle.
Cutter 1976: general survey of plant surfaces.
Frey-Wyssling 1976: cuticularization, interior and exterior cuticle, cutinization.
Holloway 1971: chemical and physical characteristics of leaf surfaces.
Hülsbruch 1966a,b: radial striation of cuticle; cuticular structure.
Lee and Priestly 1924: cuticular structure.
Litke 1966, 1968: fossil cuticles.
Mädler and Straus 1971: cuticular characters considered for computer data bank storage.
Maercker 1965a: pits in outer periclinal wall.
Priestley 1921: structure of cutin.
Romanovich 1960: cuticles of various Solanaceae, including ornamentation of trichomes.
Sitholey 1971: cutinization of whole depth of epidermal cells, hypodermis and parts of mesophyll.
Wuhrmann-Meyer, K. and Wuhrmann-Meyer, M. 1941: absorption of UV light by leaves.
Zettel 1974: cuticular characters of leaves and some fruits of various families.

PART VI : CUTICULAR THICKNESS [1]

The thickness of cuticle is a very complex subject which cannot be discussed here in great detail. It will, therefore, be useful to draw attention to the following key works: Linsbauer (1930), Pyykkö (1966), Napp-

Zinn (1973, 1974), and Hull, Morton, and Wharrie (1975).

Cuticular thickness appears to show a variable response to environmental conditions as is indicated in the following notes. Nevertheless, there is a general belief that plants from arid regions frequently (some

[1] This subject is also discussed under Ecological anatomy in Vol. II.

might say generally) have strongly cutinized leaves. Furthermore the thickness of the cuticle and the dryness of the habitat are frequently proportional to one another. Nevertheless in certain families and taxa of lower rank, the cuticle may be exceptionally thick, even when plants are not growing in a dry habitat. On the other hand plants with a thin cuticle, besides occurring in damp or wet localities, are also to be found on dry soils.

The subject has been investigated experimentally on numerous occasions of which the following are some examples.

Eberhardt (1903) experimented with many plants by subjecting them to (a) dry air and (b) humid air conditions. He reported that almost all the plants grown in dry air developed a thick cuticle, while on those grown in humid air the cuticle was thinner. Schroeter (1923) collected *Salix retusa* from an altitude of 2500 m and noted that it had a thick cuticle. When subsequently cultivated under a bell jar the leaves had a cuticle that was much thinner. Likewise Cunze (1926) observed that the leaves of many species developed a thinner cuticle when grown under a bell jar than when grown under normal conditions. However, some species of Crassulaceae developed comparable cuticles under either condition.

Several authors have reported that the cuticle of sun leaves is thicker than that of shade leaves (e.g. Dufour 1888; Kny 1909; Maximov 1929; Skoss 1955). On the other hand, Haberlandt (1914) expressed the opinion, also held by many of his contemporaries, that plants with a thick cuticle have a reduced rate of cuticular transpiration. He gave the classic examples of the Proteaceae and Epacridaceae. Furthermore, he remarks that 'even in a humid climate edaphic conditions may render the restriction of transpiration imperative.' This he thought accounted for the thick-walled epidermis in many epiphytes of the tropical rainforest, and in halophytes. Similarly, Linsbauer (1930) considered that the thickness of the cuticle varies according to the water economy of the plant. Both Haberlandt and Linsbauer reported that the upper epidermis of the leaf often has a thicker cuticle than the lower, a fact that is now well known and generally accepted. Haberlandt thought that the reason for this was that the upper surface receives more light and heat, and hence requires more effective protection against evaporation. In contrast to both Haberlandt and Linsbauer, Sitte and Rennier (1963) contend that there is no correlation between transpiration and cuticular thickness. Skoss (1955) has shown that certain plants, e.g. *Nicotiana glauca* R.

Grah., under water stress, develop up to twice the amount of cuticle as plants under an optimal water regime. Pyykkö (1966) remarks that, 'It may be that the thick cuticle and outer epidermal walls have many different functions, but it is clear from my investigation that they are associated with aridity of the habitat rather than with excessive light'. She also found in many leaves that the outer walls of epidermal cells are thicker than the cuticle itself. Further remarks of Pyykkö are worth noting:

The species of *Maytenus boaria* type are from the parts of East Patagonia with fairly high rainfall, but nearly all of them have a thick cuticle and outer epidermal wall all the same. Ecologically this type corresponds to the coriaceous leaved evergreen trees of the Mediterranean countries, which transpire copiously during the rainy season but are able to regulate water-loss efficiently during the dry period by means of their stomata. No doubt a thick cuticle also helps to prevent the leaf from collapsing when it has lost water.

Completely contrasting results on the ratio of the cuticle thickness to the thickness of the outer epidermal wall have been obtained from two different xerophytic species growing in the same vicinity. Hull, Shellhorn, and Saunier (1971) investigated the leaf anatomy of *Larrea divaricata* Cav. and reported the average ratio of cuticle to outer cell wall as 1:6, whereas Hull, Morton, and Wharrie found that the leaf of *Prosopis juliflora* var. *velutina* (Woot.) Sarg. had a cuticle-to-cell-wall ratio of 6:1. This may well indicate that both will serve the same purpose of limiting water loss, although one may be more efficient than the other. In addition to these two alternatives, Leece (1976) suggests that, 'Possibly, wax deposits have partially substituted for cuticular thickness or vice versa in different evolutionary sequences.' This might of course be due to the amount of wax included in the cuticle, the amount and orientation of epicuticular wax or the chemical composition of the wax. (See also p. 161 of Epicuticular wax).

Stober (1917) and Maximov (1929) noted that on an individual plant, the cuticle increases in thickness in passing from the radical to the cauline leaves and Zalenski (quoted by Maximov, 1929) found that the thickening increases 'the higher the point of insertion of the leaf on the stem'. However, Yapp (1912) observed on a single plant that the first leaves were more cuticularized than those that developed later. Sastre (1971) uses the thickness of the adaxial cuticle of leaves of the Ochnaceae as one of the characters in his matrices with which he attempts to resolve the possible evolution of the genera of this family.

From the above observations it appears that light, temperature, soil, and atmospheric moisture, altitude, and certain other factors not yet entirely clarified, certainly affect cuticular thickness, presumably within certain genetically inherited limits (e.g. Conde 1975, p. 460). Bearing these points in mind, and after exercising due caution, it is sometimes possible to use cuticle thickness as a character of diagnostic value, but it is more frequently of value as an indicator of climate or habitat. Occasionally it will be found in stained material examined under a light microscope, that groups of cells have a much thicker cuticle on their periclinal walls than there is on the corresponding walls of neighbouring cells. This was found to be a very distinctive and useful diagnostic character during my own investigation of the adaxial epidermal cells of certain species of *Parishia* (Anacardiaceae)

(Wilkinson 1971). In this particular example the cells with especially thick cuticle also have straight walls, whereas the neighbouring cells with thinner cuticle have undulating walls (see Plate 8(H) and Fig. 10.14, type 8).

Literature cited

Conde 1975; Cunze 1926; Dufour 1888; Eberhardt 1903; Haberlandt 1914; Hull, Morton and Wharrie 1975; Hull, Shellhorn, and Saunier 1971; Kny 1909; Leece 1976; Linsbauer 1930; Maximov 1929; Napp-Zinn 1973, 1974; Pyykkö 1966; Sastre 1971; Schroeter 1923; Sitte and Rennier 1963; Skoss 1955; Stober 1917; Wilkinson 1971; Yapp 1912.

PART VII : EPICUTICULAR WAX AND ITS MORPHOLOGY

The bloom which is to be seen on the surfaces of the leaves of many plants is due to a covering of **wax** which gives them a glaucous appearance. A similar bloom is sometimes to be seen on green stems and the surfaces of fruits. Over the past 15 years there has been an increasing awareness of the usefulness of wax morphology and biochemistry to taxonomists, plant pathologists and agricultural botanists. For taxonomists, wax morphology is particularly useful as an additional diagnostic character. The presence, absence, amount, and morphology of wax are important factors in the trapping of spores and hence of interest to plant pathologists. Likewise a knowledge of the same features is important to the agricultural botanist concerned with the entry of systemic insecticides and herbicides into leaf, stem, or fruit tissue.

Botanists became aware of the presence of wax particles on the leaves and stems of plants, partly because some kinds were of commercial use, but also in part because they were interested in the mode of secretion of epicuticular wax (Karsten 1857; Uloth 1867; Howes 1936; Wiesner 1871). De Bary (1871, 1884), however, was probably the first to attempt to classify leaf waxes on a morphological basis. He noted four types in both monocotyledons and dicotyledons. These were in the form of strata or crusts, rod-like coverings, simple granular layers, or aggregate coverings. The next important classification was not made until nearly 100 years later by Amelunxen, Morgenroth, and Picksak (1967) of which a version modified by personal observations is given below. The diagrams of Amelunxen, Morgenroth, and Picksak have been reproduced in Fig. 10.15. Barthlott and Ehler

(1977) recognize only two basic types: (1) Flat wax deposits (= Groups VI and VIII given here) and (2) Localized wax deposits (= Groups I–V given here).

Group I: wax granules (Fig. 10.15(a–c))

These are subdivided into globules, short cylinders, and warty-granules. Rather scattered wax particles, sometimes aggregated into groups, but with large visible areas of wax-free cuticle are shown by a large number of species e.g. *Melodinus orientalis* Bl. (Apocynaceae) (Place 9(A)) and over the veins of *Semecarpus vitiensis* (Seem.) Engl. (Anacardiaceae) (Plate 9(D)).

Group II: rods and filaments (Fig. 10.15(d–h))

These include rods, tubes, hooked filaments, curls, spirals, and loops. These rods are often nearly perpendicular to the surface. In cross section they are round, oval, or more-or-less angular and they may be hollow. Plate 9(I) shows a dense mass of individual rods on the abaxial surface of *Sassafras albidum* var. *molle* (Nutt.) Nees (Lauraceae), while *Salix aegyptiaca* L. (Salicaceae) has cohering groups of wax rods (Plate 10(H)). *Nyssa aquatica* Castigl. has undulating or loosely curled threads of wax (Plate 9(G)).

Group III: plates and scales (Fig. 10.15(i and j))

These may be orientated at any angle between 0° and 90°, and are illustrated by *Sophora microphylla* Ait. (Fabaceae), which has a dense covering of upright scales (Plate 9(B)), and the somewhat flatter scales of *Schefflera venulosa* var. *venulosa* Harms. (Araliaceae) (Plate 9(F)).

Group IV: aggregate coatings (Fig. 10.15(o–q))

These may have the form of granules, rods, or filaments, the last two types being frequently at a low angle and superimposed upon one another. The mature leaves of *Eucalyptus kruseana* F. Muell. (Myrtaceae) show groups of intermingling long threads (Plate 10(D)). Wax of a similar morphology is shown by *Eucalyptus globulus* Labill. and *Atriplex hortensis* L.

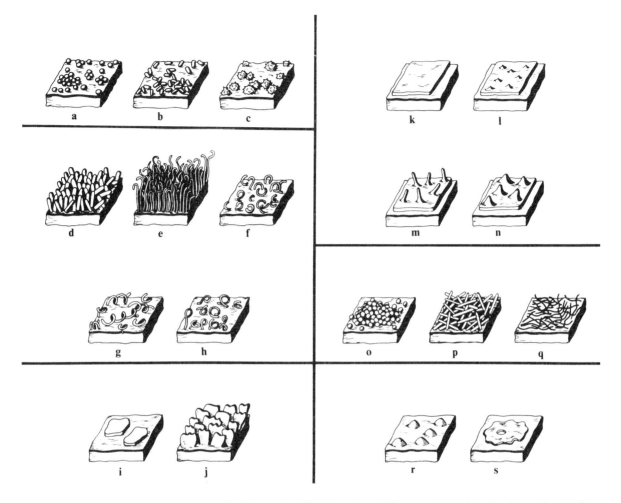

Fig. 10.15. Types of wax morphology. (a–c) granules; (d–h) rods and filaments; including hooks, curls, spirals, loops, and rings; (i, j) plates and scales; (k–n) layers and crusts; (o–q) aggregate coatings of randomly orientated granules, rods or threads; (r, s) liquids or soft wax coatings. Reproduced from Amelunxen, Morgenroth, and Picksak (1967).

(Chenopodiaceae)(Amelunxen, Morgenroth, and Picksak 1967, their Figs. 14 and 15). A dense covering of warty-granules is illustrated in *Liriodendron tulipifera* L. (Magnoliaceae)(Plate 9(C)).

Group V: mixed coatings (Plate 10(F))

The wax is of two or more types, e.g. both rods and scales on one leaf surface, as in several *Lonicera* species, here illustrated in *Lonicera korolkowii* Stapf. (Caprifoliaceae), and also several species of *Prosopis* (Fabaceae; Bleckman and Hull 1975; see comments below). According to Hall, Matus, Lamberton, and Barber (1965) and Hallam (1967), certain species of *Eucalyptus* e.g. *E. urnigera* Hook. f. also fall into this category. (See also p. 161.)

Group VI: layers and crusts (Fig. 10.15(k–n))

These may be smooth, glossy, finely warty, or have layers of rod-like or papillose appendages. The leaves of *Kalanchoë* species, e.g. *K. lugardii* Bullock (Crassulaceae), (Plate 10(A)), have a dense covering of wax, as also do the glaucous leaves of '*Hebe* × *pagei*' (Scrophulariaceae), (Plate 10(C)).

Group VII: liquid or soft wax coatings (Fig. 10.15(r and s))

These may occur as droplets or flat cakes. Apple and some other rosaceous fruits have coatings of this type, e.g. *Crataegus prunifolia* Bosc. (Plate 10(E)). Various species of *Opuntia* have each stoma surrounded by large plate of wax (Heslop-Harrison, Y. and Heslop-Harrison, J. (1968), which is itself decorated with a network of minute flakes, here illustrated for *Opuntia quimilo* K. Schum. (Plate 10(G)).

It appears that Groups I–IV, recognized by Amelunxen, Morgenroth, and Picksak should not be regarded as mutually exclusive alternatives. Wax patterns which are intermediate between Group I (Fig. 10.15(a–c)) and Group IV (Fig. 10.15(o)), sometimes exist. The aggregate coatings in Group IV (Fig. 10.15(p and q)), show prostrate threads overlapping one another, which may not be easy to distinguish from nearly upright rods and threads such as those in Fig. 10.15(d and e) (Group II). I have added Group V as a separate category, since it appears that certain taxa characteristically have two types of wax. However, Bleckman and Hull (1975) misinterpreted the aggregate group of Amelunxen, Morgenroth, and Picksak to mean 'mixed' and accordingly placed their *Prosopis* species in Group IV.

The adaxial surface of a leaf may have a different type of wax from that on the abaxial surface, e.g. *Pisum sativum* var. *alaska* L. (Martin and Juniper, 1970, their Fig. 4.21). In addition, a most interesting variation of surface wax has been observed by Hallam (1967) in *Eucalyptus polyanthemos* Schau. Over the main part of the lamina the wax is plate-like, but near the midrib there is a mixture of tubular and plate-like wax. Over the midrib itself only tubular wax is present. Clearly this species would belong in Group V. It is highly probable that other examples of this type of variation occur.

In addition to the above groups of wax types, the manner of distribution of the wax particles is sometimes also a useful character. The wax granules of *Helianthus annuus* L. are confined to areas over the anticlinal walls (Hallam and Juniper, 1971, their Fig. 7), and are predominantly located over the anticlinal walls of *Gunnera manicata* Linden (Plate 9(H)) and also (on the adaxial surface) of *Melilotus alba* Desr. (Fabaceae). An otherwise waxy cuticle sometimes has no wax particles on the guard cells of stomata. This is apparent in *Fothergilla major* Lodd. (Plate 9(E)) which has wax scales more densely aggregated near stomata, often in association with striae, rather than elsewhere on the cuticle.

The possibility of using wax morphology in classification and diagnosis has been extensively investigated on certain plants, especially *Eucalyptus*. Hallam and Chambers (1970) surveyed 315 species of *Eucalyptus* and demonstrated that the patterns of leaf wax structure revealed by the s.e.m. can, in some instances, aid the classification of this taxonomically complex genus. Their investigations yielded the following results.

Within *Eucalyptus*, wax is present in the form of plates, tubes, or both plates and tubes together. The ornamentation of the margins of the plates can frequently be correlated with taxonomic groupings or evolutionary trends suggested by other taxonomic characters. The edges of the wax plates may be entire, sinuate, crenate, digitate, or thickened. The arrangement of tube waxes on the leaf surface was also found to serve as an indicator of natural species groupings. For example, differences in the wax characters of certain taxa have confirmed that their initial classification was erroneous. Wax characters have also confirmed the existence of wrongly placed taxa, and the occurrence of large groups of taxonomically related species within the genus. With only a few exceptions, it has also been demonstrated that the type of wax remains constant, irrespective of any change in the leaf morphology, although its distribution and density may be different. The distribution of the wax, as distinct from its morphology, may also be of diagnostic value. For example, wax particles may be arranged in rows around the guard cells or on areas overlying oil glands where the cuticle is thin. Alternatively the wax particles may be in radiating clumps.

Hallam and Juniper (1971) subsequently reported for *Eucalyptus* that the pattern of wax morphology confirmed the homogeneity of the tropical *Corymbosae* or Bloodwood group. However they also state that wax morphology does not help in determining the boundaries of the existing *Eucalyptus* genus and its distinction from other closely related genera.

Differences in wax morphology can be a means of distinguishing varieties. This applies, for example, in *Brassica oleracea* L. to the variety 'Blight Proof' with green, wettable leaves and the variety 'Pale Leaf' which has water-repellant, waxy leaves. Varieties of *Ricinus communis* L. can also be distinguished by the presence or absence of a waxy bloom on the stem, petioles, and capsules (Harland 1947; see also p. 61).

Martin and Juniper (1970) warn us that wax morphology has its taxonomic limitations when they comment that 'often within a genus or family there may be general resemblances in wax structure but frequently the resemblances infringe taxonomic boundaries'. For example the wax pattern in *Pisum sativum* L. is similar to that in *Lupinus albus*, as might have been expected in two genera within the same family. However the wax in the same two species also resembles that in the wholly unrelated *Galanthus nivalis* Falk. and *Oxalis corniculata* L. The wax pattern in most glaucous Crassulaceae resembles that of *Bryophyllum tubiflorum* Harv. and *Echeveria glauca* Hort. ex Baker. Wax deposits on the leaves of some species of *Eucalyptus* resemble those on the stems of *Berberis dictyophylla* Franch. The wax of *Brassica oleracea* L. always belongs to the same type and the annulate, ridged tubes seem to be restricted to this one species. Nothing like the coiled tubes of *Chrysanthemum segetum* L. has yet been seen in any other taxon. According to Hall, Matus, Lamberton and Barber (1965), the distinctly glaucous appearance of some leaves may be due to particular types of wax arrangement. Sometimes the wax deposits grow outwards from the leaf surface. Alternatively the wax particles may show random orientation, or be more than usually dense. These types are known to increase the capacity to scatter light. On the other hand, subglaucous or non-glaucous surfaces have wax in forms that lie flat on the surface of the cuticle in such a way that each deposit is almost parallel to the cuticular surface. Alternatively the wax particles may be less plentiful on surfaces that are glaucous, or they may be orientated in a specifically defined manner.

Barber (1955) suggests that the extent to which a surface is glaucous may have a direct effect on the absorption of solar energy and he argued that a high reflection of light should confer a selective advantage in plants subjected to the intense radiation of an alpine climate. Glaucous leaves would be disadvantageous in a shaded environment. These views are supported by Thomas and Barber (1974b). Cameron (1970) studied the effect of cuticular waxes on light absorption in leaves of *Eucalyptus* spp. He found that differences in characteristics caused by the amount and orientation of waxes on the leaf cuticle caused variations in the ability of the leaves to absorb light in the 400–700 nm waveband.

The leaves of *Eucalyptus urnigera* Hook. f. characteristically have two forms of wax, both rods and flakes. Hall, Matus, Lamberton, and Barber (1965) studied the effect of altitude upon the relative amount of these two types of wax that are developed. At 2000 ft. the leaves had predominantly flaky wax and they appeared non-glaucous. Plants at 2300 ft. had a more equal mixture of flakes and rods, whilst juvenile plants at 3200 ft. had thick masses of rodlets and were consistently glaucous.

According to Barber (1955) the extent to which wax has glaucous properties is unstable in many species. For example the leaves may become less glaucous when the sun is hot. Whether this is due to melting or sublimation of the crystals is unknown.

This effect was particularly marked in *Eucalyptus risdoni* Hook. f. Barber observed that after a single hot day, leaves fully exposed to the sun became less glaucous than the leaves that remained in the shade. He also found that low temperatures seem to promote the development of wax, provided that the genotype allows this to happen. *E. risdoni* grown in a heated greenhouse during the winter remained nearly green, while plants outside exposed to mild frost became heavily glaucous. Thomas and Barber (1974a) suggest that glaucous leaves may have a selective advantage at high altitudes because of their ability to repel water. Wet leaves were killed at −2 to −4°C, whereas dry ones were not. Harland (1947) found that *Ricinus* plants with bloom were unable to fruit in cold and foggy conditions at Lima (Peru). He concluded that the presence of bloom was a physiological disadvantage under these conditions. He also reported that plants with bloom become more numerous with increasing elevation, which is concomitant with an increase in the amount of sunlight and a diminution of fog.

The effect of temperature upon the wax of leaves of *Brassica napus* L. was studied by Whitecross and Armstrong (1972). They observed marked changes in wax ultrastructure within the temperature range of 15–27°C., varying from a pattern of single, upright rodlets to one of flat, overlapping, dendritic platelets parallel to the leaf surface. Reduced light intensity apparently lowered the surface deposition of wax. However, the wax type was consistent at any particular temperature irrespective of light conditions.

An increase in the amount of cuticular wax in *Nicotiana* during drought was noted by Kurtz (1951). However, the same author (1958) studied 42 species of xerophytes growing in Arizona and reported that they did not have an excess amount of wax. Similarly Kolattukudy (1970) noted the lack of a clear correlation between surface wax and xeromorphic adaptation. (See also p. 157.)

These are only a few of the many investigations concerning the deposition of wax that have been undertaken, but it is beyond our scope to study them in further detail here. Additional particulars will be found in the literature recommended for further reading at the end of this section.

Literature cited

Amelunxen, Morgenroth and Picksak 1967; Barber 1955; Barthlott and Ehler 1977; Bary, de 1871, 1884; Bleckman and Hull 1975; Cameron 1970; Hall, Matus,

Lamberton, and Barber 1965; Hallam 1967; Hallam and Chambers 1970; Hallam and Juniper 1971; Harland 1947; Heslop-Harrison and Heslop-Harrison 1968; Howes 1936; Karsten 1857; Kolattukudy 1970; Kurtz 1951, 1958; Martin and Juniper 1970; Thomas and Barber 1947*a,b*; Worth 1867; Whitecross and Armstrong 1972; Wiesner 1871.

Suggestions for further reading

Ambronn 1888: optical properties of cuticle and included wax.

Baker and Holloway 1971: s.e.m. of waxes on plant surfaces.

Banks and Whitecross 1971: effects of temperature on ecotypes of *Eucalyptus viminalis* (Myrtaceae).

Böcher 1975: wax of *Prosopis kuntzei* (Fabaceae).

Caveness and Keeley 1970: epicuticular wax in cotton (Malvaceae).

Cowan 1975: *Eperua* (Fabaceae).

Cunze 1925: ecological significance of waxes for the water economy of the plant.

Davis 1971: s.e.m. studies of wax formations on leaves of higher plants.

Dickinson and Preece 1976: microbiology of aerial plant surfaces.

Dous 1927: theories concerning wax secretion.

Eglinton, Hamilton, Raphael, and Gonzalez 1962: hydrocarbon constituents of surface waxes: a taxonomic survey.

Fahn, Shomer, and Ben-Gera 1974: surface wax on *Citrus* juice vesicles and its importance to the canning industry.

Fisher and Bayer 1972: channels in plant cuticles as possible sites of wax precursor transport.

Hall 1967*a*: wax microchannels in white clover.

Hall 1967*b*: ultrastructure of cuticular pores and wax formation.

— and Donaldson 1963: secretion from pores of surface wax on plant leaves; ultrastructure and growth of wax on leaves of *Trifolium repens*.

Hallam 1970: growth and regeneration of waxes on leaves of *Eucalyptus*.

Herbin and Robins 1968: alkanes in plant cuticular waxes in certain genera in relation to Hutchinson's Herbaceae and Lignosae.

Martin and Batt 1958: study of surface wax and its relationship to spraying.

McNair 1929: taxonomic and climatic distribution of oils, fats and waxes on plants.

Mueller, Carr, and Loomis 1954: submicroscopic structure of plant surfaces.

Napp-Zinn 1973: review study with emphasis on wax morphology.

Preece and Dickinson 1971: ecology of leaf surface micro-organisms.

Rentschler 1974: effect of humidity on wax formation and on transpiration from leaves.

Wettstein-Knowles 1974: ultrastructure of and hypotheses upon the origin of epicuticular waxes.

PART VIII : HYDROPOTEN

Introduction

Hydropoten are multicellular, epidermal structures which serve for the absorption of water and mineral salts. They occur mainly in the abaxial epidermis of the floating leaves of a number of dicotyledonous water plants. An early reference to them was by Schilling (1894) when he described and illustrated the 'slime hairs' which he had seen on the young, unfolding leaves of many water plants such as *Brasenia peltata* Pursh, *Cabomba aquatica* Aubl. (Cabombaceae), *Nymphaea alba* L. and *Nuphar luteum* Sibth. et Smith (Nymphaeaceae), *Ranunculus fluitans* Lam., *Caltha palustris* L. (Ranunculaceae), etc. as well as in a number of monocotyledonous plants. As shown in his illustrations reproduced in Fig. 10.16(a–c) these 'hairs' are multicellular structures with an elongate

or cap-shaped terminal cell which secretes mucilage. By the time the leaves are adult, all the projecting, terminal cells have abscised and the presence of the specialized Hydropoten cells is seen in surface view. The cells are unusually small with round or oval outlines. They are surrounded by radiating, unspecialized epidermal cells (Fig. 10.16(e)). In transverse section the internal cells are mainly lens-shaped and may have lignified anticlinal walls. It was also in 1894 that Raciborski mentioned 'slime structures' on the leaves of many water plants. Mucilage hairs were again referred to and illustrated by Conard (1905) in his monograph on waterlilies.

Hydropoten of a somewhat different type were described by Perrot (1897*a,b*; 1899). He observed roundish flecks on the abaxial surface of the adult floating leaves of *Villarsia* and *Nymphoïdes peltata* (S.G. Gmel.) Kuntze both belonging to the Menyanthaceae. He discovered that these flecks are composed of islands of some 10–80 tannin-containing chlorenchymatous cells, which contrast with the surrounding chlorophyll-free epidermal cells. Furthermore, the Hydropoten cells are small and straight-walled, whereas the surrounding ordinary epidermal cells are larger and have undulate walls (Fig. 10.16(f,g)). However, Schilling (1894, pp. 313–17) observed mucilage hairs on the immature leaves of two species of *Limnanthemum*.

It was not until 1915 that the term *Hydropoten* (water drinkers) was introduced by Mayr. He described Hydropoten (of the type seen by Perrot) in both monocotyledonous and dicotyledonous plants, contrasting the 'long' ones of the former, with the 'short' ones of the latter. Among the Hydropoten which he studied in detail were those of *Trapa natans* L. (Trapaceae), *Myriophyllum spicatum* L. (Haloragaceae) and once again *Nymphoïdes peltata*. He drew attention to the fact that such structures are found in several diverse families (see also remarks on p. 165). In addition he also noted that Hydropoten cells have denser cytoplasm and considerably more chloroplasts than ordinary epidermal cells. He suggested that Hydropoten cells probably served to absorb water. However, he wrongly believed that structures resembling Hydropoten are absent from the Nymphaeaceae.

Grüss (1927*a,b*) reinvestigated the Hydropoten of *Nymphaea alba*, *N. lotus*, *Nuphar luteum*, and *Victoria regia* (= *V. amazonica*), referring to them as 'haustoria'. Grüss's opinions and terminology were mistakenly adopted in the first edition of this present work (p. 69).

Staining reactions of Hydropoten

Several authors tried to elucidate the composition and physiological function of Hydropoten cells by testing them with various stains. The following are the principal stains which were absorbed selectively by the Hydropoten cells: (1) Toluidine Blue (Drawert 1938; Lyr and Streitberg 1955; Kristen 1969, 1971); (2) Sudan III (Gessner and Volz 1951); (3) Methyl Violet (Kristen 1969, 1971); (4) Neutral Red (Goleniewska-Furmanowa 1970). Most authors agree that the chief function of these cells is to take up water and various mineral salts, such as phosphates and nitrates. Meyer (1935) on the other hand considered that they are also capable of guttation. Lüttge (1964) supplied

the lower surfaces of the leaves of *Nymphaea* with labelled $^{35}SO_4$ and found after two hours, that there were about 2·5–3·7 times more of these ions in the Hydropoten cells than in the surrounding ordinary epidermal cells. Transmission electron microscope studies by Gessner and Voltz (1951) showed that the cuticle of the outermost Hydropoten cells of *Nymphaea marliacea* is very thin. Nevertheless sieve-like perforations, which are still thinner, were apparent in the cuticle. Kristen (1969) found similar, pore-like, thin areas (of 70–140 nm) in *Nuphar luteum* (L.) Sm.

Ultrastructure of Hydropoten

In more recent years, the trend has been to examine the ultrastructure of the cell cytoplasm. Lüttge and Krapf (1969) working on *Nymphaea* spp deduced that the two innermost cells (Fig. 10.16(d)) are glandular, since they show structural characteristics in common with other glands. These characteristics are: (1) a dense cytoplasm with little vacuolation, but with numerous small vesicles; (2) numerous mitochondria; (3) relatively large nuclei; (4) cell wall protuberances; and (5) numerous pits and plasmodesmata between gland cells. The cell wall protuberances in the Hydropoten cells of *Nymphaea* and *Nuphar* are believed to indicate that they are transfer cells (see p. 122) according to Pate and Gunning (1972). These authors point out that the amplification of the plasma membrane surface area brought about by wall ingrowths varies within the range of two- to five-fold. This undoubtedly assists the unusually high intake of mineral salts and water for which these structures are noted.

Several authors comment upon the absence of Hydropoten from the aerial leaves of *Nelumbo nucifera* Gaertn., although, according to Kristen (1971), they are present in young, rolled-up, submerged leaves.

Conclusion

It appears therefore that the leaves of numerous water plants have mucilage-secreting hairs when very young, that these abscise early, and that the remaining cells take on the function of absorbing water and mineral salts. It is not clear from the literature whether or not the epidermal cells of the uniserate 'clothing hairs' (*Gliederhaare* = articulated hairs or haustoria of Grüss, 1927*b*) of *Victoria amazonica* function as

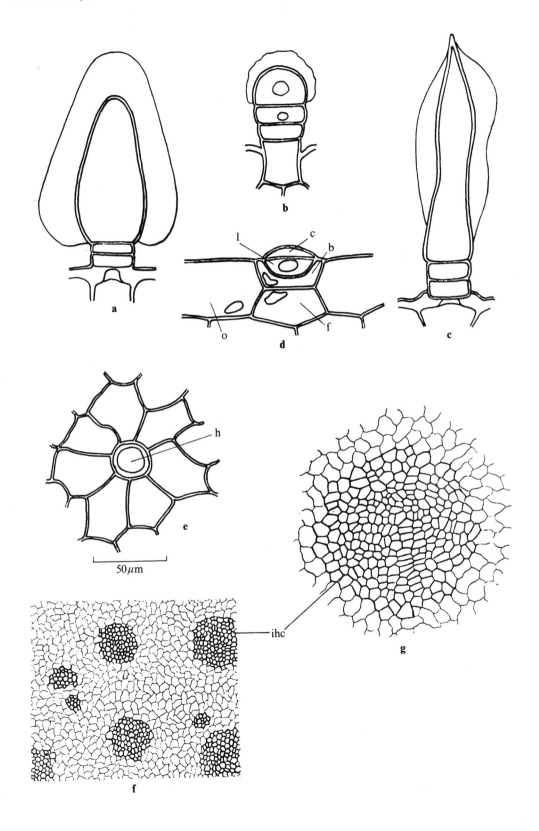

Hydropoten. However, Hydropoten are recorded in *Victoria* by Goleniewska-Furmanowa (1970, Table 2, p. 27). Their surface view appearance with both the light and scanning microscopes is very similar to that of the Hydropoten of *Nymphaea capensis* Baill. var. *zanzibarensis* (personal observation, Plate 8(G,I). Although most authors report that the Hydropoten cells occur on the abaxial surface of water plant leaves, I have seen a few on the adaxial surface of *Trapa natans* and *Victoria amazonica*.

Hydropoten cells occur in widely separated dicotyledonous families such as Nymphaeaceae (including Euryalaceae and Nupharaceae), Polygonaceae, Hippuridaceae, and Callitrichaceae, and in the water plants of certain monocotyledonous familes (e.g. Alismaceae, Hydrocharitaceae, and Limnocharitaceae) and also in water ferns.

Literature cited

Conard 1905; Drawert 1938; Gessner and Volz 1951; Goleniewska-Furmanowa 1970; Grüss 1927a and b; Kristen 1969, 1971; Lüttge 1964; Lüttge and Krapf 1969; Lyr and Streitberg 1955; Mayr 1915; Meyer, F.D. 1935; Pate and Gunning 1972; Perrot 1897a, and b, 1899; Raciborski 1894.

Suggestions for further reading

Lüttge, Pallaghy and Willert 1971: microautoradiographic investigations.
Meyer F. J. 1932: relationship of Alismatales to Ranales.
Riede 1920-21: Hydropoten in a large number of plants.
Vardar 1950: physiology of waterways in water plants with some anatomy.

Fig. 10.16. **Hydropoten.** (a–c) mucilage hairs (from Schilling 1894, magnification not given); (a) *Brasenia peltata* Pursh., (b) *Nuphar luteum* Sibth. and Smith, (c) *Cabomba aquatica* Aubl; (d) mucilage gland of *Nymphaea* sp. (× c. 892; modified from Lüttge and Krapf 1969); (e) basal cell of hair which forms Hydropoten cell surrounded by ordinary epidermal cells from *Victoria amazonica* (Poepp.) Sow. (original); (f) part of lower epidermis of '*Limnanthemum nymphaeoides*' (× 47); and (g) island of Hydropoten cells from the same species (× 227) ((f) an'd (g) from Perrot 1897b). b, bowl-shaped cell; c, cap cell; f, foot cell; h, Hydropoten cell; ihc, island of Hydropoten cells; l, lens-shaped cell; o, ordinary epidermal cell.

11

THE STEM

C. R. METCALFE

Introduction

In this Chapter the structure of stems is briefly sketched, from the time when it is first organized at the apical meristems to the beginning of secondary structure following the formation of cambium and phellogen. The transition from primary to secondary structure is often so gradual that it would not be sensible to treat them as if they were two entirely different phases in ontogenetic development. However, the most important aspect of secondary structure in systematic anatomy is wood structure and this is given separate treatment in Vol. II. The endodermis is covered in this present chapter in spite of the fact that it is more developed in and more characteristic of roots rather than stems, but it makes for better continuity to treat the subject here and to refer back to it under roots.

The mode of origin and ontogeny of the tissues that make up the structure of the mature stem must, therefore, be considered at the outset. Stem development need not be treated exhaustively, however, because it has been dealt with in the various key works and textbooks that are available, e.g. Guttenberg (1960) and Clowes (1961).

It should be remembered that the ontogeny of the stem is closely linked to that of the leaves. Indeed the distribution of the leaves on the stem (phyllotaxis) and the various types of vascular system that serve to connect the leaf to the stem are so intimately bound up with one another that the stem and leaf really need to be considered as if they are two parts of a single morphological unit. This is made abundantly clear in the contributions by Howard, beginning respectively on pp. 76 and 88.

The histogens

For many years Hanstein's (1868) histogen theory was accepted as an adequate account of the sequence of events that take place as the stem apex develops. According to this view the apical meristem or growing point, which is multicellular from the outset, was thought to become organized into three layers or zones. The outermost part of the meristem was termed the dermatogen by Hanstein. The inner solid core constituted the plerome, whilst there was a third zone, the periblem, which had the form of a hollow cylinder lying between the other two zones. It was believed that the dermatogen, periblem, and plerome each gave rise to a specific part of the mature stem. The epidermis was pictured as a derivative of the dermatogen. All tissues between the epidermis and the central cylinder (mainly cortex) were seen as a product of the periblem whereas the central cylinder, including the pith, came from the plerome.

Later on, however, it gradually became established that there is no constant relationship between Hanstein's histogens and the tissues to which they were thought to give rise; nor is it always possible to distinguish the periblem from the plerome. In consequence Hanstein's theory gave way to Schmidt's (1924) **tunica corpus** theory, according to which an outer 'tunic', comprising one to several cell layers surrounds an inner 'corpus'. The number of tunica layers is inconstant not only throughout larger taxa such as genera or families but also within the species. Indeed, according to Fahn (1974) the number of tunic layers may vary in a single plant at different stages during its development. Cell divisions in the tunic are mainly anticlinical and increase the surface area, but in the corpus the cells divide in various planes and increase the volume of the enlarging stem apex. The central apical zone of the tunic consists of larger cells than those in the surrounding lateral part. In most angiosperms the corpus comprises a zone of central mother cells situated below the large-celled part of the tunica. The central mother cells are subtended below by a lateral flank meristem and there is a rib meristem at the centre of the axis.

Fahn also refers to a second type of tunica corpus in which there is an additional cup-shaped zone, resembling a cambium, situated between the central mother cells and the rib and flank meristems. This second type of tunica corpus is referred to by Fahn as the 'Opuntia' type, but in spite of this designation it is known to occur in a few members of the Asteraceae and in *Liriodendron* and certain monocotyledons as well as in *Opuntia*, the genus from which this type of meristem takes its name. This wide distribution suggests that the Opuntia type of tunica corpus has no taxonomic significance.

Ontogenetic changes in the young stem

Immediately below the apical meristem ontogenetic changes take place rapidly. Transverse sections of a young, elongating stem show the tissues to be differentiated into a superficial covering layer, the epidermis, enclosing a primary cortex. The innermost layer of the cortex may or may not be differentiated as a morphologically recognizable endodermis (pp. 171–2). The innermost layer of the cortex, no matter whether a recognizable endodermis is present or not, is followed by the stele. This largely consists of the principal tissues that are concerned with the translocation of food, namely phloem and xylem.

In most dicotyledons the xylem and phloem are closely associated with one another and they are side by side, the phloem being towards the periphery of the axis and the xylem towards its centre. The xylem and its associated phloem are organized basically in the form of strands which are the vascular bundles. In **eustelic stems** (see p. 169) the vascular bundles are arranged in a broken circle as seen in transverse sections and they are sufficiently far apart in the circle to be easily recognized individually. However in **siphonostelic stems** vascular bundles are so closely placed that the xylem and phloem in transverse sections taken through an internode appear at first sight as closed circles. The same relative positions for the xylem and phloem are, however, maintained so that the xylem is typically towards the centre of the axis and the phloem towards the exterior. Vascular bundles showing this typical, widely occurring, side-by-side arrangement are said to be **collateral**.

Other positional arrangements of the xylem and phloem also occur, but are far less common. First, the phloem may be present on the inner as well as on the outer side of the xylem giving the **bicollateral** arrangement. Another variant is the centric arrangement in which there is either a central core of xylem surrounded by phloem (**amphicribral**) or a central core of phloem surrounded by xylem (**amphivasal**). The xylem and phloem for the most part follow the same course through the plant body and there is generally a balance in the relative proportions of the total vascular tissue that is taken up by each type of tissue. However, the amount of xylem or phloem can be reduced even to the point at which bundles consisting wholly of xylem or of phloem can occur. (See also under Anomalous structure in Vol. II).

Finally the centre of the axis is occupied by pith which occupies a very varying proportion of the total cross-sectional area of the stem in plants belonging to different taxa. In more mature stems the pith may become hollow and in a few families it is septate. Leaves soon arise from the shoot and vascular connections between the leaf and the axis form a continuum, as has been emphasized by Howard in his descriptions of the node and petiole respectively (Chapters 8 and 9).

The organization of the vascular tissue within the stem as described above applies so generally throughout the dicotyledons that it may be regarded as normal for the group. As we shall see, however, there are various deviations from the normal type that are generally referred to as anomalies and their structure as anomalous (see Vol. II). It would, however, perhaps be less misleading to regard these alternative types of vascular arrangement as variants of the normal structure rather than as anomalies.

The various anatomical patterns in young stems are at first formed wholly by the ontogenetic development of the cells and tissues derived from the apical meristem. These cells and tissues are classified as primary and should be recognized as distinct· from those that are derived from the lateral meristems that produce the cells and tissues by which the stem grows in thickness. The most important lateral meristem is the cambium which produces the incremental growth in the secondary phloem and xylem. The other familiar lateral meristem is the phellogen that gives rise to the cork (see p. 1·73).

The amount of secondary tissue in a stem is closely related to the size and woodiness of the plants. In ephemerals and small annuals there is usually little or no secondary tissue, but in the stems of large herbs as well as of shrubs, and above all of trees, the proportion of secondary tissue, and particularly of xylem, greatly exceeds that of the primary tissue. The production of secondary xylem from cambium leads to a very considerable diversity of patterns of xylem architecture and these are among the most important characters in systematic anatomy.

Vascular organization of stems in terms of stelar structure

The vascular structure of a dicotyledonous stem is usually a monostele. The concept of the stele was originally introduced by van Tieghem and Douliot (1886*a* and *b*) and, as has been pointed out by Esau (1960), the stele was originally pictured as a column in the stem (or root) made up of the vascular tissue, together with the interfascicular tissue as well as the pith and pericycle. This concept was originally suggested by the type of vascular tissue in the axis of Pteridophytes where the cortical tissues are usually conspicuously separated from the vascular tissue by an endodermis. It was subsequently found to be difficult to apply the stele concept to the angiosperm stem because the vascular system is not always clearly delimited from the cortex and epidermis by a well-defined endodermis and pericycle. However, the dicotyledonous stem usually has only one vascular cylinder made up of xylem and phloem and it is therefore stated to be **monostelic**. When a well-defined endodermis is present, the outer boundary of the stele is more clearly delimited and the original concept of the stele can more readily be applied to the dicotyledonous stem. However, in the course of time it has become customary to use the term stele even when a clearly defined endodermis is not present.

At this point a brief diversion is necessary in order to see how a monostele differs from steles of other kinds. This can be done by turning once more to the ferns and their allies where the vascular structure of the rhizome may be a protostele, siphonostele, or dictyostele.

In **protosteles** there is a solid central core of xylem encircled by a cylinder of phloem which is in turn surrounded by the pericycle and a conspicuous endodermis. **Siphonosteles** differ from protosteles in that the central part of the xylem is occupied by a solid core of parenchyma. A siphonostele is thus a **medullated protostele** in which the xylem has become cylindrical. In siphonosteles phloem may be restricted to the outer side of the xylem, when the siphonostele is said to be **ectophloic**. Alternatively phloem may be present on the inner as well as the outer side of the xylem when the siphonostele is **amphiphloic**. We have already noted that the stele is interrupted by a 'gap' occupied by parenchyma wherever a leaf is attached to the axis. When the leaf gaps are sufficiently crowded on the axis they overlap with one another to produce a **dictyostele** in which the xylem and phloem are dissected into strands, each surrounded by its own endodermis.

Of the various types of stele that have been mentioned it is usual to regard the protostele as phylogenetically the most primitive. The siphonostele is regarded as the next stage of phylogenetic advance, and it is pictured as being derived from the protostele by medullation, that is to say, by the conversion of the central part of the xylem to parenchyma. The dictyostele is still more advanced, and those who accept this view envisage that the dictyostele was evolved from the siphonostele when the leaf gaps overlapped. Recent work has shown, however, that parenchymatous gaps in siphonosteles or medullated protosteles do not necessarily arise where there are leaf gaps or branch gaps (Slade 1971).

It is currently fashionable to refer to the monosteles of dicotyledons as either siphonosteles or eusteles, and of these two terms siphonostele does not have quite the same meaning as it did when it was first used. In current usage a eustele is a hollow, cylindrical monostele in which the individual strands of xylem and phloem are spaced sufficiently widely apart to be immediately recognizable as a circle of collateral vascular bundles when viewed in transverse sections under a low power of the microscope, or even with a hand lens. A siphonostele in its modern sense is again a cylindrical monostele, but it is one in which the cut ends of the vascular strands are too close together to be individually recognizable at a low magnification. The distinction between eusteles and siphonosteles is not necessarily very clear-cut and to tell them apart may involve a personal judgment of when the vascular strands are sufficiently far apart to justify using the term eustele. Nevertheless the distinction is usually sufficiently obvious to be used as a reliable diagnostic character in systematic anatomy.

In the course of botanical history there have been controversies about the interpretation of stelar structure and its evolutionary significance (Brebner 1902; Foster and Gifford 1974, pp. 58 *et seq*.; Hill, T.G. 1906; Jeffrey 1889; Schoute 1902; Scott, D.H. 1891; Slade 1971; Tieghem, van and Douliot 1886*a* and *b*; Worsdell 1903).

Polystelic dicotyledons

Although the stems of most dicotyledons are monostelic, there are some that have a number of steles in the same way as in polystelic ferns. Each stele is provided with its own endodermis while a pith may or may not be present. In transverse sections the steles are either irregularly scattered or show an annular arrangement and at the same time they anastomose to form a network. Polystelic structure among dicotyledons has been recorded in Acanthaceae (*Justicia*), Gunneraceae (stolons), Nymphaeaceae, Parnassiaceae (at nodes), and Primulaceae (*Primula* section *auricula*). Polystelic structure has also been reported in the fruit stalks of certain Hypericaceae, Malvaceae, Meliaceae, Moraceae, Sterculiaceae, and Theaceae. A type of structure referred to as 'apparent polystely', in which there is no endodermis around the steles, is said by Solereder to occur in certain species of *Leiphamos* (Gentianaceae) and *Christisonia* (Orobanchaceae). In these last two genera there are concentric vascular bundles with central xylem. According to D.H. Scott (1891), polystely in dicotyledons developed in certain plants when they returned from a terrestrial to an aquatic mode of life.

Primary vasculature and patterns of vascular organization

It is very difficult to follow the precise sequence of the early events by which the primary vascular system is initiated immediately below the stem apex. In consequence there are different viewpoints about what actually happens. Fortunately this very complex subject has been most carefully surveyed by Esau (1965*b*). From her book we learn that there is a general consensus of opinion that the vascular system at the stem apex first becomes apparent because its component cells remain filled with dense protoplasmic contents and stand out in marked contrast to the vacuolated cells of the developing ground tissue. Vacuolation usually starts in the pith before the corresponding process starts in the cortex. The differentiating vascular tissue appears as a circle in transverse sections.

To study the cellular changes that are involved, even for a single taxon, is an exhaustive undertaking. In consequence the early stages of the differentiation of tissues at growing apices cannot at present be effectively used in systematic studies. Although, during its subsequent ontogeny, the primary vascular structure becomes much easier to study, it is still used only to a limited extent in taxonomic studies. This is because primary vascular structure has not yet been studied on a broad enough basis for its taxonomic significance to be fully assessed. Continuation of research along the lines initiated by Balfour and Philipson (1962) and Philipson and Balfour (1963) should lead to some interesting developments.

We must now pass on to consider the early vascular structure of the stem in relation to that of the leaves, and for this purpose we can take up the subject directly from Howard's (1974) publication.

Patterns of vascular organization in leaf and stem[1]

Both with eustelic and siphonostelic stems it is still a matter of dispute whether or not all of the

[1] Slightly modified from Howard (1974).

vascular tissue is to be regarded as belonging to the leaf. Some authors refer to the discrete bundles of the stem as traces, noting the ultimate departure of the conducting tissue into the leaves. Other authors, noting the continuing identity of individual bundles throughout several internodes, refer to some bundles as 'stelar' or 'cauline' (Dormer 1972). The cauline bundles are thus distinguished from their branches which constitute leaf traces. Then again, they can be interpreted as sympodia, indicating their multiple nature as branched or fused units. The primary vascular tissue can be recognized as occurring in one or another of many alternative patterns until the development of the metaxylem or secondary tissues makes the patterns obscure (Esau 1965*a,b*). The patterns which have been described as 'vascular systems' have been discovered by studying the path of the leaf traces down the stem.

Individual traces, upon association with others in the stem, may join with them in entirety or by bifurcation to establish the vascular continuity. This may occur immediately below the node or at a greater distance involving several internodes. Generally the path downward is helical and associated with a spiral phyllotaxy, but in distichous or decussate leaves the path may be straight.

Descriptions of 'eustelic' systems predominate in the literature, primarily because they can be easily studied in seedlings or herbaceous plants. Dormer (1945*b*) classified the vascular systems as 'open' or 'closed'. A closed system is essentially a reticulum or network of bundles which branch or anastomose, according to the direction the author accepts for his terminology, as they pass to or from the leaves. The open system is characterized by bundles which branch, but only exceptionally rejoin. Dormer recognized that there were intermediates, or that both types could occur in the same vascular system. The 'branches' in either system are the traces which enter the leaves, and the vascular tissue from which the branch originated is termed a stem bundle or a sympodium.

Most authors now attempt to illustrate the primary vascular system diagrammatically. This is done by depicting the stem as if it were split along one side and flattened. The vascular system can then be shown in two dimensions, with the sympodia or cauline bundles usually illustrated as heavier lines, and the traces to leaves as lighter lines. The number of sympodia may be of descriptive value, e.g. the symmetrical sixfold to symmetrical twelvefold systems (Slade 1971), as well as the nature of the branch trace pattern as cathodic or anodic into meshes or interstices. The diagrams show the open system as unconnected vertical paths of vascular tissue, and the closed system as a reticulum of anastomosing tissues. Opposite, decussate, distichous alternate, or whorled leaf arrangement can be correlated with unilacunar or multilacunar nodes. Cortical or medullary bundles can be superimposed on such a diagram (Balfour and Philipson 1962). The vascular supply to axillary buds or branches has been generally neglected in such studies, although Dormer (1972) illustrates by diagrams some of the patterns previously reported.

Philipson and Balfour (1963) reviewed and illustrated the primary vascular system patterns recognized up to that time. Subsequently Jensen's (1968) study of the Crassulaceae, which recognized 16 patterns within one family, indicates that the survey is far from complete.

Benzing (1967*a,b*) studied the primary xylem of the 'woody' Ranales in which a pseudosiphonostele (Bailey and Nast 1948) is present rather than the eustele studied by other authors. In a pseudosiphonostele there is a continuous ring of procambium and primary phloem near the stem apex, but the protoxylem is circumferentially discontinuous. What appears to be a single leaf trace to a unilacunar node or the individual traces of a trilacunar node may be, in fact, one or several stands of protoxylem exhibiting both 'open' and 'closed' systems in a variation of the application of those terms from that used by Dormer and others. By following the path of a protoxylem strand in species with unilacunar, trilacunar, and multilacunar nodes, additional diverse protoxylem strand patterns were recognized.

Benzing concluded from his evidence that the primitive node in the angiosperms was a one-trace node (one protoxylem strand), either unilacunar or trilacunar in nature. He suggested that the two-trace unilacunar node of the Clerodendrum type (Marsden and Bailey 1955) is derived.

Tissues of the mature stem

Having outlined the ontogeny and histological organization of the stem in general terms, we now consider the individual tissues that go to make up the structure of the mature stem, starting at the periphery of the stem and working inwards towards the centre.

Epidermis

The epidermis nearly always constitutes the outermost cell layer of the young stem both in herbs and woody plants. It is a continuous protective cell layer, usually perforated by stomata. In ephemerals and annuals it persists throughout the life of the axis. In more robust herbs, and also in shrubs and trees, the epidermis is sooner or later replaced by bark, the outer part of which is made up of cells with suberized walls that constitute the cork.

The delimitation of the epidermis in the leaf has already been discussed and much of what has been said for the leaf applies equally to the stem. In some stems, however, such as those of aquatic plants, the outermost cells are not always morphologically distinguishable from the subjacent cortical cells, and, where this is so, it has sometimes been claimed that the plant lacks a true epidermis (Linsbauer 1930). On the other hand, where the cells of the outermost layer are identical in appearance with those of the next one to several subjacent layers of cells, they are said to constitute a multiple epidermis, or alternatively a hypodermis, as has already been seen in the leaf. (The distinction between a multiple epidermis and a hypodermis has already been discussed on p. 65.)

The walls of the epidermal cells are often thick or sclerosed and serve to protect the living contents of the cells, whilst further protective reinforcement is provided by the waxy cuticle (see pp. 140–58). The cuticle is commonly sculptured and the various forms of sculpturing are increasingly used for diagnostic purposes (see pp. 143 *et seq.*). Epidermal cells may also serve for the deposition of crystals, opaline silica, and tannin, etc.

The morphology and histology of the trichomes and other epidermal appendages of the stem are, in most taxa, similar to those that arise from the leaf.

Primary cortex

The **primary cortex** is the ground tissue between the epidermis and the stele. It is usual to regard the endodermis (see below) as the innermost layer of the cortex.

The primary cortex is basically parenchymatous, but some parts of the tissue are commonly differentiated into mechanical elements such as collenchyma and sclerenchyma. The distribution pattern and histology of these mechanical cells provide characters that can be used for taxonomic purposes. Very often the cortex is partly photosynthetic, and this applies particularly to xerophytes. This subject, and also the anatomy of aquatic plants are dealt with under Ecological anatomy (Vol. II).

Endodermis

The term **endodermis** is applied to a layer of living cells situated where the cortex impinges on the stele, mostly in roots but sometimes in stems and leaves. The endodermis usually has the form of a closed cylinder, but a few taxa are known in which the cylinder is incomplete (Mylius 1913). The cells of which it consists are derived from the primary meristem and they are usually morphologically distinct from their neighbours, but their form may undergo ontogenetic morphological changes in any one taxon and cells that are regarded as endodermal vary in appearance in different taxa. The endodermis is occasionally, e.g. in *Escallonia rubra* and in the subterranean axis of *Saxifraga rotundifolia*, concerned with the abscission of the outer primary tissue (Mylius 1913) and, when this happens, it afterwards becomes the outermost layer of the axis. A recognizable endodermis is by no means always present in stems. In stems it is also commonly replaced by a layer of cells containing starch. An endodermis is almost universally present in young roots and the cells are characterized by bands of thickening which enter rather intimately into the composition of the transverse and radial walls. These bands are the so-called **casparian thickenings**. Many botanists use the term 'endodermis' only when casparian thickenings are present and this in effect often confines the occurrence of a true or functional endodermis to roots. However, casparian thickenings sometimes occur in stems as well. The distinction between those who use the term 'endodermis' only when casparian thickenings are present and those who employ the same term for any recognizably distinct layer of cells separating the inner boundary of the cortex from the vascular tissue largely depends on whether their approach is physiological or morphological. The great physiological interest of casparian thickenings is that they are thought to serve as impenetrable barriers in the cell walls in which they are present to the flow of liquids from the cortex to the stele or vice versa. In consequence the flow of liquid can take place only through the living protoplasts of the endodermal cells. Although casparian thickenings are normally present in roots, they also occur in the endodermal cells of stems in certain taxa. On the other hand, casparian thickenings are not necessarily confined to cells of the endodermis. For example, they occur in the exodermis of the roots of certain monocotyledons.

Casparian thickenings can usually be seen most readily when the cells of the endodermis are young.

The cell walls later become suberized and lignified, which often makes them stand out in contrast to neighbouring cells of the cortex and stele which usually have thinner walls.

It has already been noted that endodermal cells with casparian thickenings occur sparingly in stems (e.g. Codaccioni 1970). When casparian thickenings are present in stems they provide the systematic anatomist with a useful confirmatory diagnostic character. Some of the families with casparian thickenings in stems are listed on p. 216.

Those who view the endodermis as a layer of cells that are morphologically distinguishable from their neighbours at the boundary between the cortex and stele do not consider that the presence of casparian thickenings is an essential quality for a cell to be classified as endodermal. For those who adopt this attitude the position of the cells and the ease with which they can be recognized are the important characters. For purposes of systematic anatomy the ease of recognition is the most essential character of endodermal cells. Sometimes the presence of casparian thickenings affords a ready means by which they can be recognized, but it is equally probable that the cells will be more easily recognized by some other character. For example, it has already been noted that the endodermis in a stem may have the form of a starch sheath. In other taxa the size of the cells, or the thickness and chemical nature of their cell walls, may be more taxonomically significant. In making comparisons it must also be remembered that endodermal cells undergo ontogenetic changes that may alter their appearance profoundly. For example endodermal cells that are initially thin-walled, and which may or may not also show casparian thickenings, sometimes become suberized and lignified. At this stage certain individual cells in the endodermal layer may remain thin-walled and are known as passage cells.

Ontogenetic changes in the endodermis are discussed in more detail in a review article by van Fleet (1961) which includes particulars of his own investigations. Indeed van Fleet's article may be regarded as a key work from the physiological standpoint. He recognized several phases during the ontogenetic development of the endodermis in roots. To begin with, the protoendodermis is meristematic and the cells divide to add to the cortex or to give rise to cork from an endodermal phellogen. There then follows a primary phase during which fatty and phenolic substances are deposited in the casparian strips. Next there is a secondary phase during which fatty and phenolic substances are oxidized and deposited in the cell walls in cellulose lamellae. In the tertiary phase that follows, cellulose is deposited unilaterally. Finally there is a quaternary phase during which phenolic and quinoidal substances are oxidized and deposited in the cellulose walls. Incidentally van Fleet, referring to work by Barker (1953), mentions that the medullary sheath at the boundary between the vascular tissue and the pith is a homologue of the endodermis and in some plants develops characteristic endodermal features.

An extensive series of physiological investigations which have a bearing on the physiology of the endodermis were undertaken by Priestley and his colleagues at Leeds University in England (Priestley 1922, 1924, 1926, 1943; Priestley and Armstead 1922; Priestley and Ewing 1923; Priestley and North 1922; Scott and Priestley 1928).

Priestley and Ewing (1923) presented evidence to show that it is possible to induce the transformation of a non-functional endodermis (i.e. one without casparian thickenings) to a functional endodermis (with casparian thickenings) by keeping stems in darkness so as to induce etiolation. Different results were, however, obtained by Venning (1954), who found that stems of the potato plant (*Solanum tuberosum*) remained parenchymatous and devoid of casparian thickenings irrespective of whether the plants were grown in light or darkness.

It seems very strange that the endodermis can always have the one and only important physiological function that has been ascribed to it when its histological characteristics show such a range of variation in different taxa. It is interesting to note that Bond (1931), after making a detailed study of the stem endodermis in the genus *Piper*, where its structure shows a considerable range of variation, concluded that the endodermis may be no more than a vestigial structure of no great physiological significance.

Some botanists take the view that the prominence of casparian thickenings is correlated with ecological conditions. For example, according to Ginzburg (1966), casparian thickenings are particularly wide in desert plants. This conclusion was reached after studying 30 dicotyledonous species growing in various habitats in the Israeli Negev. The plants in question belong to the following families: Asteraceae (*Achillea*, *Artemisia*); Caryophyllaceae (*Gymnocarpus*); Chenopodiaceae (*Anabasis*, *Arthrocnemum*, *Chenolea*, *Hammada*, *Noaea*, *Salsola*, *Seidlitzia*, *Suaeda*); Brassicaceae (*Zilla*); Cucurbitaceae (*Citrullus*); Fabaceae (*Alhagi*, *Prosopis*, *Retama*); Polygonaceae (*Polygonum*); Plumbaginaceae (*Statice*); Resedaceae (*Ochra-*

denus); Tamaricaceae (*Reaumuria*, *Tamarix*); Zygophyllaceae (*Nitraria*, *Zygophyllum*).

It must be pointed out that there are also some plants from aquatic or marshy habitats in which the casparian thickenings are larger than usual, or conspicuous. This can be seen by referring to lists on p. 216.

Special features of the endodermis have been noted in various plants. For example, chloroplasts are present in the cells in a few plants, e.g. in some of the Chenopodiaceae (*Suaeda*) and Convolvulaceae. On the whole this character is more common amongst monocotyledons where chloroplasts are often present in the bundle sheaths (Cyperaceae and Gramineae).

Polyderm

Polyderm is a tissue of living cells that is morphologically and physiologically equivalent to the endodermis, from which it differs in showing a concentrically stratified structure. It arises from a meristem in the pericycle of roots and underground stems and consists of parenchymatous and endodermal cells which succeed one another in concentric lamellae. Each lamella consists of a layer of endodermal cells (which at first have casparian thickenings but later become lined with suberin) and a layer of parenchyma, both of which arise simultaneously. The first lamella to be formed arises on the inner flank of the endodermis. Subsequent lamellae are laid down annually on the inner flanks of the innermost of the previously formed endodermal layers. Whenever a new lamella of polyderm is laid down the outermost one becomes disorganized so that the tissue is continuously renewed. The dead polyderm tissue lacks strength or elasticity, and in submerged parts of aquatic plants the polyderm may become lacunar and then serves as aerenchyma.

The main interest of polyderm for the systematic anatomist is that it is confined to certain members of only a few families. These are Hypericaceae, Lythraceae, Melastomataceae, Myrtaceae and Onagraceae. The most comprehensive accounts of polyderm are those by Mylius (1912, 1913), but see also Luhan (1955) and Nelson and Wilhelm (1957).

Pericycle

There has been much confusion in the use of the term **pericycle** (see p. 181). It is a great convenience for systematic anatomists to continue to use the term in a purely descriptive sense because the various distribution patterns, and the nature of the component cells, of the sclerenchyma in the pericycle are diagnostically reliable. The types and distribution patterns of pericyclic sclerenchyma do not coincide very closely with any accepted system of angiosperm classification. It must also be remembered that pericyclic sclerenchyma is laid down in the course of a predetermined sequence of ontogenetic events, so the pattern varies with the age of the material being examined. Valid taxonomic conclusions can consequently be drawn only by comparing stems that are at an equivalent stage of development. The various distribution patterns, as seen in transverse sections through internodes, that were recognized and used for taxonomic and diagnostic purposes by Solereder (1908) are as follows:

(1) Pericycle remaining parenchymatous. The approximate position of the pericycle is coincident with distal ends of the primary medullary rays.
(2) Pericycle with a circle of fibre strands embedded in ground tissue composed of non-sclerosed elements.
(3) Pericycle with a nearly or completely closed circle of fibres.
(4) Pericycle with a nearly or completely closed circle made up of alternating groups of fibres and brachysclereids. This is the type of pericycle which Solereder described as showing a 'composite and continuous ring of sclerenchyma'. The families in which this type of pericyclic sclerenchyma occurs are very numerous.

Sometimes pericyclic sclerenchyma which initially has the form of a closed cylinder becomes secondarily interrupted. This is because of the pressure on the pericyclic sclerenchyma that results from the activity of the cambium. The cambial activity produces so much secondary tissue that the pericycle which is a primary tissue is forced to give way. When this happens the secondary tissue that is forced into the gaps in the initially continuous sclerenchyma ring often remains unsclerosed. In other taxa, however, the tissue that invades the gaps may itself become more or less sclerosed, and, when this is so, the 'closed' nature of the sclerenchymatous ring is maintained. In relatively thick branches of this last kind the brachysclereids often tend to occupy an increased proportion of the sclerenchyma ring. The sclerenchymatous ring may ultimately become part of the bark.

Solereder reminds us that pericyclic sclerenchyma is primarily a mechanical tissue, so its form must be correlated with the stresses to which the plant is subjected. This is no doubt one of the reasons why the pericyclic sclerenchyma and the taxonomic affinities of the plants do not go hand in hand. As an example Solereder quotes the Primulaceae, where pericyclic sclerenchyma is absent from species in which there is no need for 'flexile strength'. In the

Cucurbitaceae a strengthening ring is present in the pericycle only in species in which the stems show no great growth in thickness, the implication being that the sclerenchyma provides the necessary support when the vascular tissue is inadequate for this purpose. Finally Solereder reminds us that pericyclic sclerenchyma is often absent from succulent plants. Solereder also emphasizes that, in the fibres themselves, the length of the cells, the type of pitting, and the presence or absence of septa are of diagnostic interest. Particularly noteworthy examples that he cites include the following:

(1) In *Balanites* (Family Balanitaceae) the ring of pericyclic sclerenchyma consists of a circle of stone-cells at the inner face of which strands of fibres are present.
(2) In some families the sclerenchymatous ring includes brachysclereids with U-shaped thickenings.

Cork or phellem

Cork is a secondary protective tissue in stems and roots consisting of dead cells with suberized walls, nearly always produced by a phellogen or cork cambium. There are, however, a few dicotyledons in which cork formation without the intervention of a phellogen has been reported, e.g. in certain Lamiaceae (Lemesle 1928). This is also said to apply in some of the Epacridaceae in which the phloem becomes suberized and in the Aristolochiaceae (*Asarum*) where cork is formed in the cortex. In stems cork replaces the epidermis, and, after it has been laid down, its suberized cell walls prevent food supplies from reaching the cells on its outer side and they consequently die and become detached. The depth below the plant surface at which the phellogen is laid down varies in different kinds of plants. It is often superficial and it is so described when it arises in the sub-epidermis or outer part of the cortex. Alternatively it may arise at a deeper level in the cortex or in the pericycle. In roots, following the activity of a phellogen, which is sometimes deeply seated, the whole of the outer primary tissues may become detached in a manner recalling the removal of a glove finger from the hand. This happens for example in *Taraxacum kok-saghyz* (see under Laticifers in Vol. II). A somewhat similar situation has been described by Luhan (1955) in the roots of alpine plants belonging to certain species of Primulaceae, Rosaceae, and Gentianaceae.

The exact seat of origin of superficial cork varies from species to species. Solereder says that the position in which the cork is initiated is particularly variable amongst the Fabaceae. In other families such as the Convolvulaceae and Menispermaceae the depth of the phellogen varies below different arcs of the circumference of an individual stem. Superficial cork has been recorded in the families listed on p. 216 and this list is followed by others giving the families in which the phellogen is more deeply seated. It must be remembered, however, that during the ontogeny of a branch of a woody plant there may be a succession of phellogens at successively deeper levels and the way in which these are laid down and the nature of the cork cells to which they give rise determine the nature of the bark. We return to this topic on p. 175.

Cork sometimes arises in the xylem. References to **interxylary cork** are given by Solereder (1908), Moss (1936, 1940) and Diettert (1938). More recently Moss and Gorham (1953) have recorded its presence in 10 families, 14 genera, and 40 species of dicotyledons between which there are no close taxonomic affinities. The interxylary cork is usually formed in fissures in a perennating axis or it may be produced when annual stems die back. In the wood of *Artemisia* spp (Asteraceae) it is sometimes present as concentric layers. It was recorded by Moss and Gorham in the following families and genera: Asteraceae, (*Artemisia, Crepis, Taraxacum*); Boraginaceae (*Mertensia, Lithospermum*); Crassulaceae (*Sedum*); Geraniaceae (*Geranium*); Lamiaceae (*Salvia*); Meliaceae (*Guarea*); Onagraceae (*Epilobium*); Papaveraceae (*Corydalis*, in this genus fission may occur without the formation of cork); Polemoniaceae (*Polemonium*); Polygonaceae (*Eriogonum*); Ranunculaceae (*Aconitum, Delphinium*).

Cork in the pith, i.e. medullary cork, is rare but Solereder refers to its occurrence in *Campanula pyramidalis* and *Phyteuma limonifolium* Sibth. and Sm. Holm (1929) also records it in the hypocotyl of seedlings of a species of *Balsamocitrus*.

The part which cork plays in the formation of bark is discussed under Bark (p. 179).

The reader should note that cork is not a tissue of one uniform kind wherever it occurs. Sometimes the cell walls are thin and the lumina wide, giving the type of cork that Solereder described as spongy. If, on the other hand, the cell walls are thick, the cells may become compressed in various ways, especially in the radial direction, giving tabular cork. Then again, all or some of the cork cells may become sclerosed, thus producing stone cork. The sclerosis of the walls may be uniformly deposited, U-shaped, or restricted to the outer and inner cell walls. Sclerosis may occur on all of the cork cells, whilst in other plants it is confined to concentric layers of cells or restricted to isolated cells. In the first edition of this work sclereids were recorded in the cork of Araliaceae (*Oreopanax* and *Polyscias*); Convolvulaceae (*Erycibe, Maripa, Operculina*); Myristicaceae (*Myristica* and *Virola*). Brachysclereids were recorded in the cork in certain Combretaceae; Goodeniaceae (*Scaevola* sp); Loganiaceae (*Fagraea*); Polygalaceae; Solanaceae (*Juanulloa, Lycium, Nicotiana, Phrodus, Sessea, Solandra*). These various special types of cork cells are not at present known to be of more than limited diagnostic value, but not many broad surveys of cork cells throughout species, genera, or families have been attempted and this may account for our ignorance. Sometimes, however, the cork cells may exhibit some distinctive feature that is sufficiently restricted in its taxonomic distribution to be of diagnostic value. For example, small crystals encrust the inner walls of the cork cells of *Croton* (Euphorbiaceae). In *Liquidambar* (Hamamelidaceae) the cork cells are silicified and in *Avicennia* they are said to be lignified.

Strangely, there are species in which cells derived from the phellogen remain cellulosic. They have to be classified as cork cells because of their mode of origin, but they lack the suberin which is normally the chief characteristic of cork cells. Cork cells with cellulosic cell walls are known as **phelloid cells** and it is convenient to refer to the tissue which they form as phelloid cork. Phelloid cells may be either thin- or thick-walled and they sometimes become sclerosed. Phelloid cells and suberized cork cells sometimes alternate in concentric bands, or they may occur in groups embedded in larger masses of ordinary cork. The presence of phelloid cells has been recorded by Solereder and other authors in certain members of the following families: Burseraceae (inner cell walls also silicified); Caprifoliaceae, Combretaceae (*Quisqualis*; see also de Beuzeville and Beuzeville and Welch, 1924); Hypericaceae, Lythraceae, Melastomataceae; Myrtaceae (*Eucalyptus* and *Syzygium*), Penaeaceae, Rosaceae, Verbenaceae (*Clerodendrum inerme*; Mullan 1932-33).

Another variant is **mucilaginous cork**. This occurs according to Solereder e.g. in certain desert members of the Chenopodiaceae, Fabaceae, and Polygonaceae.

Cork tubercles, ridges etc., sometimes visible to the naked eye, have been recorded in certain Araliaceae, Bombacaceae, Cactaceae, Euphorbiaceae, Fabaceae, Rhamnaceae, Rosaceae, Rutaceae, Simaroubaceae (similar to cork warts on leaf surfaces, see p. 205).

The ultrastructure of cork cells has been described by Sitte (1962).

Relationship of periderm to cork (phellem) and phelloderm

Phelloderm is the tissue derived from and laid down on the inner side of the phellogen. It often merges into the primary cortex so that there is no obvious distinction between the primary and secondary tissue. In other plants the phelloderm can be recognized more easily because its component cells are radially arranged. The phelloderm is sometimes completely or partly sclerosed. The phelloderm and the **cork** or **phellem**, together with the phellogen from which they are initiated, constitute the **periderm**. Periderm formation is often delayed particularly in plants with reduced leaves in which the chief photosynthetic tissue is represented by the chlorenchymatous primary cortex. In *Oxylobium* (Fabaceae), in certain Menispermaceae, and in the Loranthaceae-Visceae, where it may never be formed at all, it is replaced by a cuticular epithelium consisting of the epidermis and primary cortex in which the outer cell walls become thickened or cuticularized.

Lenticels

Lenticels are local interruptions in the cork and phelloderm through which gaseous exchange with the surrounding atmosphere takes place. They can be of considerable interest in the identification of woody plants. Lenticels are histologically of two kinds. In the first of these, the more-or-less uniform component cells are at first compact, but they tend later to separate from one another to produce a spongy tissue through which gaseous diffusion takes place. In the second type of lenticel, some parts of the tissue consist of masses of somewhat powdery cells, alternating with layers of more compact cells that resemble cork cells.

Bark

In dicotyledons the term bark in a popular sense is applied to all secondary tissues external to the xylem in stems and roots of woody plants. Its morphology varies with the taxonomic affinities of the plants and also in young and old material belonging to any one taxon. The outer part of the bark consists of cork, derived from the phellogen, but variations in the depth within the axis at which the phellogen arises and differences in the ontogenetic stages by which the mature bark is produced, are responsible for the variations in the appearance of the bark of different kinds of mature trees. These topics are elaborated on

in the course of Chapter 12 on phloem.

The taxonomic value of bark structure in a general way is given in Moeller's *Baumrinden* (1882) which is a classic work on bark. The great weakness of this publication is that it refers mainly to the bark of twigs rather than to that on thick branches and trunks. The more recent treatments of bark structure by Holdheide (1951) and Parameswaran and Liese (1970) are of more interest because they deal with mature bark, but their taxonomic coverage is limited. A good example of a more detailed study of the bark of a single family is afforded by Whitmore's (1962) excellent study of the barks of the Dipterocarpaceae. This same author later examined the bark of beech, oak and sweet chestnut *Castaneae sativa* Mill. (Whitmore 1963).

Phloem

An illustrated account of the structure of the phloem, with special emphasis on diagnostic and taxonomic characters, is given in Chapter 12.

Primary and early secondary xylem

Primary xylem is the term applied to all xylem that is initiated from an apical growing point, before the cambium comes into play. It is customary to refer to early primary xylem as **protoxylem** and to late primary xylem as **metaxylem**. Stems of present-day angiosperms have endarch protoxylem situated immediately next to the pith, and the subsequent differentiation of the xylem elements is centrifugal. In roots, on the other hand, the exarch xylem is initiated at the outer boundary of the procambial cylinder and develops centripetally towards the root centre. The meristematic tissue from which the primary xylem and phloem are derived is termed **procambium** (Esau 1943*a,b*). In addition, some of the procambium cells give rise to the cambium itself. This means that the procambium and cambium represent two developmental stages of the same meristem. If the procambial meristem is entirely used up in the production of primary vascular tissue, the resulting structure is of the type that is characteristic of very small, dicotyledonous herbs.

Beginners often expect, and spend a lot of time looking for, a clearly defined boundary between protoxylem and metaxylem, and they are surprised when, as so often happens, they do not find one. This is because the distinction between protoxylem and metaxylem is largely a matter of descriptive convenience rather than of histological distinction. Proto-xylem is laid down while the young axis is still growing in length from the apical meristem, and metaxylem is formed during the rather nebulous interval after elongation has ceased and before secondary tissues are laid down by the cambium.

In correlation with its physiological function of conducting water and dissolved mineral substances upwards from the soil, it is not surprising to find that the xylem, both primary and secondary, is mostly made up of **tracheary elements**. These are elongated, relatively narrow cells, with their long axes lying parallel to that of the plant organ in which they are situated. The most important classes of tracheary elements are **vessels** (see vol. II) and **tracheids** (see vol. II). These are embedded in a matrix of narrower but still elongated fibrous cells, which are classified as **libriform fibres** when they have simple pits and as **fibre-tracheids** when the pitting is bordered. Both libriform fibres and fibre tracheids commonly have thick walls, but the thickness of the walls is further emphasized by the contrast with the comparatively narrow lumina that both types of cells possess. However, the wall thickness of both libriform fibres and fibre tracheids shows a considerable range in different taxa. Indeed the hardness of wood and the flexibility of herbaceous stems and young woody twigs is partly determined by the thickness of the fibre walls. The remaining cells of the xylem are parenchymatous. Some of the parenchyma is axially aligned (**axial parenchyma**), but the rest of it is organized into radiating plates that can be seen, in transverse sections, to divide the xylem into sectors. Some of these radiating plates extend outwards from the pith to the cambium and constitute the **primary medullary rays**. They first become apparent in young stems as inter-fascicular ground tissue in a eustele but when secondary thickening begins they are extended radially by the activity of the cambium in the manner that is more fully described under wood structure (in vol. II). As the stem increases in diameter by the addition of fresh increments of secondary xylem and phloem, additional radiating plates of parenchyma are laid down by the cambium and constitute the **secondary medullary rays**. In describing the structure of secon-dary xylem, it is usual to drop the epithet 'medullary' and to refer to the radiating plates of parenchyma as **rays**. It should be noted that in stems that are initially siphonostelic the distinction between primary and secondary rays is less obvious because rays of both kinds are more nearly equal in tangential width.

Turning now to the axially aligned parenchyma we find that it occupies a very variable proportion of the xylem in individual taxa. It is not of any great taxonomic importance in dealing with primary xylem, but in secondary wood its distribution pattern and the proportion of the ground tissue of wood that is parenchymatous have much greater taxonomic significance.

The nature of the tracheids and vessels must now be considered rather more closely. As already noted, the term tracheid is used for tracheary elements that do not communicate with one another through holes (pores) in their end walls. On the other hand the component tracheal cells that go to make up a vessel do have porous end walls. Tracheal cells that are components of a vessel are known as **vessel elements**. A vessel consists of a number of vessel elements that are vertically aligned, so that the vessel of which they form a part resembles a pipe, and it is indeed because of this pipe-like structure that the vessels are so efficient in the conduction of sap. Tracheids and vessel elements that enter into the composition of the primary xylem are generally supported by bands of wall thickening that are usually either helical (spiral), or the thickening may take the form of a series of rings (**annular thickenings**) situated at intervals around the circumference usually throughout the length of the cell. Either of these types of thickening permits the easy expansion of cells that have not ceased to elongate.

As we have already noted, when the transition from primary to secondary xylem takes place, the same mixture of sap-conducting, mechanical and storage cells is perpetuated. However, there are changes in the arrangement of the cells and in their structure. The study of the cellular organization of secondary xylem, of the ontogenetic changes by which the mature structure is achieved, as well as of the phylogenetic and taxonomic significance of the various characters that are presented, are all part of the important subject of wood structure. This is a particularly significant part of the whole subject of systematic anatomy, because of the insight into the phylogeny of the angiosperms that it gives. It also provides a practical means of identifying tim-bers, which is a matter of considerable economic im-portance. In view of these considerations it will be appropriate to leave the secondary xylem for full development in Vol. II.

Pith[1]

At the centre of the stem, surrounded by the xylem, lies the usually more-or-less cylindrical **pith** or **medulla**. The proportion of the stem that is occupied by pith shows a considerable range of variation in different taxa. In many herbaceous stems it is wide in diameter and it continues to occupy a high proportion of the stem throughout its life. This applies, for example, in numerous Asteraceae, well-known examples being the Jerusalem artichoke (*Helianthus tuberosus* L.) and the Sunflower (*Helianthus annuus* L.). In the Elder (*Sambucus nigra* L. family, Caprifoliaceae), which is a more woody species, the pith occupies a high proportion of the stems while they are still young, but as they become older the proportion of pith becomes less. In certain shoots of Elder which arise and develop very rapidly from older branches the pith reaches its maximum development (Metcalfe 1948). In most woody species the pith is small in relation to the total cross-sectional area of the stem, but there are exceptions.

Pith usually consists of parenchyma, and in mature stems its component cells may either remain alive or die. When pith consists of a mixture of dead and living cells it is described as heterogeneous. The active cells either form a network with the dead cells, or the living cells may be confined to the periphery of the pith. Pith cells may also serve to store starch or to secrete crystals and other ergastic substances. It may also become partly or wholly lignified. In spite of the fact that pith is the most deeply immersed tissue of the stem it can sometimes contain chlorophyll. A particularly surprising genus from this standpoint is *Theligonum* (now in the family Rubiaceae), for here chlorophyll is more abundant in the pith than in the more superficial cortex. In this genus the chlorophyll is especially plentiful in the peripheral part of the pith bordering on the xylem.

Attempts have been made over a period of many years to define and classify different types of pith. Some of these attempts are discussed by Solereder (1908) and he starts by referring to the work of Gris (1870) who recognized the following types of pith cells.

(1) Cells with rather thick walls in which starch is stored. This was referred to as **active pith**.
(2) Thin-walled pith cells that do not remain alive for long and become filled with air. This was termed **empty pith**.
(3) Cells containing crystals.

In some taxa the pith consists wholly of prosen-

[1] This description is based in part on notes made by Miss M. Gregory at the Jodrell Laboratory.

chymatous cells, e.g., according to Solereder, in certain species of *Myzodendron*. Alternatively there may be strands of fibres at the margin of the pith or in contact with the primary xylem. This is stated to apply, again according to Solereder, in certain Araliaceae, Corynocarpaceae, Loranthaceae, Malvaceae, Menispermaceae, Myzodendraceae, Platanaceae, Polygonaceae, Proteaceae, and Salicaceae. Sometimes the fibres form a ring at the boundary between the pith and primary xylem, e.g. in certain species of *Piper* (Piperaceae). Isolated fibres or strands of fibres are recorded by Solereder in the pith of certain Araliaceae (*Aralidium*), Asclepiadaceae, Euphorbiaceae, Geraniaceae, Lythraceae, Meliaceae, Ochnaceae (*Lophira*), Plumbaginaceae, Rutaceae, and Salicaceae. Alternatively the sclerenchyma in the pith may be represented by brachysclereids. In some plants the brachysclereids are so arranged that they form transverse diaphragms or septa (**septate pith**). For example, Solereder refers to diaphragms composed of brachysclereids in many Annonaceae (his Fig. 5, p. 35), Magnoliaceae, and Theaceae, as well as in some Convolvulaceae. When brachysclereids are present in the pith they may be organized as partial rather than as complete septa and sometimes they are present without showing any semblance of septa at all. It is also necessary to distinguish between a solid pith with transverse septa and a pith that is for the most part hollow or fistular with septa at intervals between consecutive cavities in the pith. Holm (1921) termed this last type **discoid pith** and he made a special study of it in *Juglans regia* L. The nature of the pith is recognized as having diagnostic value, but mainly at the species level.

Magócsy-Dietz (1899) has recorded diaphragms at the nodes in species of *Abelia Broussonetia, Clematis, Coronilla, Deutzia, Ficus, Forsythia, Leycesteria, Lonicera, Paulownia, Philadelphus, Symphoricarpus,* and *Vitis*.

A list of families and genera in which septate pith has been recorded is given on p. 220.

Cambium

The **cambium** need not detain us for long at this point because the manner in which it gives rise to secondary xylem is fully described under wood structure (Vol. II). The question of whether a young stem is destined to be eustelic, with separate vascular bundles, or siphonostelic, with a closed ring of xylem and phloem, is actually determined by the procambium before the cambium itself comes into play (Kostytschew 1922). As is explained under wood structure, the fully active cambium is made up of long fusiform initials that give rise to all of the cells in the xylem with the exception of the ray cells which are derived from distinctly shorter ray-initials (see, for example Fahn 1974). Ultrastructural details of dividing cambial initials have been investigated for certain species. For example, the particulars are given for *Ulmus americana* and *Tilia americana* by Evert and Deshpande (1970). The changes that take place in the walls of differentiating cambium cells are given by Kerr and Bailey (1934). Another point of interest is the question of what determines the wide range of differences in the spacing of rays in different kinds of wood. This topic has been investigated by Carmi, Sachs and Fahn (1972). It has been shown that the division of cambial cells and their differentiation are controlled by auxins and gibberellin. This has, for example, been demonstrated for *Xanthium* by Shininger (1971). The seasonal activity of cambium in relation to the production of growth rings is discussed under wood structure. It may, however, be noted here that the seasonal growth of cambium is sometimes controlled by the available moisture. It is stated, for example, by Aljaro, Guacolda, and Kummerow (1972) that *Proustia cuneifolia* (Asteraceae), a species from central Chile, sheds its leaves during droughts and, when this happens, the cambium ceases to be active and begins to divide again only when sufficient moisture becomes available. In the evergreen *Acacia caven*, from the same area in central Chile, the cambium remains active throughout the year. These are only some of the many interesting facts about cambial activity that are known, but we may justifiably leave the subject at this point because the facts are of physiological and ecological rather than of taxonomic interest. The great phylogenetic significance of the lengths of cells derived from cambial activity is discussed under wood structure (Vol. II).

Relation of organ size to tissue development

Sinnott (1936) investigated the fern *Todea hymenophylloides*, the conifer *Pinus strobus*, and an angiosperm *Datura stramonium* (Solanaceae) in order to discover whether, in stems or rhizomes differing in absolute size, the proportion occupied by each of the component tissues remains constant. He found that the pith increases more rapidly than the remainder of the stem and the cortex less rapidly, the vascular cylinder remaining intermediate between the two. No simple physiological explanation for this was found and it was concluded that the relative rates of increase of the different tissues 'may be an expression of a developmental pattern inherent in the constitution of the organism'.

Stem structure in relation to the production of flowers

Wilton and Roberts (1936) studied the structure of stems in relation to the production of flowers. They found that flowering stems of all species have certain anatomical characters in common, regardless of age or of photoperiodic classification. The flowering stem differs from the non-flowering stem in the following respects. The cambium is less active. A zone of mainly thick-walled secondary xylem elements lying next to the cambium is contrasted with the rather numerous vessels and thin-walled parenchymatous cells in the last-formed xylem of non-flowering stems. The cells of the pericycle, perimedullary region, xylem, and phloem are generally thicker-walled in flowering stems. Stems of disbudded plants have a structure resembling that of a flowering stem. This seems to indicate that the characterisic structure accompanies but is not a result of the production of flowers. The work was done on selected species from the following families: Aizoaceae, Amaranthaceae, Apocynaceae, Asteraceae, Begoniaceae, Brassicaceae, Euphorbiaceae, Geraniaceae, Linaceae, Onagraceae, Scrophulariaceae, and Solanaceae. Usually only one or two species from each family were examined, but six species of Solanaceae and five species of Asteraceae were studied.

In a subsequent investigation Wilton (1938) found that when plants reach an advanced reproductive stage the cambium is no longer active because it becomes wholly converted to xylem and phloem. It is suggested that this may account for the death of plants when the reproductive phase is over. Plants that are vegetatively vigorous have an active cambium throughout the length of their stems. The study by Wilton was based on 24 species of *Amaranthus* (Amaranthaceae), *Cannabis* (Cannabaceae), *Chrysanthemum* and *Cosmos* (Asteraceae), *Delphinium* (Ranunculaceae), and *Sidalcea* (Malvaceae).

Literature cited

Bark

Holdheide 1951; Moeller 1882; Parameswaran and Liese 1970; Whitmore 1962, 1963.

Cambium

Aljaro, Guacolda, Hoffmann, and Kummerow 1972; Carmi, Sachs, and Fahn 1972; Evert and Deshpande 1970; Fahn 1974; Kerr and Bailey 1934; Kostytschew 1922; Shininger 1971.

Cork

Beuzeville, de and Welch 1924; Diettert 1938; Holm 1929; Lemesle 1928; Luhan 1955; Moss 1936, 1940; Moss and Gorham 1953; Mullan 1932–3; Sitte 1962; Solereder 1908.

Epidermis

Linsbauer 1930

Endodermis

Barker 1953; Bond 1931; Codaccioni 1970; Fleet, van 1961; Ginzburg 1966; Mylius 1913; Priestley 1922, 1924, 1926, 1943; Priestley and Armstead 1922; Priestley and Ewing 1923; Priestley and North 1922; Scott, L. I. and Priestley 1928; Venning 1954.

Leaf and stem vasculature

Bailey and Nast 1948; Balfour and Philipson 1962; Benzing 1967a and b; Dormer 1945b, 1972; Esau 1965a,b; Howard 1974; Jensen, L.C.W. 1968; Marsden and Bailey 1955; Philipson and Balfour 1963; Slade 1971.

Stem ontogeny

Clowes 1961; Fahn 1974; Guttenberg 1960; Hanstein 1868; Schmidt, A. 1924.

Pith

Gris 1870; Holm 1921; Magócsy-Dietz 1899; Metcalfe 1948; Solereder 1908.

Polyderm

Luhan 1955; Mylius 1912, 1913; Nelson and Wilhelm 1957.

Relation of organ size to tissue development

Sinnott 1936.

Stem structure and flower production

Wilton 1938; Wilton and Roberts 1936.

Stelar structure and primary vasculature

Balfour and Philipson 1962; Brebner 1902; Esau 1943a,b, 1965a,b; Foster and Gifford 1974, pp. 58 *et seq.*; Hill. T.G. 1906; Jeffrey 1889; Philipson and Balfour 1963; Schoute 1902; Scott, D.H. 1891; Slade 1971; Tieghem, van and Douliot 1886a and b; Worsdell 1903.

Suggestions for further reading

Endodermis

Danilova and Stamboltzian 1969: physiology and ultrastructure of casparian strips in the tomato.
Guttenberg 1940: a general key work on root structure.
Kroemer 1903: an important old work; refers mainly to Monocotyledons; of no great taxonomic interest.
Rajowski 1934: a general article on the endodermis, written in Polish.

Stem ontogeny

There is a very extensive literature dealing with the initiation and differentiation of the primary tissues in the stem. The following list is only a selection:
Ball 1960: cell division in shoot apices.
Boke 1941: cell divisions in certain Cactaceae.
— 1947: shoot apex of *Vinca*.
Buvat 1944: cellular differentiation in intact plants and cuttings.
Cutter, 1957: morphogenesis of Nymphaeaceae.
— 1961: differentiation of lateral shoots.
Foster 1939: shoot apex; general.
Gifford 1954: review article.
Guttenberg 1960: reference book.
Hara 1962: shoot apex of *Daphne*.
Millington and Fisk 1956: shoot apex of *Xanthium*.
Philipson 1949: review article.
— 1954: review article.
Popham and Chan 1950: zonation in *Chrysanthemum* stem.
Vaughan 1955: stem apices of *Arabidopsis*, *Capsella* and *Anagallis*.
Wardlaw 1957: stem apex, general.
Werker and Fahn 1966: stem apex and leaf development of Chenopodiaceae.

Pith

[References marked L. are cited in the list on p. 220.]
Bailey and Smith 1942: Degeneriaceae. (L.)

Baillon 1871: general.

Beauvisage 1920: Theaceae and related families. (L.)

Bruyne, de 1922: Nymphaeaceae.

Diels 1919: Himantandraceae.

Faure 1924: Cornaceae. (L.)

Foxworthy 1903: discoid pith; Ulmaceae (L.)

Lechner 1914: *Actinidia, Saurauia, Clethra, Clematoclethra*.

Xylem: primary and early secondary

Bailey 1953*a*: wood structure and general.

Beck 1970: tracheary elements.

Bierhorst and Zamora 1965: primary xylem elements.

Bisalputra 1962: Chenopodiaceae.

Devedas and Beck 1972: *Cassia* and *Trifolium* (Fabaceae), *Geum, Potentilla, Prunus*, and *Rubus* (Rosaceae).

Esau, Cheadle, and Gill 1966: ultrastructure of tracheary elements.

Ezelarab and Dormer 1963: Ranunculaceae.

Fahn and Leshem 1963: living wood fibres.

Jacobs and Morrow 1957: auxin control of differentiation of primary vascular system.

Jeffrey 1917: classical concept of angiosperm vascular system.

O'Neill 1961: *Lupinus*.

Scott, F. M., Sjaholm, and Bowler 1960: ultrastructure of primary xylem.

12
PHLOEM
KATHERINE ESAU

Introduction

Features of dicotyledonous phloem that have been treated comparatively are almost wholly those found in the secondary phloem. The primary phloem has received much less attention in a comparative way. The metaphloem sieve elements of monocotyledons were found to be useful for the establishment of concepts about the evolutionary trends in the structure of these cells (Cheadle 1956). No studies of this kind have been carried out on the metaphloem sieve elements of dicotyledons. In contrast to metaphloem, the protophloem is relatively inaccessible to comparative studies because it is short-lived and becomes much altered after it ceases to function as a conducting tissue. Taxonomically, the modified protophloem is, in fact, more important than the younger tissue, for it acquires persisting characters that may be specific for certain taxa. These characters are usually referred to as those of the **pericyclic region**. After the term pericycle was introduced to designate the hypothetical limiting layer of the stele, the modified protophloem came to be regarded as the pericycle or part of it. Since the structure of the peripheral part of the vascular region is of taxonomic value, a simple topographical designation for the region should be available. Instead of introducing a special term for this purpose (e.g., perivascular region), the use of the familiar 'pericycle' and 'pericyclic' may be continued, with the understanding, however, that in many, and perhaps in most, dicotyledons the terms refer to the modified protophloem region or include it. Recognition of the phloic origin of the pericycle requires an ontogenetic study, an approach that seldom fits into a plan of a taxonomic investigation.

Certain features of the secondary phloem parallel those of the secondary xylem, for both kinds of tissue originate in the same meristem, the vascular cambium (see Vol. II), in which the basic organization of the secondary tissues is prefigured. At the same time, phloem and xylem show important structural differences determined by the disparate functions of the two tissues and by the difference in their position with regard to the cambium. The xylem is intensively sclerified and becomes surrounded, structurally intact, by further additions from the cambium. The active phloem usually has a high proportion of nonsclerified cells. As the tissue ages and is pushed outward by the increase in radial extent of the vascular cylinder its nonsclerified cells are deformed, displaced, and even completely crushed. The no-longer-functioning xylem preserves its taxonomically important structural features. In the phloem, only the latest-formed tissue can be used reliably for the identification of the basic structure and the interrelationship of cells typical of this tissue. In the older phloem, the basic features are obscured. Some of the new features of the nonconducting (nonfunctioning) phloem, however, such as sclerification, form of cell collapse, changes in the form and structure of rays, have a considerable taxonomic importance. The difference between xylem and phloem in the degree of sclerification among their respective elements is responsible for the greater success anatomists generally have with the xylem than with the phloem in making satisfactory preparations for microscopic studies. The relative meagreness in amount of information on the comparative structure of the phloem may be attributed, at least in part, to technical difficulties one encounters in working with this tissue.

The characters of the phloem that are potentially of taxonomic significance are those of individual cells, spatial interrelationships of cells in the tissue, modifications affecting the no-longer-conducting tissue, and relation of the phloem to the periderm (see p. 175). Thus, on a comparative basis, the phloem must be studied not only as one of the two characteristic conducting tissues but also as a component of the bark. In the following paragraphs, features of the phloem that show promise of service in taxonomic and phylogenetic investigations are discussed.

Sieve elements

Diagnostically valuable characters of sieve elements identifiable with the light microscope are: size relation to cambial cells; degree of inclination of end walls; nature of sieve plates; disparity in degree of differentiation between the sieve areas of the sieve plates and those on the lateral walls (lateral sieve areas); and the thickness of lateral walls.

Determination of sieve element length presents difficulties similar to those encountered in measurements of vessel member length, with the additional complication that maceration of the phloem tissue is not particularly useful for obtaining complete sieve elements. A more satisfactory approach is an examination of tangential sections including cambium and differentiating phloem (Plate 11(C and D)). In many species, cambial initials (and vessel members) may be used to estimate the length of sieve elements because, typically, sieve elements do not undergo intrusive elongation growth. In tangential sections, this feature may be checked by comparing sieve elements with cambial cells nearby.

A comparison of cambium and phloem in the same tangential sections may reveal the phenomenon of reduction of the potential size of sieve elements by divisions in phloem initials (phloem mother cells; compare cambial cells in Plate 11(C) with sieve elements in Plate 11(D)). This phenomenon has been called secondary septation (Zahur 1959). A preferable designation is **secondary partitioning**, for in vascular anatomy septation has the accepted connotation of belated divisions in cells with secondary walls. Secondary partitioning is secondary in the sense that it follows the additive divisions producing phloem initials from cambial initials as well as those periclinal divisions that may serve to multiply the phloem beyond the cambium. The term does not include divisions giving rise to companion cells. The partitioning divisions occur in various planes but in the mature tissue their results are most conspicuous when the partitioning walls are radial anticlinal (Plate 11(A and B), large open arrows). The cell complexes formed by secondary partitioning may combine two or more sieve elements with their companion cells or sieve elements, companion cells, and parenchyma cells. The partitioning walls vary in inclination. If they are longitudinal or nearly so, the potential width of sieve elements is decreased (Plate 12(D), arrowheads). If they are transverse or somewhat inclined, the potential length of sieve elements is decreased (Plate 12(D) at p). Secondary partitioning does not disturb the radial seriation of phloic cambial derivatives if the divisions are anticlinal (Plate 11(A and B) at large open arrows), but if the divisions occur in various planes, the radial seriation may become completely obscured.

If the sieve element is not shortened by secondary partitioning, the concept of evolutionary shortening developed with regard to cambial initials and vessel members of the xylem is applicable to sieve elements as well. In species with transverse or slightly inclined secondary partitions of phloem initials the evolutionary shortening is disguised. Lengths of sieve elements in the secondary phloem vary from less than 100 μm

to more than 700 μm (cf. Esau 1969; Zahur 1959).

The degree of inclination of the end wall is closely related to the length of the sieve element; it is greater in long than in short sieve elements. The orientation of the end wall may be expressed in terms of the angle between this wall and an imaginary horizontal wall. Vertical or extremely inclined end walls can hardly be differentiated from side walls so that the cell is truly fusiform as seen in a tangential view of the tissue (Plate 13(D and E)). The end wall may be characterized also by its length. The more inclined end walls are longer than the less inclined ones, although there is no strict correlation between length and degree of inclination because of the modifying factor of cell width.

The character of the sieve plate is affected by the length and inclination of the end wall to a considerable degree. Simple sieve plates, that is, those composed of single sieve areas (Plate 12(A and C)), commonly occur on transverse end walls; compound sieve plates, that is, those divided into several sieve areas (Plate 12 (B and D) at large arrow), are characteristic of the longer, more inclined end walls. Transverse and slightly inclined end walls, however, may also bear more than one sieve area. These relationships mean that shorter sieve elements are apt to have simple sieve plates, longer sieve elements, compound sieve plates. The number of sieve areas on inclined sieve plates commonly ranges between one and ten, but twenty and more have been recorded (Zahur 1959).

To estimate the disparity in the degree of differentiation between sieve plates and lateral sieve areas is not a simple matter except when the disparity, or lack of it, is obvious. The disparity results from differences in size of pores, a character that needs careful scrutiny under high magnification. Fortunately, however, sieve areas with larger pores show more conspicuous callose accumulations and usually occupy larger areas of the wall than do sieve areas with small pores. Thus, if pores are not clearly visible in sections, the relative size of the blue spots of callose in sections stained with lacmoid or aniline blue may be a guide to the differences in the differentiation of sieve areas of the sieve plates and those of the lateral walls (Plate 12(E), compare sa with p). In general, the disparity is greater in sieve elements with simple sieve plates and transverse end walls than in those having compound sieve plates and inclined end walls. In long sieve elements with much-inclined end walls, the sieve areas of the end wall often intergrade with those on the lateral walls so that the distinction between the two kinds of sieve area may be relatively small (Plate 13(C-E)). In short sieve elements with transverse end walls the simple sieve plates usually have conspicuously larger pores than the lateral sieve areas, and intermediate forms between the two kinds of sieve areas are lacking.

The thickness of the lateral wall in a sieve element is variable. The thickening may be set off from the cambial wall almost as clearly as a secondary wall, so that some workers interpret such a sieve element wall as secondary (cf. Esau 1969). Thin walls, and also relatively thick ones, may appear rather homogeneous except for the possible demarcation of the middle lamella. Customarily, the more or less thickened sieve element wall is called nacré or **nacreous wall** because of its glistening appearance in fresh sections. In some species, the nacreous wall is very conspicuous and is readily visible with the light microscope (Plate 14(B and C)). Its inner margin may be deeply crenulated. The thick nacreous wall develops during the differentiation of the sieve element and, at levels where it is thickest, may almost occlude the lumen of the mature, and presumably conducting, sieve element. This type of nacreous wall is a good diagnostic feature. One should have a simple descriptive terminology for indicating the presence or absence of this wall. Reference to secondary wall would not be appropriate because as yet there is no consensus as to when a sieve element has or does not have a secondary wall. 'Nacreous wall' alone is not adequate because a thick nacreous wall intergrades with a thin wall, and electron microscopy indicates that walls of different thicknesses are ultrastructurally similar. The least ambiguous approach in a comparative study would be to rely on light microscopy and to differentiate between a condition with a **nacreous wall conspicuous**, when the sieve element wall is obviously thicker than that of the adjacent parenchyma and companion cells, and one with **nacreous wall inconspicuous**, when the sieve element hardly differs from those of the associated parenchymatous cells.

Electron microscopy reveals some features in the sieve element protoplast that promise to be taxonomically significant. The **plastids** in the sieve elements examined thus far may be divided into two basic types, one, the **S-type**, typically storing starch, the other, the **P-type**, storing protein (Behnke 1972) (Plate 12(F–H)). Within the P-type one may further distinguish between the plastids frequently depositing the protein in circumscribed crystalloids (Plate 12(G)) and the plastid consistently forming a ring of fibrous protein (Plate 12(H)). The second type of P-type plastid may be called the **beta P-type** Plastid because it was first discovered in *Beta vulgaris*. To round out the terminology, the P-type without the fibrous ring may be designated as **alpha P-type** plastid. Protein-storing plastids may or may not form starch. S- or P-types of plastid appear to be characteristic of certain large taxa.

The conformation of P-protein (formerly called slime) in differentiating sieve elements may have some diagnostic value. Before the sieve element reaches maturity the P-protein forms aggregates (slime bodies) the shape and distinctness of which vary. The paracrystalline, spindle-shaped body, drawn out into a slender tail at one or both ends, is characteristic of representatives of the Fabaceae. The non-crystalline type is probably more common and may be spindle-shaped, spherical, or less precisely delimited. For the present, the paracrystalline body found in the Fabaceae (Papilionaceae) seems to be the best candidate for a diagnostic feature, especially since this kind of P-protein body persists longer as a discrete structure than does the noncrystalline type. The latter usually disperses before the sieve element reaches maturity.

A feature of the sieve element in some taxa that is readily detectable with the light microscope and may have taxonomic value is a proteinaceous inclusion that has been long referred to as an extruded nucleolus (cf. Esau 1969). This interpretation has proved to be incorrect (Esau 1978). The body arises in the cytoplasm of a sieve element in early stages of differentiation, in which the nucleus is still intact and contains a normal nucleolus; and it persists as long as the cell remains functional. The body is chemically related to the P-protein and therefore takes up cytoplasmic dyes. As seen with the electron microscope, it is partly or entirely paracrystalline. The body is a conspicuous component of the mature sieve element protoplast in a number of genera having no close taxonomic relationships one to another.

Companion cells

These parenchymatous members of the phloem are characteristic of dicotyledons. Companion cells and phloem parenchyma cells are not always clearly distinguishable, a circumstance that may lead to an impression that in a given taxon companion cells are absent. The companion cell may be one of several parenchymatous cells that arise by divisions of a sieve element precursor. There is also a functional intergrading between companion cells and phloem parenchyma cells contiguous to the same sieve element. Thus, lack of a clear cytologic differentiation may not mean that companion cells are absent. In such instances, an appropriate treatment would be to refer to 'lack of a clear morphologic specialization of companion cells' or 'companion cells not identifiable'.

The readily identifiable companion cells vary in relative size, number per sieve element, and spatial relation to the sieve element. A companion cell may be as long as the related sieve element or shorter, depending on whether the division wall separating the companion cell from the sieve element extends the entire length of the sieve element precursor or not. A companion cell which is shorter than the sieve element usually clearly reveals its identity because it appears as though carved out of the sieve element (Plate 15(C) at cc). There may be a single small companion cell or several. When there are several they are randomly distributed over the wall of the sieve element. Small companion cells may also result from a subdivision of a long companion cell precursor, which thus gives rise to a companion cell strand. Zahur (1959) found that the three ontogenetic types of companion cells (companion cell as long as the sieve element, companion cell shorter than the sieve element, and companion cells forming a strand) were characteristic of natural groupings among the dicotyledons examined.

Companion cells often display dense protoplasts, a feature helpful in their identification. Their connections with sieve elements are plasmodesmata which have a single opening on the sieve element side and several on the companion cell side of the wall (branched plasmodesma). This feature is discernible with the electron microscope. It serves as confirming evidence when a cell is suspected of being a companion cell. However, it cannot be used alone for the identification of a companion cell because similar plasmodesmata may occur between sieve elements and phloem parenchyma cells.

Phloem parenchyma cells

The parenchyma cells in the axial (or vertical) system of secondary phloem may differentiate from cambial derivatives without a subdivision of the latter into smaller cells. More commonly, the axial parenchyma cells originate through transverse or somewhat inclined divisions of cambial derivatives. The two kinds of origin of phloem parenchyma cells have led to the classification into parenchyma cells and parenchyma strands. A parenchyma strand may appear as a longitudinal file of cells which are separated from each other by transverse walls. In some taxa the partitioning

walls are positioned at various angles. The parenchyma strand then displays a patchwork pattern as seen in tangential sections of the tissue. The number of cells in a strand is variable.

In contrast to axial wood parenchyma, the axial phloem parenchyma does not occur in a variety of classifiable distributional patterns. Moreover the distinction between parenchyma cells and sieve elements in tissue transections is frequently not sharp enough to be helpful in locating the parenchyma cells precisely without a painstaking comparison of cells in serial sections. Parenchyma occurs in more or less regular tangential bands of one or more layers of cells in depth (Plates 13(A and B) and 14(A and B)) or it forms radial rows of limited extent alternating with rows containing the sieve elements. Parenchyma cells and sieve elements may be intermixed somewhat irregularly (Plate 16(D)). Parenchyma cells are seldom scattered singly (Holdheide 1951). Zahur (1959) classified the distributional patterns as banded (tangential bands) and irregular and considered the classification to be of some comparative value. Parenchyma is less abundant in the early than in the late phloem and may constitute the final layer of an annual increment.

Since the sieve elements and companion cells are more or less crushed in nonconducting phloem, phloem parenchyma becomes more prominent as the tissue ages (compare (D) and (E) in Plate 17). The cells enlarge and may participate with the rays in the dilatation phenomena that bring about the adjustment of the bark to the increase in stem circumference. Dilatation of axial parenchyma is more characteristic of some taxa than of others. Axial parenchyma may become sclerified in the old phloem. It is then transformed into sclerified or sclerotic parenchyma, sclereids (Plate 11(A) and 18(A), sc), or fibre-sclereids (Plate 18(C–E). If sclerification does not affect axial parenchyma, the cells may become crushed in the nonconducting phloem (Plate 17(E)).

In some taxa, the contents of parenchyma and related cells are useful for taxonomic descriptions. Tanniniferous compounds may be present and be abundant or scarce. Tannin-containing cells may show characteristic distributional patterns (Plate 16(B)). Parenchyma cells with specialized contents sometimes have distinctive form and merit the name of idioblasts. Among such cells are tannin sacs, oil cells, mucilage cells, and others. Laticifers of two types, non-articulated or coenocytic (some Apocynaceae, Asclepiadaceae, Moraceae) and articulated (Caricaceae, Cichorieae, *Hevea*, *Papaver*) occur in the phloem.

[For a full discussion of laticifers see Vol. II.]

Taxa vary with regard to types of crystals in the parenchyma cells of the phloem and in the distribution of these cells. (See Vol. II for general account of crystals.) Crystals in phloem may be located in somewhat specialized **crystalliferous parenchyma strands** in which the individual cells are smaller than those in ordinary parenchyma strands. Each cell usually contains a single crystal (Plate 15(D)). Crystalliferous cells frequently become sclerified. Processing of material for preparation of permanent slides may remove the crystals. Their presence is best determined in fresh material.

Rays

Because of its common ontogenetic origin with the secondary xylem, the secondary phloem shows the same basic ray characteristics as the wood. The **phloem rays** may be more or less numerous, narrow or wide, uniseriate or multiseriate, low or high, homocellular or heterocellular, and may contain secretory cells. Rays of more than one kind may occur in the same species. Phloem rays, however, assume characteristics different from those of the xylem rays after the phloem becomes part of the older bark. The rays are frequently considerably affected by the collapse of sieve elements and associated cells and by the increase in stem circumference. The wide rays commonly increase in width (dilatation) by cell division and tangential extension of cells (Plates 14(A) and 16(C)). The dividing cells may be concentrated in the central part of the ray and form a band of meristematic cells resembling cambium. The course of these rays remains radial but they flare toward the periphery of the stem. The narrow rays commonly become curved and, as seen in transections of a stem, are inclined in one direction. The degree of dilatation and displacement of rays varies in different taxa depending, in part, on the structure of the axial system and its transformation in the old bark. The presence of numerous fibres in the axial system, for example, minimizes the effect of sieve element collapse. The change in the appearance of rays in older phloem may be small if the periderm is of the type that is periodically renewed in successively deeper layers of the phloem (Plate 15(A)). In some taxa, the vascular system of the axis consists of bundles separated by wide rays which originate as wide interfascicular regions while the axis is still in a primary state of growth (Plate 17(B)). These rays are sometimes referred to as medullary rays or primary rays. Their presence gives a characteristic appearance to the vascular system, including the phloem, and is of taxonomic significance.

Sclerification of cells is less common in rays than it is in axial parenchyma, a feature probably associated with the maintenance of rays as paths of radial translocation of carbohydrates as long as the bark remains alive. If the rays are flanked by axial sclerenchyma, their peripheral cells may develop into sclerified crystalliferous cells. Such development usually occurs only in multiseriate rays. Wide rays of some genera present an exception in that they do form sclereids (Holdheide, 1951). Crystal deposition is also not a feature of rays, but exceptions occur. If crystals accumulate in rays, the rays are of the wide type.

Fibres

Phloem fibres are of considerable taxonomic importance. When present in the secondary phloem, they are conspicuous components because of their thick walls and characteristic distributional patterns. Secondary phloem fibres are absent from some taxa (*Austrobaileya scandens, Neocinnamomum delavayii, Paeonia suffruticosa*; Plates 11, 13, and 15). They are scattered singly (*Campsis radicans* (Plate 16(A)), *Litsea calicaris*) or in small tangential rows (*Liriodendron tulipifera, Magnolia kobus* Plate 14(B)) in other taxa. In still others, fibres form relatively regular tangential bands, several cells in depth, alternating with bands of the soft phloem cells (*Cananga odorata, Castanea dentata, Fraxinus americana* (Plate 16(B-D)), *Magnolia grandiflora*). The fibre bands in contiguous radial panels of the axial system usually occur at the same distance from the cambium and give the secondary phloem a stratified aspect in transverse sections. If the rays are narrow, the bands appear to be almost continuous over considerable areas of the circumference (*Castanea dentata, Fraxinus americana*; Plate 16 (B,D)). Wide rays form conspicuous breaks in the pattern formed by the fibres. Moreover, if the rays become dilated in the older phloem, a characteristic pattern develops in the bark as seen in transections: wedges of axial phloem, narrower toward the periphery, wider toward the cambium, show bands of fibres alternating with bands of more or less crushed cells (*Cananga odorata* (Plate 16(C), *Tilia cordata*). The wide rays, which are narrower toward the cambium and wider toward the periphery, alternate with the wedges of the axial system. There may be some narrow rays within the wedges of the axial system.

A feature of fibres that is rather difficult to assess taxonomically is their exact classification. Some phloem fibres correspond to the libriform fibres in the xylem in that they differentiate directly as cambial derivatives, elongate intrusively, and develop secondary walls with few inconspicuous pits (*Populus grandidentata*; Plate 18(A and B)). These are the true fibres, or **bast fibres**, of the German authors (Holdheide 1951), a term also adopted by the International Association of Wood Anatomists (1957). Since the bast fibres differentiate together with the other phloem cells, they are present in the conducting part of the phloem (*Cananga odorata*; Plate 16(C)). The second kind of fibre differentiates later in the life of the tissue, by sclerification of parenchyma cells. It is therefore not present as such in the functioning phloem (*Fraxinus americana*, *Bourreria ovata*; Plates 16(D) and 18(C)). A convenient term for this kind of fibre is **fibre-sclereid** (sclerotic fibre of the German authors: Holdheide 1951). It is a fibre in shape, length, and wall thickness, but it originates as do sclereids by sclerosis of parenchyma cells. As in sclereids, pits may be conspicuous in fibre-sclereids (Plate 18(D–E)). Fibre-sclereids elongate intrusively as do many bast fibres. The degree of intrusive elongation of both kinds of fibre varies in different taxa and may be used in characterizing the phloem. A comparison of fibres with vessel members of the same plant is a reliable method for determining whether the fibre has elongated beyond the cambial length and to what extent (cf. Esau 1969).

Both kinds of fibre may be septate, and both may occur together with sclereids. In fact, the same species may have bast fibres and fibre-sclereids. Thus, the distinction between the different kinds of sclerenchyma cells is not always obvious. Moreover, the parenchyma cells that differentiate into fibre-sclereids are not necessarily simply phloem parenchyma cells that happen to undergo sclerosis. They may be identifiable as fibre primordia near the cambium by their regular arrangement and scarcity of ergastic substances (large arrow in Plate 16(D)). The difference in timing of maturation does not invariably facilitate the recognition of fibre type. Its determination is entirely uncertain without good preservation of the cambium and the youngest phloem and requires a substantial knowledge of the character of seasonal growth in the given phloem. If the nature of the fibres is not easily determined, the presence or absence of such cells may be recorded under a group name of fibres. Whatever kinds of fibres are present, their distribution is an accessible character useful for taxonomic descriptions

of phloem.

Bast fibres often differentiate in the nonfunctioning protophloem. They may be longer than the fibres in the secondary phloem because they are initiated in an elongating part of the shoot and are thus able to extend with the surrounding tissues as well as by intrusive growth. Fibre-sclereids also may occur along the periphery of the vascular cylinder (Plate 17), often forming a continuous cylinder of sclerenchyma in combination with sclereids (pericyclic region).

Storied structure

This character is as useful in classification of phloem as it is in that of xylem. In the phloem, however, cell arrangement must be determined by the use of the latest increment of tissue because the changes occurring in nonfunctioning phloem, especially the crushing of some cells and dilatation of others, may obscure the storied structure.

Growth rings

Although occurrence of growth rings has been recorded for secondary phloem, the character is of little value for taxonomic descriptions. The identification of growth increments must be made by reference to the development of the tissue through a season. Moreover, the feature is highly susceptible to variations induced by environmental factors.

Variations in the position of the phloem

The occurrence of phloem in positions other than the typical one outside the xylem (**external phloem**) characterizes certain taxa. Some taxa have **intraxylary phloem** (internal phloem), that is, phloem located next to the pith. This tissue is mainly primary. The individual vascular bundles in plants with internal phloem are bicollateral bundles. In secondary growth, the so-called anomalous cambial activity results in distributional patterns characterized by an intermingling of xylem and phloem. Three different methods of cambial activity bring about this condition. First, in the **divided or compound xylem** cambia arise around individual vessels or groups of vessels and produce xylem and phloem in bidirectional sequence. The secondary xylem assumes the appearance of having been divided into numerous vascular cylinders. Second, in the **included (interxylary) phloem**, normally positioned cambium intermittently produces groups of cells toward the xylem which differentiate into strands of phloem. In transverse sections, the groups of unlignified phloem cells scattered in the xylem resemble holes, hence the designation **foraminate included phloem**. Third, in 'secondary growth from successive cambia', each cambium produces xylem toward the inside and phloem toward the outside so that xylem and phloem alternate in layers around the circumference (or part of it) of the axis (**concentric type** of **included phloem**). The presence of anomalous patterns can be reliably established at relatively low magnifications. Despite the indications that anomalous secondary thickening may have evolved from the normal in relation to special growth forms, the feature is useful as a taxonomic criterion when combined with other characteristics (cf. Carlquist 1961). (Anomalous structure is also discussed in Vol. II.)

Outer bark[1]

In many dicotyledons, the old phloem eventually becomes part of the outer bark, or **rhytidome**. Characters of the outer bark that distinguish different taxa are time and depth of appearance of the first periderm, frequency of origins of successive periderms, nature of cork, amount and kind of tissues isolated by the periderms, shape of periderm layers, and manner of bark abscission. The internal features of periderm and rhytidome affect the external appearance of the bark. It is possible to distinguish externally between ring bark and scaly bark; between fibrous bark containing phloem fibres and flaky bark containing numerous sclereids; between a loose bark and a coherent bark. Features of the outer bark are worthy of attention for their possible use in systematics.

Evolutionary trends in phloem structure

Since information on comparative structure of phloem is meagre, deductions regarding the evolutionary trends in this tissue must be regarded as tentative. Certain aspects of evolutionary changes in phloem are commonly treated by comparison with those established for xylem, with due consideration of the features of phloem related to the specific function of the tissue.

In the secondary phloem, a comparison of sieve elements with xylem vessels with regard to length of

[1] Bark is also mentioned on p. 175.

cells and inclination of end walls is particularly appropriate. The concept of evolutionary shortening of the cells and concomitant decrease in length and inclination of end walls is applicable to both vessel elements and sieve elements as derivatives of common precursors in the initial region of the cambium. In vessel elements the reduction in length of the end wall is correlated with a decrease in number of perforations, and in sieve elements, with a decrease in number of sieve areas. A long sieve element with a strongly inclined long end wall bearing numerous sieve areas is considered to be primitive (Plate 13(C–E)). This interpretation is supported by the predominance of this type of conducting cell in the phloem of vascular cryptogams and gymnosperms. Conversely, a short sieve element with transverse end walls bearing single sieve areas is regarded as phylogenetically advanced (Plate 17(C)).

A further feature, the degree of differentiation of sieve area pores, has been combined with those of length and form of sieve element to formulate the classification into two kinds of sieve element, the less specialized **sieve cell** and the more specialized **sieve tube member**. (The original formulation was made regarding monocotyledon metaphloem; cf. Cheadle 1956.) Sieve cells are relatively long and have much-inclined end walls bearing sieve areas indistinguishable from those on the lateral walls. A sieve tube member varies in length but tends to be shorter than a sieve cell, its end walls vary from considerably inclined to transverse, and bear sieve areas that have larger pores than those in the sieve areas on the lateral walls. One or more sieve areas on the end wall compose the **sieve plate**. The longitudinal files of sieve tube members interconnected by sieve plates are called **sieve tubes**. If the evolutionary changes from the sieve cell to the most highly evolved sieve tube member are evaluated with reference to functional specialization, an increased emphasis on longitudinal conduction may be deduced. Sieve tubes are typical of dicotyledons. Since there is a range in degree of specialization of Dicotyledon sieve elements, some may be on the borderline between a sieve cell and sieve tube member. *Austrobaileya* (Plate 13) and some Pomoideae exemplify dicotyledons having sieve elements with relatively little differentiation between the sieve areas on the end walls and those on the side walls.

The parallelism between xylem and phloem in the shortening of conducting cells is obscured in some dicotyledons by transverse secondary partitioning. As a result of this partitioning, short sieve elements and transverse sieve plates may appear in families in which the cambial initials are rather long and have inclined end walls (Plate 12(D)). Zahur's (1959) speculation that secondary partitioning indicates a phylogenetic advance is questioned by Carlquist (1961) who points out that this feature occurs in families that are relatively primitive in many other features. Significantly, the partitioning decreases the potential length and width of the conducting cells. In attempting to relate this phenomenon to functional specialization one must face the dilemma that there is no information on the possible effect of the decrease in size of conduit and the addition of sieve plates on the movement of materials in the phloem. The occurrence of thick nacreous walls in some taxa, which seemingly obstruct the path of translocate, poses a similar problem. Thus, two features of sieve elements, secondary partitioning and nacreous wall thickening, both potentially useful in systematics, cannot be meaningfully discussed with regard to their evolutionary significance.

A study of pore size in sieve areas of a large sampling of dicotyledons (Esau and Cheadle 1959) disclosed some values that fail to indicate a straightforward relation between specialization of sieve areas and a presumed increase in efficiency of the conduit. The study confirmed the generalization that simple sieve plates tend to have larger pores than scalariformly compound sieve plates. However, sieve elements with scalariform sieve plates were wider and had larger ratios of pore areas to transverse cell areas than had sieve elements with simple sieve plates. Thus, despite their large size, the pores in simple, transverse sieve plates occupy a smaller fraction of the transverse cell area than do the smaller pores in compound, inclined sieve plates. Presence of callose as a lining of the pores reverses this relation. The significance of this reversal with regard to functional specialization of sieve elements is not clear because of the possibility that the amount of callose in phloem which was injured in sampling may be greater than it is in an intact plant. The same study indicated only a minor variation in the size of pores in lateral sieve areas. From the phylogenetic aspect, the actual size of these pores is probably less significant than the contrast between the pores in the sieve plates and those in the lateral sieve areas. The contrast is greater in sieve elements with transverse sieve plates than in those with inclined sieve plates.

Zahur (1959) attempted to summarize the data on sequences in morphologic specialization of sieve elements in dicotyledons by proposing a classification of sieve elements into three categories (originally formulated by Hemenway 1913):

(1) Sieve elements long, sieve plates very oblique and bearing ten or more sieve areas.
(2) Sieve elements of medium length, sieve plates oblique and bearing two to ten sieve areas.
(3) Sieve elements short, sieve plates slightly oblique to transverse and bearing one sieve area. The shortening of sieve elements through decrease in length of initials and through secondary partitioning are both regarded as indications of evolutionary advance in this classification.

The contents of sieve elements need some consideration in a discussion of phloem phylogeny. The characteristics of sieve element contents indicate a specialization of an initially parenchymatous kind of protoplast. This specialization involves reduction of metabolic capabilities through loss of components necessary for protein synthesis, obliteration of delimitation between vacuole and cytoplasm, establishment of a high degree of continuity with adjacent sieve element protoplasts through sieve area pores, and development of features enabling the cell to interrupt the movement of materials from cell to cell by plugging the pores.

A concomitant of specialization of the sieve element protoplast is the establishment of a close functional relation between the sieve element and the associated parenchyma cells. The sieve element and companion cell association appears to be the highest expression of this relation since the companion cell is characteristic only of the angiosperms and its ontogeny and longevity are tied to those of the sieve element. To various degrees, parenchyma cells other than the companion cells are functionally related to the sieve elements. In some taxa, the morphologic distinction between companion cells and contiguous parenchyma cells is not sharp. Whether the intergrading of companion cells and parenchyma cells indicates a lower degree of specialization of companion cells has not been established by adequate surveys.

Zahur (1959) has classified the companion cells morphologically into three categories and placed them in the following order of presumed evolutionary advance:

(1) Companion cells are much shorter than the sieve element and occur singly.
(2) Companion cells are as long as the sieve element.
(3) Companion cells form a strand extending the length of the sieve element.

Zahur (1959) also considered the distributional patterns of phloem parenchyma (irregular versus banded) and sclerenchyma but recognized no decisive trends. Neither was there evidence of a correlation between type of fibre and type of sieve element. Whereas the phylogenetic relation between fibres and conducting cells in the xylem is well established, no such relation is perceptible between sieve elements and fibres of the phloem.

The phloem tissue has apparently undergone an evolutionary reduction in relation to specialized ways of life of some taxa. Thus, in some dicotyledonous parasites, phloem is poorly developed or absent entirely. The phloem of climbing roots of *Hedera helix* is said to be much reduced and that of the podostemaceous water plant *Mourera aspera* is not discernable as a differentiated tissue (cf. Esau 1969).

Literature cited

Behnke 1972; Carlquist 1961; Cheadle 1956; Esau 1969, 1978; Esau and Cheadle 1959; Hemenway 1913; Holdheide 1951; International Association of Wood Anatomists 1957; Zahur 1959.

Suggestions for further reading

Moeller 1881: comparative anatomy of barks.
Parameswaran 1971: technical properties of tropical barks.
Parameswaran and Liese 1970: comparative structure of tropical barks.
Srivastava 1964: anatomical, physiological, and chemical properties of forest tree barks.
Thorenaar 1926: comparative structure of technically useful barks.

LISTS OF FAMILIES IN WHICH CERTAIN DIAGNOSTIC FEATURES OCCUR

In the following lists the inclusion of a family signifies only that the feature occurs in certain members of the family and not necessarily in every genus. The families in which the feature is especially common are shown in bold type; those in which it is infrequent or rare are given in italics.

The lists are unavoidably incomplete and imperfect and they must inevitably remain so until the families have been described in future volumes of this second edition.

In these lists the familiar designations Leguminosae, Papilionaceae, Caesalpiniaceae, and Mimosaceae have been retained instead of including them all under Fabaceae.

Families in which epiphyllous flowers occur

Modified from Johnson (1958)

Begoniaceae. *Begonia* 2 spp
Celastraceae. *Polycardia* 2 spp
Cornaceae. *Helwingia* 2 spp
Dichapetalaceae. *Dichapetalum* 2 spp, *Tapura* 1 sp
Dulongiaceae. *Phyllonoma* 3 spp
Flacourtiaceae. *Phyllobotryon* 3 spp, *Phylloclinium* 2 spp
Gesneriaceae. *Chirita* 1 sp, *Didymocarpus* 1 sp, *Streptocarpus* 5 spp
Icacinaceae. *Leptaulus* 1 sp
Piperaceae. *Peperomia* 2 spp
Rutaceae. *Erythrochiton* 1 sp
Turneraceae. *Turnera* 10 spp

Hairs or trichomes

The lists that follow, compiled by the same authors, should be used in conjunction with the contribution by Theobald, Krahulik, and Rollins on pp. 40-53.

What was said in the first edition of this present work regarding trichomes is as applicable now as it was in 1950:

Taxonomic value of hair structure is well established. The various types are not always clearly defined, however, and their value for systematic purposes is lessened by the fact that many kinds occur in families which are generally thought to be unrelated, thus making it reasonably certain that the same type of hair must have been evolved along independent lines. Nevertheless, some kinds possess diagnostic value because of their restricted occurrence, whilst the combination of hair types is valuable in families where several or numerous kinds occur together.

In the following lists the first edition of this present work and Solereder's (1908) treatise have been used to compile the groupings. In several cases it was difficult to decide what types of trichomes were being described and thus we are certain that flaws do exist. However, these categories should prove useful until such times as the subsequent volumes in this edition deal with the families in greater detail.

In those instances where there are major subdivisions (e.g. Simple trichomes) the families may be found listed under all subdivisions of the group. This was done so that investigators would be able to place a family no matter what degree of material he might have on hand or what degree of exactness had gone into the observation.

If a particular type is especially common the family is indicated by bold face type while those in which it is rare or infrequent are indicated by italics. *The glandular or non-glandular condition is indicated by a G or N, respectively.*

Simple (unbranched)

Acanthaceae (G,N)
Aceraceae (G,N)
Actinidiaceae (G)
Adoxaceae (G)
Aizoaceae (N)
Akaniaceae (N)
Alangiaceae (G,N)
Amaranthaceae (G,N)
Anacardiaceae (G,N)
Annonaceae (N)
Apocynaceae (G,N)
Aquifoliaceae (N)
Araliaceae (N)
Aristolochiaceae (G,N)
Asclepiadaceae (G,N)
Asteraceae (G,N)
Balanophoraceae (G,N)
Begoniaceae (G,N)
Berberidaceae (G,N)
Betulaceae (N)
Bignoniaceae (G,N)
Boraginaceae (G,N)
Brassicaceae (G,N)
Brunelliaceae (N)
Bruniaceae (N)
Burseraceae (G,N)
Buxaceae (N)
Cactaceae (N)
Caesalpiniaceae (G,N)
Callitrichaceae (G)
Calycanthaceae (N)
Campanulaceae (N)
Cannabaceae (G,N)
Capparaceae (G,N)
Caprifoliaceae (G,N)
Caricaceae (G)
Caryocaraceae (N)
Caryophyllaceae (G,N)
Casuarinaceae (N)
Celastraceae (N)
Ceratophyllaceae (G,N)
Chenopodiaceae (G,N)
Chrysobalanaceae (G)
Cistaceae (G,N)
Cneoraceae (G)
Cochlospermaceae (N)
Columelliaceae (N)
Combretaceae (G,N)
Connaraceae (G,N)
Convolvulaceae (G,N)
Cornaceae (G,N)
Corylaceae (G,N)
Crassulaceae (G,N)
Cucurbitaceae (G,N)

Cunoniaceae (G)
Datiscaceae (G)
Diapensiaceae (N)
Dichapetalaceae (N)
Dilleniaceae (N)
Dipsacaceae (G,N)
Dipterocarpaceae (G,N)
Ebenaceae (G,N)
Elaeocarpaceae (G,N)
Elatinaceae (G,N)
Empetraceae (G,N)
Epacridaceae (N)
Ericaceae (G,N)
Escalloniaceae (G,N)
Eucommiaceae (N)
Eucryphiaceae (N)
Euphorbiaceae (G,N)
Eupteleaceae (N)
Fagaceae (G,N)
Flacourtiaceae (G,N)
Frankeniaceae (N)
Garryaceae (N)
Gentianaceae (G,N)
Geraniaceae (G,N)
Gesneriaceae (G,N)
Globulariaceae (G,N)
Goodeniaceae (G,N)
Goupiaceae (N)
Grossulariaceae (G,N)
Grubbiaceae (N)
Haloragaceae (G,N)
Hamamelidacae (N)
Hernandiaceae (N)
Hippocastanaceae (G,N)
Hippocrateaceae (N)
Houmiriaceae (N)
Hydrangeaceae (N)
Hydrophyllaceae (G,N)
Hypericaceae (N)
Icacinaceae (N)
Juglandaceae (G,N)
Julianiaceae (G,N)
Krameriaceae (N)
Lamiaceae (G,N)
Lauraceae (N)
Lecythidaceae (G,N)
Leitneriaceae (G,N)
Lennoaceae (G)
Lentibulariaceae (G,N)
Linaceae (G,N)
Loasaceae (G,N)
Lobeliaceae (N)
Loganiaceae (G,N)
Loranthaceae (N)
Lythraceae G,N)
Magnoliaceae (N)
Malesherbiaceae (G,N)

Malpighiaceae (G,N)
Malvaceae (G,N)
Melastomataceae (G,N)
Meliaceae (G,N)
Menispermaceae (G,N)
Mimosaceae (G,N)
Monimiaceae (N)
Moraceae (G,N)
Moringaceae (N)
Myoporaceae (G,N)
Myricaceae (G,N)
Myrsinaceae (N)
Myrtaceae (G,N)
Myzodendraceae (N)
Nyctaginaceae (N)
Nymphaeaceae (G,N)
Nyssaceae (G,N)
Ochnaceae (G,N)
Olacaceae (N)
Onagraceae (N)
Orobanchaceae (G,N)
Oxalidaceae (G,N)
Papaveraceae (N)
Papilionaceae (G,N)
Passifloraceae (G,N)
Pedaliaceae (G,N)
Penaeaceae (N)
Phytolaccaceae (N)
Piperaceae (G,N)
Pittosporaceae (G,N)
Plantaginaceae (G,N)
Platanaceae (G,N)
Plumbaginaceae (G,N)
Podostemonaceae (N)
Polemoniaceae (G,N)
Polygalaceae (N)
Polygonaceae (G,N)
Portulacaceae (G,N)
Primulaceae (G,N)
Proteaceae (N)
Ranunculaceae (G,N)
Resedaceae (N)
Rhamnaceae (N)
Rhizophoraceae (G,N)
Rosaceae (G,N)
Rubiaceae (G,N)
Rutaceae (G,N)
Sabiaceae (G,N)
Salvadoraceae (N)
Sarcolaenaceae (G,N)
Santalaceae (N)
Sapindaceae (G,N)
Sapotaceae (N)
Saxifragaceae (G,N)
Scrophulariaceae (G,N)
Scytopetalaceae (G)
Simaroubaceae (G,N)

Solanaceae (G,N)
Stackhousiaceae (G,N)
Sterculiaceae (G,N)
Stylidiaceae (G)
Symplocaceae (G,N)
Tamaricaceae (N)
Theaceae (N)
Thymelaeaceae (N)
Tiliaceae (G,N)
Tremandraceae (G,N)
Trigoniaceae (N)
Turneraceae (G,N)
Tropaeolaceae (N)
Ulmaceae (G,N)
Urticaceae (G,N)
Valerianaceae (G,N)
Verbenaceae (G,N)
Violaceae (G,N)
Vitaceae (G,N)
Vochysiaceae (N)
Zygophyllaceae (G,N)

Simple (unbranched)-short

Acanthaceae (G,N)
Aceraceae (G)
Adoxaceae (G)
Aizoaceae (N)
Alangiaceae (G,N)
Amaranthaceae (G,N)
Anacardiaceae (G,N)
Annonaceae (N)
Apiaceae (G,N)
Apocynaceae (G,N)
Aquifoliaceae (N)
Aristolochiaceae (G,N)
Asclepiadaceae (G,N)
Asteraceae (G,N)
Balanophoraceae (N)
Begoniaceae (G,N)
Bignoniaceae (G,N)
Boraginaceae (G,N)
Brassicaceae (G,N)
Burseraceae (G,N)
Cactaceae (N)
Caesalpiniaceae (G,N)
Callitrichaceae (G)
Campanulaceae (N)
Cannabaceae (G,N)
Caprifoliaceae (G,N)
Capparaceae (N)
Caryocaraceae (N)
Caryophyllaceae (G,N)
Celastraceae (N)
Chenopodiaceae (G,N)
Chrysobalanaceae (G)
Cistaceae (G,N)
Combretaceae (G,N)

Connaraceae (N)
Convolvulaceae (G,N)
Cornaceae (G,N)
Crassulaceae (G,N)
Cucurbitaceae (G,N)
Dipsacaceae (G,N)
Dipterocarpaceae (G,N)
Ebenaceae (G,N)
Elatinaceae (G)
Empetraceae (G,N)
Epacridaceae (N)
Ericaceae (G,N)
Escalloniaceae (G,N)
Eucryphiaceae (N)
Euphorbiaceae (G,N)
Fagaceae (N)
Flacourtiaceae (G,N)
Frankeniaceae (N)
Gentianaceae (N)
Geraniaceae (G,N)
Gesneriaceae (G,N)
Globulariaceae (G,N)
Goodeniaceae (G,N)
Goupiaceae (N)
Haloragaceae (G,N)
Hippocrateaceae (N)
Houmiriaceae (N)
Hydrangeaceae (N)
Hypericaceae
Icacinaceae (N)
Juglandaceae (G,N)
Julianiaceae (G)
Lamiaceae (G,N)
Lennoaceae (G)
Lentibulariaceae (G)
Linaceae (N)
Loasaceae (G,N)
Loganiaceae (G,N)
Lythraceae (G,N)
Maleshrbiaceae (N)
Malpighiaceae (G)
Malvaceae (G)
Melastomataceae (G,N)
Meliaceae (G,N)
Menispermaceae (G,N)
Mimosaceae (G,N)
Moraceae (G,N)
Myoporaceae (G,N)
Myricaceae (G)
Onagraceae (N)
Orobanchaceae (G)
Papilionaceae (G,N)
Passifloraceae (G,N)
Pedaliaceae (G)
Penaeaceae (N)
Piperaceae (G,N)
Plantaginaceae (G,N)

Podostemonaceae (N)
Polemoniaceae (G,N)
Polygonaceae (G,N)
Portulacaceae (G,N)
Primulaceae (G,N)
Ranunculaceae (G,N)
Rhamnaceae (N)
Rutaceae (G,N)
Santalaceae (N)
Sapindaceae (G,N)
Sarcolaenaceae (G,N)
Scrophulariaceae (G,N)
Selaginaceae (G,N)
Simaroubaceae (G,N)
Solanaceae (G,N)
Turneraceae (G,N)
Ulmaceae (G,N)
Urticaceae (G,N)
Valerianaceae (G,N)
Verbenaceae (G,N)

Simple (unbranched)–short, thickened

Acanthaceae (G,N)
Aizoaceae (N)
Asteraceae (G,N)
Caesalpiniaceae (G,N)
Ericaceae (G,N)
Euphorbiaceae (G,N)
Haloragaceae (G,N)
Loganiaceae (G)
Papilionaceae (G,N)
Passifloraceae (G)

Simple (unbranched)–long

Acanthaceae (G,N)
Aceraceae (G,N)
Actinidiaceae (G)
Alangiaceae (G,N)
Amaranthaceae (G,N)
Anacardiaceae (G,N)
Annonaceae (N)
Apiaceae (G,N)
Apocynaceae (G)
Araliaceae (N)
Aristolochiaceae (G,N)
Asclepiadaceae (G,N)
Asteraceae (G,N)
Balanophoraceae (G,N)
Begoniaceae (G,N)
Berberidaceae (G,N)
Bignoniaceae (G,N)
Boraginaceae (G,N)

Brassicaceae (G,N)
Bruniaceae (N)
Burseraceae (G,N)
Buxaceae (N)
Cactaceae (N)
Caesalpiniaceae (G,N)
Campanulaceae (N)
Cannabaceae (G,N)
Capparaceae (G,N)
Caprifoliaceae (G,N)
Caricaceae (G)
Caryophyllaceae (G,N)
Casuarinaceae (N)
Celastraceae (N)
Ceratophyllaceae (G,N)
Chenopodiaceae (G,N)
Cistaceae (G,N)
Cneoraceae (G)
Cochlospermaceae (N)
Combretaceae (G,N)
Connaraceae (G,N)
Convolvulaceae (G,N)
Cornaceae (G,N)
Corylaceae (G,N)
Crassulaceae (G,N)
Cucurbitaceae (G,N)
Cunoniaceae (G)
Datiscaceae (G)
Dichapetalaceae (N)
Dipsacaceae (G,N)
Dipterocarpaceae (G,N)
Ebenaceae (G,N)
Elaeocarpaceae (G,N)
Elatinaceae (G,N)
Empetraceae (G,N)
Epacridaceae (N)
Ericaceae (G,N)
Escalloniaceae (G,N)
Euphorbiaceae (G,N)
Eupteleaceae (N)
Fagaceae (G,N)
Flacourtiaceae (G,N)
Frankeniaceae (N)
Garryaceae (N)
Gentianaceae (G,N)
Geraniaceae (G,N)
Gesneriaceae (G,N)
Goodeniaceae (G,N)
Grossulariaceae (G,N)
Haloragaceae (G,N)
Hippocrateaceae (N)
Houmiriaceae (N)
Hydrangeaceae (N)
Hydrophyllaceae (G,N)
Hypericaceae
Icacinaceae (N)
Juglandaceae (N)

Julianiaceae (N)
Lamiaceae (G,N)
Lentibulariaceae (G,N)
Linaceae (G,N)
Loasaceae (G,N)
Loranthaceae (N)
Lythraceae (G,N)
Malesherbiaceae (G,N)
Malpighiaceae (G,N)
Malvaceae (G,N)
Melastomataceae (G,N)
Meliaceae (G,N)
Menispermaceae (G,N)
Mimosaceae (G,N)
Moraceae (G,N)
Myoporaceae (G,N)
Myricaceae (G)
Ochnaceae (G,N)
Onagraceae (N)
Orobanchaceae (G)
Oxalidaceae (G,N)
Papaveraceae (N)
Papilionaceae
Passifloraceae (G,N)
Pedaliaceae (G,N)
Pittosporaceae (G,N)
Plantaginaceae (G,N)
Plumbaginaceae (G,N)
Polemoniaceae (G,N)
Polygonaceae (G,N)
Portulacaceae (N)
Primulaceae (G,N)
Proteaceae (N)
Ranunculaceae (G,N)
Rhamnaceae (N)
Rhizophoraceae (G,N)
Rosaceae (G,N)
Rubiaceae (G,N)
Rutaceae (G,N)
Sabiaceae (G,N)
Sapindaceae (G,N)
Sarcolaenaceae (G,N)
Saxifragaceae (G,N)
Simaroubaceae (G,N)
Solanaceae (G,N)
Sterculiaceae (G,N)
Stylidiaceae (G)
Symplocaceae (G,N)
Tiliaceae (G,N)
Tremandraceae (G,N)
Turneraceae (G,N)
Ulmaceae (G,N)
Urticaceae (G,N)
Valerianaceae (G,N)
Verbenaceae (G,N)
Violaceae (G,N)
Vitaceae (G,N)

Zygophyllaceae (G,N)

Simple (unbranched)-long, thickened (shaggy)

Acanthaceae (G,N)
Aceraceae (G,N)
Apocynaceae (G)
Araliaceae (N)
Asclepiadaceae (G,N)
Asteraceae (G,N)
Begoniaceae (G,N)
Burseraceae (G,N)
Cactaceae (N)
Caesalpiniaceae (G,N)
Capparaceae (G,N)
Ceratophyllaceae (G)
Convolvulaceae (G,N)
Crassulaceae (G)
Cunoniaceae (G)
Datiscaceae (G)
Elatinaceae (G)
Ericaceae (G,N)
Escalloniaceae (G)
Euphorbiaceae (G,N)
Gentianaceae (G)
Geraniaceae (G)
Grossulariaceae (G)
Haloragaceae (G,N)
Linaceae (G)
Lythraceae (G,N)
Malpighiaceae (N)
Malvaceae (N)
Melastomataceae (G,N)
Menispermaceae (G)
Mimosaceae (G,N)
Ochnaceae (G)
Oxalidaceae (G)
Papaveraceae (N)
Pailionaceae (G,N)
Passifloraceae (G)
Plumbaginaceae (G)
Polygonaceae (N)
Portulacaceae (N)
Rhizophoraceae (G)
Rubiaceae (G)
Rutaceae (G)
Saxifragaceae (G,N)
Simaroubaceae (G)
Turneraceae (G,N)
Valerianaceae (N)
Violaceae (G)

2-5-armed

Acanthaceae (N)
Aceraceae (N)
Aizoaceae (N)

Amaranthaceae (N)
Araliaceae (N)
Asteraceae (N)
Begoniaceae (N)
Boraginaceae (N)
Brassicaceae
Burseraceae (N)
Caesalpiniaceae (N)
Cannabaceae (G,N)
Capparaceae (N)
Celastraceae (N)
Chenopodiaceae (N)
Chlaenaceae (N)
Cneoraceae (N)
Combretaceae (N)
Connaraceae (N)
Convolvulaceae (N)
Cornaceae (N)
Crassulaceae (N)
Ebenaceae (N)
Escalloniaceae (N)
Euphorbiaceae (N)
Flacourtiaceae (N)
Hydrangeaceae (N)
Loganiaceae (N)
Lythraceae (N)
Malpighiaceae (N)
Malvaceae (N)
Meliaceae (N)
Melianthaceae (N)
Monimiaceae (N)
Myrtaceae (N)
Papilionaceae (N)
Pittosporaceae (N)
Proteaceae (N)
Sapindaceae (N)
Sapotaceae (N)
Sarcolaenaceae (N)
Thymelaeaceae (N)
Turneraceae (N)
Verbenaceae (N)
Vitaceae (N)
Vochysiaceae (N)
Zygophyllaceae (N)

2-armed

Acanthaceae (N)
Aceraceae (N)
Aizoaceae (N)
Amaranthaceae (N)
Araliaceae (N)
Asteraceae
Begoniaceae (N)
Brassicaceae
Boraginaceae (N)
Burseraceae (N)

Caesalpiniaceae (N)
Cannabaceae (G,N)
Capparaceae (N)
Celastraceae (N)
Chenopodiaceae (N)
Cneoraceae (N)
Combretaceae (N)
Connaraceae (N)
Convolvulaceae (N)
Cornaceae (N)
Ebenaceae (N)
Escalloniaceae (N)
Euphorbiaceae (N)
Flacourtiaceae (N)
Lythraceae (N)
Malpighiaceae (N)
Meliaceae (N)
Monimiaceae (N)
Myrtaceae (N)
Papilionaceae (N)
Pittosporaceae (N)
Proteaceae (N)
Sapindaceae (N)
Sapotaceae (N)
Sarcolaenaceae
Thymelaeaceae (N)
Verbenaceae (N)
Vitaceae (N)
Vochysiaceae (N)
Zygophyllaceae (N)

2-armed-T-shaped

Acanthaceae (N)
Aceraceae (N)
Amaranthaceae (N)
Asteraceae (N)
Brassicaceae (N)
Capparaceae (N)
Cneoraceae (N)
Combretaceae (N)
Connaraceae (N)
Convolvulaceae (N)
Cornaceae (N)
Ebenaceae (N)
Malpighiaceae (N)

2-armed-U, V, Y, J-shaped

Acanthaceae (N)
Amaranthaceae (N)
Brassicaceae (N)
Cneoraceae (N)
Connaraceae (N)
Convolvulaceae (N)
Cornaceae (N)
Malpighiaceae (N)

3-5–armed

Amaranthaceae (N)
Begoniaceae (N)
Boraginaceae (N)
Brassicaceae (N)
Capparaceae (G,N)
Chenopodiaceae (N)
Convolvulaceae (N)
Crassulaceae (N)
Euphorbiaceae (G,N)
Hydrangeaceae (N)
Loganiaceae (N)
Malpighiaceae (N)
Malvaceae (N)
Melianthaceae (N)
Turneraceae (N)

Stellate

Acanthaceae (N)
Actinidiaceae
Aizoaceae (N)
Alangiaceae (G)
Amaranthaceae (N)
Anacardiaceae (G)
Annonaceae (N)
Apiaceae (N)
Araliaceae (N)
Asteraceae (N)
Begoniaceae (N)
Bixaceae (N)
Bombacaceae (N)
Boraginaceae (N)
Brassicaceae (N)
Burseraceae (N)
Caesalpiniaceae (N)
Callitrichaceae (N)
Capparaceae (G,N)
Caprifoliaceae (N)
Chenopodiaceae (N)
Chrysobalanaceae (N)
Cistaceae (N)
Clethraceae (N)
Connaraceae (N)
Convolvulaceae (N)
Crassulaceae (N)
Cunoniaceae (N)
Dilleniaceae (N)
Dipterocarpaceae (N)
Ebenaceae (N)
Elaeagnaceae (N)
Ericaceae (N)
Euphorbiaceae (G,N)
Fagaceae (N)
Flacourtiaceae (N)

Frankeniaceae (N)
Goodeniaceae (N)
Hamamelidaceae (N)
Himantandraceae (N)
Hippocrateaceae (N)
Hydrangeaceae (N)
Hydrophyllaceae (N)
Hypericaceae (N)
Icacinaceae (N)
Juglandaceae (N)
Lamiaceae (N)
Lecythidaceae (N)
Linaceae (N)
Lobeliaceae (N)
Loganiaceae (N)
Loranthaceae (N)
Lythraceae (N)
Magnoliaceae (N)
Malpighiaceae (N)
Malvaceae (N)
Melastomataceae (N)
Meliaceae (N)
Melianthaceae (N)
Mimosaceae (N)
Monimiaceae (N)
Myristicaceae (N)
Nyctaginaceae (N)
Olacaceae (N)
Papilionaceae (N)
Passifloraceae (N)
Platanaceae (N)
Polygonaceae (N)
Rhamnaceae (N)
Rhizophoraceae (N)
Rosaceae (N)
Rubiaceae (N)
Rutaceae (N)
Santalaceae (N)
Sapindaceae (N)
Solanaceae (N)
Sterculiaceae (N)
Styracaceae (N)
Theaceae (N)
Tiliaceae (N)
Tremandraceae (N)
Turneraceae (N)
Verbenaceae (N)
Vochysiaceae (N)

Stellate–rotate

Annonaceae (N)
Asteraceae (N)
Brassicaceae (N)
Capparaceae (N)
Convolvulaceae (N)

Euphorbiaceae (N)
Hydrangeaceae (N)
Malvaceae (N)
Meliaceae (N)
Solanaceae (N)
Sterculiaceae (N)
Verbenaceae (N)

Stellate-multiangulate

Aizoaceae (N)
Apiaceae (N)
Araliaceae (N)
Asteraceae (N)
Brassicaceae **(N)**
Burseraceae
Burseraceae (N)
Caesalpiniaceae (N)
Capparaceae (N)
Caprifoliaceae (N)
Chenopodiaceae (N)
Chrysobalanaceae (N)
Cistaceae (N)
Clethraceae (N)
Connaraceae (N)
Convolvulaceae (N)
Dilleniaceae (N)
Ericaceae (N)
Euphorbiaceae (N)
Hamamelidaceae (N)
Hydrangeaceae (N)
Loganiaceae (N)
Loranthaceae (N)
Malvaceae (N)
Melastomataceae (N)
Meliaceae (N)
Melianthaceae (N)
Olacaceae (N)
Papilionaceae (N)
Polygonaceae (N)
Rhamnaceae (N)
Rosaceae (N)
Rutaceae (N)
Sapindaceae (N)
Solanaceae (N)
Sterculiaceae (N)
Styracaceae (N)
Tiliaceae (N)
Tremandraceae (N)
Turneraceae (N)
Verbenaceae (N)

Stellate-porrect

Aizoaceae (N)
Asteraceae (N)
Euphorbiaceae (N)
Lamiaceae (N)

Stellate-geminate
(candelabra form)

Acanthaceae (N)
Amaranthaceae (N)
Asteraceae (N)
Capparaceae (N)
Chenopodiaceae (N)
Connaraceae (N)
Ericaceae (N)
Euphorbiaceae (N)
Goodeniaceae (N)
Loganiaceae (N)
Loranthaceae (N)
Myristicaceae (N)
Platanaceae (N)

Stellate-tufted

Actinidiaceae (N)
Araliaceae (N)
Begoniaceae (N)
Bixaceae (N)
Burseraceae (N)
Campanulaceae
Capparaceae (N)
Caprifoliaceae (N)
Cistaceae (N)
Cunoniaceae (N)
Dilleniaceae (N)
Dipterocarpaceae (N)
Ebenaceae (N)
Euphorbiaceae (G,N)
Fagaceae (N)
Flacourtiaceae (N)
Frankeniaceae (N)
Hamamelidaceae (N)
Hydrangeaceae (N)
Icacinaceae (N)
Juglandaceae (N)
Lamiaceae (N)
Lecythidaceae (N)
Linaceae (N)
Lobeliaceae (N)
Lythraceae (N)
Magnoliaceae (N)
Malvaceae (N)
Mimosaceae (N)
Monimiaceae (N)
Olacaceae (N)
Passifloraceae (N)
Polygonaceae (N)
Rhizophoraceae (N)
Rosaceae (N)
Rubiaceae (N)
Rutaceae (N)
Santalaceae (N)
Sapindaceae (N)

Saurauiaceae (N)
Solanaceae (N)
Sterculiaceae (N)
Styracaceae (N)
Theaceae (N)
Tiliaceae (N)
Turneraceae (N)
Vochysiaceae (N)

Scales

Anacardiaceae (G)
Ancistrocladaceae (G)
Annonaceae (N)
Araliaceae (G,N)
Asteraceae (G,N)
Begoniaceae (G)
Betulaceae (G)
Bignoniaceae (G)
Bixaceae (N)
Bombacaceae (G,N)
Brassicaceae (N)
Caesalpiniaceae (G)
Cannabaceae (G,N)
Capparaceae (G,N)
Caprifoliaceae (G,N)
Chrysobalanaceae (G)
Cistaceae (G)
Cochlospermaceae (G)
Combretaceae (G)
Convolvulaceae (G)
Datiscaceae (G)
Dilleniaceae (N)
Dipterocarpaceae (G)
Ebenaceae (G)
Elaeagnaceae (N)
Ericaceae (G,N)
Escalloniaceae (G)
Euphorbiaceae (G,N)
Fagaceae (G)
Flacourtiaceae (N)
Gesneriaceae (G)
Goodeniaceae (G)
Grossulariaceae (G)
Himantandraceae (N)
Hippuridaceae (N)
Hydrophyllaceae (G)
Icacinaceae (N)
Juglandaceae (G)
Loganiaceae (N)
Malvaceae (N)
Melastomataceae (N)
Meliaceae (N)
Mimosaceae (G,N)
Monimiaceae (N)
Myricaceae (G)
Myrsinaceae (G,N)

Oleaceae (G,N)
Orobanchaceae (G)
Papilionaceae (G)
Polygonaceae (G)
Rubiaceae (N)
Rutaceae (N)
Salvadoraceae (N)
Sapindaceae (G,N)
Sarcolaenaceae (G,N)
Scrophulariaceae (G)
Solanaceae (N)
Sterculiaceae (N)
Styracaceae (N)
Tiliaceae (N)
Urticaceae (G)
Verbenaceae (G,N)

Scales–sessile

Begoniaceae (G)
Betulaceae (G)
Bignoniaceae (G)
Caesalpiniaceae (G)
Combretaceae (G)
Dipterocarpaceae (G)
Ericaceae (G,N)
Escalloniaceae (G)
Euphorbiaceae (G,N)
Grossulariaceae (G)
Hippuridaceae (N)
Melastomataceae (N)
Meliaceae (N)
Myricaceae (G)
Myrsinaceae (G,N)
Oleaceae (G,N)
Orobanchaceae (G)
Papilionaceae (G)
Polygonaceae (G)
Urticaceae (N)
Verbenaceae (G,N)

Scales–peltate

Anacardiaceae (G)
Ancistrocladaceae (G)
Annonaceae (N)
Araliaceae (G,N)
Asteraceae (G,N)
Begoniaceae (G)
Betulaceae (G)
Bixaceae (N)
Bombacaceae (G,N)
Brassicaceae (N)
Cannabaceae (G,N)
Capparaceae (G,N)
Caprifoliaceae (G,N)
Chrysobalanaceae (G)
Cistaceae (G)

Cochlospermaceae (G)
Combretaceae (G)
Convolvulaceae (G)
Datiscaceae (G)
Dilleniaceae (N)
Dipterocarpaceae (G)
Ebenaceae (G)
Elaeagnaceae (N)
Ericaceae (G,N)
Escalloniaceae (G)
Euphorbiaceae (G,N)
Fagaceae (G)
Flacourtiaceae (N)
Gesneriaceae (G)
Goodeniaceae (G)
Grossulariaceae (G)
Himantandraceae (N)
Hippuridaceae (N)
Hydrophyllaceae (G)
Icacinaceae (N)
Juglandaceae (G)
Loganiaceae (N)
Malvaceae (N)
Melastomataceae (N)
Meliaceae (G)
Mimosaceae (G)
Monimiaceae (N)
Myricaceae (G)
Myrsinaceae (N)
Oleaceae (G,N)
Orobanchaceae (G)
Papilionaceae (G)
Polygonaceae (G)
Rubiaceae (N)
Rutaceae (N)
Salvadoraceae (N)
Sapindaceae (G,N)
Sarcolaenaceae (G,N)
Scrophulariaceae (G)
Solanaceae (N)
Sterculiaceae (N)
Styracaceae (N)
Tiliaceae (N)
Urticaceae (G)
Verbenaceae (G,N)

Scales–porrect

Rutaceae (N)

Dendritic (branching)

Acanthaceae (N)
Actinidiaceae (N)
Amaranthaceae (N)
Apiaceae (N)
Apocynaceae (N)
Araliaceae (N)

Asteraceae (N)
Bignoniaceae (N)
Boraginaceae (N)
Brassicaceae (N)
Caesalpiniaceae (G,N)
Capparaceae (N)
Caryophyllaceae (N)
Casuarinaceae (N)
Chenopodiaceae (N)
Connaraceae (N)
Convolvulaceae (N)
Ericaceae (N)
Euphorbiaceae (N)
Goodeniaceae (N)
Hypericaceae (N)
Lamiaceae
Loranthaceae (N)
Lythraceae (N)
Melastomataceae (N)
Mimosaceae (G,N)
Myoporaceae (N)
Myristicaceae (N)
Myrsinaceae (N)
Nyctaginaceae (N)
Olacaceae (N)
Opiliaceae (N)
Orobanchaceae (G)
Papilionaceae (N)
Platanaceae (G,N)
Polemoniaceae (N)
Primulaceae (N)
Scrophulariaceae (N)
Solanaceae (N)
Verbenaceae (N)

Dendritic–few branched

Apiaceae (N)
Apocynaceae (N)
Bignoniaceae (N)
Boraginaceae (N)
Caesalpiniaceae (G,N)
Chenopodiaceae (N)
Connaraceae (N)
Ericaceae (N)
Euphorbiaceae (N)
Fabaceae (N)
Goodeniaceae (N)
Melastomataceae (N)
Mimosaceae (N)
Myristicaceae (N)
Nyctaginaceae (N)
Opiliaceae (N)
Papilionaceae (N)
Scrophulariaceae (N)
Solanaceae (N)

Dendritic–many branched

Bignoniaceae (N)
Brassicaceae (N)
Caesalpiniaceae (G,N)
Capparaceae (N)
Chenopodiaceae (N)
Connaraceae (N)
Ericaceae (N)
Euphorbiaceae (N)
Lamiaceae (N)
Melastomataceae (N)
Mimosaceae (N)
Myristicaceae (N)
Scrophulariaceae (N)
Solanaceae (N)

Dendritic–branching terminal

Bignoniaceae (N)
Boraginaceae (N)
Brassicaceae (N)
Convolvulaceae (N)
Ericaceae (N)
Melastomataceae (N)
Mimosaceae (G)

Dendritic–branching medial

Ericaceae (N)
Melastomataceae (N)

Dendritic–branching basal

Amaranthaceae (N)
Caesalpiniaceae (G,N)
Chenopodiaceae (N)
Lamiaceae (N)
Melastomataceae (N)

Specialized types

Chalk Glands
  Plumbaginaceae
  Saxifragaceae

Pearl Glands
  Begoniaceae
  Caesalpiniaceae
  Moraceae
  Piperaceae
  Urticaceae
  Vitaceae
Salt Glands
  Acanthaceae
  Frankeniaceae
  Tamaricaceae
Snail-shaped Glands
  Burseraceae

Arachnoid Trichomes
  Asteraceae
  Chrysobalanaceae
  Crassulaceae
  Hydrophyllaceae
  Plantaginaceae
  Polemoniaceae

Calcified or Silicified trichomes
  Boraginaceae
  Campanulaceae
  Cornaceae
  Cucurbitaceae
  Gesneriaceae
  Hydrangeaceae
  Hydrophyllaceae
  Loasaceae
  Moraceae
  Polemoniaceae
  Scrophulariaceae
  Ulmaceae
  Verbenaceae

Funnel- or Cup-shaped Trichomes
  Ericaceae

Laticiferous Trichomes
  (Asteraceae-Cichorieae)

Mucilage Trichomes
  Pedaliaceae

Stinging Trichomes
  Euphorbiaceae
  Loasaceae
  Malpighiaceae
  Urticaceae

Vesicular Trichomes
  Aizoaceae
  Apiaceae
  Asteraceae
  Chenopodiaceae
  Combretaceae
  Crassulaceae
  Icacinaceae
  Melastomataceae
  Myricaceae
  Oxalidaceae
  Salicaceae

Families in which only unicellular or uniseriate trichomes have been found

Akaniaceae
Aquifoliaceae

Aristolochiaceae
Balanophoraceae
Brunelliaceae
Bruniaceae
Buxaceae
Calycanthaceae
Campanulaceae
Capparaceae
Caryocaraceae
Columelliaceae
Diapensiaceae
Dichapetalaceae
Diclidantheraceae
Epacridaceae
Eucommiaceae
Eucryphiaceae
Eupteleaceae
Flacourtiaceae
Garryaceae
Gonystylaceae
Goupiaceae
Grubbiaceae
Hippocastanaceae
Koeberliniaceae
Krameriaceae
Lauraceae
Moringaceae
Myzodendraceae
Nymphaeaceae
Onagraceae
Penaeaceae
Phytolaccaceae
Podostemonaceae
Polygalaceae
Resedaceae
Scytopetalaceae
Stackhousiaceae
Tamaricaceae
Thymelaeaceae
Trigoniaceae
Tropaeolaceae

Families with numerous types of trichomes

Acanthaceae (G,N)
Amaranthaceae (G,N)
Apiaceae (G,N)
Araliaceae (G,N)
Asteraceae (G,N)
Begoniaceae (G,N)
Boraginaceae (G,N)
Brassicaceae (G,N)
Burseraceae (G,N)
Caesalpiniaceae (G,N)
Capparaceae (G,N)

Caprifoliaceae (G,N)
Chenopodiaceae (G,N)
Cistaceae (G,N)
Connaraceae (G,N)
Convolvulaceae (G,N)
Dipterocarpaceae (G,N)
Ebenaceae (G,N)
Ericaceae (G,N)
Escalloniaceae (G,N)
Euphorbiaceae (G,N)
Fagaceae (G,N)

Flacourtiaceae (G,N)
Goodeniaceae (G,N0
Lamiaceae (G,N)
Loganiaceae (G,N)
Lythraceae (G,N)
Malvaceae (G,N)
Melastomataceae (G,N)
Meliaceae (G,N)
Mimosaceae (G,N)
Orobanchaceae (G,N)

Papilionaceae (G,N)
Polygonaceae (G,N)
Rutaceae (G,N)
Sapindaceae (G,N)
Scrophulariaceae (G,N)
Solanaceae (G,N)
Sterculiaceae (G,N)
Tiliaceae (G,N)
Turneraceae (G,N)
Verbenaceae (G,N)

LEAF CHARACTERS

I. Epidermis

Epidermis papillose
Ab = abaxial; Ad = adaxial;
X = surface unrecorded. Data
mainly from Solereder (1908).

Acanthaceae
Aceraceae Ab
Adoxaceae Ab
Aizoaceae X
Alangiaceae Ad
Anacardiaceae Ab
Annonaceae Ab
Apiaceae Ad
Apocynaceae Ab, Ad
Aristolochiaceae Ab, Ad, X
Asteraceae Ab
Begoniaceae Ad
Berberidaceae Ab
Betulaceae Ab
Bignoniaceae X
Boraginaceae Ab
Bretschneideraceae X
Brunelliaceae Ab
Bruniaceae X
Burseraceae Ab
Caesalpiniaceae B
Campanulaceae X
Capparaceae Ab
Caprifoliaceae Ab, X
Caryophyllaceae X
Celastraceae X
Cercidiphyllaceae Ab
Chloranthaceae Ab
Chrysobalanaceae Ab
Combretaceae Ab
Connaraceae Ab
Convolvulaceae X
Cornaceae Ad
Crassulaceae X
Cucurbitaceae X
Daphniphyllaceae Ab
Diapensiaceae Ab

Ebenaceae Ab
Epacridaceae Ab
Ericaceae Ab, B
Erythroxylaceae Ab
Euphorbiaceae Ab, Ad
Eupteleaceae Ab
Fagaceae Ab
Flacourtiaceae Ab, X
Garryaceae X
Geissolomataceae X
Gentianaceae X
Geraniaceae Ab
Gunneraceae X
Hamamelidaceae X
Hernandiaceae Ab
Hippocastanaceae X
Hypericaceae X
Icacinaceae Ab
Krameriaceae X
Lactoridaceae Ab
Lamiaceae X
Lardizabalaceae Ab
Lauraceae Ab
Lecythidaceae Ab
Loganiaceae Ab, Ad
Lythraceae Ad
Magnoliaceae X
Malpighiaceae Ab
Melastomataceae Ab, Ad
Meliaceae X
Menispermaceae Ab
Mimosaceae Ab, X
Moraceae X
Myoporaceae
Myricaceae Ab
Myristicaceae Ab
Myrtaceae X
Nymphaeaceae X
Nyssaceae Ab
Ochnaceae X
Olacaceae Ab
Oleaceae Ab

Oxalidaceae Ab
Papaveraceae Ab
Papilionaceae (Ab)
Passifloraceae Ab
Piperaceae X
Pittosporaceae Ad
Polemoniaceae X
Polygalaceae Ab
Polygonaceae Ad, X
Proteaceae X
Ranunculaceae X
Resedaceae X
Rhamnaceae Ab, Ad
Rosaceae Ab
Rubiaceae Ad
Rutaceae X
Santalaceae Ab, X
Sapindaceae Ab
Scrophulariaceae X
Simaroubaceae Ab
Solanaceae Ab
Sonneratiaceae Ab
Staphyleaceae Ab
Stylidiaceae Ad
Symplocaceae Ab
Tamaricaceae X
Tetracentraceae
Thymelaeaceae Ab
Ulmaceae Ab
Violaceae X
Vitaceae X
Vochysiaceae Ab
Winteraceae Ab

Epidermis mucilaginous

Aceraceae
Anacardiaceae
Annonaceae
Aquifoliaceae
Betulaceae
Bonnetiaceae
Brassicaceae

Burseraceae
Caesalpiniaceae
Caryocaraceae
Celastraceae
Chrysobalanaceae
Cistaceae
Cochlospermaceae
Connaraceae
Cornaceae
Crassulaceae
Cyrillaceae
Dichapetalaceae
Dipterocarpaceae
Elaeocarpaceae
Elatinaceae
Empetraceae
Ericaceae
Erythroxylaceae
Euphorbiaceae
Fagaceae
Flacourtiaceae
Fouquieriaceae
Geissolomataceae
Gentianaceae
Hamamelidaceae
Icacinaceae
Linaceae
Loganiaceae
Lythraceae
Malpighiaceae
Malvaceae
Melastomataceae
Meliaceae
Mimosaceae
Moraceae
Moringaceae
Myrsinaceae
Myrtaceae
Nyssaceae
Ochnaceae
Onagraceae
Papilionaceae
Passifloraceae
Pentaphylacaceae
Phytolacaceae
Polemoniaceae
Polygonaceae
Resedaceae
Rhamnaceae
Rhizophoraceae
Rosaceae
Rutaceae
Salicaceae
Sapindaceae
Sapotaceae
Schisandraceae
Simaroubaceae

Staphyleaceae
Sterculiaceae
Theaceae
Thymelaeaceae
Tiliaceae
Tremandraceae
Trigoniaceae
Tropaeolaceae
Turneraceae
Ulmaceae
Violaceae
Vochysiaceae

Epidermis including horizontally divided cells
(Two- to many-layered)

Anacardiaceae
Annonaceae
Apocynaceae
Aquifoliaceae
Araliaceae
Bignoniaceae
Bixaceae
Bombacaceae
Cactaceae (axis)
Caesalpiniaceae
Celastraceae
Connaraceae
Convolvulaceae
Cornaceae
Crassulaceae
Dichapetalaceae
Epacridaceae
Ericaceae
Globulariaceae
Goupiaceae
Hernandiaceae
Hydrangeaceae
Hypericaceae
Lythraceae
Malpighiaceae
Malvaceae
Melastomataceae
Menispermaceae
Mimosaceae
Monimiaceae
Moraceae
Myrsinaceae
Passifloraceae
Piperaceae
Pittosporaceae
Rhamnaceae
Rhizophoraceae
Rubiaceae
Rutaceae
Salvadoraceae

Santalaceae
Sapindaceae
Sapotaceae
Saxifragaceae
Scrophulariaceae
Scytopetalaceae
Simaroubaceae
Trigoniaceae
Ulmaceae
Urticaceae
Vochysiaceae

Epidermis including vertically divided cells

Araliaceae (*Panax arboreum*)
Caesalpiniaceae
Connaraceae
Hippocastanaceae (*Billia hippo-
   castanum* Peyh)
Loganiaceae (*Strychnos*)
Mimosaceae
Moraceae
Papilionaceae
Sapindaceae
Vochysiaceae

Epidermis including some tall cells
Large cells often scattered in families marked †

Apocynaceae
Asclepiadaceae
Begoniaceae
Brassicaceae †
Bruniaceae
Cistaceae
Cucurbitaceae
Elatinaceae †
Geraniaceae
Gesneriaceae
Malpighiaceae †
Melastomataceae
Menispermaceae
Myrtaceae
Resedaceae †
Sterculiaceae
Strasburgeriaceae
Styracaceae
Tremandraceae
Verbenaceae
Violaceae

Epidermis often tall and palisade-like in TS

Annonaceae
Bignoniaceae
Caesalpiniaceae

Caryophyllaceae
Celastraceae
Chrysobalanaceae
Dipterocarpaceae
Ericaceae
Hamamelidaceae
Hippocrateaceae
Hypericaceae
Lauraceae
Melastomataceae
Primulaceae
Santalaceae
Sapindaceae
Stylidiaceae
Thymelaeaceae

Epidermal cells Vesicular

Aizoaceae
Caryophyllaceae
Crassulaceae
Portulacaceae
Resedaceae

Epidermal cells Tubular
(Brown Contents)

Crassulaceae
Euphorbiaceae
Geraniaceae
Saxifragaceae
Violaceae

Epidermis small-celled in surface view

Buxaceae
Capparaceae
Chloranthaceae
Hippocrateaceae
Malpighiaceae
Myristicaceae
Salvadoraceae
Sarcolaenaceae
Vochysiaceae

Epidermis sclerotic
From first edition, list very incomplete

Apiaceae (xerophytic spp)
Euphorbiaceae (especially abaxial epidermis of *Amanoa* and *Discocarpus*)
Hypericaceae (*Clustia* etc.)
Menispermaceae (occasional)
Polygalaceae (xerophytic spp)
Resedaceae (*Ochradenus*)

Sapindaceae (*Matayba* sp)
Thymelaeaceae

II. Hypodermis

Present

Acanthaceae
Actinidiaceae
Alangiaceae
Amaranthaceae
Anacardiaceae
Ancistrocladaceae
Annonaceae
Apiaceae
Apocynaceae
Aquifoliaceae
Araliaceae
Aristolochiaceae
Asclepiadaceae
Asteraceae
Balanopsidaceae
Begoniaceae
Berberidaceae
Betulaceae
Bignoniaceae
Bixaceae
Bombacaceae
Brunelliaceae
Burseraceae
Cactaceae (axis)
Caesalpiniaceae
Campanulaceae
Canellaceae
Capparaceae
Caryocaraceae
Caryophyllaceae
Celastraceae
Chenopodiaceae
Chloranthaceae
Chrysobalanaceae
Clethraceae
Columelliaceae
Combretaceae
Connaraceae
Cornaceae
Corynocarpaceae
Crypteroniaceae
Cucurbitaceae
Cunoniaceae
Datiscaceae
Dichapetalaceae
Didieriaceae
Dilleniaceae
Dipterocarpaceae
Ebenaceae

Elaeagnaceae
Epacridaceae
Ericaceae
Euphorbiaceae
Fagaceae
Flacourtiaceae
Garryaceae
Gesneriaceae
Gomortegaceae
Hamamelidaceae
Hernandiaceae
Hippocrateaceae
Hydrangeaceae
Hypericaceae
Lamiaceae
Lardizabalaceae
Lauraceae
Lecythidaceae
Leitneriaceae
Linaceae
Loganiaceae
Loranthaceae
Lythraceae
Magnoliaceae
Malpighiaceae
Malvaceae
Marcgraviaceae
Medusagynaceae
Melastomataceae
Meliaceae
Menispermaceae
Monimiaceae
Moraceae
Myricaceae
Myristicaceae
Myrsinaceae
Myrtaceae
Nepenthaceae
Nyctaginaceae
Ochnaceae
Olacaceae
Oleaceae
Orobanchaceae
Oxalidaceae
Papilionaceae
Passifloraceae
Pelliceriaceae
Phytolaccaceae
Piperaceae
Pittosporaceae
Plantaginaceae
Polemoniaceae
Polygalaceae
Polygonaceae
Portulacaceae
Proteaceae

Rhamnaceae
Rhizophoraceae
Rosaceae
Rubiaceae
Rutaceae
Salicaceae
Salvadoraceae
Santalaceae
Sapindaceae
Sapotaceae
Sarcolaenaceae
Saururaceae
Saxifragaceae
Scrophulariaceae
Simaroubaceae
Sterculiaceae
Strasburgeriaceae
Styraceae
Symplocaceae
Tetrameristaceae
Theaceae
Thymelaeaceae
Trigoniaceae
Urticaceae
Vacciniaceae
Verbenaceae
Violaceae
Zygophyllaceae

III. Stomata

Arrangement of surrounding cells

Anomocytic

Aceraceae
Actinidiaceae
Adoxaceae
Aizoaceae
Alangiaceae
Amaranthaceae
Anacardiaceae
Apiaceae
Apocynaceae
Aquifoliaceae
Araliaceae
Aristolochiaceae
Asclepiadaceae
Asteraceae
Balanopaceae
Balanophoraceae
Balsaminaceae
Berberidaceae
Betulaceae
Bignoniaceae
Bixaceae
Boraginaceae
Bruniaceae

Burseraceae
Buxaceae
Caesalpiniaceae
Calyceraceae
Campanulaceae
Cannabaceae
Canellaceae
Capparaceae
Caprifoliaceae
Caricaceae
Caryocaraceae
Caryophyllaceae
Celastraceae
Cercidiphyllaceae
Chenopodiaceae
Chloranthaceae
Circaeasteraceae
Cistaceae
Clethraceae
Cneoraceae
Columelliaceae
Combretaceae
Convolvulaceae
Cornaceae
Crossosomataceae
Cucurbitaceae
Cyrillaceae
Datiscaceae
Diapensiaceae
Dilleniaceae
Dipsacaceae
Dipterocarpaceae
Ebenaceae
Elaeagnaceae
Elatinaceae
Empetraceae
Epacridaceae
Ericaceae
Eucommiaceae
Euphorbiaceae
Eupteleaceae
Fagaceae
Flacourtiaceae
Fouquieriaceae
Frankeniaceae
Gentianaceae
Geraniaceae
Gesneriaceae
Globulariaceae
Goodeniaceae
Goupiaceae
Grubbiaceae
Haloragaceae
Hernandiaceae
Houmiriaceae
Hydrangeaceae
Hydrophyllaceae

Hypericaceae
Icacinaceae
Juglandaceae
Julianiaceae
Lacistemaceae
Lamiaceae
Lardizabalaceae
Lecythidaceae
Lennoaceae
Lentibulariaceae
Loasaceae
Lobeliaceae
Loganiaceae
Lythraceae
Magnoliaceae
Malvaceae
Medusagynaceae
Melastomataceae
Meliaceae
Melianthaceae
Menispermaceae
Monimiaceae
Moraceae
Moringaceae
Myoporaceae
Myricaceae
Myrothamnaceae
Myrsinaceae
Myrtaceae
Myzodendraceae
Nepenthaceae
Nyctaginaceae
Nymphaeaceae
Ochnaceae
Olacaceae
Oleaceae
Onagraceae
Orobanchaceae
Papaveraceae
Papilionaceae
Passifloraceae
Pedaliaceae
Penaeaceae
Pentaphragmataceae
Phytolaccaceae
Piperaceae
Plantaginaceae
Platanaceae
Plumbaginaceae
Polemoniaceae
Polygalaceae
Polygonaceae
Primulaceae
Proteaceae
Punicaceae
Ranunculaceae
Resedaceae

Rhamnaceae
Rhizophoraceae
Rosaceae
Sabiaceae
Salvadoraceae
Santalaceae
Sapindaceae
Sapotaceae
Sarcolaenaceae
Saxifragaceae
Scrophulariaceae
Simaroubaceae
Solanaceae
Stachyuraceae
Stackhousiaceae
Sterculiaceae
Stylidiaceae
Styracaceae
Tamaricaceae
Tetrameristaceae
Theaceae
Thymelaeaceae
Tiliaceae
Tremandraceae
Tropaeolaceae
Turneraceae
Ulmaceae
Urticaceae
Valerianaceae
Verbenaceae
Vochysiaceae
Vitaceae
Zygophyllaceae

Paracytic

Acanthaceae
Actinidiaceae
Aizoaceae
Anacardiaceae
Annonaceae
Apiaceae
Apocynaceae
Araliaceae
Asclepiadaceae
Basellaceae
Bataceae
Bignoniaceae
Bixaceae
Bonnetiaceae
Byblidaceae
Cactaceae (stem)
Caesalpiniaceae
Calycanthaceae
Canellaceae
Caprifoliaceae
Casuarinaceae

Celastraceae
Chenopodiaceae
Chloranthaceae
Chrysobalanaceae
Clethraceae
Connaraceae
Convolvulaceae
Coriariaceae
Corynocarpaceae
Daphniphyllaceae
Dichapetalaceae
Dilleniaceae
Dipterocarpaceae
Elatinaceae
Ericaceae
Erythroxylaceae
Escalloniaceae
Eucryphiaceae
Euphorbiaceae
Eupomatiaceae
Flacourtiaceae
Frankeniaceae
Geraniaceae
Gesneriaceae
Globulariaceae
Gomortegaceae
Goodeniaceae
Hamamelidaceae
Hernandiaceae
Himantandraceae
Houmiriaceae
Hydrangeaceae
Hypericaceae
Icacinaceae
Julianiaceae
Krameriaceae
Lauraceae
Lecythidaceae
Linaceae
Loganiaceae
Loranthaceae
Magnoliaceae
Malpighiaceae
Melastomataceae
Menispermaceae
Mimosaceae
Monimiaceae
Myristicaceae
Myrtaceae
Nyctaginaceae
Nyssaceae
Ochnaceae
Olacaceae
Oleaceae
Opiliaceae
Oxalidaceae

Papilionaceae
Passifloraceae
Pentaphylacaceae
Phytolaccaceae
Pittosporaceae
Plumbaginaceae
Polemoniaceae
Polygalaceae
Polygonaceae
Portulacaceae
Proteaceae
Quiinaceae
Rhamnaceae
Rhizophoraceae
Rubiaceae
Rutaceae
Sabiaceae
Salicaceae
Salvadoraceae
Santalaceae
Sapindaceae
Sapotaceae
Saxifragaceae
Scrophulariaceae
Simaroubaceae
Sonneratiaceae
Sterculiaceae
Stylidiaceae
Symplocaceae
Tamaricaceae
Tetracentraceae
Tetrameristaceae
Theaceae
Trigoniaceae
Trochodendraceae
Turneraceae
Ulmaceae
Verbenaceae
Violaceae
Vochysiaceae
Winteraceae

Anisocytic

Apocynaceae
Araliaceae
Asclepiadaceae
Asteraceae
Balsaminaceae
Begoniaceae
Bignoniaceae
Bixaceae
Boraginaceae
Brassicaceae
Caryophyllaceae
Celastraceae

Clethraceae
Connaraceae
Convolvulaceae
Crassulaceae
Diapensiaceae
Dipsacaceae
Euphorbiaceae
Flacourtiaceae
Gentianaceae
Gesneriaceae
Hypericaceae
Icacinaceae
Lecythidaceae
Leguminosae
Loganiaceae
Lythraceae
Melastomataceae
Moraceae
Myoporaceae
Myrsinaceae
Olacaceae
Onagraceae
Pedaliaceae
Pentaphragmataceae
Rhamnaceae (*Cryptandra*)
Rhizophoraceae
Scrophulariaceae
Scytopetalaceae
Solanaceae
Staphyleaceae
Turneraceae
Verbenaceae
Violaceae

Cyclocytic
Only a few in each family; list
very incomplete.

Anacardiaceae
Apocynaceae
Bignoniaceae
Buxaceae
Celastraceae
Combretaceae
Connaraceae
Dipterocarpaceae
Piperaceae
Rhizophoraceae
Theaceae

Tetracytic
Rare in all families listed

Anacardiaceae
Celastraceae
Elatinaceae
Tiliaceae

Diacytic

Acanthaceae
Bignoniaceae
Caryophyllaceae
Connaraceae
Globulariaceae
Lamiaceae
Lentibulariaceae
Melastomataceae
Plantaginaceae
Scrophulariaceae
Solanaceae
Verbenaceae

Parallelocytic
Only a few in each family; list
incomplete

Anacardiaceae
Apiaceae
Apocynaceae
Cactaceae
Convolvulaceae
Euphorbiaceae
Lamiaceae
Papilionaceae
Portulacaceae
Rubiaceae

Helicocytic
Rare in all families listed

Asteraceae
Begoniaceae
Brassicaceae
Crassulaceae
Gesneriaceae
Malvaceae
Myrsinaceae
Piperaceae
Polygonaceae
Urticaceae

*Stomata surrounded by rosettes
of more or less clearly defined
subsidiary cells*

Buxaceae (*B. sempervirens* Linn.)
Celastraceae (*Mortonia*)
Menispermaceae (14 genera)
Moraceae (*Ficus*)
Papilionaceae (4 genera)
Piperaceae ('often')
Saururaceae
Thymelaeaceae (certain spp)

Stomata in Groups

Begoniaceae
Brassicaceae
Caryophyllaceae (over veins in
teeth)
Gesneriaceae (*Napeanthus*,
Gesneria sintenisii Urb.)
Himantandraceae (*Himantandra
parvifolia* Bak. f. and Norm.)
Melastomataceae (*Calycogonium*,
Leandra, *Ossaea*)
Menispermaceae (*Macrococculus*)
Moraceae (*Ficus* spp)
Ochnaceae (*Godoya*)
Papilionaceae (*Euchresta*)
Proteaceae (*Banksia*, *Dryandra*,
Lambertia, etc.)
Rubiaceae (*Pagamea*)
Sauvagesiaceae
Saxifragaceae (*Chrysosplenium*,
Saxifraga)
Simaroubaceae (*Castela*, *Soulamea*)
Theaceae (*Cleyera*)
Verbenaceae (some spp *Stachy-
tarpheta*)

Stomata in pits

Apocynaceae (*Nerium*)
Melastomataceae (*Mouriri*)
Moraceae (*Ficus* spp)
Ochnaceae (*Ouratia* sp)
Sarcolaenaceae (*Sarcolaena*,
Schizolaena)
Simaroubaceae (*Soulamea pancheri*
Brogn. & Gris)
Staphyleaceae (*Akania*)
Stylidiaceae (*Candollea*, s. in
longitudinal rows)

*Stomata restricted to adaxial leaf
surface*

A. *Floating or other leaves of
water plants*

Callitrichaceae
Menyanthaceae
Nelumbonaceae
Nymphaeaceae
Polygonaceae
Ranunculaceae
Trapellaceae

B. *Leaves of terrestrial plants*

Amaranthaceae (*Philoxerus*)
Asteraceae (*Lepidophyllum*)
Begoniaceae

Bruniaceae (*Lonchostoma,
  Pseudobaeckea, Raspalia*)
Campanulaceae (*Edraianthus*
  sp)
Dipterocarpaceae
Epacridaceae (*Leucopogon*)
Ericaceae (*Cassiope*)
Euphorbiaceae (*Euphorbia buxi-
  folia*)
Hamamelidaceae
Mimosaceae, *Pultenaea*
Onagraceae (*Epilobium crassum*
  Hook f.)
Orobanchaceae (*Lathraea*)
Papilionaceae (*Coelidium, Dillwy-
  nia, Geoffreya*)
Ranunculaceae (antarctic species
  of *Caltha*)
Saxifragaceae
Scrophulariaceae (*Hemiphragma*)
Stylidiaceae (*Candollea*)
Tamaricaceae (*Myricaria, Tamarix*)
Thymelaeaceae (*Passerina*)
Urticaceae (*Pilea serpyllifolia, Pilea
  yunnanensis* (Urb.) Britt. and
  Wils.)
Violaceae

Stomata absent
Submerged leaves of water plants,
scale leaves of saprophytes and
parasites.

Balanophoraceae
Ceratophyllaceae
Droseraceae
Ericaceae (*Monotropa*)
Haloragaceae
Lentibulariaceae
Monotropaceae
Nymphaeaceae
Pedaliaceae
Podostemonaceae
Rafflesiaceae
Ranunculaceae

Stomata unusually small
Aceraceae
Combretaceae (*Guiera,
  Calycopteris*)
Cunoniaceae (*Cunonia,
  Platylophus*)
Empetraceae
Epacridaceae
Ericaceae
Escalloniaceae (6 genera)
Euphorbiaceae (*Euphorbia*)

Melastomataceae
Papilionaceae (*Mundulea*)
Quiinaceae
Sapindaceae (at least 20 genera)
Saxifragaceae (*Euaizoonia*)

Stomata on both surfaces

Callitrichaceae
Crassulaceae
Droseraceae
Hippuridaceae
Mimosaceae
Myrtaceae
Papilionaceae
Proteaceae
Saxifragaceae (section *Dialy-
  splenium*)

Stomata on lower surface only

Aceraceae
Achariaceae
Adoxaceae
Aquifoliaceae
Begoniaceae
Bignoniaceae
Chloranthaceae
Cornaceae
Corynocarpaceae
Dichapetalaceae
Elaeocarpaceae
Epacridaceae
Erythroxylaceae
Hamamelidaceae
Juglandaceae
Linaceae
Meliaceae
Onagraceae
Portulacaceae
Rosaceae
Staphyleaceae
Tiliaceae
Turneraceae

*Stomata parallel to one another
but transverse to the midrib or
direction of axis.*

Aizoaceae (*Mesembryanthemum*
  leaf)
Balanitaceae (axis)
Bataceae (leaf and axis)
Bruniaceae (*Brunia* and *Staavia*
  leaf)
Cactaceae (axis)
Casuarinaceae (sheaths and bran-
  ches)
Chenopodiaceae (leaf in *Suaeda,*

*Salsola, Camphorosma, Echin-
  opsilon, Halegeton, Traganum,*
  axis in *Camphorosma,*
Epacridaceae (*Lysinema* leaf)
Euphorbiaceae (axis of succulent
  Euphorbia spp)
Krameriaceae (*Kramerin* spp)
Lauraceae (*Cassytha*)
Loranthaceae (leaf & axis of
  Nuytsia and axis of other
  genera)
Nepenthaceae (pro parte)
Papilionaceae
  (leaf of *Anarthrophyllum,
  Eutaxia, Latrobea,* axis of
  Daviesia, phylloclades in *Carmi-
  chaelia,* branches in *Alhagi*)
Rhamnaceae (axis of *Colletia,* acc.
  to Pfitzer)
Santalaceae (commonly on
  branches and leaves)
Staphyleaceae (axis of *Staphylea
  pinnata,* acc. to de Bary)
Tamaricaceae (leaf)

*Stomatal pores parallel to one
another and to leaf midrib or
stem axis*

Apiaceae (*Eryngium* sp)
Asteraceae (*Achillea* sp)
Bruniaceae (1–5 rows)
Cactaceae (on stem)
Callitrichaceae
Caryophyllaceae (narrow-leaved
  spp)
Chenopodiaceae
Epacridaceae
Leguminosae (sporadically in all
  sections)
Loganiaceae
Lythraceae
Melastomataceae
Myzodendraceae
Nepenthaceae
Plumbaginaceae
Polemoniaceae
Proteaceae
Rubiaceae (*Galium* sp)
Santalaceae
Saxifragaceae (linear-leaved spp)
Stylidiaceae
Trapaceae (*Trapa natans*)

IV. Hydathodes

Hydathodes present (list probably incomplete)

Aceraceae
Adoxaceae
Apiaceae
Asteraceae
Balsaminaceae
Begoniaceae
Bignoniaceae
Brassicaceae
Cabombaceae
Campanulaceae
Cannabaceae
Caprifoliaceae
Caryocaraceae
Caryophyllaceae
Celastraceae
Cephalotaceae
Chenopodiaceae
Combretaceae
Crassulaceae
Cucurbitaceae
Dipsacaceae
Elaeocarpaceae
Ericaceae (Vaccinioideae)
Fagaceae
Flacourtiaceae
Geraniaceae
Gesneriaceae
Grossulariaceae
Haloragaceae
Hamamelidaceae
Houmiriaceae
Icacinaceae
Juglandaceae
Lamiaceae
Limnanthaceae
Lobeliaceae
Malvaceae
Menispermaceae
Menyanthaceae
Moraceae
Nelumbonaceae
Nepenthaceae
Onagraceae
Orobanchaceae
Oxalidaceae
Papaveraceae
Pedaliaceae
Piperaceae
Plantaginaceae
Platanaceae
Polemoniaceae
Polygonaceae
Portulacaceae

Primulaceae
Ranunculaceae
Rhamnaceae
Rosaceae
Rubiaceae
Salicaceae
Saxifragaceae
Scrophulariaceae
Staphyleaceae
Stylidiaceae
Theaceae
Tiliaceae
Trochodendraceae
Tropaeolaceae
Ulmaceae
Urticaceae
Valerianaceae
Verbenaceae
Violaceae
Vitaceae

V. Extrafloral nectaries present

Acanthaceae
Anacardiaceae
Apocynaceae
Asclepiadaceae
Asteraceae
Balsaminaceae
Bignoniaceae
Bixaceae
Bombacaceae
Cactaceae
Caesalpiniaceae
Capparaceae
Caprifoliaceae
Caryocaraceae
Caryophyllaceae
Combretaceae
Convolvulaceae
Cucurbitaceae
Dipterocarpaceae
Dioncophyllaceae
Ebenaceae
Ericaceae
Euphorbiaceae
Flacourtiaceae
Houmiriaceae
Lecythidaceae
Loganiaceae
Lythraceae
Malpighiaceae
Malvaceae
Marcgraviaceae
Melastomataceae
Menispermaceae

Mimosaceae
Moraceae
Moringaceae
Olacaceae
Oleaceae
Papilionaceae
Passifloraceae
Pedaliaceae
Polygonaceae
Rhamnaceae
Rosaceae
Rubiaceae
Rutaceae
Salicaceae
Scrophulariaceae
Simaroubaceae
Staphyleaceae
Sterculiaceae
Tiliaceae
Turneraceae
Verbenaceae
Vochysiaceae

VI. Cork-warts present on leaves

Apocynaceae (*Carpodinus, Chilocarpus, Clitandra, Landolphia, Leuconotis, Pycnobotrya*)
Aquifoliaceae (*Ilex* spp)
Araliaceae (petioles of *Tetrapanax*)
Berberidaceae (*Berberis*, 1 sp)
Bruniaceae
Caryocaraceae (*Caryocar*)
Celastraceae (Corky outgrowths resembling lenticels; including *Salacia* spp)
Chrysobalanaceae (*Couepia*)
Corynocarpaceae (*Corynocarpus*, associated with hair bases)
Escalloniaceae (*Roussea* sp)
Hamamelidaceae (*Altingia*)
Hypericaceae (*Clusia, Garcinia*)
Loganiaceae (*Anthocleista, Fagraea*)
Marcgraviaceae (*Marcgravia, Norantea*)
Melastomataceae (*Pachyloma* sp)
Myrtaceae (*Eucalyptus*)
Piperaceae (*Peperomia*)
Rhizophoraceae (*Carallia, Cassipourea, Rhizophora*)
Rutaceae (old *Citrus* leaves)
Sonneratiaceae (*Sonneratia*)
Theaceae (*Adinandra, Anneslea, Camellia, Eurya, Ternstroemia*)

VII. Mesophyll

Homogeneous

Aizoaceae
Anacardiaceae
Annonaceae
Araliaceae
Aristolochiaceae
Asteraceae
Berberidaceae
Betulaceae
Begoniaceae
Brassicaceae
Buxaceae
Canellaceae
Campanulaceae
Capparaceae
Caryophyllaceae
Celastraceae
Chenopodiaceae
Chrysobalanaceae
Cornaceae
Cyrillaceae
Diapensiaceae
Didieeaceae
Droseraceae
Epacridaceae
Ericaceae
Fagaceae
Flacourtiaceae
Frankeniaceae
Gentianaceae
Geraniaceae
Haloragaceae
Lamiaceae
Lentibulariaceae
Linaceae
Lythraceae
Malpighiaceae
Menispermaceae
Monimiaceae
Moraceae
Myzodendraceae
Ochnaceae
Onagraceae
Orobanchaceae
Papaveraceae
Phytolaccaceae
Plantaginaceae
Plumbaginaceae
Polygalaceae
Polygonaceae
Portulacaceae
Primulaceae
Ranunculaceae
Resedaceae
Rhamnaceae

Rubiaceae
Salvadoraceae
Santalaceae
Saxifragaceae
Scrophulariaceae
Sterculiaceae
Tiliaceae
Tremandraceae
Valerianaceae
Verbenaceae
Violaceae
Vitaceae

Centric

Acanthaceae
Aceraceae
Aizoaceae
Alangiaceae
Apiaceae
Apocynaceae
Araliaceae
Aristolochiaceae
Asclepiadaceae
Bignoniaceae
Bixaceae
Boraginaceae
Brassicaceae
Bruniaceae
Campanulaceae
Capparaceae
Caryocaraceae
Caryophyllaceae
Celastraceae
Chenopodiaceae
Cistaceae
Combretaceae
Convolvulaceae
Crassulaceae
Crossosomataceae
Cucurbitaceae
Dilleniaceae
Dipsacaceae
Ebenaceae
Elaeagnaceae
Ericaceae (including *Vaccinium*)
Erythroxylaceae
Euphorbiaceae
Fagaceae
Flacourtiaceae
Fouquieriaceae
Frankeniaceae
Gentianaceae
Geraniaceae
Globulariaceae
Goodeniaceae
Haloragaceae

Hippocrateaceae
Hippuridaceae
Hydrophyllaceae
Hypericaceae
Krameriaceae
Lamiaceae
Lecythidaceae
Loasaceae
Loganiaceae
Loranthaceae
Lythraceae
Malpighiaceae
Malvaceae
Melastomataceae
Menispermaceae
Mimosaceae
Myoporaceae
Myrtaceae
Nyctaginaceae
Ochnaceae
Olacaceae
Onagraceae
Passifloraceae
Pelliceriaceae
Penaeaceae
Phytolaccaceae
Pittosporaceae
Plantaginaceae
Platanaceae
Plumbaginaceae
Polemoniaceae
Polygonaceae
Primulaceae
Proteaceae
Ranunculaceae
Resedaceae
Rhamnaceae
Rhizophoraceae
Rosaceae
Rubiaceae
Rutaceae
Santalaceae
Sapindaceae
Sapotaceae
Saxifragaceae
Scrophulariaceae
Simaroubaceae
Stackhousiaceae
Stylidiaceae
Tamaricaceae
Thymelaeaceae
Ulmaceae
Valerianaceae
Violaceae
Zygophyllaceae

Isobilateral
(sometimes merging with centric,
see previous list)

Acanthaceae
Aizoaceae
Alangiaceae
Amaranthaceae
Apocynaceae
Aristolochiaceae
Asclepiadaceae
Asteraceae
Betulaceae
Bignoniaceae
Bombacaceae
Boraginaceae
Brassicaceae
Burseraceae
Caesalpiniaceae
Campanulaceae
Capparaceae
Caryophyllaceae
Celastraceae
Convolvulaceae
Crossosomataceae
Cucurbitaceae
Dipsacaceae
Ebenaceae
Elaeagnaceae

Epacridaceae
Ericaceae
Euphorbiaceae
Fagaceae
Frankeniaceae
Gentianaceae
Geraniaceae
Globulariaceae
Goodeniaceae
Hippuridaceae
Hydrophyllaceae
Juglandaceae
Krameriaceae
Lamiaceae
Lauraceae
Linaceae (certain European spp)
Loganiaceae
Loranthaceae
Lythraceae
Malesherbiaceae
Malpighiaceae
Malvaceae
Mimosaceae
Moraceae
Myoporaceae
Myricaceae
Myrothamnaceae
Myrtaceae

Nyctaginaceae
Papaveraceae
Papilionaceae
Passifloraceae
Penaeaceae
Phytolaccaceae
Plantaginaceae
Platanaceae
Plumbaginaceae
Polemoniaceae
Polygonaceae (certain European
  spp)
Proteaceae
Ranunculaceae (certain European
  spp)
Rubiaceae (certain European spp)
Rutaceae
Salicaceae
Salvadoraceae
Santalaceae
Saxifragaceae
Scrophulariaceae
Simaroubaceae
Solanaceae
Sonneratiaceae
Thymelaeaceae
Turneraceae
Verbenaceae
Vochysiaceae

VIII. Idioblastic sclereids or sclereidal idioblasts

The information in this table should be used in conjunction with the discussion about idioblastic sclereids on pp. 58 *et seq.*

When data are included in this table without a source reference the reader should assume that the information has been taken up from the first edition of this work.

Idioblastic sclereids are so variable in form that some latitude must be allowed in interpreting the precise categories to which they belong.

Wherever a genus is mentioned the reader should remember that for many of them only a limited number of taxa have as yet been examined.

Family	Genera	Type of sclereid and location
Annonaceae	*Annona, Anaxagorea, Asteranthe, Guatteria, Heteropetalum, Popowia, Sageraea, Unona, Uvaria*	Fibres parallel to leaf surface
	Heteropetalum	Vertical fibres
	Annona, Duguetia, Guatteria, Habzelia, Unona	Branched idioblasts
Apiaceae	*Eryngium* (certain spp)	Fibres beneath hypoderm
Apocynaceae	*Bousigonia, Michrechites, Neocouma, Sclerodictyon, Trachelospermum*	Extensions of sclerenchyma surrounding the veins

Family	Genera	Type of sclereid and location
Asteropeiaceae	*Asteropeia*	Rounded brachysclereids
Begoniaceae	*Begonia*	Brachysclereids; in mesophyll and around veins; some crystalliferous, or penetrating hair bases
Bignoniaceae	*Colea, Crescentia, Phyllarthron*	Sclerenchymatous fibres
Bombacaceae	*Bombax*	Solitary sclerosed cells
Burseraceae	*Bursera, Canariellum, Canarium, Crepidospermum, Dacryodes, Garuga, Protium, Santiria, Trattinickia, Triomma*	Sclerenchymatous elements in stem cortex
Cabombaceae	*Brasenia, Cabomba*	Sclereids absent; unlie Nymphaeaceae
Capparaceae	*Boscia, Cadaba, Capparis, Courbonia, Niebuhria, Thylachium*	Polymorphic; in mesophyll. Some terminating veins. Dendrosclereids in *Boscia* (Foster 1955*a*, Bokhari and Burtt 1970)
	Boscia, Capparis	Some extending to the epidermis
	Maerua sp	Some extending externally to form hair-like structures (Rao, T.A. 1951*a*). In *Cleome aspera* secretory idioblasts as green spots (Rajagopal and Ramayya, 1968)
Caryocaraceae	*Anthodiscus, Caryocar*	Branched sclereids in stem pith and in leaf
Celastraceae	*Maurocenia, Maytenus, Microtropis, Pterocelastrus, Schaefferia*	In mesophyll
Chrysobalanaceae	*Couepia, Licania*	In mesophyll
Convolvulaceae	*Dicranostyles, Erycibe, Humbertia, Lysiostylis, Maripa, Prevostea*	Subepidermal fibres
	Ipomoea	Sclereids in mesophyll
Cornaceae	*Griselinia, Mastixia*	Sclereids in mesophyll
Cunoniaceae	*Pancheria* sp	Sclereids in mesophyll
Datiscaceae	*Octomeles*	H-shaped sclereids extending from one epidermis to the other. Branched sclereids and stone cells in cortex and pith

Family	Genera	Type of sclereid and location
Dilleniaceae Sclereids treated comprehensively by Dickison (1969, 1970)	*Davilla, Doliocarpus, Tetracera*	Sclereids with white inclusions in pith
	Curatella, Davilla, Dillenia, Doliocarpus, Tetracera	Groups of brachysclereids in pith
	Hibbertia spp	Sclereids in mesophyll
	Dillenia ochrata	Sclereids subtending trichomes (Dickison 1970)
Ebenaceae	*Diospyros*, numerous spp	Solitary and clustered pitted cells in leaf
	Diospyros spp	Osteo- and fusiform sclereids in spongy mesophyll (Rao, T.A. 1951*a*)
	Diospyros discolor	Terminal and diffuse sclereids (Rao, T. A. 1951*c*)
	Diospyros, Euclea, Maba, Oncotheca	Brachysclereids (stone-cells) in parenchyma of veins and petiole
Erythroxylaceae	*Erythroxylum* spp	Sclereids in mesophyll
Euphorbiaceae	*Acalypha, Actephila, Actinostemon, Alchornia, Bernardia, Chaetocarpus, Chondrostylis, Conceveiba, Dalechampia, Erismanthis, Pausandra, Pera, Phyllanthus, Sebastiania, Trigonostemum*	Sclereids of various kinds in mesophyll
	Amanoa, Euphorbia, Givotia, Maba, Sebastiania, Stillingia	Enlarged terminal tracheids
	Pogonophora schomburgkiana Miers	Tracheoid and secretory idioblasts of various kinds (Foster 1956)
Flacourtiaceae	*Calantica, Casearia, Homalium, Zuelania*	Fibres in mesophyll
	Erythrospermum, Patrisia, Ryania	Simple and branched sclereids in mesophyll
Frankeniaceae	*Frankenia*	Sclereids related to veins
Garryaceae	*Garrya*	Polymorphic sclereids
Gesneriaceae	*Hemiboea, Stauranthera*	Sclereids in hypodermis
	Cyrtandra	Polymorphic sclereids comprehensively treated by Bokhari and Burtt (1970)
Globulariaceae	*Globularia*	Branched sclereids, in *G. orientalis* L.
	Globularia	Solitary and grouped sclereids in stems and roots (Luhan 1954)
Goodeniaceae	*Dampiera* and other genera	Branched sclereids in leaf

Family	Genera	Type of sclereid and location
Hamamelidaceae	*Hamamelis virginiana*	Brachysclereids in leaf
	Bucklandia, Rhodoleia	Sclereids short and gnarled
	Eustigma, Hamamelis	Sclereids columnar
	Dicoryphe stipulacea J. St. Hil.	Sclereids thick-walled, irregular
	Distylium, Loropetalum, Sycopsis	Sclereids thin-walled, irregular
	Bucklandia, Eustigma, Hamamelis, Rhodoleia	Some sclereids terminal (Foster 1947, 1955a)
Houmiriaceae	*Sacoglottis*	Sclereids extending from one epidermis to the other
Ixonanthaceae	*Irvingia*	Sclereids perivascular but ending blindly in mesophyll
	Ochthocosmus	Sclereids in mesophyll: some spirally thickened
Linaceae	*Hugonia*	Sclereids in mesophyll
Loganiaceae (see Potaliaceae)	*Strychnos* spp	Branched stone-cells
Loranthaceae	spp with evergreen leaves	Branched stone-cells
	Dendrophthoë sp.	Astrosclereids in stem, leaf, perianth and fruits; crystals in mature sclereids (Rao, T. A. 1951a, Rao, A. R. and Malaviya 1962)
Malvaceae	*Goethea, Pavonia* (some spp)	Solitary sclereids in mesophyll
Marcgraviaceae	All genera	Sclereids abundant in mesophyll and stem cortex (Roon 1967, Rao, A. N. 1971)
Melastomataceae	*Memecylon, Mouriri* (69 spp. examined)	Sclereids terminal; ophiuroid and polymorphic (Foster 1946, 1947, 1955a, Morley 1953, Rao, T. A. 1950b, 1951a,b, 1957
	Graffenrieda, Meriania	Sclereids solitary; with reticulate thickening
	Anplectrum, Graffenrieda, Medinilla, Melastoma, Meriania, Ochthocharis, Pachyloma	Sclereids with local thickenings as in Menispermaceae
	Aciotis, Bellucia, Henriettea, Sonerila	Spiral tracheids in mesophyll
	Gravesia, Henrietella, Huberia, Lavoisiera, Leandra, Macairea, Medinilla, Memecylon, Miconia, Mircrolicia, Mouriri, Ossaea,*	Sclereids polymorphic

Family	Genera	Type of sclereid and location
	Plethiandra * of considerable taxonomic value according to Solereder	
Meliaceae	*Dysoxylum* spp, *Khaya senegalensis* A. Juss.	Sclereids in mesophyll
Menispermaceae	*Abuta, Adeliopsis, Anamirta, Anomospermum, Arcangelisia, Burasaia, Chlaenandra, Chondrodendrum, Coscinium, Detandra, Heptacyclum, Limacia*	Sclereids polymorphic; some locally thickened as in some Melastomataceae
Menyanthaceae	*Limnanthemum, Liparophyllum, Villarsia*	Astrosclereids; similar to those in Nymphaeaceae
Moraceae	*Balanostreblus, Ficus, Sahagunia;* few spp. only	Sclereids in mesophyll (Katsumata 1971, 1972)
Myristicaceae	*Gymnacranthera*	Fibres in mesophyll
	Iryanthera	Astrosclereids in mesophyll
Myrsinaceae (further details on p. 79)	*Weigeltia*	Fibres below adaxial and abaxial epidermis
	Aegiceras corniculatum	Polymorphic sclereids and smaller 'Palosclereids' (Rao, A.N. 1971)
Nelumbonaceae	*Nelumbo*	Sclereids absent as in Cabombaceae, but differing from Nymphaeaceae where present
Nymphaeaceae	*Euryale, Nuphar, Nymphaea, Victoria*	Sclereids ranging from stellate to girder and H-shaped; some covered with minute crystals cf Cabombaceae, Nelumbonaceae; sclereids absent from roots (Malaviya 1962)
Nyssaceae	*Nyssa*	Traversing entire thickness of lamina
Ochnaceae	*Blastemanthus, Cespedesia, Elvasia, Hilairella, Luxemburgia, Poecilandra, Trichovaselia*	Continuous layer of sclereids subjacent to adaxial epidermis; sometimes jacketing small veins and projecting into mesophyll (Rao, T.A. 1950*b*, 1951*a*)
Olacaceae	*Eganthus, Endusa, Heisteria, Minquartia, Ochanostachys, Scorodocarpus*	Sclereids in mesophyll
Oleaceae	*Ligustrum, Linociera, Noronhia, Notelaea, Olea*	Sclereids polymorphic
	Olea	Dense network (Arzee 1953a)
	Osmanthus fragrans	Columnar with ramified ends (Griffith 1968)

Family	Genera	Type of sclereid and location
	Schrebera swietennoides Roxb	Sclereids vesicular (Rao, T. A. 1949, 1951*a*)
	Linociera	Mostly terminal, but sometimes diffuse (Rao, T.A. 1950*a*, 1951*a*); includes osteosclereids
	Olea	Polymorphic; types of sclereids fully described; branched sclereids sometimes criss-cross (Rao, T. A. 1951*a*)
	Olea	More branched in compact than in lacunar tissue (Rao, T. A. and Kulkarni 1952)
Oliniaceae	*Olinia*	Branched sclereids in stem cortex
Orobanchaceae	*Kopsiopsis*	Sclereids in mesophyll
Pandaceae	*Centroplacus*	Sclereids polymorphic; in mesophyll
Papilionaceae	*Ammodendron, Andira, Bossiaea, Bowdichia, Ormosia, Platymiscium Pultenaea, Swartzia* etc.	Fibres or branched sclereids in mesophyll
	Buchenroedera	Sclerosed parenchyma in mesophyll
Passifloraceae	*Mittostema* 1 sp	Pitted cells in mesophyll
	Passiflora, few spp	Sclereids in mesophyll
Pelliceriaceae	*Pelliceria*	Fibres parallel to leaf surface; grouped stone-cells in spongy tissue
Penaeaceae	*Penaea*	Fibres sometimes spirally thickened
Plumbaginaceae	*Aegialitis, Limoniastrum, Limonium*	Branched sclereids in mesophyll
	Limonium (70 spp examined)	Sclerenchyma, diffuse and terminal; vesicular and parenchymatous sclereids, osteo-sclereids, astrosclereids, and filiform sclereids all recorded (Bokhari 1970)
Polygalaceae	*Moutabea*	Terminal sclereids (Foster 1947, 1955*a*)
Polygonaceae	*Polygonum equisetiforme*	Bundles of subepidermal fibres
Potaliaceae (see also Loganiaceae)	*Anthocleista, Fagraea, Potalia*	Spp with mesophyll of sclereids in fleshy leaves, and in stem pith
Proteaceae	*Hakea*	Columnar sclereids in mesophyll
	Adenanthos, Grevillea, Hakea, Isopogon, Petrophila, Roupala, Stenocarpus	Sclereids palisade-like

Family	Genera	Type of sclereid and location
	Adenanthos, Bellendena, Hakea, Isopogon, Leucospermum, Nivenia, Sorocephalus, Xylomelum	Sclereids almost unbranched
	Isopogon	Astrosclereids with narrow rays
	Leucospermum	Sclereids vesicular and polymorphic (Rao, T.A. 1950*b*, 1951*a*)
Rhizophoraceae	*Bruguiera, Rhizophora*	Sclereids H-shaped in aerial parts of roots and also in stems and leaf (Rao, T.A. 1951*a*)
	Poga oleosa	Solitary, sclerosed cells, in mesophyll
Rubiaceae	*Chomelia, Hippotis, Pauridiantha, Pentagonia, Stilpnophyllum, Tammsia*	Forming superficial network, sometimes connected to veins
Rutaceae (Further particulars on p. 000)	*Boronella, Boronia*	Polymorphic sclereids in mesophyll; some terminal (Foster 1955*a,b*)
Sapotaceae	*Amorphospermum, Bumelia, Chrysophyllum, Labourdonnaisia, Madhuca, Manilkara, Mimusops, Pouteria, Sideroxylon, Synsepalum*	Fibres in mesophyll sometimes spirally thickened
	Mimusops spp	Sclereids terminal with branches to mesophyll (Rao, T. A. 1950*b*, 1951*a*)
Scytopetalaceae	All genera, but not in every sp	Sclereids spreading below epidermis
Simaroubaceae	*Eurycoma, Hyptiandra, Mannia, Odyendea, Perriera, Quassia, Simaba, Simaruba, Simmanibopsis*	Polymorphic sclereids in mesophyll; for *Hannoa* (*Quassia*) see Foster 1955*a*
	Picramnia, Picrasma, Samadera	Terminal sclereids with branches in mesophyll
	Harrisonia	No sclereids recorded
Sonneratiaceae	*Sonneratia*	'Curved, lignified spicules' in pneumatophores, and in mesophyll
Tetrameristaceae	*Tetramerista*	Few sclereids in midrib and central part of mesophyll
Theaceae	*Adinandra, Anneslea, Camellia, Cleyera, Eurya, Franklinia, Freziera, Gordonia, Nabiasodendron, Pyrenaria, Schima, Ternstroemia, Ternstroemiopsis, Visnea*	Sclereids in cortex of petiole, in mesophyll and sometimes in stem pith; those in *Ternstroemia* stellate (Rao, T.A. 1951*a*)
	Cleyera, Eurya, Freziera, Ternstroemia, Visnea	Sclereids forming diaphragms in stem pith

Family	Genera	Type of sclereid and location
	Camellia	Sclereids polymorphic (Foster 1944) (Barua and Wight 1958; Barua and Dutta 1959)
Theophrastaceae	*Clavija* (pro parte), *Deherainia*, *Jacquinia*, *Theophrasta*	Fibres subjacent to adaxial and abaxial epidermis
Trochodendraceae	*Trochodendron*	Sclereids large, very polymorphic Foster 1945*b*)
Turneraceae	*Turnera hilaireana* Urban	Sclereids in mesophyll
Verbenaceae	*Clerodendrum splendens* G. Don	Brachysclereids in groups of 2–3 in mesophyll; groups of 3–20 or more in midrib and petiole; some contain stellate, rod-shaped or cubical crystals (Inamdar 1968*c*)
	Nyctanthes	Polymorphic scleriods (Rao, T.A., 1947, 1951*a*)
Winteraceae	*Bubbia*, *Drimys* section *Wintera*, *Zygogynum*	Branched sclereids in mesophyll
	Few spp of *Belliolum*, *Bubbia*	Some mesophyll cells with reticulate thickenings
	Belliolum (6 spp), *Bubbia* (17 spp), *Drimys* section *Tasmannia*	Larger and terminal veins jacketed by sclereids (Bailey and Nast 1944)
	Drimys section *Wintera*	Large veins jacketed by sclereids; terminal veins not jacketed (Bailey and Nast 1944)
	Zygogynum	Sclereids polymorphic (Bailey and Nast 1944)

IX. Kranz structure

This type of structure (see also pp. 72–3) occurs in many monocotyledons, notably in certain grasses and sedges (Metcalfe 1960, 1971).

Aizoaceae
  (*Mollugo* and *Trianthema*; Sabnis 1919–21)
Amaranthaceae
  (*Amaranthus*; Laetsch 1968, Tregunna and Downton 1967, Cookston and Moss 1970)
  (*Amaranthus*, *Froelichia*, *Gomphrena*)

Asteraceae
  (*Centaurea* spp; Heinricher 1884)
  (*Pectis*; Solereder 1908, Schöch and Kramer 1971)
Boraginaceae
  (*Heliotropium*; Solereder 1908, Sabnis 1919–21, Downton, Bisalputra, and Tregunna 1969)
Brassicaceae
  (*Farsetia*; Sabnis 1919–21)
  (*Diplotaxis*, 2 spp; Moser 1934)
Caryophyllaceae
  (*Dianthus*, *Silene*; Heinricher 1884, Crookston and Moss 1970)

  (*Polycarpaea*; Sabnis 1919–21)
Chenopodiaceae
  (*Atriplex*, 79 spp; Moser 1934, Laetsch 1968; not in *A. patula* group, Frankton and Bassett 1970)
  (*Atriplex*; Downton, Bisalputra, and Tregunna 1969)
  (*Atriplex*, *Bassia*, *Kochia*, *Salsola*; Crookston and Moss 1970)
Dipsacaceae
  (*Scabiosa* sp; Heinricher 1884)
Euphorbiaceae
  (*Euphorbia*; Crookston and Moss 1970)
  (*Chamaesyce* sp = *Euphorbia*;

Schöch and Kramer 1971)
Nyctaginaceae
  (*Boerhaavia, Bougainvillea,*
    Phaeoptilium; Solereder 1908)
  (*Boerhaavia*, 2 spp; Sabnis
    1919-21)
Oleaceae
  (A few spp. Moser 1934)
Papaveraceae
  (A few spp; Moser 1934)
Papilionaceae
  (*Genista, Spartium*; Haberlandt
    1914)
  (*Crotalaria, Indigophora*; Sabnis
    1919-21)
Polemoniaceae
  (Some spp of all genera
    examined *except Bonplandia*;
    Solereder 1908)
Portulacaceae
  (Some spp; Crookston and Moss
    1970)
Resedaceae
  (*Ochradenus, Oligomeris*; Moser
    1934)
Sapindaceae
  (*Cardiospermum*; Sabnis
    1919-21)
Scrophulariaceae
  (*Pentstemon*; Heinricher 1884)
Zygophyllaceae
  (*Tribulus*; Solereder 1908)
  (*Fagonia, Tribulus, Zygo-*
    phyllum; Sabnis 1919-21)

X. Veins

Vertically transcurrent

Aceraceae
Actinidiaceae
Anacardiaceae
Ancistrocladaceae
Annonaceae
Betulaceae
Berberidaceae
Bixaceae
Boraginaceae
Brassicaceae
Brunelliaceae
Burseraceae

Caesalpiniaceae
Capparaceae
Caprifoliaceae
Celastraceae
Chrysobalanaceae
Cistaceae
Clethraceae
Combretaceae
Connaraceae
Convolvulaceae
Cornaceae
Cunoniaceae
Dipsacaceae
Dipterocarpaceae
Ebenaceae
Epacridaceae
Ericaceae
Escalloniaceae
Eucryphiaceae
Fagaceae
Flacourtiaceae
Gentianaceae
Geraniaceae
Globulariaceae
Grossulariaceae
Hippocastanaceae
Hydrangeaceae
Hypericaceae
Ixonanthaceae
Lamiaceae
Lardizabalaceae
Lauraceae
Lythraceae
Malvaceae
Melastomataceae
Menispermaceae
Mimosaceae
Moraceae
Myricaceae
Myristicaceae
Nyssaceae
Ochnaceae
Papilionaceae
Picrodendraceae
Pittosporaceae
Platanaceae
Plumbaginaceae
Polygalaceae
Proteaceae
Ranunculaceae
Rhamnaceae

Rhizophoraceae
Rosaceae
Sapindaceae
Sapotaceae
Sarcolaenaceae
Saxifragaceae
Simaroubaceae
Sterculiaceae
Styracaceae
Symplocaceae
Tiliaceae
Tremandraceae
Verbenaceae
Violaceae
Vochysiaceae

Sheathed by large parenchy-
matous cells in some species of:

Aizoaceae
Amaranthaceae
Aristolochiaceae
Asteraceae
Begoniaceae
Boraginaceae
Campanulaceae
Caryophyllaceae
Ceratophyllaceae
Chenopodiaceae
Cucurbitaceae
Euphorbiaceae
Goodeniaceae
Hippuridaceae
Lamiaceae
Mimosaceae
Nyctaginaceae
Polemoniaceae (in *Gilia* cells
  with U-shaped thickenings)
Primulaceae
Punicaceae
Rhamnaceae
Rhizophoraceae
Rosaceae
Salvadoraceae
Sarcolaenaceae
Scrophulariaceae
Stylidiaceae
Simaroubaceae
Thymelaeaceae
Ulmaceae
Zygophyllaceae (*Tribulus*; cell
  walls thick and pitted)

STEM CHARACTERS

I. Endodermis Conspicuous

Acanthaceae (*Andrographis, Barleria, Thunbergia*)

Actinidiaceae

Aizoaceae

Amaranthaceae (Frequently conspicuous)

Aristolochiaceae

Asclepiadaceae

Asteraceae (Species of *Aster, Lactuca, Layia, Raoulia* (walls thickened)

Boraginaceae (*Heliotropium* 1 sp; cells with granular contents)

Brassicaceae (Species of *Brassica, Capsella, Kernera, Lepidium, Nasturtium, Sinapis* and other genera)

Byblidaceae (*Byblis*; as starch sheath)

Caesalpiniaceae (*Cercis siliquastrum*)

Campanulaceae (*Asyneuma, Campanula, Wahlenbergia*)

Caryophyllaceae (*Arenaria, Corrigiola, Dianthus, Lychnis, Saponaria*)

Chenopodiaceae

Cistaceae (*Helianthemum*; sometimes less conspicuous in *Cistus*)

Convolvulaceae (*Convolvulus*; root of 1 sp; suberized)

Dipsacaceae (*Dipsacus*; cell walls thick)

Elatinaceae (*Elatine*)

Epacridaceae (*Needhamia, Oligarrhena*, one species of each; cells suberized)

Gentianaceae (Frequent)

Geraniaceae (*Geranium pratense*, in young stems)

Gesneriaceae (*Achimenes, Aeschynanthus, Ramondia*)

Haloragaceae (Especially aquatic spp)

Hippuridaceae (*Hippuris*)

Hydrophyllaceae (Frequent)

Hypericaceae (*Hypericum elodes*; young stems of terrestrial species; (*Kielmeyera*)

Lamiaceae (Walls thick or thin; very widespread in the family)

Lobeliaceae (Usually conspicuous)

Melastomataceae (Often thin-walled, but cells with U-shaped thickenings in species of *Anerincleistus, Calophysa, Dichaetanthera, Kibessa, Memecylon, Mouriri*; suberized in *Gravesia* and *Medinella*)

Menyanthaceae (*Nymphoïdes* (formerly *Limnanthemum*)

Nyctaginaceae (*Boerhaavia*; cells thick-walled)

Onagraceae (*Epilobium pedunculare* Hook. f and *E. pubens* A. Rich.)

Penaeaceae (*Penaea*; cells large)

Pentaphragmataceae (*Pentaphragma*; cells thin-walled)

Piperaceae (Endodermis very variable in *Piper*; Bond 1931)

Plantaginaceae (*Littorella*; walls thickened and pitted)

Polemoniaceae (*Collomia* and *Gilia*)

Polygonaceae (Recorded in about 20 genera)

Portulacaceae (*Montia australis*)

Primulaceae (usually conspicuous)

Rosaceae (Species of *Alchemilla, Fragaria, Gillenia, Kerria, Neviusa, Potentilla, Poterium, Rubus*)

Rubiaceae (*Cephaelis, Hoffmannia, Manettia*)

Scrophulariaceae (*Antirrhinum, Euphrasia, Gratiola, Herpestis, Linaria, Scrophularia, Veronica*)

Solanaceae (*Browallia, Nicotiana, Solanum*)

Stackhousiaceae (Cells with granular contents and thick walls)

Tropaeolaceae (*Tropaeolum majus* L.).

Violaceae (Large cells with suberized walls in *Viola cunninghamii*)

II. Families with Casparian Thickenings in stems

Acanthaceae (*Asystaria, Dianthera, Eranthemum, Jacobinia*)

Adoxaceae

Amaranthaceae (*Alternanthera* sp)

Asteraceae (*Cichorium, Cosmos, Erigeron, Felicia, Tagetes, Xanthium*)

Campanulaceae (*Asyneuma, Campanula, Wahlenbergia*)

Caryocaraceae (*Caryocar*; young roots)

Celastraceae (*Euonymus*; Codaccioni 1970)

Ceratophyllaceae (*Ceratophyllum*)

Convolvulaceae (*Convolvulus arvensis*; Codaccioni 1970)

Elaeagnaceae (*Elaeagnus*; upper part of hypocotyl)

Lamiaceae (in genera such as *Ballota, Galeopsis, Prunella, Scutellaria*, and *Stachys*. for further particulars see Briquet 1903, Codaccioni 1970, Courtot and Baillaud 1960, Lemesle 1928)

Lythraceae (*Adenaria, Cuphea, Lagerstroemia, Peplis*)

Melastomataceae (*Sonerila*)

Penaeaceae (*Penaea* sp)

Pentaphragmataceae

Piperaceae (*Piper* in part; see Bond 1931)

Plantaginaceae

Plumbaginaceae (root)

Polemoniaceae (*Phlox paniculata, Polemonium caeruleum*)

Rubiaceae (Tribe Galieae and *Theligonum*)

Scrophulariaceae (*Antirrhinum, Euphrasia, Gratiola, Herpestis, Linaria*)

Solanaceae (*Jaborosa*; root)

Tropaeolaceae

Verbenaceae (*Verbena* spp; obscure in *Phryma*)

Vitaceae (in root, but rare)

III. Cork

Superficial

Acanthaceae

Aceraceae

Achariaceae

Actinidiaceae

Akaniaceae
Amaranthaceae
Anacardiaceae
Ancistrocladaceae
Annonaceae
Apiaceae
Apocynaceae
Aquifoliaceae
Araliaceae
Aristolochiaceae
Asclepiadaceae
Asteraceae
Balanopaceae
Basellaceae
Begoniaceae
Berberidaceae
Betulaceae
Bignoniaceae
Bombacaceae
Bonnetiaceae
Boraginaceae
Bruniaceae
Burseraceae
Buxaceae
Cactaceae
Caesalpiniaceae
Calycanthaceae
Campanulaceae
Canellaceae
Cannabaceae
Capparaceae
Caprifoliaceae
Caryocaraceae
Celastraceae
Cercidiphyllaceae
Chenopodiaceae
Chrysobalanaceae
Cistaceae
Combretaceae
Connaraceae
Convolvulaceae
Coriariaceae
Cornaceae
Corynocarpaceae
Crassulaceae
Crossosomataceae
Curcurbitaceae
Cunoniaceae
Datiscaceae
Diapensiaceae
Dichapetalaceae
Dilleniaceae
Dipsacaceae
Dipterocarpaceae
Ebenaceae
Elaeagnaceae
Ericaceae

Elaeocarpaceae
Erythroxylaceae
Escalloniaceae
Eucommiaceae
Euphorbiaceae
Fagaceae
Flacourtiaceae
Fouquieriaceae
Garryaceae
Geissolomataceae
Geraniaceae
Gesneriaceae
Gomortegaceae
Goodeniaceae
Grubbiaceae
Hamamelidaceae
Hernandiaceae
Himantandraceae
Hippocrateaceae
Houmiriaceae
Hydrophyllaceae
Hypericaceae
Icacinaceae
Juglandaceae
Julianiaceae
Lamiaceae
Lardizabalaceae
Lauraceae
Lecythidaceae
Leitneriaceae
Lobeliaceae
Loganiaceae
Loranthaceae
Magnoliaceae
Malesherbiaceae
Malpighiaceae
Malvaceae
Marcgraviaceae
Medusagynaceae
Melastomataceae
Meliaceae
Menispermaceae
Mimosaceae
Monimiaceae
Moraceae
Myoporaceae
Myricaceae
Myristicaceae
Myrsinaceae
Myrtaceae
Nyctaginaceae
Nyssaceae
Ochnaceae
Olacaceae
Oleaceae
Oliniaceae
Opiliaceae

Oxalidaceae
Papilionaceae
Pedaliaceae
Pelliceriaceae
Pentaphylacaceae
Phytolaccaceae
Piperaceae
Pittosporaceae
Plantaginaceae
Platanaceae
Plumbaginaceae
Polemoniaceae
Polygalaceae
Polygonaceae
Primulaceae
Proteaceae
Rhamnaceae
Rhizophoraceae
Rosaceae
Rubiaceae
Rutaceae
Sabiaceae
Salicaceae
Salvadoraceae
Santalaceae
Sapindaceae
Sapotaceae
Saxifragaceae
Scrophulariaceae
Scytopetalaceae
Simaroubaceae
Solanaceae
Sonneratiaceae
Stachyuraceae
Staphyleaceae
Sterculiaceae
Symplocaceae
Tamaricaceae
Tetracentraceae
Theaceae
Thymelaeaceae
Tiliaceae
Trigoniaceae
Trochodendraceae
Turneraceae
Ulmaceae
Urticaceae
Verbenaceae
Vitaceae
Vochysiaceae
Winteraceae
Zygophyllaceae

Deep seated

Acanthaceae
Aceraceae

Actinidiaceae
Aizoaceae
Anacardiaceae
Ancistrocladaceae
Apiaceae
Apocynaceae
Asteraceae
Bataceae
Berberidaceae
Bignoniaceae
Boraginaceae
Brassicaceae
Burseraceae
Buxaceae
Campanulaceae
Capparaceae
Caprifoliaceae
Caryophyllaceae
Casuarinaceae
Celastraceae
Chenopodiaceae
Cistaceae
Clethraceae
Columelliaceae
Combretaceae
Convolvulaceae
Crassulaceae
Cucurbitaceae
Cyrillaceae
Diapensiaceae
Dilleniaceae
Dipsacaceae
Ebenaceae
Empetraceae
Epacridaceae
Ericaceae
Escalloniaceae
Euphorbiaceae
Eupteleaceae
Fouquieriaceae
Frankeniaceae
Gesneriaceae
Goodeniaceae
Grossulariaceae
Hydrangeaceae
Hypericaceae
Krameriaceae
Lamiaceae
Lauraceae
Loasaceae
Loganiaceae
Lythraceae
Malpighiaceae
Melastomataceae
Melianthaceae
Menispermaceae
Myrtaceae

Nepenthaceae
Nyctaginaceae
Oleaceae
Onagraceae
Papilionaceae
Penaeaceae
Plantaginaceae
Polemoniaceae
Polygonaceae
Primulaceae
Proteaceae
Punicaceae
Ranunculaceae
Rosaceae
Rubiaceae
Sapindaceae
Sargentodoxaceae
Saxifragaceae
Scrophulariaceae
Solanaceae
Styracaceae
Tamaricaceae
Tetrameristaceae
Theaceae
Tropaeolaceae
Urticaceae
Valerianaceae
Verbenaceae
Vitaceae
Vochysiaceae
Zygophyllaceae

IV. Little or no sclerenchyma in the pericyclic region

Acanthaceae
Aizoaceae
Apiaceae
Apocynaceae
Araliaceae
Balsaminaceae
Begoniaceae
Boraginaceae
Cactaceae
Campanulaceae
Clethraceae
Convolvulaceae
Cornaceae
Cyrillaceae
Dipsacaceae
Ericaceae
Euphorbiaceae
Grossulariaceae
Hydrangeaceae
Hydrophyllaceae

Lamiaceae
Lobeliaceae
Loganiaceae
Lythraceae
Malpighiaceae
Nymphaeaceae
Pittosporaceae
Plantaginaceae
Portulacaceae
Primulaceae
Rubiaceae
Salicaceae
Saxifragaceae
Scrophulariaceae
Stackhousiaceae
Turneraceae
Valerianceae
Verbenaceae

V. Width of primary medullary rays

Broad

Acanthaceae (Steles)
Aceraceae
Adoxaceae
Aizoaceae
Akaniaceae
Amaranthaceae
Apiaceae
Apocynaceae
Aquifoliaceae
Araliaceae
Aristolochiaceae
Asclepiadaceae
Asteraceae
Balanophoraceae (rhizome)
Balsaminaceae
Basellaceae
Bataceae
Begoniaceae
Berberidaceae
Bignoniaceae
Brassicaceae
Byblidaceae
Cactaceae
Caesalpiniaceae
Calyceraceae
Capparaceae
Caricaceae
Caryophyllaceae
Casuarinaceae
Celastraceae
Chenopodiaceae
Coriariaceae
Corynocarpaceae

Cucurbitaceae
Didiereaceae
Dilleniaceae
Droseraceae (scape)
Ericaceae (*Monotropa*)
Euphorbiaceae
Eupomatiaceae
Fagaceae
Garryaceae
Geraniaceae
Goodeniaceae
Grossulariaceae
Haloragaceae (steles)
Hydrangeaceae
Hydrophyllaceae
Hydrostachyaceae
Icacinaceae
Lamiaceae
Lardizabalaceae
Lauraceae
Lecythidaceae
Lennoaceae
Limnanthaceae
Loasaceae
Loganiaceae
Loranthaceae
Magnoliaceae
Medusagynaceae
Melianthaceae
Menispermaceae
Menyanthaceae
Mimosaceae
Monimiaceae
Moraceae
Myricaceae
Myrsinaceae
Myzodendraceae
Nyctaginaceae
Nymphaeaceae (steles)
Orobanchaceae
Oxalidaceae
Papaveraceae
Papilionaceae
Phytolaccaceae
Piperaceae
Plantaginaceae
Platanaceae
Plumbaginaceae
Polygonaceae
Portulacaceae
Primulaceae
Proteaceae
Ranunculaceae
Rhizophoraceae
Rosaceae
Sabiaceae
Salvadoraceae

Sargentodoxaceae
Saururaceae
Saxifragaceae
Solanaceae
Tamaricaceae
Tropaeolaceae
Ulmaceae
Urticaceae
Valerianaceae
Verbenaceae
Violaceae
Vitaceae
Zygophyllaceae

Narrow

Acanthaceae
Aceraceae
Actinidiaceae
Alangiaceae
Anacardiaceae
Apocynaceae
Asclepiadaceae
Asteraceae
Balanopsidaceae
Betulaceae
Bignoniaceae
Bixaceae
Bonnetiaceae
Boraginaceae
Brassicaceae
Brunelliaceae
Bruniaceae
Burseraceae
Buxaceae
Caesalpiniaceae
Calycanthaceae
Campanulaceae
Canellaceae
Cannabaceae
Capparaceae
Caprifoliaceae
Caryocaraceae
Caryophyllaceae
Casuarinaceae
Celastraceae
Cercidiphyllaceae
Chlaenaceae
Chrysobalanaceae
Cistaceae
Clethraceae
Cneoraceae
Combretaceae
Connaraceae
Convolvulaceae
Cornaceae
Crassulaceae
Crossosomataceae

Crypteroniaceae
Cunoniaceae
Daphniphyllaceae
Datiscaceae
Diapensiaceae
Dipsacaceae
Ebenaceae
Elaeagnaceae
Elaeocarpaceae
Empetraceae
Epacridaceae
Ericaceae
Erythroxylaceae
Escalloniaceae
Eucommiaceae
Eucryphiaceae
Euphorbiaceae
Fagaceae
Flacourtiaceae
Frankeniaceae
Geissolomataceae
Gentianaceae – Gentianoideae
Geraniaceae
Gesneriaceae
Globulariaceae
Gomortegaceae
Goodeniaceae
Grossulariaceae
Grubbiaceae
Haloragaceae
Hamamelidaceae
Hippocastanaceae
Hydrangeaceae
Hydrophyllaceae
Hypericaceae
Juglandaceae
Krameriaceae
Lamiaceae
Lauraceae
Lecythidaceae
Leitneriaceae
Linaceae
Loasaceae
Lobeliaceae
Loganiaceae
Loranthaceae
Lythraceae
Malesherbiaceae
Malpighiaceae
Marcgraviaceae
Melastomataceae
Meliaceae
Melianthaceae
Mimosaceae
Monimiaceae
Moringaceae
Myoporaceae

Myricaceae
Myristicaceae
Myrothamnaceae
Myrsinaceae
Myrtaceae
Nepenthaceae
Nyssaceae
Ochnaceae
Olacaceae
Oleaceae
Oliniaceae
Onagraceae
Oxalidaceae
Papilionaceae
Passifloraceae
Pedaliaceae
Pelliceriaceae
Penaeaceae
Pentaphylacaceae
Plantaginaceae
Polemoniaceae
Polygalaceae
Polygonaceae
Portulacaceae
Primulaceae
Punicaceae
Quiinaceae
Resedaceae
Rhamnaceae
Rhizophoraceae
Rosaceae
Rubiaceae
Rutaceae
Salicaceae
Santalaceae
Sapindaceae
Sapotaceae
Sarcolaenaceae
Schisandraceae
Scrophulariaceae
Scytopetalaceae
Simaroubaceae
Solanaceae
Sonneratiaceae
Stachyuraceae
Stackhousiaceae
Staphyleaceae
Styracaceae
Symplocaceae
Tetrameristaceae
Theaceae
Thymelaeaceae
Tremandraceae
Turneraceae

Urticaceae
Valerianaceae
Verbenaceae
Violaceae
Winteraceae
Zygophyllaceae

Narrow and broad mixed

Actinidiaceae
Anacardiaceae
Annonaceae
Araliaceae
Begoniaceae
Betulaceae
Bombacaceae
Boraginaceae
Brassicaceae
Cannabaceae
Caryophyllaceae
Chloranthaceae
Crassulaceae
Dipterocarpaceae
Elaeagnaceae
Eupteleaceae
Hernandiaceae
Himantandraceae
Hydrophyllaceae
Icacinaceae
Lactoridaceae
Loasaceae
Loranthaceae
Malvaceae
Moraceae
Orobanchaceae
Passifloraceae
Pedaliaceae
Pittosporaceae
Primulaceae
Sabiaceae
Saxifragaceae
Scytopetalaceae
Simaroubaceae
Solanaceae
Staphyleaceae
Tetracentraceae
Tiliaceae
Trigoniaceae
Trochodendraceae
Ulmaceae

VI. Diaphragms in Pith

F = fistular; I = diaphragms sometimes incomplete; D = diaphragms of discoid type. (For description of types see p. 000.)

Actinidiaceae (Lechner 1914)

Annonaceae; diaphragms very common (see first edition of this work; Wyk and Canright (1952)

Asteraceae; certain species of *Senecio*; F (Solereder 1908)

Begoniaceae; D (Holm 1921)

Brassicaceae; *Diplotaxis* Solereder 1908)

Caprifoliaceae; F, in *Lonicera*

Convolvulaceae; *Maripa cayennensis* Meissn. (Solereder 1908)

Daphniphyllaceae; F (Solereder 1908)

Degeneriaceae (Bailey and Smith 1942)

Dilleniaceae; *Wormia*; D (Solereder 1908)

Fouquieriaceae; F (Solereder 1908)

Himantandraceae; nests and transverse plates of sclereids (Foxworthy 1903)

Juglandaceae; *Juglans cinerea* and *J. nigra* (Foxworthy 1903; Harlow 1930); *Pterocarya* sp (Holm 1921)

Magnoliaceae; *Magnolia*, sclerenchymatous in some spp, diaphragms sometimes incomplete (Foxworthy 1903); *Liriodendron*, diaphragms heterogeneous

Myristicaceae (Wyk and Canright 1952)

Nymphaeaceae; De Bruyne 1922 and other authors)

Nyssaceae (Faure 1924; Foxworthy 1903; Holm 1921; some disagreement amongst authors)

Olacaceae; *Brachynema*, plates of stone-cells (Foxworthy 1903; Solereder 1908)

Oleaceae; *Forsythia*, *Jasminum* (Foxworthy 1903)

Pedaliaceae; *Pedalium*, F (Solereder 1908)

Phytolaccaceae; *Phytolacca decandra*, F Solereder 1908)

Rosaceae; *Prinsepia*, F (Solereder
  1908)
Scrophulariaceae; *Paulownia*
  (Foxworthy 1903)
Theaceae; *Adinandra*, *Cleyera*,
Eurya, *Freziera*, *Ternstroemia*,
  Visnea, I (Beauvisage 1920)
Ulmaceae; *Celtis*, F (Foxworthy
  1903)
Winteraceae; *Drimys howeana*
  F. Muell., I(Solereder 1908).
  Recent work indicates that
  sclerenchyma is common in the
  pith of this family.

BIBLIOGRAPHY

Aalders, L. E. and Hall, I. V. (1962). New evidence on the cytotaxonomy of *Vaccinium* species as revealed by stomatal measurements from herbarium specimens. *Nature, Lond.* 196, 694.

Acqua, C. (1887). Sulla distribuzione dei fasci fibrovascolari nel loro passaggio dal fusto alla foglia. *Malpighia* 1, 277–82.

Adamson, R. S (1912). On the comparative anatomy of the leaves of certain species of *Veronica. J. Linn. Soc., Bot.* 40, 247–74.

Agardh, J. G. (1850). Ueber die Nebenblätter (Stipulae) der Pflanzen. *Flora* 33, 758–61.

Agthe, C. (1951). Ueber die physiologische Herkunft des Pflanzennektars. Thesis, Zurich. (*Ber. schweiz. bot. Ges.* 61, 240–77.)

Ahmad, K. J. (1962). Cuticular striations in *Cestrum. Curr. Sci.* 31, 388–90.

— (1964a). On stomatal abnormalities in the Solanaceae. *Sci. Cult.* 30, 349–51.

— (1964b). Cuticular studies with special reference to abnormal stomatal cells in *Cestrum. J. Indian bot. Soc.* 43, 165–77.

Airy Shaw, H. K. (1971). Notes on Malesian and other Asiatic Euphorbiaceae. New and noteworthy species of *Macaranga* Thou. *Kew Bull.* 25, 473–553.

Aljaro, M. E., Avila, A. G., Hoffmann, A., and Kummerow, J. (1972). The annual rhythm of cambial activity in two woody species of the Chilean 'matorral'. *Am. J. Bot.* 59, 879–85.

Allaway, W. G. and Milthorpe, F. L. (1976). Structure and functioning of stomata. In *Water deficits and plant growth*. (ed. T. T. Kozlowski) Vol. 4. Academic Press, London.

Alvin, K. and Boulter, D. [1975] (1974). A controlled method of comparative study for Taxodiaceae leaf cuticles. *J. Linn. Soc., Bot.* 67, 277–86.

Amar, M. (1904). Sur le rôle de l'oxalate de calcium dans la nutrition des végétaux. *Annls Sci. nat., Bot.*, (ser. VIII) 19, 195–291.

Ambronn, H, (1888). Ueber das optische Verhalten der Cuticula und der verkorkten Membranen. *Ber. dt. bot. Ges.* 6, 226–30.

Amelunxen, F., Morgenroth, K., and Picksak, T. (1967). Untersuchungen an der Epidermis mit dem Stereoscan-Elektronenmikroskop. *Z. Pflanzenphysiol.* 57, 79–95.

Anderson, D. B. (1928). Struktur und Chemismus der Epidermis-aussenwände von *Clivia nobilis. Jb. wiss. Bot.* 69, 501–15.

— (1935). The structure of the walls of the higher plants. *Bot. Rev.* 4, 52–76.

Anderson, L. C. (1972). Systematic anatomy of *Solidago* and associated genera (Asteraceae). *Brittonia* 24, 117. [Abstract.]

Anderson, W.R. (1972). A monograph of the genus *Crusea* (Rubiaceae). *Mem. N. Y. Bot. Gard.* 22 (4), 1–128.

Andersson, L. (1977). The genus *Ischnosiphon* (Marantaceae). *Op. Bot.* 43, 1–113.

Anheisser, R. (1900). Uber die aruncoide Blattspreite. *Flora* 87, 64–94.

Arbeitsgruppe 'Cuticulae' der C.I.M.P. (1964). Entwurf für einheitliche diagnostische Beschreibung von Kutikulen. *Fortschr. Geol. Rheinld Westf.* 12, 11–24.

Arber, A. (1941a). Tercentary of Nehemiah Grew (1641–1712). *Nature, Lond.* 147, 630–2.

— (1941b). The Relation of Nehemiah Grew and Marcello Malpighi. *Chronica Bot.* 6, 391–2.

— (1941c). Nehemiah Grew and Marcello Malpighi. *Proc. Linn. Soc. Lond.* 218–38.

— (1941d). The interpretation of leaf and root in the angiosperms. *Biol. Rev.* 16, 81–105.

Arbo, M. M. (1972). Estructura y ontogenia de los nectarios foliares del género *Byttneria* (Sterculiaceae). *Darwiniana* 17, 104–58.

— (1973). Los nectarios foliares de *Megatritheca* (Sterculiaceae). *Darwiniana* 18, 272–6. [English summary.]

Arens, T. (1968).. Radialstrukturen in den Stomata von *Ouratea spectabilis* (Mart.) Engl. *Protoplasma* 66, 403–11.

Areschoug, F. W. C. (1897). Über die physiologischen Leistungen und die Entwickelung des Grundgewebes des Blattes. *Lunds univ. årsskr.* 33 II (10), 1–46.

Arnal, C. (1962). La notion de noeud. *Bull. Soc. bot. Fr.* 104–11.

Arraes, M. A. B. (1970). Significação estatistica do indice estomatico em especies do genero *Cassia* L. *Bol. cear; Agron. bras.* 11, 31–8.

Artschwager, E. (1943). Contribution to the morphology and anatomy of guayule (*Parthenium argentatum*). *U.S.D.A. Tech. Bull.* 842, 1–33.

Arzee, T. (1953a). Morphology and ontogeny of foliar sclereids in *Olea europaea*. I. Distribution and structure. *Am. J. Bot.* 40, 680–7.

— (1953b). Morphology and ontogeny of foliar sclereids in *Olea europaea*. II. Ontogeny. *Am. J. Bot.* 40, 649–744.

Arzt, T. (1933). Untersuchungen über das Vorkommen einer Kutikula in den Blättern dikotyler Pflanzen. *Ber. dt. bot. Ges.* 51, 470–500.

Ascherson, P. (1878). Galls on *Salix comifera* Wang. and *Acacia fistula* Schweinf. *Verh. bot. Ver. Prov. Brandenb.* XX, 44–45.

Auer, S. (1962). Untersuchungen zur Epidermis-Entwicklung an Keimblättern von *Pulsatilla vulgaris. Z. Bot.* 50, 128–53.

Aufrecht, S. (1891). Beiträge zur Kenntnis der Extrafloralen Nektarien. Thesis, Zurich.

Aylor, D. E., Parlange, J.-Y., and Krikorian, A. D. (1973). Stomatal mechanics. *Am. J. Bot.* 60, 163–71.

Baas, P. (1970). Anatomical contributions to plant anatomy I. Floral and vegetative anatomy of *Eliaea* from Madagascar and *Cratoxylum* from Indo-Malesia (Guttiferae). *Blumea* 18, 369–91.

— (1972a). Anatomical contributions to plant taxonomy II. The affinities of *Hua* Pierre and *Afrostyrax* Perkins et Gilg. *Blumea* 20, 161–92.

— (1972b). The vegetative anatomy of *Kostermansia malayana* Soegeng. *Reinwardtia* 8, 335–44.

— (1973b). The wood anatomical range in *Ilex* (Aquifoliaceae) and its ecological and phylogenetic significance. *Blumea* 21, 193–258.

— (1974). Stomatal types in the Icacinaceae. Additional observations on genera outside Malaysia. *Acta bot. neerl.* **23** (3), 193–200.

— (1975). Vegetative anatomy and the affinities of Aquifoliaceae, *Sphenostemon*, *Phelline*, and *Oncotheca*. *Blumea* **22**, 311–407.

Bolton, A. J., and Catling, D. M. (1976). *Wood structure in biological and technological research.* Leiden Botanical Series No. 3. Leiden University Press.

Bailey, I. W. (1922). The anatomy of certain plants from the Belgian Congo, with special reference to myrmecophytism. In Wheeler. The ants collected by the American Museum Congo expedition. *Bull. Am. Mus. nat. Hist.* **45**, 585–621.

— (1953*a*). The anatomical approach to the study of genera. *Chronica Bot.* **14** (3), 121–5.

— (1953*b*). Evolution of the tracheary tissue of land plants. *Am. J. Bot.* **40**, 4–8.

— (1954). *Contributions to plant anatomy.* Chronica Botanica, Waltham, Mass., U. S. A.

— (1956). Nodal anatomy in retrospect. *J. Arnold Arbor.* **37**, 269–87.

— and Howard, R. A. (1941). The comparative morphology of the Icacinaceae I-IV. *J. Arnold Arbor.* **22**, 125–32, 171–87, 432–42, 556–68.

— and Nast, C. G. (1944). The comparative morphology of the Winteraceae V. Foliar epidermis and sclerenchyma. *J. Arnold Arbor.* **25**, 342–8.

— — (1945). Morphology and relationships of *Trochodendron* and *Tetracentron*. 1. Stem, root and leaf. *J. Arnold Arbor.* **26**, 143–54.

— — (1948). Morphology and relationships of *Illicium*, *Schisandra* and *Kadsura* I. Stem and leaf. *J. Arnold Arbor.* **29**, 77–89.

— and Smith, A. C. (1942). Degeneriaceae, a new family of flowering plants from Fiji. *J. Arnold Arbor.* **23**, 356–65.

— and Swamy, B. G. L. (1949). Morphology and relationships of *Austrobaileya*. *J. Arnold Arbor.* **30**, 211–26.

— and Tupper, W. W. (1918). Size variations in tracheary cells: I. a comparison between the secondary xylems of vascular cryptogams, gymnosperms and angiosperms. *Proc. Am. Acad. Arts Sci.* **54**, 149–204.

Baillon, H. (1871). *The natural history of plants*, Vol. 1. Reeve, London.

Baker, E. A. (1970). The morphology and composition of isolated plant cuticles. *New Phytol.* **69**, 1053–8.

— (1971). Chemical and physical characteristics of cuticular membranes. In *Ecology of leaf surface micro-organisms* (eds. T. F. Preece and C. H. Dickinson) Sect. I, pp. 55–65. Academic Press, London.

— (1974). The influence of environment on leaf wax development in *Brassica oleracea* var. *gummifera*. *New Phytol.* **73**, 955–66.

— and Holloway, P. J. (1971). SEM of waxes on plant surfaces. *Micron* **2**, 364–80.

Baker, R. T. (1919). *The hardwoods of Australia and their economics.* Tech. Educ. Ser. No. 23. Government NSW, Sydney.

Balfour, E. E. and Philipson, W. R. (1962). The development of the primary vascular system of certain dicotyledons. *Phytomorphology* **12**, 110–43.

Ball, E. (1960). Cell divisions in living shoot apices. *Phytomorphology* **10**, 377–96.

Bandulska, H. (1923). A preliminary paper on the cuticular structure of certain dicotyledonous and coniferous leaves from the middle Eocene flora of Bournemouth. *J. Linn. Soc. Bot.* **46**, 241–69.

— (1924). On the cuticles of some recent and fossil Fagaceae. *J. Linn. Soc., Bot.* **46**, 427–41.

— (1926). On the cuticles of some fossil and recent Lauraceae. *J. Linn. Soc., Bot.* **47**, 383–425.

— (1928). A Cinnamon from the Bournemouth Eocene. *J. Linn. Soc., Bot.* **48**, 139–47.

— (1931). On the cuticles of some recent and fossil Myrtaceae. *J. Linn. Soc. Bot.* **48**, 325, 657–71.

Banks, J. C. G. and Whitecross, M. I. (1971). Ecotypic variation in *Eucalyptus viminalis* Labill. I. Leaf surface waxes, a temperature × origin interaction. *Austrl. J. Bot.* **19**, 327–34.

Baranova, M. A. (1968). Stomatography and Taxonomy. *Bot. Zh. SSSR* **53**, 383–91. [In Russian.]

— (1972). Systematic anatomy of the leaf epidermis in the Magnoliaceae and some related families. *Taxon* **21** (4), 447–69.

Barber H. N. (1955). Adaptive gene substitutions in Tasmanian eucalypts. I. Genes controlling the development of glaucousness. *Evolution* **9**, 1–14.

Barker, W. G. (1953). Proliferative capacity of the medullary sheath region in the stem of *Tilia americana*. *Am. J. Bot.* **40**, 773–8.

Barros, M. A. A. de (1961*a*) Ocorrência das domácias nas Angiospermas. *Anais Esc. sup. Agric 'Luiz Queiroz'* **18**, 113–30.

— (1961*b*). Domácias nas Angiospermas-Variações na forma e na localizaçao (1). *Anais Esc. sup. Agric. 'Luis Queiroz'* **18**, 132–46.

Barth, F. (1896). Anatomie comparée de la tige et de la feuille des Trigoniacées et des Chailletiacées (Dichapetalées). *Bull. Herb. Boiss.* **4**, 481–520.

Barthlott, W. and Ehler, N. (1977). Raster-Elektronenmikroskopie der Epidermis-Oberflächen von Spermatophyten. *Trop. subtrop. Pflanzenwelt* **19**, 105 pp.

Barton, L. V. (1967). *Bibliography of seeds.* Columbia University Press, New York.

Barua, P. K. and Dutta, A. C. (1959). Leaf sclereids in the taxonomy of *Thea* camellias. II. *Camellia sinensis* L. *Phytomorphology* **9**, 372–82.

— and Wight, W. (1958). Leaf sclereids in the taxonomy of *Thea* camellias. I. Wilson's and related camellias. *Phytomorphology* **8**, 257–64.

Bary, A. de (1871). Ueber die Wachsüberzüge der Epidermis. *Bot. Ztg.* **29**, 129–39, 145–54, 161–76, 566–71, 573–85, 589–600, 605–19.

— (1877). *Vergleichende Anatomie des Vegetationsorgane der Phanerogamen und Farne.* Engelmann, Leipzig.

— (1884). *Comparative anatomy of the vegetative organs of the Phanerogams and ferns.* Clarendon Press, Oxford.

Bazavaluk, V. G. (1936). Métamorphose des glandes pétiolaires chez les *Prunus*. *Sov. Bot.* **1**, 81–98. [In Russian.]

Beauvisage, L. (1920). *Contribution à l'étude anatomique de la famille des Ternstroemiacées.* Arrault et Cie, Tours.

Beccari, O. (1884). *Piante ospitatricii ossia piante formicarie della Malesia e della Papuasia, Malesia.* Sordo-Muti, Genoa.

— (1904). *Wandering in the great forests of Borneo.* Archibald Constable, London.

Beck, C. B. (1970). The appearance of gymnospermous structure. Addendum. *Biol. Rev.* **45**, 379-400.

Beer, M. and Setterfield, G. (1958). Fine structure in thickened primary walls of collenchyma cells of celery petioles. *Am. J. Bot.* **45**, 571-80.

Behnke, H. D. (1972). Sieve-tube plastids in relation to angiosperm systematics – an attempt toward classification by ultrastructural analysis. *Bot. Rev.* **38**, 155-97.

Beille, L. (1947). Anatomie comparative du genre *Coffea* et de quelques Rubiacées-Ixorées. In *Les caféiers du globe* (ed. A. Chevalier) Vol. III, pp. 23-81. Paul Lechevalier, Paris.

Belin-Depoux, M. (1969). Contribution à l'étude des hydathodes I. Remarques sur la type 'à épithème' chez les Dicotylédones. *Revue gén. Bot.* **76**, 631-57.

— and Clair-Maczulajtys, D. (1974). Introduction à l'étude des glandes foliaires de *l'Aleurites moluccana* Willd. (Euphorbiacée). I. La glande et son ontogénèse. *Revue gén. Bot.* **81**, 335-51.

— (1975). Introduction à l'étude des glandes foliaires de *l'Aleurites moluccana* Willd. (Euphorbiacée). II. Aspects histologiques et cytologiques de la glande pétiolaire fonctionnelle. [Short English summary.] *Revue gén. Bot.* **82**, 119-55.

Bellini, R. (1909). Nettari extranuziali nella *Paulownia imperialis* Sieb. e Zucc. *Annali Bot.* **7**, 515-16.

Belt, T. (1874). *The naturalist in Nicaragua*. Murray, London.

Benzing, D. H. (1967a). Developmental patterns in stem primary xylem of woody Ranales. I. Species with unilacunar nodes. *Am. J. Bot.* **54**, 805-13.

— (1967b). Developmental patterns in stem primary xylem of woody Ranales. II. Species with trilacunar and multilacunar nodes. *Am. J. Bot.* **54**, 813-20.

Bequaert, J. G. (1921-2). Ants and their diverse relations to the plant world. In Wheeler, The ants collected by the American Museum Congo expedition. *Bull. Am. Mus. nat. Hist.* **45**, 333-84.

Berg, O. C. (1861). *Anatomischer Atlas zur pharmazeutischen Waarenkunde*. Rudolf Gaertner, Berlin.

Bergen, J. Y. (1904). Transpiration of sun leaves and shade leaves of *Olea europaea* and other broad-leaved evergreens. *Bot. Gaz.* **38**, 285-96.

Bernhard, F. (1964). Les glandes pétiolaires de certaines Euphorbiacées dérivent de méristèmes identiques à ceux des lobes foliaires. *C. r. hebd. Séanc. Acad. Sci.*, Paris **258**, 6213-5.

— (1966). Contribution à l'étude des glandes foliaires chez les Crotonoidées (Euphorbiacées.) *Mem. Inst. Fond. Afr. Noire*, **75**, 67-156.

Betts, M. W. (1920). Notes from Canterbury College Mountain Biological Station. No. 7. The rosette plants, Part II *Trans. N. Z. Inst.* **52**, 253-75.

Beuzeville, W. A. W. de and Welch, M. B. (1924). A description of new species of *Eucalyptus* from southern New South Wales. *J. Proc. R. Soc. N. S. W.* **58**, 177-80.

Bhattacharyya, B. and Maheshwari, J. K. (1971a). Studies on extrafloral nectaries of the Leguminales I. Papilionaceae, with a discussion on the systematics of the Leguminales. *Proc. Indian natn. Sci. Acad.* **37B**, 11-30.

— — (1971b). Studies on extrafloral nectaries of the Leguminales II. The genus *Cassia* Linn. (Caesalpiniaceae). *Proc. Indian natn, Sci, Acad.* **37B**, 74-90.

— — (1973). Studies on extrafloral nectaries of the Legu-

minales III. Mimosaceae. *J. Ind. Bot. Soc.* **52**, 267-98.

Bierhorst, D. W. and Zamora, P. M. (1965). Primary xylem elements and element associations of angiosperms. *Am. J. Bot.* **52**, 657-710.

Birch, T. (1660-87). *History of the Royal Society of London*, 4 Vols. London.

Bisalputra, T. (1962). Anatomical and morphological studies in the Chenopodiaceae. III. The primary vascular system and nodal anatomy. *Aust. J. Bot.* **10**, 13-24.

Björkman, O., Gauhl, E., and Nobs, M. A. (1969). Comparative studies of *Atriplex* species with and without carboxylation photosynthesis and their first generation hybrid. *Carnegie Inst. Washington, Yearb.* **68**, 620-33.

— Nobs, M. A., and Berry, J. A. (1971). Further studies on hybrids between C3 and C4 species of *Atriplex*. *Carnegie Inst. Wahington, Yearb.* **70**, 507-11.

Blake, S. T. (1972). *Idiospermum* (Idiospermaceae), a new genus and family for *Calycanthus australiensis*. *Contrib. Queensland Herb.* **12**, 1-37.

Bleckmann, C. A. and Hull, H. M. (1975). Leaf and cotyledon surface ultrastructure of five *Prosopis* species. *J. Arizona Acad. Sci.* **10**, 98-105.

Blyth, A. (1958). Origin of primary extraxylary stem fibres in Dicotyledons. *Univ. Calif. Publs Bot.* **30**, 145-232.

Bobisut, O. (1910). Über den Funktionswechsel der Spaltöffnungen in der Gleitzone der *Nepenthes*- Kannen. *Sber. Akad. Wiss. Wien (Math.-Nat. Kl.)* **119** (1), 3-10.

Böcher, T. W. (1972). Comparative anatomy of three species of the apophyllous genus *Gymnophyton*. *Am. J. Bot.* **59**, 494-503.

— (1975). Structure of the multinodal photosynthetic thorns in *Prosopis kuntzei* Harms. *Biol. Skr., k. danske Vidensk. Selsk.* **20** (8), 1-44.

— and Lyshede, O. B. (1968). Anatomical studies in xerophytic apophyllous plants. I. *Monttea aphylla, Bulnesia retama* and *Bredemeyera colletioides*. *Biol. Skr.* **16** (3), 1-44.

— — (1972). Anatomical studies in xerophytic apophyllous Plants. II. Additional species from South American shrub steppes. *Biol. Skr.* **18** (4), 1-137.

Böhmker, H. (1917). Beiträge zur Kenntniss der floralen und extrafloralen Nektarien. *Beih. bot. Zbl.* **33**, 169-247.

Boke, N. H. (1940). Histogenesis and morphology of the phyllode in certain species of *Acacia. Am. J. Bot.* **27**, 73-89.

— (1941). Zonation in the shoot apices of *Trichocereus spachianus* and *Opuntia cylindrica. Am. J. Bot.* **28**, 656-64.

— (1947). Development of the shoot apex and floral initiation in *Vinca rosea* L. *Am. J. Bot.* **34**, 433-9.

— (1961). Determinate shoot meristems in the Cactaceae. *Recent Adv. Bot.* **1**, 759-61.

Bokhari, M. H. (1970). Morphology and taxonomic significance of foliar sclereids in *Limonium. Notes R. bot. Gdn Edinb.* **30**, 43-53.

— and Burtt, B. L. (1970). Studies in the Gesneriaceae of the Old World XXXII: foliar sclereids in *Cyrtandra. Notes R. bot. Gdn Edinb.* **30**, 11-21.

Bollinger, R. (1959). Entwicklung und Struktur der Epidermisaussenwand bei einigen Angiospermenblättern. *J. Ultrastruct. Res.* **3**, 105-30.

Bond, G. (1931). The stem-endodermis in the genus *Piper*, *Trans. R. Soc. Edinb.* **56**, 695-724.

Bondeson, W. (1952). Entwicklungsgeschichte und Bau der

Spaltöffnungen bei den Gattungen *Trochodendron* Sieb. et Zucc, *Tetracentron* Oliv. und *Drimys* J. R. et G. Forst. *Acta Horti Bergiani* **16** (5), 169–217.

Bongers, J. M. (1973). Epidermal leaf characters of the Winteraceae. *Blumea* **21**, 381–411.

— Jansen, W. T., and Staveren, M. G. M. van (1973). Epidermal variation and peculiarities in the Winteraceae, Celastraceae and Icacinaceae. *Acta bot. neerl*. **22**, 250–1.

Bonnett, H. T. and Newcomb, E. H. (1966). Coated vesicles and other cytoplasmic components of growing root hairs of radish. *Protoplasma* **62**, 59–75.

Bonnier, G. (1878). Les Nectaires. Étude critique, anatomique et physiologique. *Annls Sci. nat. Bot. sér.* 6, **8**, 5–212.

— (1879). Etude anatomique et physiologique des nectaires. *C. r. hebd. Séanc. Acad. Sci., Paris* **88**, 662–5.

Bornemann, J. G. (1856). *Über organische Reste der Fetten-kohlengruppe Thuringens*. Engelmann, Leipzig.

Borodin, J. P. (1870). Ueber den Bau des Blattspitze einiger Wasserpflanzen. *Bot. Ztg.* **28**, 841–51.

Bower, F. O. (1884). On the structure of the stem of *Rhynchopetalum montanum* (Fresen). *J. Linn. Soc., Bot.* **20**, 440–6.

— (1938). *Sixty years of botany in Britain (1875–1935)*. Macmillan, London.

Brändlein, H. (1907). Systematische anatomischen Untersuchungen der Blatte der *Samydaceen* Benth. & Hook. Thesis, Erlangen.

Brazier, J. D. (1968). The contribution of wood anatomy to taxonomy. *Proc. Linn. Soc., Lond.* **179**, 271–4.

— (1975). The changing pattern of research in wood anatomy. *J. Microsc.* **104**, 53–64.

Brebner, G. (1902). On the anatomy of *Danaea* and other Marattiaceae. *Ann. Bot.* **16**, 517–52.

Bremekamp, C. E. B. (1947). A monograph of the genus *Streblosa* Korthals (Rubiaceae). *J. Arnold Arbor.* **28**, 145–85.

Brenner, W. (1900). Untersuchungen an einigen Fettpflanzen. *Flora* **87**, 387–439.

Briquet, J. (1920). Sur la présence d'acarodomaties foliaires chez les Cléthracées. *C. r. Soc. Phys. Hist. Nat. Genève* **35**, 12–15.

Brocheriou, J. and Belin-Depoux, M. (1974) [1975]. Contribution à l'étude ontogénique des poches sécrétrices des feuilles de quelques Myrtacées. *Phytomorphology* **24**, 321–38.

Brodie, P. B. (1842). Notice on the occurrence of plants in the plastic clay of the Hampshire coast. *Proc. geol. Soc.* **3**, 592.

Brongniart, A. (1834). Nouvelles recherches sur la structure de l'épiderme des végétaux. *Annls Sci. nat., Bot.*, 2nd ser. **1**, 65–71.

Brown, W. L. (1960). Ants, acacias and browsing mammals. *Ecology* **41**, 587–92.

Bruyne, C. de (1922). Idioblastes et diaphragmes des Nymphéacées. *C. r. hebd. Séanc. Acad. Sci., Paris* **175**, 452–5.

Buch, O. (1870). Über Sclerenchymzellen. Thesis, Breslau.

Bukvic, N. (1912). Die thylloiden Verstopfungen der Spaltöffnungen und ihre Beziehungen zur Korkbildung bei den Cactaceen. *Öst. Bot. Z.* **62**, 401–6.

Bünning, E. (1956). *The growth of leaves*. (ed. F. J. Milthorpe). Butterworth, London.

— and Sagromsky, H. (1948). Die Bildung des Spaltöffnungsmusters in der Blattepidermis. *Z. Naturf.* **3b**, 203–16.

Burck, W. (1910). Contribution to the knowledge of water secretion in plants. *Proc. Kon. Akad. Van Wet. te Amsterdam* **12**, 306–21.

Burgestein, A. (1887). Materialen zu einer Monographie betreffend die Erscheinungen der Transpiration der Pflanzen. *Zoolog. bot. Ges., Wien* **37**, 691–782.

Burström, H. (1961). Development of stomata in submerged leaves. *K. fysiogr. Sällsk. Lund Förh.* **31**, 25–30.

Buvat, R. (1944). Recherches sur la dédifférenciation des cellules végétales: I. Plantes entières et boutures. *Annls Sci. nat., Bot.*, 11th ser. **5**, 1–130.

Cameron, R. J. (1970). Light intensity and the growth of *Eucalyptus* seedlings. II. The effect of cuticular waxes on the light absorption in leaves of *Eucalyptus* species. *Austr. J. Bot.* **18**, 275–84.

Cammerloher, H. (1929). Zur Kenntnis von Bau und Funktion extrafloraler Nektarien. *Biologia gen.* **5**, 208–302.

Candolle, A. P. De (1825). *Prodromus systematis Naturalie. Regni Vegetabilis* Par. II.

— (1827). *Organographie végétale*. 2 Vols. Deterville Paris.

— (1841). *Vegetable organography* (English version) 2 Vols. Houlston and Stoneman, London.

Candolle, C. De (1866). Mémoire sur la famille des Piperacées. *Mém. Soc. Phys. Hist. nat., Genève* **18**, 1–32.

— (1868). Théorie de la feuille. *Arch. Sci. Bibl. Univ.* **32**, 32–64.

— (1879). Anatomie comparée des feuilles de quelques familles de dicotylédones. *Mém. Soc. Phys. Hist. nat.* Genève **26**, 427–80.

— (1890). Recherches sur les inflorescences épiphylles. *Mém. Soc. Phys. Hist. Nat. Genève.* Vol. supplement (celebrating the Centennial of the foundation of the Society) No. 6.

Canright, J. E. (1955). The comparative morphology and relationships of the Magnoliaceae. IV. Wood and nodal anatomy. *J. Arnold Arbor.* **36**, 115–40.

Carlquist, S. (1957). Leaf anatomy and ontogeny in *Argyroxiphium* and *Wilkesia* (Compositae). *Am. J. Bot.* **44**, 696–705.

— (1958). Structure and ontogeny of glandular trichomes of Madiinae (Compositae). *Am. J. Bot.* **45**, 675–82.

— (1959a). The leaf of *Calycadenia* and its glandular appendages. *Am. J. Bot.* **46**, 70–80.

— (1959b). Glandular structures of *Holocarpha* and their ontogeny. *Am. J. Bot.* **46**, 300–8.

— (1961). *Comparative plant anatomy*. Holt, Rinehart and Winston, New York.

— (1969). Morphology and anatomy. In *A short history of botany in the United States*, issued at the XIth international Botanical Congress, Seattle, pp. 49–67. Hafner, New York.

— (1975). *Ecological strategies of xylem evolution*. University of California Press, Berkeley.

Carmi, A., Sachs, T. and Fahn, A. (1972). The relation of ray spacing to cambial growth. *New Phytol.* **71**, 349–53.

Carolin, R. C. (1954). Stomatal size, density and morphology in the genus *Dianthus*. *Kew Bull.* **9**, (2), 251–8.

— (1971). The trichomes of the Goodeniaceae. *Proc. Linn. Soc. N.S.W.* **96**, 8–22.

Carpenter, S. B. and Smith, N. D. (1975). Stomatal distribution and size in southern Appalachian hardwoods. *Can. J. Bot.* **53**, 1153-6.

Carr, S. G. M., Milkovits, L., and Carr, D. J. (1971). Eucalypt phytoglyphs; the microanatomical features of the epidermis in relation to taxonomy. *Aust. J. Bot.* **19**, 173-90.

Carroll, C. R. and Janzen, D. H. (1973). Ecology of foraging by ants. *Ann. Rev. Ecol. Syst.* **4**, 231-57.

Carruthers, W. (1902). On the life and work of Nehemiah Grew. *J. R. microsc. Soc.* 129-41.

Caspary, R. (1848). *De nectariis*, Elberfeld.

Caveness, F. E. and Keeley, P. E. (1970). *Meloidogyne incognita* and production of leaf epicuticular wax in cotton. *J. Nematol.* **2**, 412.

Chafe, S. C. (1970). The fine structure of the collenchyma cell wall. *Planta* **90**, 12-21.

— and Wardrop, A. B. (1973) [1972]. Fine structural observations on the epidermis II. The cuticle. *Planta* **109**, 39-48.

Chakravarty, H. L. (1937). Physiological anatomy of the leaves of Cucurbitaceae. *Philip. J. Sci.* **63**, 409-31.

Chalk, L. and Davy, J. B. (1932-9). *Forest trees and timbers of the British Empire*, Parts 1-4. Clarendon Press, Oxford.

Chandra, V., Kapoor, S. L., Sharma, P. C., and Kapoor, L. D. (1969). Epidermal and venation studies in Apocynaceae-I. *Bull. bot. Surv. India* **11** (3 and 4), 286-9.

Chappet, A. (1969). Diplome d'études supérieures. Faculté de Dijon.

Chavan, A. R. and Bhatt, R. P. (1962). Anatomy of the extrafloral glands in the leaf of *Blastania fimbristipula* Kotschy and Peyr. *Sci. Cult.* **28**, 137.

— and Deshmukh, Y. S. (1960). The ontogeny of extrafloral nectaries in the genus *Gmelina*. *J. Indian bot. Soc.* **39**, 410-14.

Cheadle, V. I. (1956). Research on xylem and phloem – progress in fifty years. *Am. J. Bot.* **43**, 719-31.

Chemin, E. (1920). Observations anatomiques et biologiques sur le genre *Lathraea*. *Annls. Sci. nat., Bot.*, Ser 10 **2**, 125-272.

Chevalier, A. (1942). Les Caféiers du globe. *Encycl. biol.* **XXII**, fasc. II.

— and Chesnais, F. (1941*a*). Sur les domaties des feuilles des Juglandées. *C. r. hebd. Séanc. Acad. Sci., Paris* **213**, 389-92.

— — (1941*b*). Nouvelles observations sur les domaties des Juglandacées. *C. r. hebd. Séanc. Acad. Sci., Paris* **213**, 597-601.

Chowdhury, K. A. (1964). Growth rings in tropical trees and taxonomy. *J. Indian bot. Soc.* **43**, 334-42.

— (1968). *History of botanical researches in India, Burma and Ceylon. X. Wood anatomy*. Aligarh Muslim University Press.

— and Ghosh, S. S. (1958-63). *Indian woods*, Vols. I and II. Delhi.

— Rao, V. S., and Mitra, G. C. Anatomy 1939-50. In *Progress of science in India*, Section VI, pp. 71-92.

C.I.M.P. (Commission Internationale de Microflore du Paléozoique) (1964). Entwurf für eine einheitliche diagnostische Beschreibung von Kutikulen. *Fortschr. Geol. Rheinld Westf.* **12**, 11-24.

Clos D. (1879). Indépendance, développement, anomalies des stipules; bourgeons à écailles stipulaires. *Bull. Soc. Bot. Fr.* **26**, 189-93.

Clowes, F. A. L. (1961). *Apical meristems*. Blackwell, Oxford.

Codaccioni, M. (1970). Présence de cellules de type endodermique dans la tige de quelques Dicotylédones. *C. r. hebd.*

Séanc. Acad. Sci., Paris **271**, 1515-17.

Cohn, F. (1850). De Cuticula *Linnaea* **7**, 337-407.

Col, A. (1904). Recherches sur la disposition des faisceaux dans la tige et les feuilles de quelques dicotylédones. *Ann. Sci. Nat.*, Bot., ser. 8, **20**, 1-288.

Conard, H. S. (1905). *The waterlilies. A monograph of the genus* Nymphaea, Vol. 4. Carnegie Inst., Washington D.C.

Conde, L. F. (1975). Anatomical comparisons of five species of *Opuntia* (Cactaceae). *Ann. Mo. Bot. Gd* **62**, 425-73.

Condit, I. J. (1969). Ficus: *the exotic species*. Berkeley, Univ. Calif. Div. Agric. Sci.

Cook, O. F. (1911). Dimorphic leaves of cotton and allied plants in relation to heredity. *U.S.D.A. Bur. Pl. Ind. Bull.* **221**, 1-59.

Corner, E. J. H. (1949). The Durian Theory or the Origin of the Modern Tree. *Ann. Bot. N.S.* **13**, 376-414.

Cotthem, W. R. J. van (1968). Vergelijkend-morfologische studie van de stomata bij de Filicopsida. Thesis, Ghent, Belgium. (See also *Bull. Jard. bot. natn Belg.* **40**, 81-239.)

— (1970). A classification of stomatal types. *Bot. J. Linn. Soc.* **63**, 235-46.

— (1971). Vergleichende morphologische Studien über Stomata und eine neue Klassifikation ihrer Typen. *Ber. dt. bot. Ges.* **84**, 141-68.

— (1973). A new classification of the ontogenetic types of stomata. *Bot. Rev.* **39** (1), 71-138.

Cotti, T. (1962). Ueber die quantitative Messung der Phosphatase Aktivität in Naktarien. *Ber schweiz. bot. Ges.* **72**, 306-31.

Cowan, J. M. (1950). *The Rhododendron leaf, a study of the epidermal appendages*. Oliver and Boyd, Edinburgh.

Cowan, R. S. (1975). A monograph of the genus *Eperua* (Leguminosae: Caesalpinioideae). *Smithsonian Contrib. Bot.* No 28, 1-45.

Cristóbal, C. L. (1971). Mirmecofilia en *Byttneria* (Sterculiaceae). *Kurtziana* **6**, 271-4.

— and Arbo, M. M. (1971) Sobre las especies de *Ayenia* (Sterculiaceae), nectarios foliares. *Darwiniana* **16**, 603-12. [English summary]

Croizat, L. (1940). A comment on current notions concerning the leaf, stipule and budscale of the Angiosperms. *Lingnan Sci. J.* **19**, 49-66.

— (1960). *Principia botanica*. Published by the author, Caracas, Venezuela.

— (1973). Quelques réflexions sur la morphologie, la morphogénèse et la symétrie. *Adansonia*, sér. 2 **13**, 351-82.

Cronquist, A. (1968). *The evolution and classification of flowering plants*. Houghton Mifflin, Boston.

Crookston, R. K. and Moss, D. N. (1970). The relation of carbon dioxide compensation and chlorenchymatous vascular bundle sheaths in leaves of Dicots. *Pl. Physiol., Lancaster* **46**, 564-7.

Cumming, N. M. (1925). Notes on strand plants. 1. *Atriplex babingtonia* Woods. *Trans. Proc. bot. Soc. Edinb.* **29**, 171-5.

Cunze, R. (1925). Untersuchungen über die ökologische Bedeutung des Wachses im Wasserhaushalt der Pflanzen. *Beih. bot. Zbl.* **42** (1), 160-89.

Curtis, J. D. and Lersten, N. R. (1973). Seasonal changes in foliar gland structure and function in *Populus deltoides*. *Am. J. Bot.* **60** (Suppl.), 6.

— — (1974). Morphology, seasonal variation and function of resin glands on buds and leaves of *Populus deltoides*

(Salicaceae). *Amer. J. Bot.* **61**, 835–45.

Curtis, L. C. (1943). Deleterious effects of guttated fluids on foliage. *Am. J. Bot.* **30**, 778–81.

— (1944). The influence of guttation fluid on pesticides. *Phytopathology* **34**, 196–205.

Cusset, G. (1965). Les nectaires extra-floraux et la valeur de la feuille des Passifloracées. *Rev. gén. Bot.* **72**, 145–216.

— (1970). Remarques sur les feuilles de dicotylédones. *Boissiera* **16**, 1–210.

Cutler, D. F. (1969). *Anatomy of the monocotyledons, Vol. IV Juncales* (ed. C. R. Metcalfe). Clarendon Press, Oxford.

— (1975). Anatomical notes on the leaf of *Eleutharrhena* Forman and *Pycnarrhena* Miers (Menispermaceae). *Kew Bull.* **30**, 41–8.

Cutter, E. G. (1957). Studies of morphogenesis in the Nymphaeaceae. II. Floral development in *Nuphar* and *Nymphaea*; bracts and calyx. *Phytomorphology* **7**, 57–73.

— (1961). Formation of the lateral members of the shoot. *Recent Adv. Bot.* **1**, 820–3.

— (1976). Aspects of the structure and development of the aerial surfaces of higher plants. In *Ecology of leaf surface microorganisms* (eds. T. F. Preece and C. H. Dickinson). Academic Press, London.

Danilova, M. F. and Stamboltzian, E. U. (1969). On the structure of the 'casparian strip'. *Bot. J. U.S.S.R.* **54**, 1288–91.

Darlington, A. (1968). *The pocket encyclopaedia of plant galls.* Blandford Press, London.

Darwin, F. (1876). On the glandular bodies on *Acacia sphaerocephala* and *Cecropia peltata* serving as food for ants. *J. Linn Soc., Lond.* **15**, 398–409.

Daumann, E. (1967). Zur Bestäubungs- und Verbreitungsökologie dreier *Impatiens*-Arten. *Preslia* **39**, 43–58.

Dave, Y. S. and Patel, N. D. (1975). A developmental study of extrafloral nectaries in slipper spurge (*Pedilanthus tithymaloides*). Euphorbiaceae. *Am. J. Bot.* **62**, 808–12.

Davidson, C. (1973). An anatomical and morphological study of Datiscaceae. *Aliso* **8**, 49–110.

Davies, W. J. and Kozlowski, T. T. (1974). Stomatal responses of five woody angiosperms to light intensity and humidity. *Can. J. Bot.* **52**, 1525–34.

Davis, D. G. (1971). Scanning electron microscopic studies of wax formations on leaves of higher plants. *Can. J. Bot.* **49**, 543–6.

Davis, E. L. (1961). Medullary bundles in the genus *Dahlia* and their possible origin. *Am. J. Bot.* **48**, 108–13.

Davis, J. J. (1883). Nectar glands on leaves. *Bot. Gaz.* **8**, 339–40.

Décamps, O. (1974). Types stomatiques chez les Renonculacées. *C. r. hebd. Séanc. Acad. Sci., Paris* **279** (18), 1527–29.

Decrock, E. (1901). Anatomie des Primulacées. *Annls Sci. nat.* **13**, 1–99.

Dehay, C. L. F. (1935). L'appareil libéro-ligneux foliaire des Euphorbiacées. *Annls Sci. nat.*, Bot., ser. 10 **17**, 148–296.

— (1942). Notice sur les Titres et travaux scientifiques de Charles Dehay Arras, pp. 1–57.

Deinega, V. (1898). Beiträge zur Kenntniss der Entwicklungsgeschichte des Blattes und der Anlage der Gefässbündel. *Flora* **85**, 439–98.

Delpino. F. (1886). Funzione mirmecofila nel Regno Vegetale (part 1, **106**, Bologna) *Memorie R. Accad. Sci. Ist. Bologna*, ser. IV, **VII**, 215–323, **VIII**, 601–50.

Desch, H. E. (1941, 1954). *Manual of Malayan timbers.* Malayan Forest Records No. 15, Vols I and II. Kuala Lumpur.

Devadas, C. and Beck, C. B. (1972). Comparative morphology of the primary vascular stystems in some species of Rosaceae and Leguminosae. *Am. J. Bot.* **59**, 557–67.

Dewar, T. (1933). The histology of the leaves of *Digitalis thapsi*. *Q. Jl. Pharm. Pharmac.* **6**, 443–53.

— (1934a). The histology of the leaves of *Digitalis lutea*. *Q. Jl. Pharm. Pharmac.* **7**, 1–22.

— (1934b). The histology of the leaves of *Digitalis lanata* Ehrh. *Q. Jl. Pharm. Pharmac.* **7**, 331–45.

— and Wallis, T. E. (1935). Digitalis leaf. The macroscopical and microscopical characters, potencies and constituents of certain species. *Pharm. J. ser. 4*, **81**, 565–6.

Dickinson, C. H. and Preece, T. F. (1976). *Microbiology of aerial plant surfaces.* Academic Press, London, New York and San Francisco, 669 pp.

Dickinson, T. A. and Sattler, R. (1974). Development of the epiphyllous inflorescence of *Phyllonoma integerrima* (Turcz.) Loes.: implications for comparative morphology. *Bot. J. Linn. Soc.* **69**, 1–13.

Dickison, W. C. (1969). Comparative morphological studies in Dilleniaceae, IV. Anatomy of the node and vascularization of the leaf. *J. Arnold Arbor.* **50**, 384–410.

— (1970). Comparative morphological studies in Dilleniaceae V. Leaf anatomy. *J. Arnold Arbor.* **51**, 89–113.

— (1973). Nodal and leaf anatomy of *Xanthophyllum* (Polygalaceae). *Bot. J. Linn. Soc.* **67**, 103–15.

Diels, L. (1919). Über die Gattung *Himantandra*, ihre Verbreitung und ihre systematische Stellung. *Bot. Jb.* **55**, 126–34.

Diettert, R. A. (1938). The morphology of *Artemisia tridentata* Nutt. *Lloydia* **1**, 3–74.

Dilcher, D. L. (1974). Approaches to the identification of angiosperm leaf remains. *Bot. Rev.* **40** (1), 1–157.

— and Zeck, C. A. (1968). A study of the factors controlling variation of cuticular characters. *Indiana Acad. Sci.* **78**, 1115 [Abstract].

Dop, P. (1927). Les glandes florales externes des Bignoniacées. *Bull. Soc. Hist. Nat. Toulouse* **56**, 189–98.

— (1928). Les glandes de *Clerodendron foetidum* Bunge. *Bull. Soc. Hist. Nat. Toulouse*, **57**, 167–9.

Dormer, K. J. (1945a). On the absence of a plumule in some leguminous seedlings. *New Phytologist* **44**, 25–8.

— (1945b). An investigation of the taxonomic value of shoot structure in angiosperms with especial reference to Leguminosae. *Ann. Bot. N. S.* **9**, 141–53.

— (1972). *Shoot organization in vascular plants.* Chapman and Hall, London.

Dorsey, M. J. and Weiss, F. (1920). Petiolar glands of the plum. *Bot. Gaz.* **69**, 391–405.

Dous, E. (1927). Über die Wachsausscheidungen bei Pflanzen; ein Studium mit dem Oberflächenmikroskop. *Bot. Archiv.* **19**, 461–73.

Downton, W. J. S., Bisalputra, T., and Tregunna, E. B. (1969). The distribution and ultrastructure of chloroplasts in leaves differing in photosynthetic carbon metabolism. II. *Atriplex rosea* and *Atriplex hastata* (Chenopodiaceae). *Can. J. Bot.* **47**, 915–9.

— and Tregunna, E. B. (1968). Carbon dioxide compensation – its relation to photosynthetic carboxylation reactions, systematics of the Gramineae and leaf anatomy. *Can.*

J. Bot. **46**, 207-15.

Drawert, H. (1938). Elektive Färbung der Hydropoten an fixierten Wasserpflanzen. Ein Beitrag zur protoplasmatischen Anatomie fixierter Gewebe. *Flora* **132**, 234-52.

Duchaigne, A. (1953). Sur la transformation du collenchyme en sclérenchyme chez certaines Ombellifères. *C. r. hebd. Séanc. Acad. Sci., Paris* **236**, 839-41.

— (1954a). Nouvelles observations sur la sclérification du collenchyme chez les Ombellifères. *C. r. hebd. Séanc. Acad. Sci., Paris* **238**, 375-7.

— (1954b). La sclérification du collenchyme chez les Labiées. *Bull. Soc. Bot. Fr.* **101**, 235-7.

— (1955). Les divers types de collenchymes chez les dicotylédones, leur ontogénie et leur lignification. *Ann. Sci. nat.,* Bot. **16**, 455-79.

Duchartre, P. (1859). Recherches physiologiques, anatomiques et organogéniques sur la Colocis des Anciens, *Colocasia antiquorum* Schott. *Ann. Sci. nat., Bot.*, ser. 4 **12**, 232-79.

— (1868). Observations physiologiques et anatomiques faites sur une colocase de la Chine. *Bull. Soc. Bot. Fr.* **5**, 267-71.

Dufour, J. (1886). Notices microchimiques sur le tissu épidermique des végétaux. *Bull. Soc. Vaud. Sci. nat.* **22**, 134-42.

Dunn, D. B., Sharma, G. K., and Campbell, C. C. (1965). Stomatal patterns of dicotyledons and monocotyledons. *Am. Midl. Nat.* **74**, 185-95.

Dupont, S. (1962). Observations sur les types stomatiques des Ficoidaceae. *Bull. Soc. Hist. nat. Toulouse* **97** (1-2), 93-8.

Dutta, S. C. and Mukerji, B. (1952). *Pharmacognosy of Indian leaf drugs*. Calcutta.

Eberhardt, P. (1903). Influence de l'air sec et de l'air humide sur la forme et sur la structure des végétaux. *Annls Sci. nat., Bot.*, ser. 8 **18**, 61-153.

Eckerson, S. H. (1908). The number and size of the stomata. *Bot. Gaz.* **46**, 221-4.

Eckhardt, Th. (1957). *Neue Hefte zur Morphologie*. Weimar.

Edelstein, W. (1902). Zur Kenntnis der Hydathoden an den Blättern des Holzgewächse. *Bull. Acad. Imp. St. Petersb.* **17**, 59-64.

Edwards, W. N. (1935). The systematic value of cuticular characters in recent and fossil angiosperms. *Biol. Rev.* **10**, 442-59.

Eggeling, W. J. (1952). The indigenous trees of the Uganda Protectorate. 2nd edn. I. R. Dale, Entebbe.

Eglinton, G., Hamilton, R. J., Raphael, R. A., and Gonzalez, A. G. (1962). Hydrocarbon constituents of the wax coatings of plant leaves: a taxonomic survey. *Nature, Lond.* **193**; 739-742.

Elias, T. S. (1972). Morphology and anatomy of foliar nectaries of *Pithecellobium macradenium* (Leguminosae). *Bot. Gaz.* **133**, 38-42.

— Rozich, W. R., and Newcombe, L. (1975). The foliar and floral nectaries of *Turnera ulmifolia* L. *Am. J. Bot.* **62**, 570-6.

Elsler, E. (1907). Das extraflorale Nektarium und die Papillen der Blattunterseite bei *Diospyros discolor* Willd. *Sber. Akad. Wiss. Wien (Math.- Nat. Kl.)* **116** (1), 1563-90.

El-Sharkawy, M. A., Loomis, R. S., and Williams, W. A. (1967). Apparent reassimilation of respiratory carbon dioxide by different plant species. *Physiologia Pl.* **20**, 171-86.

Emberger, L. (1952). Tige, racine feuille. *Année biol.* **28**, 107-25.

Engler, A. (1897). *Rutaceae* in *Die natür L. Pflanzenfam.* Vol. III, pp. 4, 165; f. 96F.

Esau, K. (1936). Ontogeny and structure of collenchyma and of vascular tissues in celery petioles. *Hilgardia* **10**, 431-67.

— (1943a) Vascular differentiation in the vegetative shoot of *Linum*. II. The first phloem and xylem. *Am. J. Bot.* **30**, 248-55.

— (1943b). Origin and development of primary vascular tissue in seed plants. *Bot. Rev.* **9**, 125-206.

— (1945). Vascularization of the vegetative shoots of *Helianthus* and *Sambucus*. *Am. J. Bot.* **32**, 18-29.

— (1953). *Plant anatomy* (1st edn). John Wiley, New York; Chapman and Hall, London.

— (1960). The development of inclusions in sugar beets infected with beet-yellows virus. *Virology* **11**, 317-28.

— (1965a). *Plant anatomy* (2nd edn). John Wiley, New York.

— (1965b). *Vascular differentiation in plants*. Holt, Rinehart and Winston, New York.

— (1969). In Zimmermann and Ozenda's *Encyclopaedia of plant anatomy*. Band V. Teil 2. *The phloem*. Gebrüder Borntraeger, Berlin.

— (1977). *Anatomy of seed plants* (2nd edn). John Wiley, London.

— (1978). The protein inclusions in sieve elements of cotton (*Gossypium hirsutum* L.) *J. Ultrastruct. Res.* **63**, 224-35.

— and Cheadle, V. I. (1959). Size of pores and their contents in sieve elements of dicotyledons. *Proc. natn. Acad. Sci. U.S.A.* **45**, 156-62.

— — and Gill, R. H. (1966). Cytology of differentiating tracheary elements. I. Organelles and membrane systems. *Am. J. Bot.* **53**, 756-64.

Espinosa, R. (1932). Ökologische Studien über Kordillerenpflanzen (morphologisch und anatomisch dargestellt). *Bot. Jb.* **65**, 120-211.

Ettingshausen, C. R. von (1845a). Über die Nervation der Papilionaceen. *Sber. Akad. Wiss. Wien (Math.- Nat. Kl.)* **12**, 600-33.

— (1845b). Über die Nervation der Blätter und blattartigen Organe bei den Euphorbiaceen, mit besonderer Rücksicht auf die vorweltlichen Formen. *Sber. Akad. Wiss Wien (Math.-Nat. Kl.)* **12**, 138-54.

— (1857). Über die Nervation der Blätter bei den Celastrineen. *Denkschr. Akad. Wiss. Wien* **13**, 43-83.

— (1861). *Die Blattskelete der Dicotyledonen*. K. K. Hof, Vienna.

Evert, R. F. and Deshpande, B. P. (1970). An ultrastructural study of cell division in the cambium. *Am. J. Bot.* **57**, 942-61.

Eymé, J. (1963). Observations cytologiques sur les nectaires de trois Renonculacées. (*Helleborus foetidus* L., *H. niger* L. et *Nigella damascena* L.). *Botaniste* **46**, 137-79.

Ezelarab, G. E. and Dormer, K. J. (1963). The organization of the primary vascular system in the Ranunculaceae. *Ann. Bot.* **27**, 23-38.

Fahn, A. (1963). The fleshy cortex of articulated Chenopodiaceae. J. Indian bot. Soc. (Maheshwari Commemorative volume) **42A**, 39–45.

— (1967). *Plant anatomy*. Pergamon Press, Oxford.

— (1974). *Plant anatomy* (2nd ed). Pergamon Press, Oxford.

— and Bailey, I. W. (1957). The nodal anatomy and the primary vascular cylinder of the Calycanthaceae. *J. Arnold Arbor.* **38**, 107–17.

— and Dembo, N. (1964) [1965]. The structure and development of the epidermis in articulated Chenopodiaceae. *Israel J. Bot.* **13**, 177–92.

— and Leshem, B. (1963). Wood fibres with living protoplasts. *New Phytol.* **62**, 91–8.

— and Rachmilevitz, T. (1970). Ultrastructure and nectar secretion in *Lonicera japonica*. *J. Linn. Soc., Lond.* (*Bot.*) **63** (Suppl. *New research in plant anatomy*) pp. 51–6.

— — (1975). An autoradiographical study of nectar secretion in *Lonicera japonica* Thunb. *Ann. Bot.* **39**, 975–6.

— Shomer, I. and Ben-Gera, I. (1974). Occurrence and structure of epicuticular wax on the juice vesicles of *Citrus* fruits. *Ann. Bot.* **38**, 869–72.

Faure, A. (1924). Étude organographique, anatomique et pharmacologique de la famille des Cornacées (groupe des Cornales). Thesis, Lille.

Faust, W. Z. and Jones, S. B., Jr. (1973). The systematic value of trichome complements in the North American group of *Vernonia* (Compositae). *Rhodora* **75**, 517–28.

Ferguson, D. K. (1974). The significance of the leaf epidermis for the taxonomy of *Cocculus* (Menispermaceae). *Kew Bull.* **29**, 483–92.

Field, D. V. (1967). An investigation into the anatomy of several genera of the Stapelieae of the family Asclepiadaceae with a view to improving the definition of generic limits. Thesis, London.

Figier, J. (1968). Localisation infrastructurale de la phosphomonestérase acide dans la stipule de *Vicia faba* L. au niveau du nectaire. Rôles possibles de cet enzyme dans les mécanismes de la sécrétion. *Planta* **83**, 60–79.

— (1971). Étude infrastructurale de la stipule de *Vicia faba* L. au niveau du nectaire. *Planta* **98**, 31–49.

— (1972*a*). Localisation infrastructurale de la phosphatase acide dans les glandes pétiolaires d'*Impatiens holstii*. Rôles possibles de cette enzyme au cours des procéssus sécrétoires. *Planta* **108**, 215–26.

— (1972*b*). Les cytosomes dans le nectaire stipulaire de *Vicia faba* L. et dans la glande pétiolaire de *Mercurialis annua* L. Étude infrastructurale, cytochimique et radioautographique. *Botaniste* **55**, 289–310.

— (1972*c*). Etude infrastructurale des glandes pétiolaires d'*Impatiens holstii*. *Botaniste* **55**, 311–38.

Fisher, D. A. and Bayer, D. E. (1972). Thin sections of plant cuticles demonstrating channels and wax platelets. *Can. J. Bot.* **50**, 1509–1511.

Fisher, Sir R. A. and Yates, F. (1963). *Statistical tables for biological, agricultural and medical research workers*. Oliver and Boyd, Edinburgh.

Flachs, K. (1916). Über die Verbreitung des äquifazialen Blattbaues in der australischen Flora. Thesis, Kgl. Bayer. Ludwig-Maximilians Univ., Munich.

Fleet, D. S. van (1961). Histochemistry and function of the endodermis. *Bot. Rev.* **27**, 165–220.

Florin, R. (1921). Über Cutikularstrukturen der Blätter bei einigen rezenten und fossilen Coniferen. *Ark. Bot.* **16** (6), 1–32.

— (1933). Studien über die Cycadales des Mesozoikums

nebst Erörterungen über die Spaltöffnungsapparate des Bennettitales. *K. svenska Vetenska-Akad. Handl.* **12** (5), 1–134.

— (1958). On Jurassic taxads and conifers from Northwestern Europe and Eastern Greenland. *Acta Horti Bergiana* **17**, 257–402.

Forbes, H. O. (1880). Notes from Java. *Nature, Lond.* **22**, 148.

— (1885). *A naturalist's wanderings in the Eastern Archipelago*. Sampson, Low, Marston, Searle, and Rivington, London.

Forsdike, J. L. (1946). Determination of adulterants of drugs consisting of leaves. *Q. J. Pharm. Pharmacol.* **19**, 270–9.

Foster, A. S. (1939). Problems of structure, growth and evolution in the shoot apex of seed plants. *Bot. Rev.* **5**, 454–70.

— (1944). Structure and development of sclereids in the petiole of *Camellia japonica* L. *Bull. Torrey bot. Club* **71**, 302–26.

— (1945*a*). Origin and development of sclereids in the foliage leaf of *Trochodendron aralioides* Sieb. and Zucc. *Am. J. Bot.* **32**, 456–68.

— (1945*b*). The foliar sclereids of *Trochodendron aralioides* Sieb. and Zucc. *J. Arnold Arbor.* **25**, 155–62.

— (1946). Comparative morphology of the foliar sclereids in the genus *Mouriria* Aubl. *J. Arnold Arbor.* **27**, 253–71.

— (1947). Structure and ontogeny of the terminal sclereids in the leaf of *Mouriria huberi* Cogn. *Am. J. Bot.* **34**, 501–14.

— (1949). *Practical plant anatomy* (2nd ed.). Van Nostrand, New York.

— (1950). Morphology and venation of the leaf in *Quiina acutangula* Ducke. *Am. J. Bot.* **37**, 159–71.

— (1952). Foliar venation in angiosperms from the ontogenetic standpoint. *Am. J. Bot.* **39**, 752–66.

— (1955*a*). Comparative morphology of the foliar sclereids in *Boronella* Baill. *J. Arnold Arbor.* **36**, 189–98.

— (1955*b*). Structure and ontogeny of terminal sclereids in *Boronia serrulata*. *Am. J. Bot.* **42**, 551–60.

— (1956). Plant idioblasts; remarkable examples of cell specialization. *Protoplasma* **46**, 184–93.

— (1959*a*). The phylogenetic significance of dichotomous venation in angiosperms. *Proc. IX int. Bot. Congr.* Vol. 2, pp. 119–20.

— (1959*b*). The morphological and taxonomic significance of dichotomous venation in *Kingdonia uniflora* Balfour f. et W. W. Smith. *Notes R. bot. Gdn Edinb.* **23**, 1–12.

— (1961*a*). The floral morphology and relationships of *Kingdonia uniflora*. *J. Arnold Arbor.* **42**, 397–410.

— (1961*b*). The phylogenetic significance of dichotomous venation in angiosperms. *Recent Adv. Bot.* **2**, 971–5.

— (1966). Morphology of anastomoses in the dichotomous venation of *Circaeaster*. *Am. J. Bot.* **53**, 588–99.

— (1968). Further morphological studies on anastomoses in the dichotomous venation of *Circaeaster*. *J. Arnold Arbor.* **49**, 52–72.

— (1970). Types of blind vein-endings in the dichotomous venation of *Circaeaster*. *J. Arnold Arbor.* **51**, 70–88.

— (1971). Additional studies on the morphology of blind vein-endings in the leaf of *Circaeaster agrestis*. *Am. J. Bot.* **58**, 263–72.

— and Gifford, E. M. (1974). *Comparative morphology of vascular plants* (2nd edn). W. H. Freeman, San Francisco.

Foxworthy, E. W. (1903). Discoid pith in woody plants. *Proc. Indiana Acad. Sci.* 191–4.

Fraine, E. de (1913). The anatomy of the genus *Salicornia*. *J. Linn. Soc., Lond. (Bot.)* 41, 317–48.

Franke, W. (1960*a*). Über Beziehungen der Ektodesmen zur Stoffaufnahme durch Blätter. I. Mitteilung Beobachtungen an *Plantago major* L. *Planta* 55, 390–423.

— (1960*b*). Über die Beziehungen der Ektodesmen zur Stoffaufnahme durch Blätter II. Mitteilung Beobachtungen an *Helxine soleirolii* Req. *Planta* 55, 533–41.

— (1971) [1972]. Über die Natur der Ektodesmen und einen Vorschlag zur Terminologie. *Ber. dt. bot. Ges.* 84, 533–7.

Frankton, C. and Bassett, I. J. (1970). The genus *Atriplex* (Chenopodiaceae) in Canada. II. Four native western annuals: *A. argentea, A. truncata, A. powellii,* and *A. dioica. Can. J. Bot.* 48, 981–9.

Frei, É. (1955). Die Innervierung der floralen Nektarien dikotyler Pflanzenfamilien. *Ber. schweiz. bot. Ges.* 65, 60–114.

Freund, H. (1951, 1970). *Handbuch der Mikroskopie in der Technik.* Band V, Teil 1. *Mikroskopie des Holzes und des Papiers.* xxi, 456 pp. (There is also a second edition of Band V, Teil 1, published in 1970.) Band V, Teil 2, xxiv, 891 pp.
   Band VI. *Mikroskopie der Textilfasern und Textilen.* Teil 1. *Die Preparationsmethoden der Faserstoffe und ihre mikroskopischen Untersuchungsverfahren,* xxviii, 442 pp. Teil 2. *Die Mikroskopie der einzelnen Fasergruppen,* xxvi, 400 pp. Umschau Verlag, Frankfurt.

Frey, A. (1926). Die submikroskopische Struktur des Zellmembranen. *Jb. wiss. Bot.* 65, 195–223.

Frey-Wyssling, A. (1933). Über die physiologische Bedeutung der extrafloralen Nektarien von *Hevea brasiliensis. Ber. schweiz. bot. Ges.* 42, 109–22.

— (1935). *Die Stoffausscheidung der höheren Pflanzen.* Springer, Berlin.

— (1941). Die Guttation als allgemeine Erscheinung. *Ber. schweiz. bot. Ges.* 51, 321–5.

— (1955). The phloem supply to the nectaries. *Acta bot. neerl.* 4, 358–69.

— (1976). The plant cell wall. In *Encyclopaedia of plant anatomy.* Gebrüder Borntraeger, Berlin.

— and Häusermann, E. (1960). Deutung der gestaltlosen Nektarien. *Ber. schweiz. bot. Ges.* 70, 150–62.

Fritsch, F. E. (1903). The use of anatomical characters for systematic purposes. *New Phytol.* 2, 177–84.

Fryns-Claessens, E. and Cotthem, W. van (1973). A new classification of the ontogenetic types of stomata. *Bot. Rev.* 39, 71–138.

Funke, G. L. (1929). On the biology and anatomy of some tropical leaf joints. *Annls Jard. bot. Buitenz.* 40, 45–74.

Furuya, M. (1953). Problèmes de l'organogènèse dans rameau axillaire végétatif de Dicotylédones. *J. Fac. Sci. Tokyo Univ. (section 3, Botany)* 6, 159–207.

Gadkari, P. D. (1964). Further studies in stomatal frequencies of cotyledonary leaves of Indian cotton. *Indian Cott. Grow. Rev.* 18, 222–47.

Gamble, J. S. (1902). *A manual of Indian timbers.* Sampson, Low, and Marston, London.

Gangadhara, M. and Inamdar, J. A. (1975). Action of growth regulators on the cotyledonary stomata of *Cucumis sativus* L.: structure and ontogeny. *Biologia Pl.* 17, 292–303.

Gardiner, W. (1881). The development of the water-glands in the leaf of *Saxifraga crustata. Q. Jl. microsc. Sci.* 21, 407–14.

— (1883). On the significance of water glands and nectaries. *Proc. Camb. phil. Soc.* 5, 35–40.

Gassner, G. (1955).*Mikroskopische Untersuchung pflanzlicher Nahrungs und Genussmittel.* Gustav Fischer, Stuttgart.

Gaudet, J. (1960). Ontogeny of the foliar sclereids in *Nymphaea odorata. Am. J. Bot.* 47, 525–32.

Gaulhofer, K. (1908). Die Perzeption der Lichtrichtung im Laubblätte mit Hilfe der Randtüpfel, Randspalten und der windschiefen Radialwände. *Sber. Akad. Wiss. Wien (Math-Nat. K.)* I, 117, 153–90.

Gay, A. P. and Hurd, R. L. (1975). The influence of light on stomatal density in the tomato. *New Phytol.* 75, 37–46.

Gessner, F. and Volz, G. (1951). Die Kutikula der Hydropoten von *Nymphaea. Planta* 39, 171–4.

Gibson, R. J. H. (1888). *The history of the science of biology.* Liverpool.

Gifford, E. M., Jr. (1954). The shoot apex in angiosperms. *Bot. Rev.* 20, 477–529.

Gill, A. M. and Tomlinson, P. B. (1971). Studies on the growth of red mangrove (*Rhizophora mangle* L.). 2. Growth and differentiation of aerial roots. *Biotropica* 3, 63–77.

Gindel, J. (1969). Stomatal numbers and size as related to soil moisture in tree xerophytes in Israel. *Ecology* 50, 263–7.

Ginzburg, C. (1966). Xerophytic structures in the roots of desert shrubs. *Ann. Bot., N.S.* 30, 403–18.

Gleason, H. A. (1922). Vernonieae. *N. Am. Flora* 33, 47–110.

Glück, H. (1919). *Blatt-und blutenmorphologische Studien.* Gustav Fischer, Jena.

Goatley, J. L. and Lewis, R. W. (1966). Composition of guttation fluid from rye, wheat and barley seedlings. *Pl. Physiol., Lancaster* 41, 373–5.

Goebel, K. (1897). Morphologische und biologische Bemerkungen. 7. Ueber die biologische Bedeutung des Blatthöhlen bei *Tozzia* und *Lathraea. Flora* 83, 449–53.

Gogelein, A. J. F. (1968). A revision of the genus *Cratoxylum* Bl. (Guttiferae). *Blumea* 15, 453–75.

Goleniewska-Furmanowa, M. (1970). Comparative leaf anatomy and alkaloid content in the Nymphaeaceae Bentham and Hooker. *Monographiae bot.* 31, 5–56.

Gorenflot, R. (1971). Intérêt taxonomique et phylogénique des caractères stomatiques (application à la tribu des Saxifragées). *Boissiera* 19, 181–92.

Goris, M. A. (1910). Contribution à l'étude des Anacardiaceae de la tribu des Mangiférées.*Annls Sci. nat. (ser. 9)* 11, 1–29.

Govier, R. N., Brown, J. G. S., and Pate, J. S. (1968). Hemiparasitic nutrition in angiosperms. II. Root haustoria and leaf glands of *Odontites verna* (Bell.) Dum. and their relevance to the abstraction of solutes from the host. *New Phytol.* 67, 963–72.

Grambast, N. (1954). Sur la structure et le développement de l'appareil stomatique dans le genre *Ficus. Rev. gén. Bot.* 61, 607–31.

Gravis, A. (1934), (1936). Théorie des traces foliaires. *Mém. Acad. r. Belg. Cl. Sci. (ser. 2)* 12, 59 pp.

Green, J. R. (1914). *A history of botany in the United Kingdom from the earliest times to the end of the nineteenth century.* Dent, London.

Greenish, H. G. and Collin, E. (1904). *An anatomical atlas of vegetable powders.* J. and A. Churchill; London.

Greensill, N. A. R. (1902). Structure of leaf of certain species of *Coprosma. Trans. N.Z. Inst.* 35, 342–55.

Gregory, T. C. (1915). The taxonomic value and structure of

the peach leaf glands. *Cornell Univ. Agric. Exp. Station Bull.* **365**, 183–224.

Grew, N. (1672). *The anatomy of vegetables begun. With a general account of vegetation founded thereon.* London.

— (1675). *The comparative anatomy of trunks, together with an account of their vegetation grounded there-upon.* London.

— (1679). *Anatomie des plantes qui contient une description exacte de leurs parties et leurs usages et qui fait voir comment elles se forment, et comment elles croissent.* 2nd edn. Paris.

— (1682). *The anatomy of plants with an idea of a philosophical history of plants and several other lectures read before the Royal Society.* London.

Griffith, M. M. (1968). The structure and development of the foliar sclereids in *Osmanthus fragrans* Lour. *Phytomorphology* **18**, 75–9.

Gris, A. (1870). Mémoire sur la moelle des plantes ligneuses. *Nouv. Archs Mus. Hist. nat. Paris* **6**, 201–302.

Groom, P. (1893). On *Dischidia rafflesiana* (Wall.). *Ann. Bot.* **7**, 223–42.

— (1894). On the extra-floral nectaries of *Aleurites. Ann. Bot.* **8**, 228–30.

— (1897). On the leaves of *Lathraea squamaria* and some allied Scrophulariaceae. *Ann. Bot.* **11**, 385–95.

Grüss, J. (1927*a*). Die Luftblätter des Nymphaeaceen. *Ber dt. bot. Ges.* **45**, 454–8.

— (1927*b*). Die Haustoren der Nymphaeaceen. *Ber. dt. bot. Ges.* **45**, 459–66.

Guédès, M. (1972*a*). Contribution à la morphologie du phyllome. *Mém. Mus. natn. Hist. nat., Paris (N.S., ser. B)* **21**, 1–179.

— (1972*b*). Leaf morphology in some *Lachemilla*, with a reassessment of leaf architecture. *Adv. Front. Pl. Sci.* **29**, 183–221.

— (1975). Intrusive hair sclereids in *Jovetia* (Rubiaceae). *Bot. J. Linn. Soc., Lond.* **71**, 141–4.

Guérin, P. (1906). Sur les domaties des feuilles de Dipterocarpées. *Bull. Soc. bot. Fr.* **53**, 186–92.

— and Guillaume, G. (1908). Falsification des feuilles de Belladonne. *Bull. Sci. pharmac.* **15**, 213.

Gunckel, J. E., Thimann, K. V., and Wetmore, R. H. (1949). Studies in development in long shoots and short shoots of *Gingko biloba* L. IV. Growth habit, shoot expression and mechanism of its control. *Am. J. Bot.* **36**, 309–16.

— and Wetmore, R. H. (1946). Studies of development in long and short shoots of *Gingko biloba* L. I. The origin and pattern of development of cortex, pith and procambium. *Am. J. Bot.* **33**, 285–95.

Gunning, B. E. S. and Pate, J. S. (1969). Plant cells with wall ingrowths, specialized in relation to short distance transport of solutes – their occurrence, structure and development. *Protoplasma* **68**, 107–33.

— —, and Briarty, L. G. (1968). Specialized 'transfer cells' in minor veins of leaves and their possible significance in phloem translocation. *J. cell Biol.* (suppl. to *J. biophys. biochem. Cytol.*) **37**, C7–C12.

— —, and Green, L. W. (1970). Transfer cells in the vascular system of stems: taxonomy, association with nodes, and structure. *Protoplasma* **71**, 147–71.

Gupta, B. (1961). Correlation of tissues in leaves. I. Absolute vein-islet numbers and absolute veinlet termination numbers. *Ann. Bot. (N.S.)* **25**, 65–70.

Guttenberg, H. von (1940). In Linsbauer's *Handbuch der Pflanzenanatomie.* Abt. II. Teil 3. *Samenpflanzen.* Band VIII. *Der primäre Bau der Angiospermenwurzel.* 369 pp. Gebrüder Borntraeger, Berlin.

— — (1959). Die physiologische Anatomie der Spaltöffnungen. In W. Ruhland (ed.), *Encyclopaedia of Plant Physiology,* **17**, 339–415., Berlin.

— (1960). *Grundzüge der Histogenese höherer Pflanzen. I. Die Angiospermen.* (Handbuch der Pflanzenanatomie VIII, 3.) Gebrüder Borntraeger, Berlin.

— (1971). In Zimmerman and Ozenda's *Encyclopaedia of Plant Anatomy.* Band V. Teil 5. *Bewegungsgewebe und Perzeptionsorgane.* pp. 332. Gebrüder Borntraeger, Berlin.

Guyot, M. (1966). Les stomates des Ombellifères. *Bull. Soc. bot. fr.* **113**, 244–73.

Haberlandt, G. (1894). Anatomisch-physiologische Untersuchungen über das tropische Laubblatt. I. Über wassersecernirende und-absorbirende Organe. *Sber. Akad. Wiss. Wien. (Math-Nat. Kl.)* **103**, 489–583.

— (1895). Anatomische-physiologische Untersuchungen über das tropische Laubblatt, II. Über wasser secerni rende und secer-absorbirende Organe. *Sber. Akad. Wiss. Wien (Math-Nat. Kl.)* **104** (Abt I), 55–116.

— (1897). Zur Kenntniss der Hydathoden. *Jb. wiss. Bot.* **30**, 511–28.

— (1898). Bemerkungen zur Abhandlung von Otto Spanjer. *Bot. Zg* **56**, 177–81.

— (1910). Wagers Einwände gegen meine Theorie der Lichtperzeption in den Laubblättern. *Jb. wiss. Bot.* **47**, 377–90.

— (1914). *Physiological plant anatomy.* (Translated from the 4th German edition by Montagu Drummond.) Macmillan, London.

— (1924). *Physiologische Pflanzenanatomie.* I. W. Engelmann, Leipzig.

— (1926). Über den Blattbau der *Crataegomespili* von Bronvaux und ihrer Eltern. *Sber. preuss. Akad. Wiss. Phys.-Math. Kl.* **17**, 170–208.

— — (1934*a*). Blattepidermis und Palisadengewebe der *Crataegomespili* und ihrer Eltern. *Sitz. Preuss. Akad. Wiss., Phys.-Math. Kl.* (12), 176–9.

— — (1934*b*). Über die Sonnen- und Schattenblätter der *Crataegomespili* und ihrer Eltern. *Sitz. Preuss. Akad. Wiss., Phys-Math. Kl.* (20), 363–76.

— (1935). Über den Blattbau sexueller Bastarde zwischen Mispel und Weissdorn. *Sber. preuss. Akad. Wiss.* (6), 118–43.

Habernicht, L. (1823). Ueber die tropfbare Absonderung des Wassers aus den Blättern der *Calla aethiopica. Flora* **34**, 529–36.

Hagemann, W. (1970). Studien zur Entwicklungsgeschichte der Angiospermenblätter. Ein Beitrag zur Klärung ihres Gestaltungsprinzips. *Bot. Jb.* **90**, 297–413.

Hall, B. N. (1762). *Dissertatio Botanica sistens Nectarium Florum.* Upsala.

Hall, D. M. (1967*a*). Wax microchannels in the epidermis of white clover. *Science* **158**, 505–06.

— (1967*b*). The ultrastructure of wax deposits on plant leaf surfaces. II. Cuticular pores and wax formation. *J. Ultrastruct. Res.* **17**, 34–44.

— and Donaldson, L. A. (1963). The ultrastructure of wax deposits on plant leaf surfaces. 1. Growth of wax on leaves of *Trifolium repens. J. Ultrastruct. Res.* **9**, 259–67.

— Matus, A. I., Lamberton, J. A. and Barber, H. N. (1965).

Infra-specific variation in wax on leaf surfaces. *Aust. J. biol. Sci.* **18**, 323–32.

Hall, J. P. and Melville, C. (1951). Veinlet termination number: a new character for the differentiation of leaves. *J. Pharm Pharmac.* **3**, 934–41.

—— (1954). Veinlet termination numbers – some further observations. *J. Pharm. Pharmac.* **6**, 129–33.

Hallam, N. D. (1967). An electron microscope study of the leaf waxes of the genus *Eucalyptus* L'Héritier. Thesis, University of Melbourne.

— (1970). Growth and regeneration of waxes on the leaves of *Eucalyptus*. *Planta* **93**, 257–68.

— and Chambers, T. C. (1970). The leaf waxes of the genus *Eucalyptus* L'Héritier. *Aust. J. Bot.* **18**, 335–86.

— and Juniper, B. E. (1971). The anatomy of the leaf surface. *The Ecology of leaf surface micro-organisms* (ed. Preece and Dickinson) pp. 3–37. Academic Press, London.

Hallé, F. (1966). Étude biologique et morphologique de la tribu des Gardéniées (Rubiacées). *Mém. Off. Rech. Sci. Tech. Outre-Mer.* No. 22, 1–146.

— (1971). Architecture and growth of tropical trees exemplified by the Euphorbiaceae. *Biotropica* **3**, 56–62.

— (1974). Architecture of trees in the rain forest of Morobe District, New Guinea. *Biotropica* **6**, 43–50.

— and Oldeman, R. A. (1970). *Essai sur l'architecture et la dynamique de croissance des arbres tropicaux.* Masson et Cie, Paris.

—— and Tomlinson, P. B. (1978). *Tropical Trees and Forests: an architectural analysis* Springer-Verlag, Berlin.

Hamilton, A. G. (1896). On domatia in certain Australian and other plants. *Proc. Linn. Soc. N.S.W.* **21**, 758–92.

Hanstein, J. (1868). Die Scheitelzellgruppe im Vegetationspunkt der Phanerogamen. *Festschr. Niederrhein Ges. Natur- und Heilkunde*, 109–34.

Hara, N. (1957). On the types of the marginal growth in dicotyledonous foliage leaves. *Bot. Mag., Tokyo* **70**, 108–14.

— (1962). Structure and seasonal activity of the vegetative shoot apex of *Daphne pseudomezereum*. *Bot. Gaz.* **124**, 30–42.

Hardy, A. D. (1912). The distribution of leaf glands in some Victorian acacias. *Victorian Nat.* **29**, 26–32.

Hare, C. L. (1943). The anatomy of the petiole and its taxonomic value. *Proc. Linn. Soc., Lond.* **155**, 223–9.

Harland, S. C. (1947). An alteration in the gene frequency in *Ricinus communis* L. due to climatic conditions. *Heredity, London.* **1**, 121–5.

Harlow, W. M. (1930). The formation of chambered pith in the twigs of butternut and black walnut. *J. Forestry* **28**, 739–41.

Harms, H. (1917). Ueber eine Meliacee mit blattbürtigen Blüten. *Ber. dt. bot. Ges.* **35**, 338–48.

Harris, T. M. (1926). Note on a new method for the investigation of fossil plants. *New Phytol.* **25**, 58–60.

— (1956). The fossil plant cuticle. *Endeavour* **15**, 210–14.

— (1961–9). *The Yorkshire Jurassic flora.* Vol. I (1961); Vol. II (1964), Vol. III (1969). British Museum (Natural History), London.

—, Millington, W., and Miller, J. (1974). *The Yorkshire Jurassic flora*, Vol. IV. British Museum (Natural History), London.

Hartog van ter Tholen, R. M. and Baas, P. (1979). Epidermal characters of the Celastraceae (*sensu lato*). Acta Bot.

Neerl. **27**, 355–8.

Hasselberg, G. B. F. (1937). Zur Morphologie des vegetativen Sprosses der Loganiaceen. *Symb. bot. Upsal.* II, 3, 1–170.

Haupt, H. (1902). Zur Sekretionsmechanik der extrafloralen Nektarien. *Flora* **90**, 1–41.

Häusermann, E. and Frey-Wyssling, A. (1963). Phosphatase-Aktivität in Hydathoden. *Protoplasma* **57**, 371–80.

Heilbronn, M. (1916). Die Spaltöffnungen von *Camellia japonica* L. (*Thea japonica* Nois.). *Ber. dt. bot. Ges.* **34**, 22–31.

Heimann, M. (1950). Einfluss periodischer Beleuchtung auf die Guttationsrhythmik. (Untersuchungen an *Kalanchoë blossfeldiana*.) *Planta* **38**, 157–95.

Heinrich, G. (1973). Die Feinstruktur der Trichom-Hydathoden von *Monarda fistulosa*. *Protoplasma* **77**, 271–8.

Heinricher, E. (1884). Ueber isolateralen Blattbau mit besonderer Berücksichtigung der europäischen, speciell der deutschen Flora. Ein Beitrag zur Anatomie und Physiologie der Laubblätter. *Jb. wiss. Bot.* **15**, 502–67.

— (1910). Beiträge zur Kenntnis der Anisophyllie. *Annls Jard. bot. Buitenz.* supple. 3, 2, 649–64; taf. xx-xxv.

Heiser, B. B. (1949). Study in the evolution of the sunflower species *Helianthus annuus* and *H. bolanderi*. *Univ. Calif. Publs. Bot.* **23**, 157–208.

Heitzelman, C. E. and Howard, R. E. (1948). The comparative morphology of the Icacinaceae. V. The pubescence and the crystals. *Am. J. Bot.* **35**, 42–52.

Hemenway, A. F. (1913). Studies on the phloem of the dicotyledons. II. The evolution of the sieve- tube. *Bot. Gaz.* **55**, 236–43.

Henslow, J. S. (1832). On the examination of a hybrid *Digitalis*. *Trans. Camb. phil. Soc.* **4**, 257–78.

Herbin, G. A. and Robins, P. A. (1968). Studies on plant cuticular waxes. II. Alkanes from members of the genus *Agave* (Agavaceae), the genera *Kalanchoë, Echeveria, Crassula* and *Sedum* (Crassulaceae) and the genus *Eucalyptus* (Myrtaceae) with an examination of Hutchinson's sub-division of the angiosperms into Herbaceae and Lignosae. *Phytochemistry* **7**, 257–68.

Heslop-Harrison, Y. and Heslop-Harrison, J. (1968). Scanning electron microscopy of leaf surfaces. *Proc. I.I.T. SEM. Symp.* 119–26.

Hetschko, A. (1908). Über den Insektenbesuch bei einigen *Vicia*-Arten mit extrafloral-Nektarien. *Wien. ent. Zg.* **27**, 299–305.

— (1916). Ueber den Insektenbesuch bei *Vicia Faba* L. *Wien ent. Zg.* **25**, 123–5.

Hickey, L. J. (1973). Classification of the architecture of dicotyledonous leaves. *Am. J. Bot* **60**, 17–33.

— and Wolfe, J. A. (1975). The bases of angiosperm phylogeny: vegetative morphology. *Ann. Mo. bot. Gdn* **68**, 538–89.

Hill, A. W. (1931). A hybrid Daphne (*D. petraea* Leybold X *D. cneorum* L.). *Ann. Bot.* **45**, 229–31.

Hill, J. (1770). *The construction of timber.* London.

Hill, T. G. (1906). Stelar theories. *Sci. Prog. Lond.* **2**, 1–18.

Hocquette, M. (1954). Anatomie. In *Histoire de botanique en France.* pp. 125–46. Issued at the VIIIth International Botanical Congress, Paris.

Hohn, K. (1950). Untersuchungen über Hydathoden und deren Funktion. *Abh. Akad. wiss. Lit. Mainz* **2**, 11–42.

Holdheide, W. (1951). Anatomie mitteleuropäischer Gehölzrinden (mit mikrophotographischem Atlas). In *Handbuch der Mikroskopie in der Technik* (ed. H. Freund) Vol. V, Teil, I, 193-367. Umschau Verlag. Frankfurt.

Holloway, P. J. (1971). The chemical and physical characteristics of leaf surfaces. In *Ecology of leaf surface microorganisms.* (eds. T. F. Preece and C. H. Dickinson.) Section I, chap. 2, pp. 39-53. Academic Press, London.

Holm, T. (1910). *Grindelia squarrosa* (Pursh.) Duval. *Merck's Rep.* **19**, 310-12. (See *Bot. Zbl.* **116**, 383 (1911).))

— (1921). Morphological study of *Carya alba* and *Juglans nigra. Bot. Gaz.* **72**, 375-89.

— (1929). Medullary cork in *Balsamocitrus.* An anatomical study. *Am. J. Sci.* (ser. 5) **18**, 505-8.

Hooke, R. (1665). *Micrographia or some physiological descriptions of minute bodies made by magnifying glasses with observations and inquiries thereupon.* London. [Also Facsimile edition 1961. New York and London.]

Horner, H. T., Jr. and Lersten, N. R. (1968). Development structure and function of secretory trichomes in *Psychotria bacteriophila* (Rubiaceae). *Am. J. Bot.* **55**, 1089-99.

Houard, C. (1922). *Les zoocécidies des Plantes d'Afrique, D'Asie et d'Océanie.* Librarie Scientifique Jules Hermann, Paris.

— (1933). *Les zoocécidies des plantes de l'Amérique du Sud et du l'Amérique centrale.* 34-6. Hermann, Paris.

Hove, C. van and Kagoyre, K. (1974). A comparative study of stipular glands in nodulating and non-nodulating species of Rubiaceae. *Ann. Bot.* **38**, 989-91.

Howard, J. E. (1862). *Illustrations of the Nueva Quinologia of Pavon.* Lovell Reeve & Co., London.

Howard, R. A. (1962). The vascular structure of the petiole as a taxonomic character. Proc. 15th Intn. hort. Cong., Nice 1958, pp. 7-13.

— (1969). The ecology of an elfin forest in Puerto Rico. 8. Studies of stem growth and form and of leaf structure. *J. Arnold. Arbor.* **50**, 225-67.

— (1970a). The ecology of an elfin forest in Puerto Rico. 10. Notes on two species of *Marcgravia. J. Arnold Arbor.* **51**, 41-55.

— (1970b). Some observations on the nodes of woody plants with special reference to the problem of the 'split lateral' versus the 'common gap'. In *New research in plant anatomy* (eds. N. K. B. Robson, D. F. Cutler, and M. Gregory) *Bot. J. Linn. Soc.* **63**, Supplement 1, pp. 195-214. Academic Press, London.

— (1974). The stem-node-leaf continuum of the Dicotyledoneae. *J. Arnold Arbor.* **55**, 125-81.

Howes, F. N. (1936). Sources of vegetable wax. *Kew Bull,* 503-26.

Hryniewiecki, B. (1912a). Ein neuer Typus der Spaltöffnungen bei den Saxifragaceen. *Bull. int. Acad. Sci. Lett. Cracovie (Cl. Sci, math. nat., ser. B)* 52-73; with plates I to VI.

— (1912b). Anatomische Studien über die Spaltöffnungen bei den Dikotylen. *Bull. int. Acad. Sci. Lett. Cracovie (Cl. Sci. math. nat., ser. B.)* 585-605.

Hull, H. M., Morton, H. L., and Wharrie, J. R. (1975). Environmental influences on cuticle development and resultant foliar penetration. *Bot. Rev.* **41**, 421-52.

—, Shellhorn, S. J., and Saunier, R. E. (1971). Variations in creosote bush (*Larrea divaricata*) epidermis. *J. Ariz. Acad. Sci.* **6**, 196-205.

Hüller, G. (1907). Beiträge zur vergleichenden Anatomie der Polemoniaceen. *Beih. bot. Zbl.* **21**, (I), 173-244.

Hülsbruch, M. (1966a). Zur Radialstreifung cutinisierter Epidermis-aussenwände I [English Summary]. *Z. Pflanzenphysiol.* **55**, 181-97.

— (1966b). Cutin als konstruktive Wandsubstanz–nicht nur Akkruste oder Inkruste. *Ber. dt. bot. Ges.* **79**, 87-91.

Hülsbruch, W. (1932). Beiträge zur Kenntnis der Gattung *Dysophylla* und einiger anderer Labiaten. *Flora, N.S.* **26**, 329-62.

Hunter, G. E. and Austin, D. F. (1967). Evidence from trichome morphology of interspecific hybridization in *Vernonia.* Compositae. *Brittonia* **19**, 38-41.

Idle, D. B. (1969). Scanning electron microscopy of leaf surface replicas and the measurement of stomatal aperture. *Ann. Bot.* (*N.S.*) **33**, 75-6.

Inamdar, J. A. (1968a). Development of stomata in vegetative and floral organs of some Caryophyllaceae. *Aust. J. Bot.* **16**, 445-9.

— (1968b). Epidermal structure and ontogeny of stomata in some Nyctaginaceae. *Flora* **158B**, 159-66.

— (1968c). Anatomical studies in *Clerodendrum splendens* G. Don. *Proc. Indian Acad. Sci. B* **67**, 8-17.

— (1969a). Structure and ontogeny of foliar nectaries and stomata in *Bignonia chamberlaynii* Sims. *Proc. Indian Acad. Sci. B.* **70**, 232-40.

— (1969b). Epidermal structure and stomatal ontogeny in some Polygonales and Centrospermae. *Ann. Bot.* **33**, 541-52.

—, Bhatt, D. C., Patel, R. C., and Dave, V. H. (1973). Structure and development of stomata in vegetative and floral organs of some Passifloraceae. *Proc. Indian nat. Sci. Acad. B* **39**, 553-60.

— and Gangadhara, M. (1975). Effects of growth regulators on stomatal structure and development in the cotyledons of *Lagenaria leucantha* (Duch.) Rusby. *Aust. J. Bot.* **23**. 13-25.

International Association of Wood Anatomists. (1957) *International glossary of terms used in wood anatomy. Trop. Woods* No. 107, 1-36.

International Association of Wood Anatomists Bulletin (1975). No. 2, p. 36.

International Wood Collector's Society Bulletin. (Editor: Eleanor P. Frost, 148 Summer St., Lanesborough, Ma 01237, U.S.A.).

Iterson, van G. Jr. (1937). A few observations on the hairs of the stamens of *Tradescantia virginica. Protoplasma* **27**, 190-211.

Ivanoff, S. S. (1963). Guttation injuries of plants. *Bot. Rev.* **29**, 202-229.

Jacobs, D. L. (1946). Shoot segmentation in *Anacharis densa. Am. Midl. Nat.* **35**, 283-6.

Jacobs, M. (1966). On domatia – the viewpoints and some facts. *Proc. K. ned. Akad. Wet.* (sec. C) **69**, 275-316.

Jacobs, W. P. and Morrow, I. B. (1957). A quantitative study of xylem development in the vegetative shoot system of *Coleus. Am. J. Bot.* **44**, 823-42.

Jähnichen, H. von (1969). Revision zu Originalen strukturbieternder Blätter aus der Lausitzer und Niederrheinischen Braunkole (P. Menzel 1906 und 1913). *Geologie* **18**, 77-112.

Jalan, S. (1962). The ontogeny of the stomata in *Schisandra grandiflora* Hook. f. & Thoms. *Phytomorphology* **12**, 239-42.

Janda, C. (1937). Die extranuptialen Nektarien der Malvaceen.

Öst. Bot. Z. **86**, 81–130.

Jansen, W. T. and Baas, P. (1973). Comparative leaf anatomy of *Kokoona* and *Lophopetalum* (Celastraceae). *Blumea* **21**, 153–78.

Janssonius, H. H. (1952). *Key to the Javanese woods on the basis of anatomical features.* E. J. Brill, Leiden.

Janzen, D. H. (1966). Coevolution of mutualism between ants and acacias in Central America. *Evolution* **20**, 249–75.

— (1967*a*). Fire, vegetation structure and the ant × acacia interaction in Central America. *Ecology* **48**, 25–35.

— (1967*b*). Interaction of the bull's-horn acacia (*Acacia cornigera* L.) with an ant inhabitant (*Pseudomyrmex ferruginea* F. Smith) in Eastern Mexico. *Kans. Univ. Sci. Bull.* **47**, 315–58.

— (1969). Allelopathy by myrmecophytes: the ant Azteca as an allelopathic agent of *Cecropia*. *Ecology* **50**, 147–53.

— (1974*a*). Swollen-thorn acacias of Central America. *Smithson. Contr. Bot.* No. 13, 1–131.

— (1974*b*). Epiphytic myrmecophytes in Sarawak: mutualism through the feeding of plants by ants. *Biotropica* **6**, 237–59.

Jeffrey, E. C. (1899). The morphology of the central cylinder in the angiosperms. *Trans. Canad. Inst.* **6**, 599–636.

— (1917). *The anatomy of woody plants.* Chicago University Press.

— (1925). The origin of parenchyma in geological time. *Proc. nat. Acad. Sci. Washington* **11**, 106–10.

Jensen, L. C. W. (1968). Primary stem vascular patterns in three sub-families of the Crassulaceae. *Am. J. Bot.* **55**, 553–63.

Jensen, W. A. (1962). *Botanical histochemistry; principles and practices.* Freeman, San Francisco.

Johansen, D. A. (1940). *Plant microtechnique.* McGraw-Hill, New York.

Johnson, Sister C. and Brown, W. V. (1973). Grass leaf ultrastructural variations. *Am. J. Bot.* **60**, 727–35.

Johnson, M. A. (1958). The epiphyllous flowers of *Turnera* and *Helwingia*. *Bull. Torrey bot. Club* **85**, 313–23.

— and Truscott, F. H. (1956). On the anatomy of *Serjania*. I. Path of the bundles. *Am. J. Bot.* **43**, 509–18.

Juniper, B. E. and Cox, G. C. (1973). The anatomy of the leaf surface: the first line of defence. *Pestic. Sci.* **4**, 543–61.

Kalmán, F. and Gulyás, S. (1974). Ultrastructure and mechanism of secretion in extrafloral nectaries of *Ricinus communis* L. *Acta biol., Szeged.* **20**, 57–67.

Kanehira, R.(1921*a*). *Identification of the important Japanese woods by anatomical characters.* Tokyo. [In English.]

— (1921*b*). *Anatomical characters and identification of Formosan woods, with critical remarks from the climatic point of view.* Government of Formosa. [In English.]

— (1926). *Anatomical characters and identification of the important woods of the Japanese Empire.* Department of Forestry, Government Research Institute, Taihoku, Formosa. [In Japanese.]

Kaplan, D. R. (1970*a*). Comparative foliar histogenesis in *Acorus calamus* and its bearing on the phyllode theory of monocotyledonous leaves. *Am. J. Bot.* **57**, 331–61.

— (1970*b*). Comparative development and morphological interpretation of the 'rachis-leaves' in Umbelliferae. In *New research in plant anatomy* (eds. N. K. B. Robson, D. F. Cutler, and M. Gregory) pp. 101–25 + plates. Acade-

mic Press, London. (Published as supplement I to *Bot. J. Linn. Soc. Lond.* **63**.

Kapoor, S. L., Sharma, P. C., Chandra, V. and Kapoor, L. D. (1969) [1972]. Epidermal and venation studies in Apocynaceae II. *Bull. bot. Surv. India* **11**, 372–6.

Karsten, H. (1857). Ueber die Entstehung des Harzes, Wachses, Gummis und Schleims durch die assimilierende Thätigkeit der Zellmembran. *Bot. Zeit.* **15**, 313–21.

Kato, N. (1966-7*a*). On the variation of node types in woody plants I. *J. Jap. Bot.* **41**, 101–7.

Kato, N. (1966-7*a*). On the variation of nodal types in woody plants. II. *J. Jap. Bot.* **42**, 161–8.

Katsumata, F. (1971). Shape of idioblasts in mulberry leaves with special reference to the classification of mulberry trees. *J. seric. Sci., Tokyo* **40**, 313–22.

— (1972). Relationships between the length of styles and the shape of idioblasts in mulberry leaves, with special reference to the classification of mulberry trees. *J. seric. Sci., Tokyo* **41**, 387–95.

Kaufmann, K. (1927). Anatomie und Physiologie der Spaltöffnungsapparate mit verholzten Schliesszellmembranen. *Planta* **3**, 27–59.

Kaul, M. L. H. (1970). Studies on ecotypes and ecads of *Mecardonia dianthera*. III. Ecological distribution, phenology, dispersal and stomata. *Proc. nat. acad. Sci. India.* B. **39**, 178–84.

Kenda, G. (1952). Stomata in Antheren I. Anatomischer Teil. *Phyton, Horn* **4**, 83–96.

Kerr, A. F. G. (1912). Notes on *Dischidia rafflesiana* Wall and *Dischidia nummularia* Br. *Scient. Proc. R. Dubl. Soc.* **13**, 293–309.

Kerr, T. and Bailey, I. W. (1934). The cambium and its derivative tissues. X Structure, optical properties and chemical composition of the so-called lamella. *J. Arnold Arbor.* **15**, 327–49.

Ketellapper, H. J. (1963). Stomatal physiology. *Ann. Rev. Plant Physiol.* **14**, 249–70.

Keuchenius, P. E. (1916). Beiträge zur Anatomie von *Hevea brasiliensis*. *Annls Jard. bot. Buitenz.* **29**, 109–11.

Kirchmayr, H. (1908). Die extrafloralen Nektarien von *Melampyrum* vom physiologisch-anatomischen Standpunkt. *Sber. Akad. Wiss. Wien (math.-naturwiss. Kl.).* I, **117**, 439–52.

Knapheisowna, G. (1927). Beiträge zur Verteilung und Anatomie der Sekretionsorgane an den Blättern der *Prunus*-Arten. *Acta Soc. Bot. Pol.* **4**, 106–13.

Knecht, G. N. and O'Leary, J. W. (1972). The effect of light intensity on stomate number and density of *Phaseolus vulgaris* L. leaves. *Bot. Gaz.* **133**, 132–4.

— and Orton, E. R. (1970). Stomate density in relation to winter hardiness of *Ilex opaca* Ait. *J. Am. Soc. hort. Sci.* **95**, 341–5.

Kny, L. (1909). Innerer Bau des Sonnen- und Schattenblätter der Rotbuche (*Fagus sylvatica* L.). *Bot. Wandtafeln. Tafel* cxii und cxiv, 502–513.

Kolattukudy, P. E. (1970). Plant waxes. *Lipids* **5**, 259–75.

Körnicke, M. (1918). Über die extrafloralen Nektarien auf den Laubblättern einiger Hibisceen. *Flora* **111–12**, 526–40.

Korn, R. W. (1972). Arrangement of stomata on the leaves of *Pelargonium zonale* and *Sedum stahlii*. *Ann. Bot.* **36**, 325–33.

— and Frederick, G. W. (1973). Development of D-type

stomata in leaves of *Ilex crenata* var. *connexa. Ann. Bot.* **37**, 647–56.

Kostytschew, S. (1922). Der Bau und das Dickenwachstum der Dikotylenstamme. *Ber. dt. bot. Ges.* **40**, 297–305.

Krafft, K. (1907). Systematisch-anatomische Untersuchung der Blattstruktur bei den Menispermaceen. Thesis, Erlangen, Stuttgart.

Kralj, D. and Sušnik, F. (1967). Reconnaissance de la polyploidie chez le houblon *Humulus lupulus* L. *Biol. Věst.* **15**, 67–71. [Serbo-Croat.]

Kraus, G. (1872). Ueber eigenthumliche Sphaerocrystalle in der Epidermis von *Cocculus laurifolius. Jb. wiss. Bot.* **8**, 421–6.

Kraus, K. (1880). Die Lebensdauer der immergrunen Blätter. *Sber. naturf. Ges., Halle.* 1–15.

Kristen, H. (1969). Licht- und electronenmikroskopische Untersuchungen an den Hydropoten von *Nuphar lutea, Nymphoides peltata, Sagittaria macrophylla* und *Salvinia auriculata. Flora* **A159**, 536–58. [English Summary]

— (1971). Licht-und elektronenmikroskopische Untersuchungen zur Entwicklung der Hydropoten von *Nelumbo nucifera. Ber. dt. bot. Ges.* **84**, 211–24.

Kritikos, P. G. and Steinegger, E. (1948). Heteroploidie Versuche an Arzneipflanzen. 6. Spaltöffnungsindex bei diploiden und tetraploiden *Lobelia inflata. Pflanzen Pharm. Acta Helv.* **23**, 343–52.

— — (1949a). Spaltöffnungsindex verschiedener *Lobelia* Spezies. *Pharm. Acta Helv.* **24**, 100–7.

— — (1949b). Heteroploidie-Versuche an Arzneipflanzen. 8 Spaltöffnungsindex bei diploiden und tetraploiden *Lobelia syphilitica. Pharm. Acta Helv.* **24**, 45–50.

Kroemer, K. (1903). Wurzelhaut, Hypodermis und Endodermis der Angiospermenwurzel. *Bibl. Bot. Heft.* **59**, 151 pp.

Kropfitsch, M. (1951a). Apfelgas. Wirkung auf Stomatazahl. *Protoplasma* **40**, 256–65.

— (1951b). UV-Bestrahlung und Stomatazahl. *Protoplasma* **40**, 266–74.

Kugler, H. (1928). Über invers- dorsiventrale Blätter. *Planta* **5**, 89–134.

Kumar, A. (1976). A new type of nodal organization in angiosperms. *Acta bot. indica* **4**, 76–7.

Kundu, B. C. and De, A. (1968). Taxonomic position of the genus *Nyctanthes. Bull. bot. Surv. India* **10**, 397–408.

Kuntze, O. (1891). Beiträge zur vergleichenden Anatomie der Malvaceen. *Bot. Zbl.* **45**, 161–8, 197–202, 229–34, 261–8, 293–9.

Kurer, G. A. (1917). Kutikularfalten und Protuberanzen an Haaren und Epidermen und ihre Verwendung zur Differenzialdiagnose offizineller Blätter. Thesis, Zurich.

Kurr, J. G. (1833). *Untersuchungen über die Bedeutung der Nektarien in den Blumen.* Stuttgart.

Kurt, J. (1929). Ueber die Hydathoden der Saxifrageae. *Beih. bot. Zbl.* **46** (1), 203–46.

Kurtz, E. B. (1951). Studies in metabolism of lipids in plants. Thesis, California Institute of Technology.

— (1958). Survey of some plant waxes of Southern Arizona. *J. Am. Oil Chemists Soc.* **35**, 465–7.

Kutík, J. (1973). The relationships between quantitative characteristics of stomata and epidermal cells of leaf epidermis. *Biologia Pl., Tchécosl.* **15**, 324–8.

Laetsch, W. M. (1968). Chloroplast specialization in dicotyledons possessing the C_4 dicarboxylic acid pathway of photosynthetic CO_2 fixation. *Am. J. Bot.* **55**, 875–83.

Lamarlière, L. Géneau de (1906). Sur les membranes cutinisées des plantes aquatiques. *Rev. gén. Bot.* **18**, 289–95.

Lange, R. T. (1969). Concerning the morphology of isolated plant cuticles. *New Phytol.* **68**, 423–5.

Larsen, C. M. (1961). Développement des stomates des peupliers au cours d'une année sèche. *Physiol. Plant.* **14**, 877–9.

Lavier-George, L. (1936). Recherches sur les épidermes foliaires des *Philippia* de Madagascar; utilisation de leurs caractères comme bases d'une classification. *Bull. Mus. Hist. nat. Paris,* (ser. 2) **8**, 173–99.

— (1937). Epidermes foliaires et appareils aquifères de l'*Hottonia palustris* L. *Ann. Sci. nat. Bot.* (ser. 10) **19**, 299–308.

Lawrence, G. H. M. (1951). *Taxonomy of Vascular Plants.* Macmillan, New York.

— (1952). Morphology and the taxonomist. *Phytomorphology.* **2**, 30.

Lechner, S. (1914). Anatomische Untersuchungen über die Gattungen *Actinidia, Saurauia, Clethra* und *Clematoclethra* mit besonderer Berücksichtigung ihrer Stellung im System. *Beih. bot. Zbl.* **32** (1) 431–67.

Lee, A. T. (1948). The genus *Swainsona. Contr. N.S.W. Herb.* **1**, 131–271.

Lee, B. and Priestley, J. H. (1924). The plant cuticle. I. Its structure, distribution and function. *Ann. Bot.* **38**, 525–45.

Leece, D. R. (1976). Composition and ultrastructure of leaf cuticle from fruit trees, relative to differential foliar absorption. *Aust. J. Plant Physiol.* **3**, 833–47.

Leick, E. (1927). Untersuchungen über den Einfluss des Liches auf die Öffnungsweite unterseitiger und oberseitiger Stomata desselben Blattes. *Jahb. wiss. Bot.* **67**, 771–848.

Lemesle, R. (1928). Contributions à l'étude structurale de quelques Labiées extra-européennes. *Bull. Soc. bot. Fr.* **75**, 18–25.

Léonard, J. (1957). Genera des Cynometreae et des Amherstieae africaines (Leguminosae-Caesalpinioideae). Essai de blastogénie appliquée à la systématique. *Acad. R. Belgique, Classe des Sciences, Mémoires* **30**(2), 1–314.

Lepeschkin, W. W. (1906). Zur Kenntnis des Mechanisms der aktiven Wasserausscheidung der Pflanzen. *Bot. Zbl.* **19**, 409–52.

Lepeschkin, W. (1921). Recherches sur les organes du bord des jeunes feuilles. *Bull. Soc. bot. Genève.* (ser 2) **13**, 226–35.

Lerch, G. (1964). Kallose in der Epidermis höherer Pflanzen. *Flora,* **154**, 36–52.

Lersten, N. R. (1972). Stipular glands and trichomes in relation to the bacterial leaf nodule symbiosis in *Psychotria* (Rubiaceae). *Brittonia* **24**, 123 (Abstract).

— (1974a). Morphology and distribution of colleters and crystals in relation to the taxonomy and bacterial leaf nodule symbiosis of *Psychotria* (Rubiaceae). *Am. J. Bot.* **61**, 973–81.

— (1974b). Colleter morphology in *Pavetta, Neorosea* and *Tricalysia* (Rubiaceae) and its relationship to the bacterial nodule symbiosis. *Bot. J. Linn. Soc.* **69**, 125–36.

— and Carvey, K. A. (1974). Leaf anatomy of ocotillo (*Fouquieria splendens*; Fouquieriaceae) especially vein endings and associated veinlet elements. *Can. J. Bot.* **52**, 2017–21.

— and Curtis, J. D. (1974). Colleter anatomy in red mangrove,

Rhizophora mangle (Rhizophoraceae). *Can. J. Bot.* **52**, 2277-8.

— and Peterson, W. H. (1974). Anatomy of hydathodes and pigment disks in leaves of *Ficus diversifolia* (Moraceae). *Bot. J. Linn. Soc.* **68**, 109-13.

Letouzey, R., Hallé, N., and Cusset, G. (1969). *Phyllobotryae* (Flacourtiaceae) d'Afrique centrale, variations morphologiques et biologiques, conséquences taxonomiques. *Adansonia* (*ser. 2*) **9**, 515-37.

Levin, D. A. (1973). Role of trichomes in plant defense. *Quart. Rev. Biol.* **48**, 3-15.

Levin, F. A. (1929). The taxonomic value of vein islet areas based upon a study of the genera *Barosma, Cassia, Erythroxlyon* and *Digitalis*. *Q. J. Pharm. Pharmac.* **2**, 17-43.

Levitt, J. (1974). The mechanism of stomatal movement – once more. *Protoplasma* **82**, 1-17.

Lewis, S. M. (1968). The gross morphology of the glandular trichome of the leaf of Jonathan apple. *Bot. Gaz.* **129**, 89-91.

Lignier, M. O. (1887). Recherches sur l'anatomie comparée des Calycanthées, des Mélastomacées et des Myrtacées. *Arch. Sci. Nord. de la France* **4**, 455.

— (1888). De la forme du système libéro-ligneux foliaire chez les phanérogames. *Bull. Soc. Linn. Normandie* **IV**, (ser. 2). 81-92.

Link, H. F. (1840). *Icones selectae anatomico-botanicae*. Berlin.

Linsbauer, K. *Handbuch der Pflanzenanatomie*. The various contributions to Linsbauer's *Handbuch der Pflanzenanatomie* that refer to the anatomy of Dicotyledons are entered in this work under the names of the authors of the individual parts. See also Zimmermann and Ozenda's *Encyclopaedia of plant anatomy*. The authors concerned include Frey (1929), Guttenberg, H. von. (1940, 1966, 1968, 1971); Linsbauer, K. (1930); Netolitzky (1926), (1932); Sperlich, A. (1925).

— (1930). In Linsbauer's *Handbuch der Pflanzenanatomie*. Abt. 1. Teil 2. Band IV. *Die Epidermis*. Gebrüder Borntraeger, Berlin.

Lippmann, E. (1925). Ueber das Vorkommen der verschiedenen Arten der Guttation und einige physiologische und ökologische Beziehungen. *Bot. Arch.* **11**, 361-464.

Litke, R. (1966). Kutikularanalytische Untersuchungen im Niederlausitzer Unterflöz. *Paläont. Abh.* (*Abt. B. Paläobot.*) **2**, 193-426.

— (1968). Pflanzenreste aus dem Untermiozän in Nordwestsachsen. *Palaeontographica* **123B** 173-83.

Lloyd, F. E. (1942). *The carnivorous plants*. Chronica Botanica Co., Waltham, Mass.

Lobreau-Callen, D. (1973). Le pollen des Icacinaceae II. Observations en microscopie électronique, corrélations, conclusions. *Pollen et spores* **15**, 47-89.

Loftfield, J. V. G. (1921). The behaviour of stomata. *Publ. Carnegie Inst.* **314**.

Lohr, P. J. (1919). Untersuchungen über die Blattanatomie von Alpen und Ebenenpflanzen. *Rec. Trav. bot. Néerl.* **16**, 1-62.

Lorougnon, G. (1966). Recherches sur quelques représentants tropicaux de groupes végétaux tempérés. *Adansonia* **6**, 289-300.

Louguet, P. (1974). Les mécanismes du mouvement des stomates; étude critique des principales théories classiques et modernes et analyse des effets du gaz carbonique sur

le mouvement des stomates du *Pelargonium* × *hortorum*, à l'obscurité. *Physiol. Vég.* **12**, 53-81.

Lucas, G. L. (1968). Icacinaceae. *Flora Trop. East Africa* 1-17.

Lück, H. B. (1966). Sur l'indice stomatique. *Naturalia Monspeliensis bot.* **17**, 145-56.

Luhan, M. (1954). Über das Vorkommen von Sklerenchym-Idioblasten bei *Globularia*-Arten. *Ber. dt. bot. Ges.* **67**, 346-55.

— (1955). Das Abschlussgewebe der Wurzeln unserer Alpenpflanzen. *Ber. dtsch. bot. Ges.* **68**, 87-92.

Lundstroem, A. N. (1887). Pflanzenbiologische Studien II. Die Anpassungen der Pflanzen an Thiere (I. von Domatien, pp. 3-72). *Nova Acta Regiae Soc. Sci. Upsaliensis* (*ser 3*) **13** (10), 1-88.

— (1888). Domatia. *J. R. Microsc. Soc.* **(1)**, 87.

Lüttge, U. (1961). Über die Zusammensetzung des Nektars und den Mechanismus seiner Sekretion. I. *Planta* **56**, 189-212.

— (1962a). Über die Zusammensetzung des Kektars und den Mechanismus seiner Sekretion. II Mitteilung. Der Kationengehalt des Naktars und die Bedeutung des Verhältnisses Mg^{++}/Ca^{++} im Drüsengewebe für die Sekretion. *Planta* **59**, 108-14.

— (1962b). Über die Zusammensetzung des Nektars und den Mechanismus seiner Sekretion. III Mitteilung. Die Rolle der Rückresorption und der spezifischen Zuckersekretion. *Planta* **59**, 175-94.

— (1964). Mikroautoradiographische Untersuchungen über die Funktion der Hydropoten von *Nymphaea*. *Protoplasma* **59**, 157-62.

— (1971). Structure and function of plant glands. *Ann. Rev. Plant physiol.* **22**, 23-44.

— and Krapf, G. (1969). Die Ultrastruktur der *Nymphaea*-Hydropoten in Zusammenhang mit ihrer Funktion als Salz-transportierende Drüsen. *Cytobiologie* **1**, 121-31. [*Biol. Abstr.* 51 (1970) No. 127086.]

—, Pallaghy, C. K., and Willert, K. von (1971). Microautoradiographic investigations of sulfate uptake by glands and epidermal cells of waterlily (*Nymphaea*) leaves with special reference to the effect of poly-L-lysine. *J. Membrane Biol.* **4**, 395-407.

Lyr, H. and Streitberg, H. (1955). Die Verbreitung von Hydropoten in verschiedenen Verwandtschaftskreisen der Wasserpflanzen. *Wiss. Z. Univ. Halle, Math-nat. Reihe* **4**, 471-84.

McNair, J. B. (1929). The taxonomic and climatic distribution of oils, fats and waxes in plants. *Am. J. Bot.* **16**, 832-41.

Mädler, K. and Straus, A. (1971). Ein System der Blattformen mit spezieller Anwendung für die Bestimmung neogener Blattreste (Miozän und Pliozän). *Bot. Jb.* **90**, 562-74.

Maekawa, F. (1948). *Folia orixata*, a new type of phyllotaxis and its significance in phyllotaxis evolution. *Bot. Mag., Tokyo* **61**, 7-10.

— (1952). Topo-morphological investigations on the relation between stem and leaves and their bearing on the phylogenetic systematics of vascular plants. Part I, *J. Fac. Sci. Tokyo Univ.* (*sect. III, Bot.*) **6**, 1-28.

Maercker, U. (1965a). Über vermeintliche Poren in Epidermisaussenwänden von *Cocculus laurifolius* und *Camellia japonica*. *Z. Pflanzenphysiol.* **53**, 86-9. [English summary.]

—— (1965b). Über das Vorkommen von Stomata in der Epidermis bunter Perianthblätter. *Z. Pflanzenphysiol.* **53**, 422-8.

Magócsy-Dietz, A. (1899). Das Diaphragma in dem Marke der dicotylen Holzgewächse. *Math. naturw. Ber. Ung.* **17**, 181–226.

Maheshwari, J. K. (1954). The structure and development of extrafloral nectaries in *Duranta plumieri* Jacq. *Phytomorphology* **4**, 208–11.

— and Chakrabarty, B. (1966). Foliar nectaries of *Clerodendrum japonicum* (Thunb.) Sweet. *Phytomorphology* **16**, 75–80.

Maheshwari, P. (1929). Origin and development of internal bundles in the stem of *Rumex crispus*. *J. Indian. bot. Soc.* **8**, 89–117.

— (1930). Contributions to the morphology of *Boerhaavia diffusa* II. *J. Indian bot. Soc.* **9**, 42–61.

Majumdar, G. P. (1941). The collenchyma of *Heracleum sphondylium* L. *Proc. Leeds phil. Soc.* **4**, 25–41.

— (1957). The shoot of higher plants: its morphology and phylogeny. *J. Asiat. Soc., Calcutta* **23**, 39–62.

Malaviya, M. (1962). A study of sclereids in three species of *Nymphaea*. *Proc. Indian Acad. Sci.* **56**, 232–6.

Malpighi, M. (1675, 1679). *Anatome plantarum*. 2 vols. Printed by John Martyn for the Royal Society, London.

Mani, M. S. (1964). *Ecology of plant galls*. Junk, The Hague.

Mankevich, O. I. (1971). Influence de la ploidie sur les caractères anatomiques des plantes décoratives. *Vest. Akad. Navuk B SS R* (V), 88–91. [Russian]

Markgraf, F. (1964). Gedanken zur neueren morphologisch-systematischen Forschung. *Ber. dt. bot. Ges.* **77**, 17–22.

Marsden, M. P. F. and Bailey I. W. (1955). A fourth type of nodal anatomy in dicotyledons, illustrated by *Clerodendron trichotomum* Thunb. *J. Arnold Arbor.* **36**, 1–51.

Martens, P. (1931*a*). Dépouillement cuticulaire et phénomènes osmotiques dans les poils staminaux de *Tradescantia*. *Bull. Soc. r. Bot. Belg.* **64**, 108–11.

— (1931*b*). Recherches sur la cuticule I. Phénomènes cuticulaires et phénomènes osmotiques dans les poils staminaux de *Tradescantia*. *Cellule* **41**, 15–48.

— (1933*a*). Recherches sur la cuticule. II. Dépouillement cuticulaire spontané sur les pétales de *Tradescantia*. *Bull. Soc. Bot. Belg.* **66**, 58–64.

— (1933*b*). Origine et rôle des plissements superficiels sur l'épidermes des pétales floraux. *C. R. Acad. Sci. Paris* **197**, 785–87.

— (1934*a*). Recherches sur la cuticule III: structure, origine et signification du relief cuticulaire. *Protoplasma* **20**, 483–515.

— (1934*b*). Recherches sur la cuticule. IV Le relief cuticulaire et la différenciation épidermique des organes floraux. *Cellule* **43**, 287–320.

— (1934*c*). Nouvelles observations sur la cuticule des épidermes floraux. *C. r. hebd. Séanc. Acad. Sci., Paris* **199**, 309–11.

— (1935). Recherches sur la cuticule V. Différenciation épidermique et cuticulaire chez *Erythraea centaurium*. *Annls Sci. nat.* (ser. 10) **17**, 5–33.

Martin, J. T. and Batt, R. F. (1958). Studies on plant cuticle. I. The waxy coverings of leaves. *Ann. Appld. Biol.* **46**, 375–92.

— and Juniper, B. E. (1970). *The cuticles of plants*. Edward Arnold, Edinburgh.

Martinet, M. (1872). Organes de sécrétion des végétaux. *Annls Sci. nat.* (*Bot. ser.* 5) **14**, 91–232.

Masuda, T. (1933). Studies on the elongation of petioles in some dicotyledons. *Bot. Mag. Tokyo* **47**, 347–70.

Maximov, N. A. (1929). *The plant in relation to water. A study of the physiological basis of drought resistance.* (English translation by R. H. Yapp.) George Allen and Unwin, London.

— (1931). The physiological significance of the xeromorphic structure of plants. *J. Ecol.* **19**, 272–82.

Mayr, F. (1915). Hydropoten an Wasser- und Sumpfpflanzen. *Beih. bot. Zbl.* I, **32**, 278–371.

Meidner, H. and Mansfield, T. A. (1968). *Physiology of stomata*. McGraw-Hill, London.

Melville, R. (1962). A new theory of the Angiosperm flower. I. *Kew Bull.* **16**, 1–50.

— (1976). The terminology of leaf architecture. *Taxon* **25**, 549–61.

Mercer, F. V. and Rathgeber, N. (1962). *Int. Congr. electron Microscopy, Philadelphia* Vol. 2. Academic Press, New York.

Mersky, M. L. (1973). Lower Cretaceous (Potomac group) angiosperm cuticles. *Am. J. Bot.* **60** suppl., 17–18.

Messager, A. (1886). Les glandes du pétiole. *Rev. Hort.* 36–8.

Metcalfe, C. R. (1933). A note on the structure of the phyllodes of *Oxalis Herrerae* R. Knuth and *O. bupleurifolia* St. Hil. *Ann. Bot.* **47**, 355–9.

— (1938*a*). The morphology and mode of development of the axillary tubercles and root tubers of *Ranunculus ficaria. Ann. Bot.* (*N.S.*) **2**, 145–57.

— (1938*b*). Extra-floral nectaries on *Osmanthus* leaves. *Kew Bull.* 254–6.

— (1942). A short history of the Jodrell Laboratory. *Chron. Bot.* 174–6.

— (1946). The systematic anatomy of the vegetative organs of the angiosperms. *Biol. Rev.* **21**, 159–72.

— (1948). The elder tree (*Sambucus nigra* L.) as a source of pith, pegwood and charcoal, with some notes on the structure of the wood. *Kew Bull.* 163–9.

— (1951). The anatomical structure of the Dioncophyllaceae in relation to the taxonomic affinities of the family. *Kew Bull.* 351–68.

— (1953). The anatomical approach to the classification of the flowering plants. *Sci. Prog.* **41**, 42–53.

— (1954). An anatomist's views on Angiosperm classification. *Kew Bull.* 427–40.

— (1959). A vista in plant anatomy. In *Vistas in Botany* **1** (ed. W. B. Turrill) pp. 76–98. Pergamon Press, London.

— (1960). *Anatomy of the Monocotyledons I. Gramineae*. Clarendon Press, Oxford.

— (1961). The anatomical approach to systematics. In *Recent advances in botany* **1**, pp. 146–50. Toronto University Press.

— (1967) [1968]. Some current problems in systematic anatomy. *Phytomorphology* **17**, 128–32.

— (1971). *Anatomy of the Monocotyledons V. Cyperaceae*. Clarendon Press, Oxford.

— (1972) In *Dictionary of scientific biography* (Edited under the auspicies of American Council of Learned Societies) Vol. V, pp. 534–6. Charles Scribners' Sons, New York.

— (1973). Metcalfe and Chalk's Anatomy of the Dicotyledons and its revision. *Taxon* **22**, 659–68.

— (1976). History of the Jodrell Laboratory as a centre for systematic anatomy. In *Wood structure in biological and technological research* (eds. P. Baas, A. J. Bolton, and

D. M. Catling) pp. 1–19. Leiden University Press.
— and Chalk, L. (1950). *Anatomy of the dicotyledons*, 2 vols. Clarendon Press, Oxford.

Mettenius, G. H. (1856). *Filices Horti Botanici Lipsiensis.* Verlag von Leopold Voss, Leipzig.

— (1865). Über die Hymenophyllaceae. *Abh. sächs Akad. Wiss. (Math.-Phys. Kl.)* 7, 403–504.

Meyen, F. J. F. (1837). Über die Epidermis der Gewächse. *Wiegm. Arch.* 3 (1), 221.

Meyer, F. J. (1932). Die Verwandtschaftsbeziehungen der Alismataceen zu den Ranales im Lichte der Anatomie. *Engler's bot. Jahrb.* 65, 53–59.

— (1935). Zur Frage der Funktion der Hydropoten. *Ber. dt. bot. Ges.* 53, 542–6.

— (1962). In *Encyclopaedia of plant anatomy. Das tropische Parenchym* (eds. W. Zimmerman and P. Ozenda). Gebrüder Borntraeger, Berlin.

Meyer, M. (1938). Die submikroskopische Struktur der kutinisierten Zellmembranen. *Protoplasma* 27, 552–8.

Michele, D. W. de and Sharpe, P. J. H. (1974). A parametric analysis of the anatomy and physiology of the stomata. *Agric. Meteorol.* 14, 229–41.

Miehe, H. (1911*a*). Ueber die javanische *Myrmecodia* und die Beziehung zu ihren Ameisen. *Biol. Zbl.* 31, 733–86.

— (1911*b*). Javanische Studien II. Untersuchungen über die javanische *Myrmecodia. Abh. Sächs. Akad. Wiss (Math.-Phys. Kl.)* 32, 312–61.

Millington, W. F. and Fisk, E. L. (1956). Shoot development of *Xanthium pennsylvanicum*: I. The vegetative plant. *Am. J. Bot.* 43, 655–65.

Minden, M. von. (1899). Beiträge zur anatomischen und physiologischen Kenntnis Wasser-secernierender Organe. *Bibl. Bot.* 9 (46), 1–76.

Möbius, M. (1897). Beitrag zur Anatomie der *Ficus*-blätter. *Ber. senckenb. naturf. Ges.* 117–38.

Moeller, J. (1882). *Anatomie der Baumrinden*. Julius Springer, Berlin.

— (1889). *Lehrbuch der Pharmacognosie.* Alfred Holder, Wien.

Mohl, H. von (1842). Über die Cuticula der Gewächse. *Linnaea* 16, 401–16.

— (1845). Ueber das Eindringen der Cuticula in die Spaltöffnungen. *Bot. Ztg.* 3, 1–6.

— (1847). Untersuchung der Frage: Bildet die Cellulose die Grundlage sämmtlicher vegetabilischen Membranen? *Bot. Ztg.* 5, 497–505, 521–9, 545–53.

— (1852). *Principles of the anatomy and physiology of the vegetable cell.* (English translation by Henfrey.) John van Voorst, London.

Molisch, H. (1916). Beiträge zur Mikrochemie der Pflanze 2. Über orangefarbige Hydathoden bei *Ficus javanica. Ber. dt. bot. Ges.* 34, 66–72.

Moll, J. W. and Janssonius, H. H. (1906–36). *Mikrographie des Holzes der auf Java vorkommenden Baumarten.* 6 vols. E. J. Brill, Leiden.

Moncontié, C. (1969). Les stomates des Plantaginacées. *Rev. gén. Bot.* 76, 491–529.

Money, L. L., Bailey, I. W., and Swamy, B. G. L. (1950). The morphology and relationships of the Monimiaceae. *J. Arnold Arbor.* 31, 372–404.

Monod, T. and Schmitt, C. (1968). Contribution à l'étude des pseudo-galles formicaires chez quelques Acacias africains. *Bull. Inst. fond. Afr. noire. A.* 30, 953–1027.

Morini, F. (1886). Contributo all'anatomia ed alla fisiologia dei nettarii estranuziali. *Memorie R. Accad. Sci. Ist. Bologna (ser. 4)* 7, 325–31.

Morley, T. (1953), The genus *Mouriri* (Melastomaceae). *Univ. Calif. Publs Bot.* 26, 223–312.

Mortlock, C. (1952). The structure and development of the hydathodes of *Ranunculus fluitans* Lam. *New Phytol.* 51, 129–38.

Morvillez, F. (1919). Recherches sur l'appareil conducteur foliaire des Rosacées, des Chrysobalanées et des Légumineuses. Thesis, Fac. Sci. Lille.

Moser, H. (1934). Untersuchungen über die Blattstruktur von *Atriplex*-Arten und ihre Beziehungen zur Systematik. *Beih. bot. Zbl. B* 52, 378–88.

Moss, E. H. (1936). The ecology of *Epilobium angustifolium* with particular reference to rings of periderm in the wood. *Am. J. Bot.* 23, 14–20.

— (1940). Interxylary cork in *Artemisia* with reference to its taxonomic significance. *Am. J. Bot.* 27, 762–8.

— and Gorham, A. L. (1953). Interxylary cork and fission of stems and roots. *Phytomorphology* 3, 285–94.

Mouton, J. A. (1970). Architecture de la nervation foliaire. *C. r. 92e. Congr. natl. Soc. Savantes* 3, 165–76.

Mueller, L. E., Carr, P. H., and Loomis, W. E. (1954). The submicroscopic structure of plant surfaces. *Am. J. Bot.* 41, 593–600.

Mueller, S. (1966*a*). The taxonomic significance of cuticular patterns within the genus *Vaccinium* (Ericaceae). *Am. J. Bot.* 53, 633.

— (1966*b*). Cuticular patterns as a taxonomic tool within the genus *Vaccinium. Ass. southeast. biol. Bull.* 13, 42.

Mühldorf, A. (1922). Ein neuer xeromorpher Spaltöffnungsapparat bei Dicotylen. *Öst. bot. Z.* 71, 50–4.

Mullan, D. P. (1932–33). Observations on the biology and physiological anatomy of some Indian halophytes. *J. Indian bot. Soc.* 11, 103–18, 285–302; *J. Indian bot. Soc.* 12, 165–82, 235–53.

Müller, C. (1890). Ein Beitrag zur Kenntniß der Formen des Collenchyms. *Ber. dt. bot. Ges.* 8, 150–66.

Munting, A. (1672). *Waare oeffening der platen.* Amsterdam. [Cited from Ivanoff, S.S. 1963.]

— (1696). *Naauwkeurige beschryving der aardgewassen.* Leyden.

Mylius, G. (1912). Das Polyderm. *Ber. dt. bot. Ges.* 30, 363–5.

— (1913). Das Polyderm. Eine vergleichende Untersuchung über die physiologischen Scheiden Polyderm, Periderm und Endodermis. *Bibl. Bot.* 18, Heft 79, 1–119.

Nägeli, C. and Cramer, C. (1858). *Pflanzenphysiologische Untersuchungen.* Heft 2 by Nägeli, C. *Die Stärkekörner.* Zurich.

Nakazawa, K. (1956). The vascular course of Piperales. I, Chloranthaceae. *Jap. J. Bot.* 15, 199–207.

Napp-Zinn, K. (1973, 1974). *Anatomie des Blattes II. Blattanatomie der Angiospermen. A. Entwicklungsgeschichtliche und topographische Anatome des Angiospermenblattes.* 2 vols. In *Handbuch der Pflanzenanatomie.* VIII, 2A. Gebruder Borntraeger, Berlin.

Nast, C. G. and Bailey, I. W. (1946). Morphology of *Euptelea* and comparison with *Trochodendron. J. Arnold Arbor.* 27, 186–92.

Nathorst, A. G. (1907–12). Paläeobotanische Mitteilungen, 1–11. *K. Svensk. Vetenskapsakad. Handl. Ny Fiöligd* 42,

(5)–48 (2).

Neese, P. (1916). Zur Kenntnis der Struktur der Niederblätter und Hochblätter einiger Laubhölzer. *Flora* **109**, 144–87.

Neischlova, E. and Kaplan, J. (1975). Types de stomates et complexes stomatiques. *Biol. Ceskol.* **30**, 315–23.

Nelson, P. E. and Wilhelm, S. (1957). Some anatomic aspects of the strawberry root. *Hilgardia* **26**, 631–42.

Nesmeyanova, A. B. (1968). Quelques données sur les nectaires de *Hibiscus cannabinus* L. *Uzbek. biol. Zh.* **12** (1), 38–40. [Russian, English summary.]

Nestler, A. (1896). Untersuchungen über die Ausscheidung von Wassertropfen an den Blättern. *Sber. Akad. Wiss. Wien* (*Math.-Naturw.*) **55**, 521–51.

— (1899). Die Sekrettropfen an den Laubblättern von *Phaseolus multiflorus* Willd. und den Malvaceen. *Ber. dt. bot. Ges.* **17**, 332–7.

Netolitzky, F. (1926). *Anatomie der Angiospermen Samen.* In *Handbuch der Pflanzenanatomie* II (ed. K. Linsbauer) Abt. 2, Teil 2, Band X. Gebrüder Borntraeger, Berlin.

— (1932). In *Handbuch der Pflanzenanatomie* (ed. K. Linsbauer) Abt. 1. Teil 2. *Hautgewebe.* Band IV. *Die Pflanzenhaare.* Gebrüder Borntraeger, Berlin.

Neumann-Reichart, E. (1917). Anatomisch-physiologische Untersuchungen über Wasserspalten. *Beitr. all. Bot.* **1**, 301–40.

Nieuwenhuis von Uexküll-Güldenband, M. (1907). Extraflorale Zuckerausscheidungen und Ameisenschutz. *Annls Jard. bot. Buitenz.* (ser. 2) **6**, 195–328.

— (1914). Sekretionskanäle in den Cuticularschichten der extrafloralen Nektarien. *Rec. Trav. bot. néerl.* **11**, 291–311.

Odell, M. E. (1932). The determination of fossil angiosperms by the characteristics of their vegetative organs. *Ann. Bot.* **46**, 941–63.

Odhelius, J. L. (1774). Om naturlig cristalliferadt Socker (*in nectario Impatientis Balsaminae*). *Kongl. Vetenskaps Academiens Handlingar* **35**, 359–60.

Ogura, Y. (1937). Disarticulation of the branches in *Bladhia* (Myrsinaceae). *Bot. Mag. Tokyo* **51**, 158–67.

Ojehomon, O. O. (1968). The development of the inflorescence and extrafloral nectaries of *Vigna unguiculata* (L.) Walp. *J. West Afr. Sci. Assoc.* **13**, 93–110.

Oliver, F. W. (1935). Dukinfield Henry Scott 1854–1934. *Ann. Bot.* **49**, 823–40.

O'Neill, T. B. (1961). Primary vascular organization of *Lupinus* shoot. *Bot. Gaz.* **123**, 1–9.

Ozenda, P. (1949). Recherches sur les dicotylédones apocarpiques. Contribution à l'étude des angiospermes dites primitives. *Publ. Lab. Biol. Ecole norm supér.* Masson and Cie.

Öztig, Ö. F. (1940). Beiträge zur Kenntnis des Baues der Blattepidermis bei den *Mesembrianthemum* im besonderen den extrem xeromorphen Arten. *Flora, N. S.* **134**, 105–44.

Paganelli Cappelletti, E. M. (1975). Studio morfologico al microscopio elettronico a scansione di foglie di *Atropa belladonna* L. *In. Bot. Ital.* **7**, 24–5.

Paliwal, G. S. (1967). Ontogeny of stomata in some Cruciferae. *Can. J. Bot.* **45**, 495–500.

— (1969). Stomata in certain angiosperms: their structure, ontogeny and systematic value. In *Recent advances in the anatomy of tropical seed plants*, pp. 63–73. Hindustan Publishing Corp, Delhi, India.

—— (1972). Stomata. *Vistas in Plant Sci.* **2**, 23–48.

— and Bhandari, N. N. (1962). Stomatal development in some Magnoliaceae. *Phytomorphology* **12**, 409–12.

Pant, D. D. (1965). On the ontogeny of stomata and other homologous structures. *Plant Sci. Ser.* (*Allahabad*) **1**, 1–24.

— and Gupta, K. L. (1966). Development of stomata and foliar structure of some Magnoliaceae. *J. Linn. Soc.* (*Bot.*) **59**, 265–77.

— and Mehra, B. (1962). Path of bundles in the stem of *Bougainvillaea. Curr. Sci.* **31**, 295–6.

— — (1964). Nodal anatomy in retrospect. *Phytomorphology* **14**, 384–87.

— — (1965). Ontogeny of stomata in some Rubiaceae. *Phytomorphology.* **15**, 300–10.

Paoli, G. (1929). Strane abitazioni di una formica su acacie della Somalia. *Riv. Coleott. Ital.* (III Maggio) **5**, 474–85.

— (1930). Contributo allo studio dei rapporti fra le Acacie et le Formice. *Mem. Soc. Entomol. Ital.* **9**, 131–95.

Parameswaran, N. (1971). Über die Struktur der tropischen Baumrinde und ihre Verwertungsmöglichkeiten. *Forstarchiv.* **41**, 193–8.

— and Liese, W. (1970). Mikroskopie der Rinde tropischer Holzarten. In *Handbuch der Mikroskopie in der Technik* (2nd edn.) (ed. Freund, H.) Vol. V, 1, pp. 227–306. Umschau Verlag. Frankfurt.

Parija, P. and Samal, K. (1936) Extrafloral nectaries in *Tecoma capensis* Lindl. *J. Indian Bot. Soc.* **15**, 241–46.

Parkin, J. (1904). The extrafloral nectaries of *Hevea brasiliensis* Muell. Arg. (The para rubber tree). An example of bud scales serving as nectaries. *Ann. Bot.* **18**, 217–26.

— (1924). Stomata and phylogeny. *Ann. Bot.* **38**, 1–16.

Pataky, S. (1969). Comparison of the leaf epidermis of *Salix alba* L. in different regions of the leafy crown. *Acta biol. Szeged* **15**, 29–36.

Pate, J. S. and Gunning, B. E. S. (1969). Vascular transfer cells in angiosperm leaves. A taxonomic and morphological survey. *Protoplasma* **68**, 135–56.

— — (1972). Transfer cells. *Ann. Rev. Plant Physiol.* **23**, 173–96.

Patel, J. D. (1978). How should we interpret and distinguish subsidiary cells? *Bot. J. Linn. Soc.* **77**, 65–72.

Patel, R. C. and Inamdar, J. A. (1971). Structure and ontogeny of stomata in some Polemoniales. *Ann. Bot.* **35**, 389–409.

— — (1974). Studies in the trichomes and nectaries of some Gentianales. In *Biology of the land plants* (ed. Puri, V. et al.), pp. 328–40. Sarita Prakishan Meerut, India.

Paula, J. E. de, (1974). Estomatos de Guttiferae: estudo morfologico, dimensional e quantitativo. *Acta amazonica* **4** (3) 23–40. [Portuguese, with English Summary.]

Payer, J. B. (1857). *Traité d'organogénie végétale comparée de la fleur.* 2 Vols. Paris.

Payne, W. W. (1969). A quick method for clearing leaves. *Ward's Bull.* **61**, 4–5.

— (1970). Helicocytic and allelocytic stomata; unrecognized patterns in the Dicotyledoneae. *Am. J. Bot.* **57**, 140–7.

Pazourek, J. (1970). The effect of light intensity on the stomatal frequency in leaves of *Iris hollandica* hort., var. Wedgwood. *Biol. Plant.* **12**, 208–15.

Pearson, R. S. and Brown, H. P. (1932). *Commercial timbers of India*, 2 vols. Government of India Central Publications Branch, Calcutta.

Pease, V. A. (1917). Duration of leaves in evergreens. *Am. J. Bot.* **4**, 145–60.

Penzig, O. and Chiabrera, C. (1903). Contributo alla conoscenza delle piante acarofile. *Malpighia* **17**, 429–87.

Percival, M. S. (1961). Types of nectar in Angiosperms. *New Phytol.* **60**, 235–81.

Perrin, A. (1970). Organisation infrastructurale en rapport avec les processus de sécrétion, des poils glandulaires (trichome-hydathodes) de *Phaseolus multiflorus*. *C. R. Acad. Sci. Paris, D.* **270**, 1984–7.

— (1971). Présence de 'cellules de transfert' au sein de l'épithème de quelques hydathodes. *Z. Pflanzenphysiol.* **65**, 39–51.

— (1972*a*). Organisation et nature de l'inclusion cristalline des organites du type 'crystal-containing body' rencontrés dans les cellules de l'épitheme des hydathodes de *Cichorium intybus* L. et *Taraxacum officinale* Weber. *Protoplasma* **74**, 213–225.

— (1972*b*). Contribution à l'étude de l'organisation et du fonctionnement des hydathodes; recherches anatomiques, ultrastructurales et physiologiques. Thesis. Univ. Claude Bernard, Lyon, France.

— and Zandonella, P. (1971). Présence d'invaginations nucléaires dans les cellules de quelques nectaires floraux et hydathodes. *Planta* **96**, 136–44.

Perrot, É. (1897*a*). Anatomie comparée des Gentianacées aquatiques (*Menyanthes* Griseb.). *Bull. Soc. bot. Fr.* **44**, 340–53.

— (1897*b*). Sur une particularité de structure de l'épiderme inférieur de la feuille chez certaines Gentianées aquatiques. *J. de Bot.* **11**, 195–201.

— (1899). Anatomie comparée des Gentianacées. *Ann. Sci. nat. bot.* ser. 8, **7**, 105–292.

Petaj, N. V. (1916). Ekstrafloralni nektarijii lišéu pajasena (*Ailanthus glandulosa* Desf.). *Rad. Jugoslavenske Akademije zanestii i umjetnostii*. **215**, 59–81.

Peters, T. (1912). Zur Anatomie des Phyllopodiums von *Acacia*. *Inaug. Diss*. Kiel (Braunschweig.)

Petit, L. (1886). Sur l'importance taxonomique du pétiole. *C. R. Acad. Sci., Paris* **103**, 767–9.

— (1887). Le pétiole des dicotylédones au point de vue de l'anatomie comparée et de la taxinomie. *Mem. Soc. Sci. Phys. Nat. Bordeaux* III **3**, 217–404.

— (1889). Nouvelles recherches sur le pétiole des phanérogames. *Acta Soc. Linn. Bordeaux* **43**, 11–60.

Philipson, W. R. (1948). Studies in the development of the inflorescence. *Ann. Bot.* (N.S.) **12**, 147–56.

— (1949). The ontogeny of the shoot apex in dicotyledons. *Biol. Rev.* **24**, 21–50.

— (1954). Organization of the shoot apex in dicotyledons. *Phytomorphology* **4**, 70–5.

— (1968). The abaxial stipules of *Plagianthus divaricatus* J. R. and G. Forst. *New Zealand J. Bot.* **6**, 518–21.

— and Balfour, E. E. (1963). Vascular patterns in dicotyledons. *Bot. Rev.* **29**, 382–404.

— and Philipson, M. N. (1968). Diverse nodal types in *Rhododendron*. *J. Arnold Arbor.* **49**, 193–224.

Pierre, L. (1896). Plantes du Gabon. *Bull. Mens. Soc. Linn. Paris* **2**, 1249–56.

Pijl, L. van der (1951). On the morphology of some tropical plants. *Gloriosa, Bougainvillea, Honckenya, Rottboellia*. *Phytomorphology* **1**, 185–88.

— (1955). Some remarks on myrmecophytes. *Phytomorphology* **5**, 190–200.

Pisek, A., Knapp, H. and Ditterstorfer, J. (1970). Maximale Öffnungsweite und Bau der Stomata, mit Angaben über ihre Grösse und Zahl. *Flora* **159**, 459–79.

Popham, R. A. and Chan, A. P. (1950). Zonation in the vegetative stem tip of *Chrysanthemum morifolium* Bailey. *Am. J. Bot.* **37**, 476–84.

Porsch, O. (1903). Zur Kenntnis des Spaltöffnungsapparates submerser Pflanzenteile. *Sber. Akad. Wiss. Wien math.-nat. Kl. 1*, **112**, 97–138.

Post, D. M. (1958). Studies in Gentianaceae. I. Nodal anatomy of *Frasera* and *Swertia perennis*. *Bot. Gaz.* **120**, 1–14.

Poulsen, V. A. (1875). Om nogle trikomer og nektarier. *Vidensk. Meddr Dansk Naturh. Foren. Kjøbenhavn* No. 16–19, 242–83 [French summary].

— (1877). Das extraflorale Nektarium bei *Batatas edulis*. *Bot Zeit.* **35**, 780–782.

— (1918). Planteanatomiske Bidrag II. iii Det extraflorale Nektarium ·hos *Carapa guyanensis* Aubl. *Vidensk. Meddr Dansk Naturh. Foren. Kjøbenhavn* **69**, 329–343.

Prantl, K. (1883). Studien über Wachstum, Verzweigung und Nervatur der Laubblätter, insbesondere der Dicotylen. *Ber. dtsch. bot. Ges.* **1**, 280–88.

Preece, T. F. and Dickinson, C. H. (1971). *Ecology of leaf surface micro-organisms*. Academic Press, London.

Preston R. D., and Duckworth, R. B. (1946). The fine structure of the walls of collenchyma cells in *Petasites vulgaris* L. *Proc. Leeds Phil. Lit. Soc. Sci. Sec.* **4**, 345–51.

Priestley, J. H. (1921). Suberin and cutin. *New Phytol.* **20**, 17–29.

— (1922). Physiological studies in plant anatomy I. Introduction. *New Phytol.* **21**, 58–61.

— (1924). The fundamental fat metabolism of the plant. *New Phytol.* **23**, 1–19.

— (1926). Light and growth II. On the anatomy of etiolated plants. *New Phytol.* **25**, 145–70.

— (1943). The cuticle in angiosperms. *Bot. Rev.* **9**, 593–616.

— and Armstead, D. (1922). Physiological studies in plant anatomy. II. The physiological relation of the surrounding tissue to the xylem and its contents. *New Phytol.* **21**, 62–80.

— and Ewing, J. (1923). Physiological studies in plant anatomy VI, Etiolation. *New Phytol.* **22**, 30–43.

— and North, E. E. (1922). Physiological studies in plant anatomy. III. The structure of the endodermis in relation to its function. *New Phytol.* **21**, 113–39.

Pyykkö, M. (1966). The leaf anatomy of East Patagonian xerophytic plants. *Ann. bot. fennici* **3**, 453–632.

Rachmilevitz, T. and Fahn, A. (1975). The floral nectary of *Tropaeolum majus* L. The nature of the secretory cells and the manner of nectar secretion. *Ann. Bot.* **39**, 721–8.

Raciborski, M. (1894). Beiträge zur Kenntnis der Cabombaceae und Nymphaeaceae. *Flora* **79**, 92–108.

Radlkofer, L. (1896). Sapindaceae in Engler and Prantl's *Die natürl. Pflanzenfam.* ii, 5.

— (1934). Sapindaceae in Engler's *Das Pflanzenreich* iv, 165.

Ragonese, A. M. (1960). Ontogenia de los distintos tipos de tricomas de *Hibiscus rosa-sinensis* L. (Malvaceae). *Darwiniana* **12**, 58–66.

— (1973). Systematic anatomical characters of the leaves of *Dimorphandra* and *Mora*, Leguminosae, Caesalpinoidae. *Bot. J. Linn. Soc.* **67**, 255–74.

Rajagopal, T. and Ramayya, N. (1968). Occurrence of idioblastic 'cell-sacs' in the leaf epidermis of *Cleome aspera* Koen. ex DC. with observations on their taxonomic significance, structure and development. *Curr. Sci.* **37**, 260–2.

Rajkowski, S. (1934). Badania histologiczne i morfologiczne nad środskárnią w łodygach roślin kwriatowych (Histologische und morphologische Untersuchungen über die Endodermis in den Stengeln der Blütenpflanzen). *Acta Soc. Bot. Pol.* **11**, 19–50.

Ramayya, N. (1972). Classification and phylogeny of the trichomes of angiosperms. In *Research trends in plant anatomy* (eds. A. K. M. Ghouse and M. Yunus) pp. 91–102. Tata McGraw-Hill, Bombay.

—— and Bahadur, B. (1968). Morphology of the 'squamellae' in the light of their ontogeny. *Curr. Sci.* **37**, 520–2.

—— and Rajagopal, T. (1968). Foliar epidermis as taxonomic aid in 'The Flora of Hyderabad' I. Portulacaceae and Aizoaceae. *J. Osmania Univ. (sci.)*, Golden Jubilee vol., 147–60.

Rangaswamy Ayyangar, K. (1962). Observations on the effect of sound waves on the stomatal frequency and index of *Urginea indica* Kunth. *Proc. 49th. Indian Sci. Congr.* (Cuttak) Vol. III, p. 293.

Rao, A. N. (1963). Reticulate cuticle on leaf epidermis in *Hevea brasiliensis* Muell. *Nature, Lond.* **197**, 1125–6.

—— (1971). Morphology and morphogenesis of foliar sclereids in *Aegiceras corniculatum. Israel J. Bot.* **20**, 124–32.

Rao, A. R. and Malaviya, M. (1962). The distribution, structure and ontogeny of sclereids in *Dendropthoe falcata* (L. f.) Ettings. *Proc. Indian Acad. Sci. B* **55**, 239–43.

—— and Rao, C. K. (1972). Root sclereids of *Syzygium cumini* (L.) Skeels. *Proc. Indian Acad. Sci.* **75**, (Sec. B), 177–90.

Rao, K. R. and Purkayastha, S. K. (1972). *Indian Woods* Vol, III. *Leguminosae to Combretaceae.* Manager of Publications, Delhi.

Rao, L. N. (1926). A short note on the extra-floral nectaries in *Spathodea stipulata. J. Indian bot. Soc.* **5**, 113–16.

Rao, T. A. (1947). On the occurrence of sclerosed palisade cells in the leaf of *Nyctarthes arbor tristis* L. *Curr. Sci.* **16**, 122–3.

—— (1949). Foliar sclereids in the Oleaceae. 1. On the occurrence of sac-like spicular cells in the leaf of *Schrebera swietennoides.* Roxb. *J. Indian bot. Soc.* **28**, 251–4.

—— (1950a). Foliar sclereids in the Oleaceae. 2. Occurrence of terminal foliar sclereids in some species of the genus *Linociera.* Swartz. *J. Indian bot. Soc.* **29**, 220–4.

—— (1950b). Studies in the foliar sclereids in dicotyledons. (2) On sclereids in species of *Leucospermum* (Proteaceae), *Mimusops* (Sapotaceae) and *Memecylon* (Melastomaceae). *J. Univ. Bombay* **19** (3), 25–31.

—— (1951a). Studies on foliar sclereids. A preliminary survey. *J. Indian bot. Soc.* **30**, 28–39.

—— (1951b). Studies on foliar sclereids in Dicotyledons V. Structure of the terminal sclereids in the leaf of *Memecylon heyneanum.* Benth. *Proc. Indian Acad. Sci. B* **34**, 329–34.

—— (1951c). Studies on foliar sclereids in Dicotyledons I. Structure and ontogeny of sclereids in the leaf of *Diospyros discolor* Willd. *Proc. Indian Acad Sci. B* **34**, 92–8.

—— (1957). Comparative morphology and ontogeny of foliar sclereids in seed plants. II. *Linociera. Proc. nat. Inst. Sci. India* **23B**, 152–64.

—— and Bhupal, O. P. (1973). Typology of sclereids. *Proc. Indian Acad. Sci. B,* **77**, 41–55.

—— and Dakshni, K. M. M. (1963). Systematics of *Memecylon* (*Melastomaceae*) – a preliminary survey based on the sclereid morphology. *Proc. Indian Acad. Sci. B* **58**, 28–35.

—— and Kulkarni, G. Y. (1952). Foliar sclereids in the Oleaceae III Ontogeny of the sclereids in 2 species of the genus

Olea L. *J. Univ. Bombay* **20** (5), 52–7.

Raschke, K. (1975). Stomatal action. *Ann. Rev. Plant Physiol.* **26**, 309–40.

Rathay, E. (1889). Ueber extrafloralen Nektarien. *Verh. O. zool. bot. Vereines* **39**, 14–21.

Raunkiaer, C. (1934). *The life forms of plants and statistical plant geography.* (Raunkiaer's collected papers.) Clarendon Press, Oxford.

Rea, M. W. (1921). Stomata and hydathodes in *Campanula rotundifolia* L. and their relation to environment. *New Phytol.,* **20**, 56–72.

Reams, W. M. (1953). The occurrence and ontogeny of hydathodes in *Hygrophila polyspermata* T. Anders. *New Phytol.* **52**, 8–13.

Record, S. J. (1930–45). Notes concerning the formation and early history of the International Association of Wood Anatomists. See *Trop. Woods* **22**, 1; **24**, 1–5; **27**, 20–3; **29**, 29–31; **30**, 41–3; **32**, 22; **33**, 29–30; **36**, 1–12; **37**, 46; **41**, 16; **84**, 25.

—— (1934a). Rôle of wood anatomy in taxonomy. *Trop. Woods* **37**, 1–9.

—— (1934b). *Identification of the timbers of temperate North America.* John Wiley, New York; Chapman and Hall, London.

—— (1936). Classifications of various anatomical features of dicotyledonous woods. *Trop. Woods* **47**, 12–27.

—— and Hess, R. W. (1943). *Timbers of the New World.* New Haven.

Reed, E. L. (1917). Leaf nectaries of *Gossypium. Bot. Gaz.* **63**, 229–31.

—— (1923). Extrafloral glands of *Ricinus communis. Bot. Gaz.* **76**, 102–6.

Regel, E. (1843). Beobachtungen über den Ursprung und Zweck der Stipeln. *Linnaea* **17**, 193–234.

Rehder, A. (1911–18). *The Bradley Bibliography: a guide to the literature of woody plants.* Cambridge, Mass.

Reinke, J. (1875). Beiträge zur Anatomie der an Laubblättern, besonders an den Zähnen derselben vorkommenden Sekretionsorgane. *Jb. wiss. Bot.* **10**, 117–78.

Renaudin, S. (1966). Sur les glandes de *Lathraea clandestina* L. *Bull. Soc. bot. Fr.* **113**, 379–85.

—— and Garrigues, R. (1967). Sur l'ultrastructure des glandes en bouclier de *Lathraea clandestina* L. et leur rôle physiologique. *C. r. hebd. Séanc. Acad. Sci., Paris* **264**, 1984–7.

Renner, O. (1907). Beiträge zur Anatomie und Systematik der Artocarpeen und Conocephaleen, insbesondere der Gattung *Ficus. Bot. Jb.* **39**, 319–448.

Rentschler, I. (1974). Elektronenmikroskopische Untersuchungen an wachsüberdeckten Spaltöffnungen. *Planta* **117**, 153–61.

Rettig, E. (1904). Ameisenpflanzen - Pflanzenameisen. *Beih. bot. Zbl.* **17**, 89–122.

Reule, H. (1937). Vergleichendanatomische Untersuchungen in der Gattung *Mesembryanthemum* L. *Flora (N.S.)* **31**, 400–24.

Rickson, F. R. (1969). Developmental aspects of the shoot apex, leaf, and Beltian bodies of *Acacia cornigera. Am. J. Bot.* **56**, 195–200.

—— (1971). Glycogen plastids in Müllerian body cells of *Cecropia peltata* – a higher green plant. *Science* **173**, 344–7.

Riede, W. (1920–1). Untersuchungen über Wasserpflanzen. *Flora* **114**, 1–118.

Rippel, A. (1919). Der Einfluss der Bodentrockenheit auf

den anatomischen Bau der Pflanzen, insbesondere von *Sinapis alba* L. und die sich daraus ergebenden physiologischen und entwicklungsgeschichtlichen Fragen. *Beih. bot. Zbl.* 36 (1), 187–260.

Rocchetti, R. (1905). Richerche sugli acarodomazi. *Contrzioni Biol. veg.* IV fasc. I, 7–36.

Roe, K. E. (1971). Terminology of hairs in the genus *Solanum*. *Taxon* 20, 501–8.

Roelofsen, P. A. (1952). On the submicroscopic structure of cuticular cell walls. *Acta bot. neerl.* 1, 99–114.

— (1959). The plant cell-wall. In *Handbuch Pflanzenanat*, III/4. Gebrüder Borntraeger, Berlin.

Roland, J. C. (1961). Compléments relatifs à la connaissance des stéréomes. I. Etude des tissus de soutien de la feuille de l'Olivier: *Olea europea* L. (Oléacées). *Bull. Soc. Bot. Fr.* 108, 393–413.

— (1964). Infrastructure des membranes du collenchyme. *C. r. hebd. Séanc. Acad. Sci. Paris* 259, 4331–4.

— (1965a). Edification et infrastructure de la membrane collenchymateuse. Son remaniement lors de la sclérification. *C. r. hebd. Séanc. Acad. Sci. Paris* 260, 950–3.

— (1965b). Premières observations sur l'infrastructure du protoplasme au cours de la mise en place et de la différenciation des collocytes de *Sambucus nigra* L. *C. r. hebd. Séanc. Acad. Sci., Paris* 260, 2293–6.

— (1967). Recherches en microscopie photonique et en microscopie électronique sur l'origine et la différenciation des cellules du collenchyme. *Annls Sci. nat. Bot. (ser. 12)* 8, 141–214.

Rollins, R. C. (1941). Monographic study of *Arabis* in Western North America. *Rhodora* 43, 289–325, 348–411, 425–81.

— (1944). Evidence of natural hybridity between guayule (*Parthenium argentatum*) and mariola (*Parthenium incanum*). *Am. J. Bot.* 31, 91–9.

— (1958). The genetic evaluation of a taxonomic character in *Dithyrea* (Cruciferae). *Rhodora* 60, 145–52.

— and Shaw, E. A. (1973). *The genus* Lesquerella (*Cruciferae*) *in North America*. Harvard University Press, Cambridge.

Romanovich, E. A. (1960). Obsobennosti anatomicheskago stroeniya epidermisa lista u predstavitelei Solanaceae. *Bot. Zhur. SSSR* 45, 259–66.

Roon, A. C. de (1967). Foliar sclereids in the Marcgraviaceae. *Acta bot. neerl.* 15, 585–623.

Ross, H. and Hedicke, H. (1927). *Die Pflanzengallen (Cecidien Mittel und Nordeuropas)*. Jena. (See especially pp. 60–2.)

— and Suessenguth, K. (1925–6). Das Apikalorgan der Blätter von *Lafoënsia*. *Flora* 120, 1–18.

Rost, T. L. (1969) [1970]. Vascular pattern and hydathodes in leaves of *Crassula argentea* (Crassulaceae). *Bot. Gaz.* 130, 267–70.

Roth, I. (1949). Zur Entwicklungsgeschichte des Blattes, mit besonderer Berücksichtigung von Stipular- und Ligularbildungen. *Planta* 37, 299–336.

— (1952). Beiträge zur Entwicklungsgeschichte der Schildblätter. *Planta* 40, 350–76.

— (1957a). Zur Histogenese der dorsalen 'Ligula' von *Thalictrum*. *Ost. bot. Z.* 104, 165–72.

— (1957b). Das Dorsalmeristem der Deckblätter von *Beta trigyna*. *Flora* 144, 635–46.

— (1958). Über die Entstehung des Kollenchyms von *Beta trigyna*. *Öst. bot. Z.* 105, 88–101.

— (1974a). Anatomia de las hojas de plantas de los páramos venezolanos; 1. *Hinterhubera imbricata* (Compositae). *Acta bot. venez.* 9, 381–98. [English summary].

— (1974b). Morfologia, anatomia y desarrollo de la hoja pinnada y de las glándulas laminales en *Passiflora* (Passifloraceae). *Acta bot. venez.* 9, 363–80. [English summary].

— and Clausnitzer, I. (1972). Desarrollo de los hidatodos en *Sedum argenteum*. *Acta bot. venez.* 7, 207–17.

Rowson, J. M. (1943a). The significance of the stomatal index as a differential character. I. A statistical investigation of the stomatal indices of *Senna* leaflets. *Q. Jl Pharm. Pharmac.* 16, 24–31.

— (1943b). The significance of the stomatal index as a differential character. II. The identification of the leaves of English and of Indian Belladonna: the characterisation of *Coca* B.P.C. *Q. Jl Pharm. Pharmac.* 16, 255–64.

— (1946). The significance of the stomatal index as a differential character. III. Studies on the genera *Atropa, Datura, Digitalis, Phytolacca* and in polyploid leaves. *Q. Jl Pharm. Pharmac.* 19, 136–43.

Royen, P. van and Steenis, C. G. G. J. van (1952). *Eriandra*, new genus from New Guinea. *J. Arnold Arbor.* 33, 91–5.

Ryder, V. L. (1954). On the morphology of leaves. *Bot. Rev.* 20, 263–76.

Sabnis, T. S. (1919–21). The physiological anatomy of the plants of the Indian desert. A series of short articles in *J. Ind. Bot. Soc.*, 1–2.

Sachs, Julius von (1882). *Textbook of botany, morphological and physiological* (2nd edn.) (Transl. and annotated by A. W. Bennett & W. T. Thiselton-Dyer.) Oxford.

— (1906). *History of botany, 1530–1860*, pp. 228–41. (Transl. Henry E. F. Garnsey, reviewed by Isaac Bayley Balfour.) Oxford.

Sagromsky, H. (1949). Weitere Beobachtungen zur Bildung des Spaltöffnungsmusters in der Blattepidermis. Zur Frage der Gruppenbildung. *Zeitschr. Naturforschung* 4b, 360–7.

Saha, B. (1952). Morphology of the leaves of *Phyllarthron commorense* DC. *Bull. bot. Soc. Bengal* 6, 25–31.

Sahasrabudhe, S. and Stace, C. A. (1974). Developmental and structural variation in the trichomes and stomata of some Gesneriaceae. *New Botanist (India)*. 1, 46–62.

Sainte-Hilaire, A. de (1847). *Leçons de Botanique*, Paris.

Salisbury, E. J. (1909). The extrafloral nectaries of the genus *Polygonum*. *Ann. Bot.* 23, 229–41.

— (1927). On the causes and ecological significance of stomatal frequency, with special reference to the woodland flora. *Phil. Trans. R. Soc.* 216B, 1–65.

— (1932). The interrelations of soil, climate and organism, and the use of stomatal frequency as an integrating index of the water relations of the plant. *Beih. Bot. Zbl.* 99, 402–20. Festschrift O. Drude.

Samantarai, B. and Kabi, T. (1953). Secondary growth in petioles and the partial shoot theory of the leaf. *Nature, Lond.* 172, 37.

Sampson, F. B. and McLean, J. (1965). A note on the occurrence of domatia on the underside of leaves in New Zealand plants. *New Zealand J. Bot.* 3, 104–112.

Sargent, C. (1976a). *In situ* assembly of cuticular wax. *Planta* 129, 123–26.

— (1976b). The occurrence of a secondary cuticle in *Libertia elegans* (Iridaceae). *Ann. Bot.* 40, 355–59.

— (1976c). Studies on the ultrastructure and development of the plant cuticle. Thesis, University of London.

— and Gay, J. L. (1977). Barley epidermal apoplast structure and modification by powdery mildew contact. *Physiol. Plant Pathol.* 11, 195–205.

Sass, J. E. (1958). *Botanical Microtechnique*. (3rd ed.) Iowa State Univ. Press. Ames.

Sastre, C. (1971). Recherche sur les Ochnacées. V. Essai de taxonomie numérique et schéma évolutif du genre *Sauvagesia* L. *Sellowia* 23, 9–44.

— and Guédès, M. (1974). Les hydathodes à épithème chlorophyllien de *Sauvagesia erecta* L. (Ochnacées). *C. R. Acad. Sci. Paris* 279, 49–52.

Satter and Galston, A. W. (1973). Leaf movements: Rosetta stone of plant behaviour? *Bioscience* 23, 407–16.

Sattler, R. L. (1966) [1967]. Towards a more adequate approach to comparative morphology. *Phytomorphology* 16, 417–29.

Sax, K. and Sax, H. J. (1937). Stomatal size and distribution in diploid and polydiploid plants. *J. Arnold Arbor.* 18, 164–72.

Scherffel, A. (1928). Die Hydathoden von *Lathraea squamaria* L. und deren epiphytisches Bakterium *Microbacterium lathreae* Mihi. *Mag. Tud. Akad Math. Termeszett Ertesito*.

Schieferstein, R. H. and Loomis, W. E. (1959). Development of the cuticular layers in angiosperm leaves. *Am. J. Bot.* 46, 625–35.

Schilling, A. J. (1894). Anatomisch-biologische Untersuchungen über die Schleimbildung der Wasserpflanzen. *Flora* 78, 281–360.

Schimper, A. F. W. (1903). *Plant geography upon a physiological basis.* (Transl. W. R. Fisher.) Revised and edited by Percy Groom and Isaac Bailey Balfour. Clarendon Press, Oxford.

Schittengruber, B. (1953a). Stomata auf weissen Blattflecken. *Öst. bot. Z.* 100, 652–6.

— (1953b–54). Die Stomataverteilung an Blättern von *Pulmonaria*-Arten. *Phyton (Austria)* 5, 128–32.

Schleiden, M. J. (1846). Ueber die fossilen Pflanzenreste des Jenaischen Muschelkalks. Schmied & Schleid. *Geognost. Verhält. Saalth. Jena* 2.

— (1861). *Grundzuge der wissenschaftlichen Botanik.* Vol. IV. W. Engelmann, Leipzig.

Schmidt, A. (1924). Histologische Studien an phanerogamen Vegetationspunkten. *Bot. Arch.* 8, 345–404.

Schmidt, H. (1930). Zur Function der Hydathoden von *Saxifraga*. *Planta* 10, 314–44.

Schmidt, L. (1831). Beobachtungen über die Ausscheidung von Flüssigkeit aus der Spitze des Blattes des *Arum colocasia*. *Linnaea* 6, 65–75.

Schmucker, T. and Linnemann, G. (1951). Geschichte der Anatomie des Holzes. In Hugo Freund's *Handbuch der Mikroskopie in der Technik*. Band V. Teil 1. Umschau Verlag, Frankfurt am Main.

Schnell, R. (1960). Sur une galle foliaire d'un *Wendlandia* (Rubiacée) du Cambodge en rapport avec la question des domaties. *J. Agr. trop. bot. appl.*, 7, 539–59.

— (1963a). Organes marginaux et organes portés par la surface du limbe. *Mém. Soc. bot. Fr.* 104–8.

— (1963b). Sur une structure anormale de *Casuarina equisetifolia* L. et la question des 'halomorphoses'. *Bull Inst. fr. Afr. noir* 25A, 301–6.

— (1966a). Remarques morphologiques sur les 'myrmécophytes'. *Mém. Soc. bot. Fr.* 121–32.

— (1966b). Contribution a l'étude des genres guyanoamazoniens *Tococa* Aubl. et *Maieta* Aubl. (Mélastomacées) et de leurs poches foliaires. *Adansonia* (ser 2) 6, 525–32.

— and Beaufort, F. G. de (1966). Contribution à l'étude des plantes à myrmécodomaties de l'Afrique intertropicale. *Mém. Inst. Fond. Afr. Noire* 75, 1–66.

— Cusset, G. (1963). Glandularisation and foliarisation. *Bull. Jard. bot. État Bruxelles* 33, 525–30.

— — and Quenum, M. (Mlle) (1963). Contribution à l'étude des glandes extra-florales chez quelques groupes de plantes tropicales. *Rev. gén. bot.* 70, 269–342.

— — Tchinaye, V., and Tô Ngoc Anh (1968). Contribution à l'étude des 'acarodomaties'. La question aisselles de nervures. *Rev. gén. Bot.* 75, 5–64.

Schnepf, E. (1959). Untersuchungen über Darstellung und Bau der Ektodesmen und ihre Beeinflussbarkeit durch stoffliche Faktoren. *Planta* 52, 644–708.

— (1965). Licht- und elektronenmikroskopische Beobachtungen an den Trichomhydathoden von *Cicer arietinum*. *Z. Pflanzenphysiol.* 53, 245–54. [English summary]

— (1968). Zur Feinstruktur der schleimsezernierenden Drüsenhaare auf der Ochrea von *Rumex* und *Rheum*. *Planta* 79, 22–34.

Schöch, E. and Kramer, D. (1971). Korrelation von Merkmalen der C_4 Photosynthese bei Vertretern verschiedener Ordnungen der Angiospermen. *Planta* 101, 51–66.

Schoch, P. G. (1972). Variation de la densité stomatique de *Capsicum annuum* L. en fonction du rayonnement global. *C. R. Acad. Sci. Paris, D.* 274, 2496–8.

Schofield, E. K. (1968). Petiole anatomy of the Guttiferae and related families. *Mem. N. Y. Bot. Gard.* 18 (1), 1–55.

Schoute, J. C. (1902). *Die Stelar-Theorie*. Groningen.

Schremmer, F. (1969). Extranuptiale Nektarien. Beobachtungen an *Salix eleagnos* Scop. und *Pteridium aquilinum* (L.) Kuhn. *Öst. bot. Z.* 117, 205–22.

Schroeter, C. (1923). *Das Pflanzenleben der Alpen. Eine Schilderung des Höchgebirgsflora*. 336 pp. Albert Ranstein, Zurich.

Schumacher, W. and Halbsguth, W. (1939). Über den Anschluss einiger höherer Parasiten an die Siebröhren der Wirtspflanzen. Ein Beitrag zum Plasmodesmenproblem. *Jb. wiss. Bot.* 87, 324–55.

Schumann, K. (1891). Rubiaceae. In Engler's *Die natürlichen Pflanzenfamilien*, Vol. IV, pp. 1–156. [See p.2, Fig. 1B and p. 3, line 11.]

Schürmann, B. (1959). Über den Einfluss der Hydratur und des Lichtes auf die Ausbildung der Stomata-Initialen. *Flora* 147, 471–520.

Schuster, W. (1908). Die Blattaderung des Dicotylenblattes und ihre Abhängigkeit von äusseren Einflüssen. *Ber. dt. bot. Ges.* 26, 194–237.

— (1910). Zur Kenntnis der Aderung des Monocotylenblattes. *Ber. dt. bot. Ges.* 28, 268–78.

Schwendt, E. (1907). Zur Kenntnis der extrafloralen Nektarien. *Beih. bot. Zbl.* 22 (1), 245–86.

Scott. D. H. (1889). On some recent progress in our knowledge of the anatomy of plants. *Ann. Bot.* 4, 147–61.

— (1891). Origin of polystely in dicotyledons. *Ann. Bot.* 5, 514–17.

— (1925). German reminiscences of the early eighties. *New Phytol.* 24, 9–16.

— and Sargant, E. (1893). On the pitchers of *Dischidia rafflesiana* (Wall). *Ann. Bot.* 7, 243–69.

Scott, F. M. (1963). Root hair zone of soil-grown roots. *Nature, Lond.* 199, 1009–10.

— (1966). Cell wall surface of the higher plants. *Nature, Lond.* 210, 1015–17.

—, Bystrom, B. G., and Bowler, E. (1963). Root hairs, cuticle, and pits. *Science* **140**, 63–4.

— Hamner, K. C., Baker, E., Bowler, E. (1958). Electron microscope studies of the epidermis of *Allium cepa*. *Am. J. Bot.* **45**, 449–61.

—, Sjaholm, V. and Bowler, E. (1960). Light and electron microscope studies of the primary xylem of *Ricinus communis*. *Am. J. Bot.* **47**, 162–73.

Scott, L. I. and Priestley, J. H. (1928). The root as an absorbing organ. I. A reconsideration of the entry of water and salts in the absorbing region. *New Phytol.* **27**, 126–40.

Sen, S. (1958). Stomatal types in the Centrospermae. *Curr. Sci.* **27**, 65–7.

Sharma, G. K. and Dunn, D. B. (1968). Effect of environment on the cuticular features in *Kalanchoë fedschenkoi*. *Bull. Torrey Bot. Cl.* **95**, 464–73.

—— (1969). Environmental modifications of leaf surface traits in *Datura stramonium*. *Can. J. Bot.* **47**, 1211–16.

Sheldrake, A. R. and Northcote, D. H. (1968). Some constituents of xylem sap and their possible relationship to xylem differentiation. *J. Exp. Bot.* **19**, 681–89.

Shininger, T. L. (1971). The regulation of cambial division and secondary xylem differentiation in *Xanthium* by auxins and gibberellin. *Plant Physiol.* **47**, 417–22.

Simon-Moinet, L. (1965). Homologies foliaires et florales chez l'*Impatiens Balsamina* L. Recherches préliminaires. *C. R. Acad. Sci., Paris* **260**, 2047–50.

Sinclair, C. B. and Sharma, G. K. (1971). Epidermal and cuticular studies of leaves. *J. Tennessee Acad. Sci.* **46**, 2–11.

Singh, P. and Kundu, B. C. (1962). Occurrence of special types of thickenings in the stomata of *Digitalis* spp. *J. Sci. Indust. Res.* **21c**, 107–8.

Sinnott, E. W. (1914). The anatomy of the node as an aid in the classification of angiosperms. *Am. J. Bot.* **1**, 303–22.

— (1936). The relation of organ size to tissue development. *Am. J. Bot.* **23**, 418–21.

— and Bailey I. W. (1914). Investigations on the phylogeny of the Angiosperms. 3. Nodal anatomy and morphology of stipules. *Am. J. Bot.* **1**, 441–53.

— and Durham, G. B. (1923). A quantitative study of anisophylly in *Acer*. *Am. J. Bot.* **10**, 278–87.

Sitholey, R. V. (1971). Observations on the three-dimensional structure of the leaf cuticle in certain plants. *Ann. Bot.* **35**, 637–9.

— and Pandey, Y. N. (1971). Giant stomata. *Ann. Bot.* **35**, 641–2.

Sitte, P. (1955). Der Feinbau verkorkter Zellwand. *Mikroskopie* **10**, 178–200.

— (1962). Zum Feinbau der Suberinschichten im Flaschenkork. *Protoplasma* **54**, 555–9.

— and Rennier, R. (1963). Untersuchungen an cuticularen Zellwandschichten. *Planta* **60**, 19–40. [English summary.]

Skene, M. (1924). *The biology of flowering plants.* Sidgwick and Jackson, London.

Skoss, J. D. (1955). Structure and composition of plant cuticle in relation to environmental factors and permeability. *Bot. Gaz.* **117**, 55–72.

Skutch, A. F. (1946). A compound leaf with annual increments of growth. *Bull. Torrey Bot. Cl.* **73**, 542–46.

Slade, B. F. (1952). Cladode anatomy and leaf trace systems in the New Zealand brooms. *Trans. R. Soc. New Zealand* **80**, 81–96.

— (1971). Stelar evolution in vascular plants. *New Phytol.* **70**, 879–84.

Slatyer, R. O. (1967). *Plant-water relationships.* Academic Press, New York.

Slavik, B. (1974). Guttation. In *Methods of studying plant water relations,* Chapman and Hall, London.

Smith, H. B. (1937). Number of stomata in *Phaseolus vulgaris* studied with the analysis of variance technique. *Am. J. Bot.* **24**, 384–387.

Solereder, H. (1899). *Systematische Anatomie der Dicotyledonen.* Ferdinand Enke, Stuttgart.

Solereder, H. (1908). *Systematic anatomy of the Dicotyledons.* (Transl. L. A. Boodle and F. E. Fritsch, revised by D. H. Scott.) 2 vols. Clarendon Press, Oxford.

Spackman, W. and Swamy, B. G. L. (1949). The nature and occurrence of septate fibers in dicotyledons. *Am. J. Bot.* **36**, 804. [Abstract]

Spanjer, O. (1898). Untersuchungen ueber die Wasserapparate der Gefässpflanzen. *Bot. Zeit.* **56**, 35–81.

Sperlich, A. (1939). *Das tropische Parenchym.* B. Exkretionsgewebe, p. 170. In Linsbauer, *Handbuch der Pflanzenanatomie. IV.* Gebrüder Borntraeger; Berlin.

Spinner, H. (1936). Stomates et altitude. *Ber. schweiz. bot. Ges.* **46**, 12–27.

Springensguth, W. (1935). Physiologische und ökologische Untersuchungen über extraflorale Nektarien und die sie besuchend Insekten. *Sitz. Abh. Naturf. Ges. Rostok* **5**, 31–110.

Spruce, R. (1908). *Notes of a botanist on the Amazon and Andes.* (Edited and condensed by A. R. Wallace.) 2 vols. London. [See especially pp. 384–412.]

Srivastava, L. M. (1964). Anatomy, chemistry and physiology of bark. *Internat. Rev. Forest Res.* **1**, 203–77.

— (1966). Histochemical studies on lignin. *Tappi* **49**, 173–83.

Stace, C. A. (1963). Cuticular patterns as an aid to plant taxonomy. Thesis, University of London.

— (1965a). Cuticular studies as an aid to plant taxonomy. *Bull. Br. Mus. (Nat, Hist.) Bot.* **4** (1), 1–78.

— (1965b). The significance of the leaf epidermis in the taxonomy of the Combretaceae. I. A general view of tribal, generic and specific characters. *J. Linn. Soc. Lond. Bot.* **59**, 229–52.

— (1966). The use of epidermal characters in phylogenetic considerations. *New Phytol.* **65**, 304–18.

Stahl, E. (1896). Über bunte Laubblätter; ein Beitrag zur Pflanzenbiologie II. *Ann. Jard. bot. Buitenzorg.* **13**, 137–215.

Staveren, M. G. C. van, and Baas, P. (1973). Epidermal leaf characters of the Malesian Icacinaceae. *Acta bot. néerl.* **22**, 329–59.

Stearn, W. T. (1966). In *Supplement to the Dictionary of gardening,* pp. 318–22. (Ed. P. Synge.) Oxford University Press.

Stebbins, G. L. (1950). *Variation and evolution in plants.* Columbia University Press, New York and Oxford University Press, London.

— and Jain, S. K. (1960). Developmental studies of cell differentiations in epidermis of monocotyledons. I. *Allium, Rhoeo* and *Commelina. Dev. Biol.* **2**, 409–26.

— and Khush, G.S. (1961). Variation in the organization of the stomatal complex in the leaf epidermis of monocotyledons and its bearing on their phylogeny. *Am. J. Bot.* **48**, 51–9.

Steenis, C. G. G. J. van (1968). Occurrence of domatia as a systematic character. *Flora Malesiana Bull.* **5**, 1568–70.

Stern, W. L. (1957). Guide to institutional wood collections. *Trop. Woods* 106, 1–27.

— (1967). *Index xylariorum*: institutional wood collections of the world. *Regnum Vegetabile* 49, 1–36.

— (1973). The wood collection, what should be its future? *Arnoldia* 33, 67–80.

— (1974). Comparative anatomy and systematics of woody Saxifragaceae, *Escallonia. Bot. J. Linn. Soc.* **68**, 1–20.

— (1976). Multiple uses of institutional wood collections. *Curator* 19, 265–70.

— (1978). *Index Xylariorum*. Institutional wood collections of the world, 2. *Taxon* **27**, 233–69.

— Sweitzer, E. M., and Phipps. R. E. (1970). Comparative anatomy and systematics of woody Saxifracaceae. *Ribes* In: *New Research in plant anatomy* (eds. Robson, N. K. B., Cutler, D. F., and Gregory, M.). Academic Press. *Bot. J. Linn. Soc.* **63**, Suppl. 1, 215–37.

Stevens, A. B. P. (1956). The structure and development of the hydathodes of *Caltha palustris* L. *New Phytol.* **55**, 339–45.

Stober, J. P. (1917). A comparative study of winter and summer leaves of various herbs. *Bot. Gaz.* **63**, 89–109.

Stocking, C. R. (1956). Guttation and bleeding. In W. Ruhland, *Handbuch der Pflanzenphysiologie*, III, *Pflanze und Wasser*. Springer, Berlin.

Stone, D. E. (1961). Ploidal level and stomatal size in the American hickories. *Brittonia* **13**, 293–302.

Stork, H. E. (1956). Epiphyllous flowers. *Bull. Torrey Bot. Cl.* **83**, 338–41.

Strain, R. W. (1933). A study of vein endings in leaves. *Am. Mid. Natur,* **14**, 367–73.

Strasburger, E. (1866). Ein Beitrag zur Entwicklungsgeschichte der Spaltöffnungen. *Jb. wiss. Bot.* **5**, 297–342.

Stromberg, A. (1956). On the question of classification of stomatal types in leaves of dicotyledonous plants. *Tbilisi Sci. Stud. Chemical-Pharmaceutical Inst.* 8, 51–67. [In Russian.]

Styer, C. H. (1977). Comparative anatomy and systematics of the Moutabeae (Polygalaceae).*J. Arnold Arbor.* **58**, 109–45.

Subramanyam, K. and Rao, T. A. (1949). Foliar sclereids in some species of *Memecylon* L. *Proc. Indian Acad. Sci. B* **30**, 291–8.

Sugiyama, M. (1972). A vascular system of 'node to leaf' in *Magnolia virginiana* L. *J. Jap. Bot.* **47**, 313–20.

Swamy, B. G. L. (1953d). The morphology and relationships of the Chloranthaceae. *J. Arnold Arbor.* **34**, 375–408.

— and Bailey, I. W. (1949). The morphology and relationships of *Cercidiphyllum. J. Arnold Arbor.* **39**, 187–210.

— — (1950). *Sarcandra*, a vessel-less genus of the Chloranthaceae. *J. Arnold Arbor.* **31**, 117–29.

Takhtajan, A. (1969). *Flowering plants: origin and dispersal.* (Transl. from Russian by C. Jeffrey.) Oliver and Boyd, Edinburgh.

Tarnavschi, I. T. and Paucă-Comănescu, M. (1972). Morphological variation of leaf epidermis depending on station in several herbaceous species.*Revue Roumaine Biol. (sér Bot.)* 17, 299–309.

Taub, S. (1910). Beiträge zur Wasserausscheidung und Intumeszenzbildung bei Urticaceen. *Sber. Akad. Wiss. Wien (Math. Nat. Kl.)* I 119, 683–708.

Terraciano, A. (1897-9). I nettarii estranuziali delle Bombacaceae. In Borzì, *Contrzioni Biol. veg.* II, 139–91.

Teuscher, H. (1967). *Dischidia pectenoides. Natn. hort. Mag.* **46**, 36–40.

Theobald, W. L. (1967a). Comparative morphology and anatomy of *Chlaenosciadium* (Umbelliferae). *J. Linn. Soc., Lond. (Bot.)* 60, 75–84.

— (1967b). Anatomy and systematic position of *Uldinia* (Umbelliferae). *Brittonia* 19, 165–73.

Thomas, D. A. and Barber, H. N. (1974a). Studies on leaf characteristics of a cline of *Eucalyptus urnigera* from Mount Wellington, Tasmania. I. Water repellency and the freezing of leaves. *Austr. J. Bot.* **22**, 501–12.

— — (1974b). Studies on leaf characteristics of a cline of *Eucalyptus urnigera* from Mount Wellington, Tasmania II. Reflection, transmission and absorption of radiation. *Austr. J. Bot.* **22**, 701–7.

Thomas, H. H. (1932). The old morphology and the new. *Proc. Linn. Soc., Lond.* **145**, 17–32.

Thorenaar, A. (1926). Onderzoek naar bruikbare kenmerken ter identificatie van boomen naar hun bast. Dissertation. H. Voenman and Zonen, Wageningen.

Thouvenin, M. (1890). Recherches sur la structure des Saxifragacées. *Ann. Sci. nat. Bot.* (ser. 7) **12**, 1–174.

Thurston, E. L. and Lersten, N. R. (1969) (1970). The morphology and toxicology of plant stinging hairs. *Bot. Rev.* **35**, 393–412.

Tichoun, Y. T. (1923). Morphologie comparative des feuilles à l'état adulte. Thesis, University de Nancy.

Tieghem, P. E. L. van (1891). *Traité de Botanique.* Paris.

— and Douliot, H. (1886a). Sur les tiges à plusieurs cylindres centraux. *Bull. Soc. Bot. Fr.* **33**, 213–16.

— (1886b). Sur la polystélie. *Ann. Sci. nat. Bot.* (ser. 7) **3**, 275–322.

Timmerman, H. A. (1927). Stomatal numbers; their value for distinguishing species. *Pharm. J.* **118**, 241–3.

Titman, P. W. and Wetmore, R. H. (1955). The growth of long and short shoots in *Cercidiphyllum. Am. J. Bot.* **42**, 364–72.

Tobler, J. (1957). In Zimmermann and Ozenda's *Encyclopaedia of Plant Anatomy.* Band IV, Teil 6. *Die mechanischen Elemente und das mechanische System.* Gebrüder Borntraeger, Berlin.

Tomlinson, P. B. (1961). *Anatomy of the Monocotyledons. II. Palmae.* Clarendon Press, Oxford.

— (1969). *Anatomy of the Monocotyledons. III. Commelinales – Zingiberales.* Clarendon Press, Oxford.

— (1974). Development of the stomatal complex as a taxonomic character in the monocotyledons. *Taxon* **23**, 109–28.

— and Gill, A. M. (1973). Growth habits of tropical trees: some guiding principles. In *Tropical forest ecosystems in Africa and S. America: a comparative review* (ed. B. J. Meggers *et al.*). Smithsonian Institution Press, Washington, D.C.

Tô Ngoc Anh (1966). Sur la structure anatomique et l'ontogénèse des acarodomaties et les interprétations morphologiques qui paraissent s'en dégager. *Adansonia* 6, 147–51.

Tortorelli, L. A. (1956). *Maderas y bosques argentinos.* Editorial Acme, S.A.C.I., Buenos Aires.

Tregunna, E. B. and Downton, J. (1967). Carbon dioxide compensation in members of the Amaranthaceae and some related families. *Can. J. Bot.* **45**, 2385–7.

—, Smith, B. N., Berry, J. A., and Downton, J. S. (1970). Some methods for studying the photosynthetic taxonomy of the angiosperms. *Can. J. Bot.* **48**, 1209–14.

Trelease, W. (1881). The foliar nectar glands of *Populus*. *Bot. Gaz.* **6**, 284–90.

Treub, M. M. (1883). Sur le *Myrmecodia echinata* Gaudich. *Annls Jard. bot. Buitenz.* **3**, 129–59.

— (1888). Nouvelles recherches sur le myrmecodia de Java (*Myrmecodia tuberosa* Beccari) [non Jack.] *Annls Jard. bot. Buitenz.* **7**, 191–212.

— (1889). Les bourgeons floraux du *Spathodea campanulata* Beauv. *Annls Jard. bot. Buitenz.* **8**, 38–46.

Treviranus, L. Chr. (1835–8). *Physiologie der Gewächse*, Vol. I. 1835; Vol. II 1838. Bonn.

Trimen, H. (1881). *Cinchona ledgeriana* Moens. *J. Bot.* **19**, 323–5.

Troll, W. (1932). Morphologie der schildförmigen Blätter. *Planta* **17**, 153–314.

— (1939). *Vergleichende Morphologie der höheren Pflanzen*. Lieferung 3, 1426–676. Berlin.

— and Meyer, H. J. (1955). Entwicklungsgeschichtliche Untersuchungen über das Zustandekommen unifazialer Blattstrukturen. *Planta* **46**, 286–360.

Tschirch, A. (1885). Beiträge zur Kenntnis des mechanischen Gewebesystems der Pflanzen. *Jb. wiss. Bot.* **16**, 303–35.

— (1889). *Angewandte Pflanzenanatomie*. Wien und Leipzig.

— and Oesterle, O. (1893–1900). *Anatomischer Atlas der Pharmakognosie und Nahrungsmittelkunde*. Part I. Leipzig.

Tswett, M. (1907). Recherches anatomiques sur les hydathodes des Lobéliacées. Nouveau type de stomates aquifères. *Rev. gén. Bot.* **19**, 305–16.

Tucker, S. C. (1964). The terminal idioblasts in magnoliaceous leaves. *Am. J. Bot.* **51**, 1051–62.

— and Hoefert, L. L. (1968). Ontogeny of the tendril in *Vitis vinifera*. *Am. J. Bot.* **55**, 1110–19.

Tumanyan, S. (1963). Anatomy of the dicotyledon leaf and its value for systematics. *Izv. Akad. Nauk. armyan. SSR* (*Biol. Nauki*) **16** (11), 3–12.

Tyler, A. A. (1897). The nature and origin of stipules. *Ann. N. Y. Acad. Sci.* **10**, 1–49.

Uhlarz, H. (1975). Über die strittige Homologie sogennanter Stipulardrüsen bei einigen *Euphorbia*-Arten der Sektion *Euphorbium*. *Plant Syst. Evol.* **124**, 229–50.

Uloth, W. (1867). Ueber Wachsbildung im Pflanzenreich. *Flora*, **50**, 385–92.

Unger, F. (1855). *Anatomie und Physiologie der Pflanzen*. Pest, Wien, Leipzig.

— (1858). Beiträge zur Physiologie der Pflanzen. VII Ueber die Allgemeinheit wässeriger Ausscheidungen und deren Bedeutung für das Leben der Pflanzen. *Sber. Akad. Wiss. Wien* **28**, 111–34.

— (1861). Beiträge zur Physiologie der Pflanzen. Über die kalkausscheidenden Organe der *Saxifraga crustata* Vest. *Sber. Akad. Wiss. Wien* **43**, 519–24.

Untawale, A. G. and Mukherjee, P. K. (1969) [1970]. Structure and development of glands in *Jatropha gossypifolia* Linn. *J. Indian bot. Soc.* **48**, 359–62.

Uphof, J. C. T. (1942). Ecological relations of plants with ants and termites. *Bot. Rev.* **8**, 563–98.

—, Hummel, K., and Staesche, K. (1962). *Plant hairs*. In *Handb. Pflanzenanat.*, Vol. IV. pt 5. Gebrüder Borntraeger, Berlin.

Vardar, Y. (1950). Untersuchungen über die Wasserbewegungen in untergetauchten Pflanzen. *Rev. Fac. Sci. Univ. Istanbul* **15**, 1–59.

— and Bütün, G. (1974) [1975]. Studies on the water

economy of macchia elements in Turkey. I. Structural relations of sun and shade leaves of some typical macchia elements. *Ber. dt. bot. Ges.* **87**, 581–8.

Vaughan, J. G. (1955). The morphology and growth of the vegetative and reproductive apices of *Arabidopsis thaliana* (L.) Heynh., *Capsella bursa-pastoris* (L.) Medic. and *Anagallis arvensis* L. *J. Linn. Soc. Bot.* **55**, 279–301.

— (1970). *The structure and utilization of oil seeds*. Chapman and Hall, London.

Venning, F. D. (1949). Stimulation by wind motion of collenchyma formation in celery petioles. *Bot. Gaz.* **110**, 511–14.

— (1954). The relationship of illumination to the differentiation of a morphologically specialized endodermis in the axis of the potato, *Solanum tuberosum* L. *Phytomorphology* **4**, 132–9.

Vesque, J. (1881). L'anatomie des tissus appliquée à la classification des plantes. *Nouv. Archs Mus. Hist. nat., Paris* (ser. 2) **4**, 1–56.

— (1885). Caractères des principales familles gamopétales tirés de l'anatomie de la feuille. *Annls Sci. nat. bot.* (ser. VII) **1**, 183–360.

— (1889). De l'emploi des caractères anatomiques dans la classification des végétaux. *Bull. Soc. Bot. Fr.* **36**, XLI-LXXVII.

Villiers, J.-F. (1973). Icacinaceae. *Flore du Gabon* No. 20, 3–100.

Vines, S. H. (1925). Reminiscences of German botanical laboratories in the seventies and eighties of the last century. *New Phytol.* **24**, 1–8.

Vink, W. (1970). The Winteraceae of the Old World. I. *Pseudowintera* and *Drimys* — morphology and taxonomy. *Blumea* **18**, 225–354.

Vis, J. H. (1958). The histochemical demonstration of acid phosphatase in nectaries. *Acta bot. neerl.* **7**, 124–30.

Vogl, A. E. (1887). *Anatomisches Atlas zur Pharmacognosie*, Plates 16–60. Urban and Schwarzenberg, Vienna.

Volkens, G. (1883). Ueber Wasserausscheidung in liquider Form an den Blättern höherer Pflanzen. *Jb. Konig. botan. Gartens bot. Mus. Berlin* **2**, 166–209.

— (1884). Die Kalkdrüsen der Plumbagineen. *Ber. dt. bot. Ges.* **8**, 334–42.

Vouk, V. (1916). Dodatak istrazivanjima 'O gutaciji i hidatodama kod Oxalis-vrsta'. *Rad. Jugosl. Akad. Znan. Umjetn.* **215**, 55–58.

— and Njegovan, M. (1949). Svijetla pjega na lotosovom listu. (A light spot on the leaf of *Nelumbium*.) *Acta bot. Inst. bot., Zagreb* **12–13**, 195–206. [English summary.]

Waddle, R. M. and Lersten, N. R. (1973). Morphology of discoid floral nectaries in Leguminosae, especially tribe Phaseoleae (Papilionoideae). *Phytomorphology* **23**, 152–61.

Wagner, A. 1892. Zur Kenntnis des Blattbaues der Alpenpflanzen und dessen biologische Bedeutung. *Sitzber. kais. Akad. Wiss. Wien* **101**, 487–547.

Waldner, M. (1877). Die Kalkdrüsen der Saxifragen. *Mitt. naturw. Vereins f. Steiermark* **1**, 9–17.

Walker, N. E. and Dunn, D. B. (1967). Environmental modification of cuticular characteristics of Alaska pea plants. *Trans. Missouri Acad. Sci.* **1**, 17–24.

Walker, W. S. (1960). The effects of mechanical stimulation and etiolation on the collenchyma of *Datura stramonium*.

Am. J. Bot. **47**, 717-24.

Wallis, T. E. (1946). *Textbook of pharmacognosy.* J. and A. Churchill, London.

— (1949). Sclerenchyma in the diagnosis and analysis of vegetable powders. *J. Pharm. Pharmac.* **1**, 505-13.

— (1952). Mint and its adulterants. *Pharm. Weekblad* **87**, 684-92.

— (1957). Crystals in the leaf of *Lobelia inflata* Linn. *J. Pharm. Pharmac.* **9**, 663-5.

— (1960). *Textbook of pharmacognosy.* (4th edn.) J. and A. Churchill, London.

— (1966). *Analytical microscopy. Its aims and methods in relation to foods, water, spices and drugs.* (3rd. edn.) J. and A. Churchill, London.

— (1967). *Textbook of pharmacognosy.* (5th edn.) J. and A. Churchill, London.

— and Dewar, T. (1933). Buchu and the leaves of other species of *Barosma. Q.J. Pharm. Pharmac.* **6**, 347-62.

— and Forsdike, J. L. (1938). Palisade ratio. Its value for detecting certain adulterants of Belladonna Leaf and Stramonium, especially *Scopolia carniolica* and *Solanum nigrum. Q. J. Pharm. Pharmac.* **11**, 700-8.

— and Saber, A. H. (1933). The quantitative determination of foreign leaves in powdered drugs. *Q. J. Pharm. Pharmacol.* **6**, 655-68.

Warden, J. (1971-2). Leaf-plantlet meristems ('leaf embryos') of *Bryophyllum*, revision of terminology. *Portugalliae Acta Biol. A.* **12**, 97-100.

Wardlaw, C. W. (1957). On the organization and reactivity of the shoot apex in vascular plants. *Am. J. Bot.* **44**, 176-85.

Wardrop, A. B. (1969). The structure of the cell wall in lignified collenchyma of *Eryngium* sp. (Umbelliferae) *Aust. J. Bot.* **17**, 229-40.

Watari, S. (1934). Anatomical studies of some leguminous leaves with special reference to the vascular system in petioles and rachises. *J. Fac. Sci. Univ. Tokyo (Sec. III)* **4**, 225-365.

— (1936). Anatomical studies on the vascular system in the petioles of some species of *Acer*, with notes on the external morphological features. *J. Fac. Sci. Imp. Univ. Tokyo (Sec. III)* **5**, 1-73.

— (1939). Anatomical studies on the leaves of some saxifragaceous plants, with special reference to the vascular system. *J. Fac. Sci. Univ. Tokyo,* (sect. III, Bot.) **5**, 195-316.

Watson, L. (1962). The taxonomic significance of stomatal distribution and morphology in Epacridaceae. *New Phytol.* **61**, 36-40.

—, Pate J. S., and Gunning, B. E. S. (1977). Vascular transfer cells in leaves of Leguminosae-Papilionoideae. *Bot. J. Linn. Soc.* **74**, 123-30.

Watson, R. W. (1942). Effect of cuticular hardening on the form of epidermal cells. *New Phytol.* **41**, 223-9.

Weber, F. (1951). *Impatiens*-Nektar. *Phyton. Austria* **3**, 110-111.

Weber, H. (1949a). Über die Bildung von Spaltöffnungsgruppen an Spressachsen. *Ber. dt. bot. Ges.* **62**, 114-15.

— (1949b). Morphologische und anatomische Studien an höheren Pflanzen. I. Über die Verteilung der Spaltöffnungen an den Sprossachsen krautiger Pflanzen. *S. B. Heidelberger Akad. Wiss. Math.-Nat. Kl.* **6**, 161-88.

— (1955). Haben die Marcgraviaceen, 'Inulinblätter'? *Ber. dt. bot. Ges.* **68**, 408-12.

— (1956). Über die Blütenstände und die Hochblätter von *Norantea* Aubl. (Marcgraviaceae). *Beitr. Biol. Pfl.* **32**, 313-29.

— (1959). Negative idioblasts. *Phyton (Austria)* **8**, 130-1.

Weberling, F. (1955a). Morphologische und entwicklungsgeschichtliche Untersuchungen über die Ausbildung des Unterblattes bei dikotylen Gewächsen. *Beiträge Biol. Pfl.* **32**, 27-105.

— (1955b). Die Stipularbildungen der Coriariaceae. *Flora* **142**, 629-30.

— (1956). Untersuchungen ueber die rudimentären Stipeln bei den Myrtales. *Flora* **143**, 201-18.

— (1957a) Über das Vorkommen rudimentärer Stipeln bei den Lecythidaceae (s.1.)und Sonneratiaceae. *Flora* **145**, 72-7.

— (1957b) Weitere Untersuchungen zur Morphologie des Unterblattes bei den Dikotylen I. Balsaminaceae. II. Plumbaginaceae. *Beitr. Biol. Pfl.* **33**, 17-32.

— (1957c). Morphologische Untersuchungen zur Systematik der Caprifoliaceen. *Abh. Math-Nat. Kl. Akad. Wiss. Mainz.*

— (1963). Ein Beitrag zur systematischen Stellung der Geissolomataceae, Penaeaceae, und Oliniaceae sowie der Gattung *Heteropyxis* (Myrtaceae). *Bot. Jb.* **82**, 119-28.

— (1966a). Additional notes on the myrtaceous affinity of *Kania eugenioides* Schltr. *Kew Bull.* **20**, 517-20.

— (1966b). Zur systematischen Stellung der Gattung *Heptacodium* Rehd. *Bot. Jb.* **85**, 253-8.

— (1967). Nebenblattbildungen als systematisches Merkmal. *Naturwiss. Rundschau* **20**, 518-25.

— (1968a). Bemerkungen über das Vorkommen rudimentärer Stipeln. I. Cyrillaceae und die Gattung *Cyrillopsis* Kuhlm. II *Alzatea* Ruiz. und Pav. (Lythraceae) und *Tristania* R. Br. *Acta bot. neerl.* **17**, 282-7.

— (1968b). Über die Rudimentärstipeln der Resedaceae. *Acta bot. neerl* **17**, 360-72.

— (1970a). Weitere Untersuchungen zur Morphologie des Unterblattes bei den Dikotylen. V. Piperales. *Beitr. Biol. Pfl.* **46**, 403-34.

— (1970b). Weitere Untersuchungen zur Morphologie des Unterblattes bei den Dikotylen. VI. Polygonaceae. *Beitr. Biol. Pfl.* **47**, 127-40.

— (1970c). Die vermeintlichen Stipulardornen bei *Zanthoxylum* L. und *Fagara* L. (Rutaceae) sowie bei *Acanthopanax* Miq. (Araliaceae). *Ber. Oberhess. Ges. Natur-Heilk. Giessen (N.F.) Naturw. Abt.* **37**, 141-7.

— (1974a,b). Weitere Untersuchungen zur Morphologie des Unterblattes bei den Dikotylen VII. Polygalales; VIII. *Koeberlinia* Zucc. *Beitr. Biol. Pflanzen* **50**, 277-89.

— (1975). Über die Beziehungen zwischen Scheidenlappen und Stipeln. *Bot. Jahb. Syst.* **96**, 471-91.

— (1976). Die Pseudostipeln der Sapindaceae. *Abhandlung math-naturwiss. Kl. Akad. Wiss. Lit. Mainz* No. 2, 5-27.

— and Leenhouts, P. W. (1965) [1966]. Systematisch-morphologische Studien an Terebinthales-Familien (Burseraceae, Simaroubaceae, Meliaceae, Anacardiaceae, Sapindaceae). *Abh. Akad. Wiss. Mainz Math- naturwiss Kl.* **10**, 497-584.

Weingart, W. (1920), Extranuptiale Nektarien an einem *Phyllocactus. Monatschr. Kakteenkunde* **30**, 136-8.

— (1923). *Crassula schmidtii* Regel. *Zeit. Sukkulentenkunde* **1**, (Part 2), 23-6.

Weiss, A. (1865). Untersuchungen über die Zahlen und Grössenverhältnisse der Spaltöffnungen. *Jahb. wiss. Bot.*

4, 124–96.

Wergin, W. P., Elmore, C. D., Hanny, B. W., and Ingber, B. F. (1975). Ultrastructure of the subglandular cells from the foliar nectaries of cotton in relation to the distribution of plasmodesmata and the symplastic transport of nectar. *Am. J. Bot.* **62**, 842–9.

Werker, E. and Fahn, A. (1966). Vegetative shoot apex and the development of leaves in articulated Chenopodiaceae. *Phytomorphology* **16**, 393–401.

Wessel, P. and Weber, O. (1855). Neuer Beitrag zur Tertiärflor des Niederrheinischen Braunkohlformation. *Palaeontographica* **4**, 111–78.

Wetmore, R. H. and Barghoorn, E. S. (1974). The Harvard Wood Collection. *International Association of Wood Anatomists Bulletin* No. 2, 17–21.

— and Garrison, R. (1961). The growth and organization of internodes. In *Recent Advances in Botany*, pp. 827–32. University of Toronto Press.

— — (1966). The morphological ontogeny of the leafy shoot. In *Trends in Plant Morphogenesis*, pp. 187–99. (ed. E. G. Cutter), Longmans, Green and Co, London.

— and Steeves, T. A. (1971). Morphological introduction to growth and development. *Plant physiology, a treatise,* Vol. 6A, pp. 3–166. (Ed. F. C. Steward.) Academic Press, New York.

— Barghoorn, E. S., and Stern, W. L. (1974). The Harvard University wood collection in the rejuvenation of systematic wood anatomy. *Taxon,* **23**, 739–45.

Wettstein, R. von (1888). Über die Kompositen der österreichisch-ungärischen Flora mit zuckerabscheidenden Hüllschüppen. *Sber. Akad. Wiss. Wien, Math. Naturw. Klasse* **97**, 570–589.

Wettstein-Knowles, P. von (1974). Ultrastructure and origin of epicuticular wax tubes. *J. Ultrastruct. Res.* **46**, 483–98.

Wheeler, M. C. (1913). Observations on the Central American acacia ants. *Trans 2nd. Congr. Entom. Oxford* **II**, 109–39.

— (1942). Studies of neotropical ant-plants and their ants. *Bull. Mus. Comp. Zool. Harvard* **90**, 1–262.

Whiffin, T. (1972). Observations on some upper Amazonian formicarial Melastomataceae. *Sida* **5**, 33–41.

Whitecross, M. I. and Armstrong, D. J. (1972). Environmental effects of epicuticular waxes of *Brassica napus* L. *Austr. J. Bot.* **20**, 87–95.

Whitmore, T. C. (1962). Studies in systematic bark morphology. II. General features of bark construction in Dipterocarpaceae. *New Phytol.* **61**, 208–220.

— (1963). Studies in systematic bark morphology. IV. The bark of beech, oak and sweet chestnut. *New Phytol.* **62**, 161–9.

Wiehler, H. (1975). *Besleria* L. and the re-establishment of *Gasteranthus* Benth. (Gesneriaceae). *Selbyana* **1**, 150–6.

Wiesner, J. (1871). Beobachtungen über die Wachsüberzüge der Epidermis. *Bot. Zeit.* **29**, 769–774.

— (1890). *Anatomie und Physiologie der Pflanzen* (3rd edn). Wien.

Wigand, A. (1854). *Botanische Untersuchungen. III. Intercellularsubstanz und Cuticula*, pp. 67–82. Braunschweig.

Wildeman, É. de (1938). *Dioscorea* alimentaires et toxiques (morphologie et biologie). *Mém. Inst. R. colon. Belge (sér. 8)* **7**(2), 1–262.

Wilkinson, H. P. (1971). Leaf anatomy of various Anacardiaceae with special reference to the epidermis and some contribution to the taxonomy of the genus *Dracontomelon*

Blume. Thesis, University of London.

— (1978). Leaf anatomy of the tribe Coscinieae Hook. f. and Thoms. (Menispermaceae). *Kew Bull.* **32**, 347–60.

Willis, J. C. (1966). *A dictionary of the flowering plants and ferns.* Revised by H. K. Airy Shaw. Cambridge University Press.

Wilson, C. L. (1924). Medullary bundle in relation to the primary vascular system in Chenopodiaceae and Amaranthaceae. *Bot. Gaz.* **78**, 175–99.

— (1942). The telome theory and the origin of the stamen. *Am. J. Bot.* **29**, 759–64.

Wilson, J. K. (1923). The nature and reaction of water from hydathodes. *Cornell Univ. Agric. Exp. Stn.* **55**, 3–11.

Wilton, O. C. (1938). Correlation of cambial activity with flowering and regeneration. *Bot. Gaz.* **99**, 854–64.

— and Roberts, R. H. (1936). Anatomical structure of stems in relation to the production of flowers. *Bot. Gaz.* **98**, 45–64.

Winkler, H. (1913). Die Pflanzenwelt des Tropen. In *Das Leben der Pflanzen*, Vol. VI, Ameisen und Pflanzen, pp. 383–98. Stuttgart.

— and Zimmer, C. (1912). *Eine akademische Studienfahrt nach Africa*. Breslau.

Wolkinger, F. (1969). Morphologie und systematische Verbreitung der lebenden Holzfasern bei Sträuchern und Bäumen. I. Zur Morphologie und Zytologie. *Holzforschung* **23**, 135–43.

Wood, D. (1970). The role of marginal hydathodes in foliar water absorption. *Trans. Bot. Soc. Edinburgh* **41**, 61–4.

Worsdell, W. C. (1903). The stelar theory. *New Phytol.* **2**, 140–44.

Wuhrmann-Meyer, K. and Wuhrmann-Meyer, M. (1941). Untersuchungen über die Absorption ultravioletter Strahlen durch Kutikular- und Wachsschichten von Blättern I. *Planta* **32**, 43–50.

Wulff, T. (1898). Stüdien über verstopfte Spaltöffnungen. *Öst. bot. Z.* **48**, 201–9, 252–58, 298–307.

Wyk, R. W. vander, and Canright, J. E. (1956). The anatomy and relationships of the Annonaceae. *Trop. Woods.* **104**, 1–24.

Wylie, R. B. (1946). Relations between tissue organization and vascularization in leaves of certain tropical and subtropical dicotyledons. *Am. J. Bot.* **33**, 721–6.

— (1952). The bundle sheath extension in leaves of dicotyledons. *Am. J. Bot.* **39**, 645–51.

Yamazaki, T. (1965). Phylogeny of the leaf in the dicotyledons. *Bot. Mag. Tokyo* **78**, 332–43. [Japanese.]

Yapp, R. H. (1912). *Spiraea ulmaria* and its bearing on the problem of xeromorphy in marsh plants. *Ann. Bot.* **26**, 815–70.

Yin, H. C. (1941). Studies on the nyctinastic movement of the leaves of *Carica papaya. Am. J. Bot.* **28**, 250–61.

Yuncker, T. G. and Gray, W. D. (1934). Anatomy of Hawaiian Peperomias. *Bernice P. Bishop Mus. Occasional Papers* **10**, (20), 1–19.

Zahur, M. S. (1959). Comparative study of the secondary phloem of 423 species of woody dicotyledons belonging to 85 families. *Mem. Cornell Univ. Agr. Exp. Sta.* **358**, 160 pp.

Zalensky, V. (1904). Material for the study of the quantitative anatomy of different leaves on the same plant. *Mem. Polytechn. Inst. Kiev* **4**, 1–203.

Zandonella, P. (1967). Stomates des nectaires floraux chez les Centrospermales. *Bull. Soc. Bot. Fr.* **114**, 11–20.

Zettel, J. (1974). Mikroskopische Epidermiskennzeichen von Pflanzen als Bestimmungshilfen I–IV. *Mikrokosmos* **63**, 106–11; 136–9; 177–81; 201–6.

Ziegenspeck, H. (1941). Der Bau der Spaltöffnungen. III. Eine phyletischphysiologische Studie. *Report Nov. spec. regni Veg. Beih.* **123**, 1–56.

— (1949). Zur Phylogenie der Hydathoden. *Phyton (Austria)* **1**, 302–18.

Ziegler, H. (1955). Phosphataseaktivität und Sauerstoffverbrauch des Nektariums von *Abutilon striatum* Dicks. *Naturwissenschaften* **42**, 259–60.

— (1965). Die Physiologie pflanzlicher Drüsen. *Ber. dt. bot. Ges.* **78**, 466–77.

Zimmermann, J. G. (1932). Über die extrafloralen Nektarien der Angiospermen. *Beih. bot. Zbl.* **49**, 99–196.

Zörnig, H. and Weiss, G. (1925). Beiträge zur Anatomie des Laubblattes offizineller und pharmäzeutisch gebräuchlicher Compositen-Drogen. *Arch. Pharm., Berl.* **263**, 451–70.

Zubkova, I. G. (1975). Ways of anisocytic type stomatal apparatus formation in dicotyledons. *Bot. Zh. SSSR* **60**, 322–30. [Russian; English summary.]

NAME INDEX

The Subject Index is on p. 257.

This index contains the names of the pioneers mentioned in Chapter 1, the names of the authors of the chapters, and the first authors from the literature cited in the text. The page numbers refer to the lists at the end of each chapter or part; for further reference details consult the Bibliography on pp. 222–49.

'Suggestions for further reading' are also listed at the end of each chapter or part but are not included in this index.

SUBJECT INDEX

All genera and higher taxonomic units are indexed except for those in the 'Lists of families' on pp. 190–221. A key to this section is provided under the index heading 'Diagnostic features'.

Figures and drawings are indicated by page numbers in **bold type** as are plate references. The plates follow p. 276.

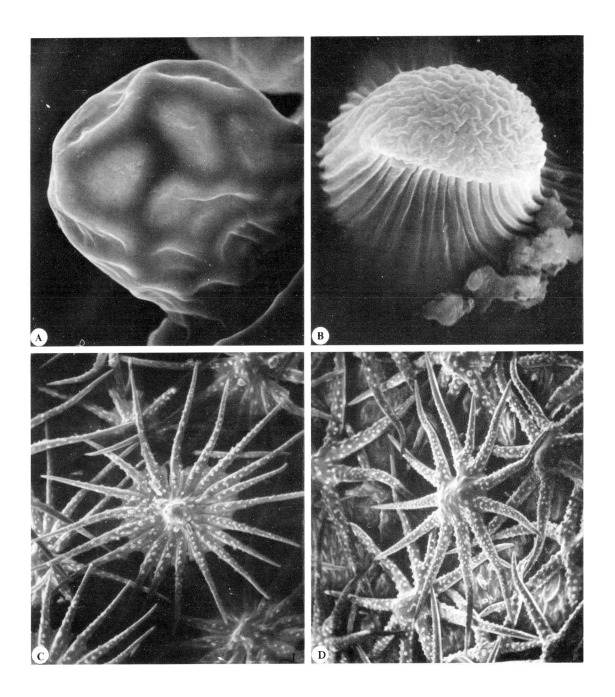

Plate 1. Trichomes (A) *Uldinia ceratocarpa* Burbidge (Apiaceae), × 2500. (B) *Tordylium apulum* L. (Apiaceae), × 5000. (C) *Lesquerella fendleri* S. Wats. (Brassicaceae), × 200. (D) *Lesquerella engelmanni* S. Wats. (Brassicaceae), × 140.

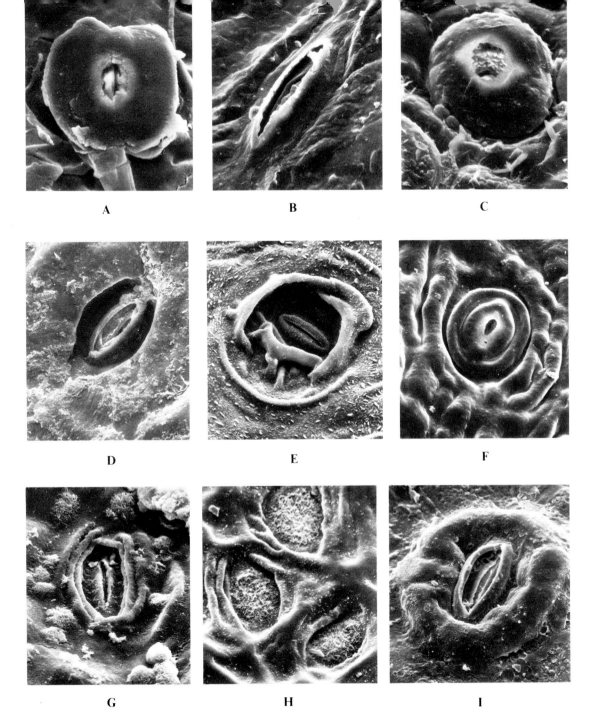

Plate 2. Various stomata. (A) a raised, roundish stoma of *Coscinium fenestratum* (Gaertn.) Colebr. (Menispermaceae), lacking a raised stomatal rim, × 1662. (B) an elongate stoma of *Sedum aizoon* L. (Crassulaceae), with raised rim and very long, narrow aperture, × 807. (C) a round stoma of *Hyaenanche globosa* (Gaertn.) Lamb and Vahl. (Euphorbiaceae), entirely represented by the grossly developed, raised rim and round aperture, × 853. (D) sunken stoma of *Buchanania obovate* Engl. (Anacardiaceae), with raised stomatal rim, long, narrow aperture, and also overlapping peristomatal rim, × 777. (E) a slightly sunken stoma of *Schefflera venulosa* var. *venulosa* Harms. (Araliaceae), surrounded by curved ridges of cuticle on the subsidiary cells; stomatal rim raised, aperture long and narrow, × 1000. (F) round stoma of *Prunus laurocerasus* L. showing two stomatal rims and short ellipsoidal aperture, × 600. (G) a stoma of *Swintonia robinsonii* Ridl. (Anacardiaceae) with long, narrow aperture, barely perceptable outer stomatal rim, and also 2–3 peristomatal rims, × 1543. (H) sunken stomata of *Drimys lanceolata* Baill. (Winteraceae), obscured by alveolar material, × 565. (I) stoma of *Ceriops decandra* (Griff.) Hou (Rhizophoraceae), showing two outer stomatal rims (pore not visible) and a stout peristomatal rim, × 543. All tilted at 30°.

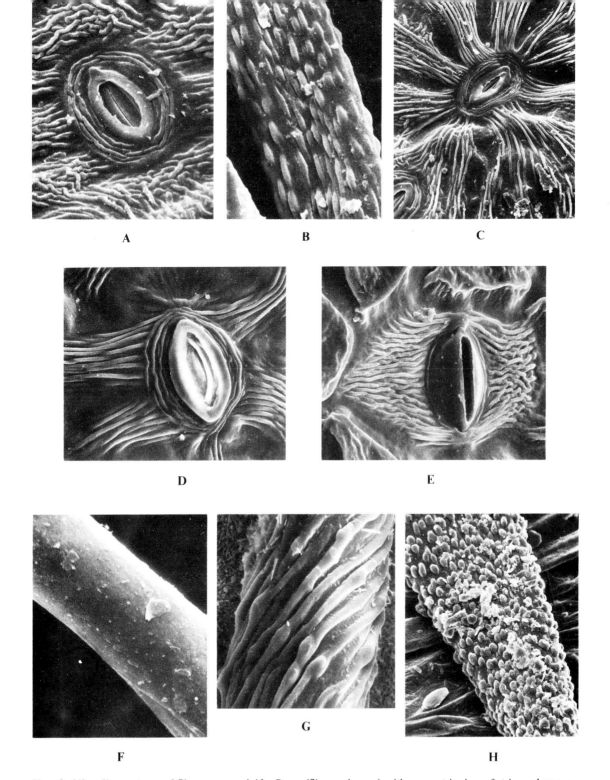

Plate 3. (A) ordinary stoma of *Picrasma quassioides* Benn. (Simaroubaceae) with concentric rings of striae and suggestion of radiating striae outside concentric ones, × 1000. (B) linear, warty ornamentation on hair of *Heeria argentea*
(E. Mey.) O. Kuntze (Anacardiaceae), × 1500. (C) primary or giant stoma of *Picrasma quassioides* Benn. with very long,
radiating striae, × 600. (D) stoma of *Nyssa sylvatica* Marsh (Nyssaceae), showing concentric rings of striae and also
extended wings of striae, × 840. (E) stoma of *Coriaria nepalensis* Wall. (Coriariaceae) with striae extended as lateral
wings; stomatal rim raised into funnel-like structure with long and narrow aperture, × 863. (F–H) parts of hairs. (F)
smooth hair of *Anamirta cocculus* Wight and Arn. (Menispermaceae), × 1800. (G) spiral, somewhat nodular ridges on
hair of *Garrya elliptica* Dougl. ex Lindl. (Garryaceae), × 900. (H) dense covering of warty blobs on hair of *Hydrophyllum canadense* L. (Hydrophyllaceae), × 1800. All at tilt 30°.

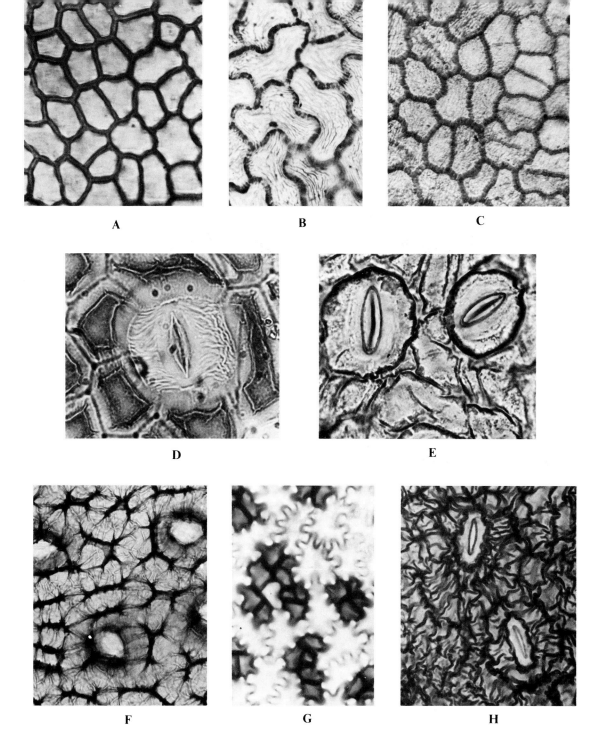

Plate 4. Light micrographs of various cuticular features. (A) *Schinus longifolia* Speg., with straight to slightly undulate, non-pitted anticlinal, and smooth periclinal walls, × 400. (B) *Rhus succedanea* L., with undulate (type 3), densely pitted anticlinal, and finely striate periclinal walls, × 700. (C) *Schinus terebinthifolius* Raddi, showing densely pitted anticlinal walls and granular periclinal walls, in which the granules are arranged in rows giving the appearance of striae, × 325. (D) *Coriaria nepalensis* Wall., showing long, navicular-shaped outer poral rim, alae of striae over paracytic subsidiary cells, and ordinary epidermal cells with raised central region, × 600. (Compare with s.e.m. photograph, Plate 3(E).) (E) *Pseudosmodingium perniciosum* Engl. with peristomatal rims of cuticle and randomly orientated wrinkle-striae, × 700. (F) *Hevea brasiliensis* Muell. Arg., showing reticulum of buttressed ridges; stomata out of focus below cuticular ramparts, × 400. (Compare with s.e.m. photograph, Plate 6(C).) (G) *Parishia maingayi* Hook. f. showing two types of cells: (a) those with darkly staining thicker cuticle and straight anticlinal walls, and (b) thin-cuticled cells with undulate (type 8) anticlinal walls, × 450. (Compare the latter with s.e.m. photograph, Plate 8(H). (H) *Choerospondias axillaris* (Roxb.) Burtt and Hill, with undulated, reticulate striation, some forming peristomatal rims, × 700. (See also Plate 6(G).) A–C, G,H, Anacardiaceae; D, Coriariaceae; F, Euphorbiaceae.

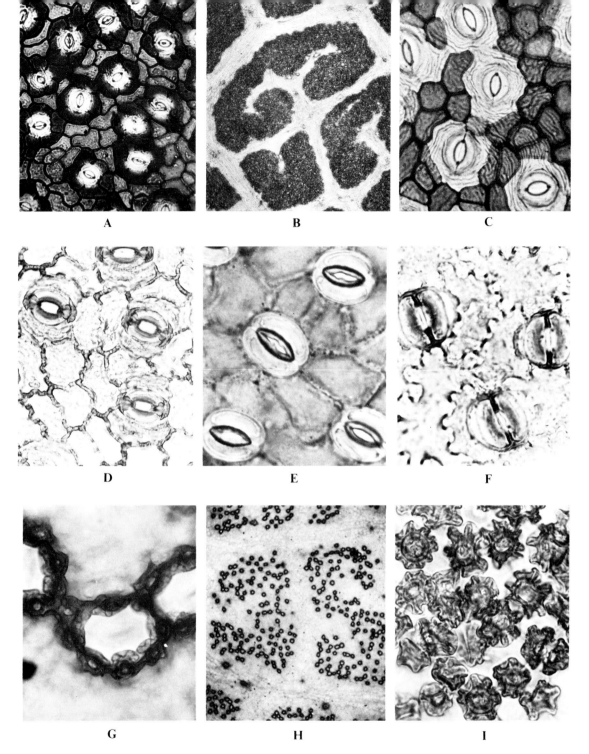

Plate 5. Light micrographs of various cuticular features. (A) *Lithraea molleoides* Engl. showing staurocytic–tetracytic–cyclocytic subsidiary cells with thicker cuticle than that of the ordinary epidermal cells, × 280. (B) *Semecarpus verniciferus* Hay and Kaw, with densely papillose areolae contrasting with the non-papillose vein cuticle, × 70. (C) *Antrocaryon klaineanum* Pierre, showing cyclocytic subsidiary cells with thinner cuticle than the surrounding ordinary epidermal cells, × 400. (D) *Illicium philippinensis* Merr. (Illiciaceae) showing massive cuticular thickening of the guard cell polar regions ('suggesting a head with a broad-brimmed hat', Baranova 1972), and paracytic subsidiary cells (see also Fig. 10.4 (f)) × 354. (E) *Rhus potaninii* Maxim., stomata of a common type without any striking cuticular features except a distinct outer stomatal rim (thinner cuticle over pore visible below) and with anomocytic surrounding cells, × 645. (F) *Melodinus orientalis* B1. (Apocynaceae), stomata with polar T pieces, × 600. (G) *Melanochyla auriculata* Hook. f., showing papillose subsidiary cells; ordinary epidermal cells not papillose (see also Fig. 10.4 (1)), × 600. (H) The same, × 70, areolae made distinct by papillose subsidiary cells. (I) *Melanochyla bracteata* King smooth-lobed papillae, × 600. All abaxial. A–C, E, G–I all Anacardiaceae.

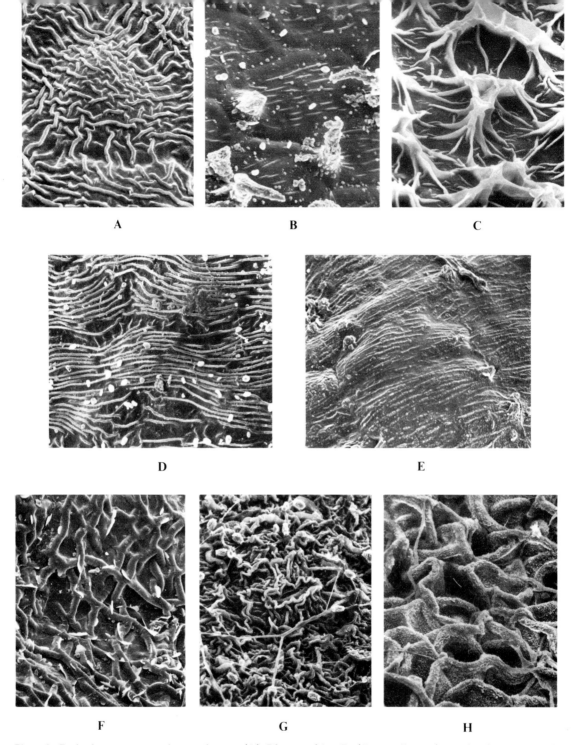

Plate 6. Cuticular ornamentation on leaves. (A) *Rhus typhina* L. (Anacardiaceae), randomly orientated striae on adaxial cell with radiating striae across anticlinal wall boundaries, × 1215, tilt 30°. (B) *Philippia myriadenia* Baker (Ericaceae), short striae confined to each cell, × 585, tilt 30°. (C) *Hevea brasiliensis* Muell Arg. (Euphorbiaceae), a reticulum of buttressed ridges, × 890, tilt 45°. (D) *Philippia capitata* Baker (Ericaceae), long, parallel striae continuous over several cells, × 588, tilt 30°. (E) *Philippia andringitensis* H. Perrier de la Bâthie, (Ericaceae), fine warty striae, × 240, tilt 30°. (F) *Lewisia cotyledon* B.L. Robinson (Portulacaceae), reticulum of rounded ridges (abaxial surface), × 625, tilt 45°. (G) *Choerospondias axillaris* (Roxb.) Burtt and Hill; dense, complicated network of undulate striae, thrown into ridges and forming a peristomatal rim (see also Plate 4(H)), × 600, tilt 30°. (H) *Oxythece leptocarpa* Miq. (Sapotaceae), reticulum of tall ridges with deep sulci at sides, × 607, tilt 45°.

Plate 7. Various papillae. (A) several domes per cell on the adaxial surface of *Dillwynia hispida* Lindl. (Fabaceae), × 350, tilt 45°. (B) simple, cone-like papillae on the abaxial surface of *Rhododendron parryae* Hutch. (Ericaceae), × 650, tilt 30°. (C) tall, striate papillae of *Rhus semialata* Murr. (Anacardiaceae), × 1800, tilt 30°. (D) tall papillae with globular branches on *Willughbeia grandiflora* Dyer ex Hook. f. (Apocynaceae), × 570. (G) part of same, × 1850, both tilted at 45°. (E) tall, striated papillae of *Semecarpus densiflorus* (Merrill) van Steenis (Anacardiaceae), × 1000, tilt 45°. (F) lobed papillae of *Semecarpus vitiensis* (A. Gray) Engl. (Anacardiaceae), × 570, tilt 45°. (I) one papilla from same specimen, × 1650, tilt 30°. (H) globular, striate papilla connected by radiating ridges on *Idesia polycarpa* Maxim. (Flacourtiaceae), × 1800, tilt 15°. B–I all abaxial.

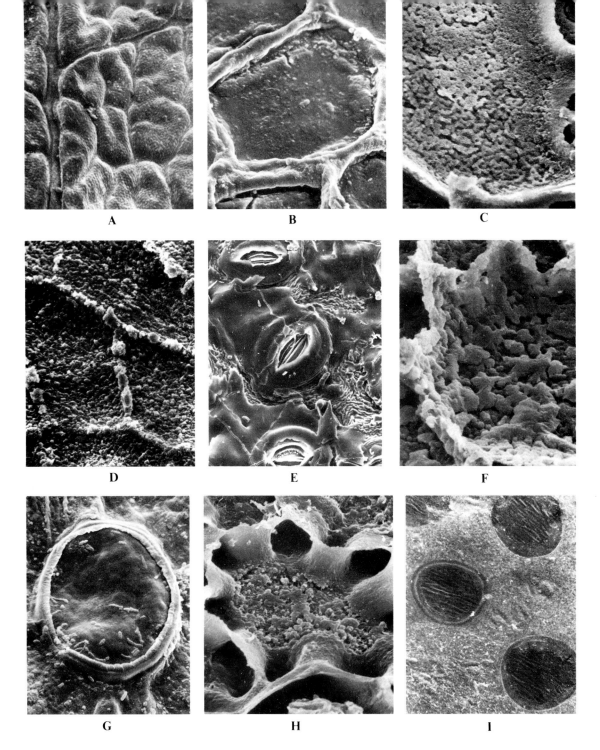

Plate 8. (A) sunken veins on adaxial surface of leaf of *Picrasma quassioides* Benn. (Simaroubeceae), × 27. (B) inner surface of cuticle of *Lithraea caustica* (Molina) Hooker and Arnott. (Anacardiaceae), showing smooth surface and anticlinal walls without pitting, × 845. (C) inner surface of cuticle of *Buchanania obovata* Engl. (Anacardiaceae), showing crustose surface, × 1725. (D) inner surface of *Lannea stuhlmannii* (Engl.) Engl. (Anacardiaceae), showing granular surface and low, pitted anticlinal walls, × 850. (E) outer, abaxial surface of *Viburnum vernicosum* L.S. Gibbs (Caprifoliaceae), showing varnish, partly removed in places revealing striae underneath, × 275. (F) inner surface of cuticle of *Buchanania arborescens* (B1.) B1. (Anacardiaceae), showing flocculent-granular periclinal wall and anticlinal walls projecting at corner junctions, × 1800. (G) outer surface of abaxial cuticle of *Victoria amazonica* (Poepp.) Sow. (Nymphaeaceae) showing a Hydropote, × 1160. (H) inner surface of adaxial cuticle of *Parishia maingayi* Hook. f. (Anacardiaceae) showing coarsely granular periclinal wall and relatively tall, very undulate anticlinal walls, × 1688. (I) inner surface of abaxial cuticle of *Nymphaea capensis* Baill. var *zanzibarensis* (Nymphaeaceae), showing very smooth surface, absence of anticlinal wall (cuticular flanges) and three Hydropoten with striate surfaces, × 564. All at tilt of 30°.

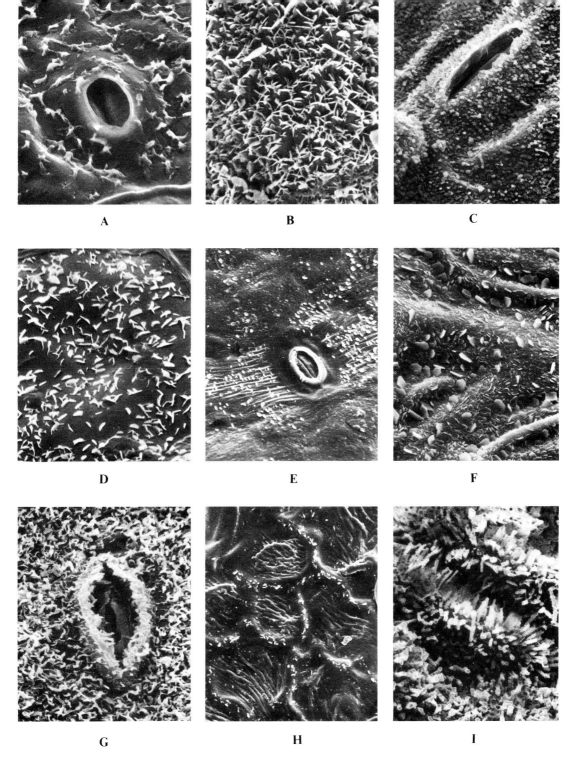

Plate 9. Some examples of leaf wax types and their distribution. (A) short, branched ridges of wax on *Melodinus orientalis* Bl. (Apocynaceae), × 1832. (B) upright scales of wax on *Sophora microphylla* Ait. (Fabaceae), × 2646. (C) warty–crustose wax particles on *Liriodendron tulipifera* L. (Magnoliaceae), × 1655. (D) small, scattered, variously shaped flakes of wax, over and adjacent to a vein on *Semecarpus vitiensis* (Seem.) Engl. (Anacardiaceae), × 1700. (E) scales of wax predominantly aggregated over striae adjacent to a stoma of *Fothergilla major* Lodd. (Hamamelidaceae), × 600. (F) scales of wax on *Schefflera venulosa* var. *venulosa* Harms. (Araliaceae), × 1740. (G) undulating or loosely curled filaments of wax on *Nyssa aquatica* Castigl. (Nyssaceae), × 1600. (H) fine flakes of wax predominantly located over the anticlinal walls of *Gunnera manicata* Linden (Gunneraceae), × 600. (I) rods of wax on *Sassafras albidum* var. *molle* (Nutt.) Nees (Lauraceae), × 2700. H at tilt 15°, remainder all at 30°. B,C, adaxial, all others abaxial.

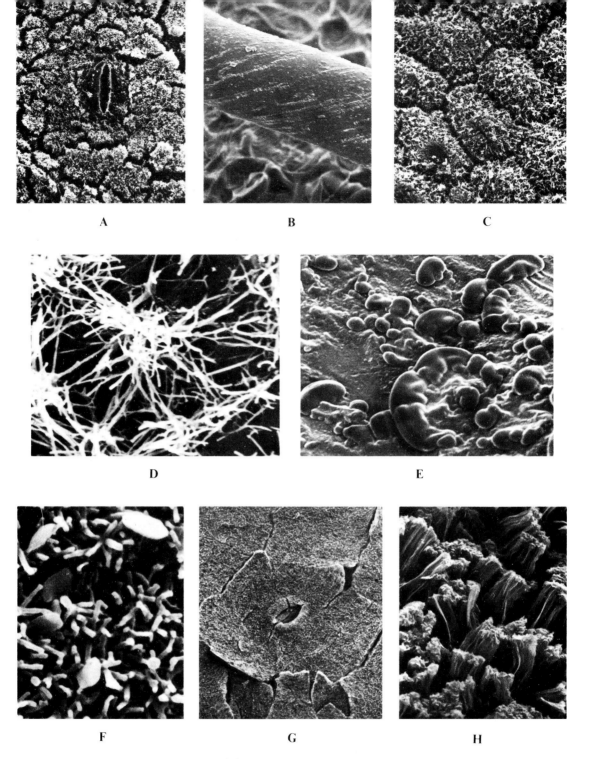

Plate 10. Some examples of wax types. (A) crustose wax on abaxial surface of leaves of *Kalanchoë lugardii* Bullock (Crassulaceae), × 500. (B) minute wax plates spirally arranged on a hair of *Gunnera tinctoria* Mirb. (Gunneraceae), × 550. (C) crustose wax on adaxial surface of leaves of '*Hebe × pagei*' (Scrophulariaceae), × 575. (D) threads of wax on abaxial surface of *Eucalyptus kruseana* F. Muell. (Myrtaceae) leaf, × 3000. (E) soft wax on the fruit of *Crataegus prunifolia* Bosc. (Rosaceae), × 810. (F) two types of wax, rods and scales on the adaxial surface of *Lonicera korolkowii* Stapf (Caprifoliaceae), × 6000, (G) wax plate around a stoma on the stem of *Opuntia quimilo* K. Schum. (Cactaceae), × 215. (H) wax rods cohering in groups on the abaxial surface of *Salix aegyptiaca* L. (Salicaceae), × 1200. B and H at tilt 45°, all others at 30°.

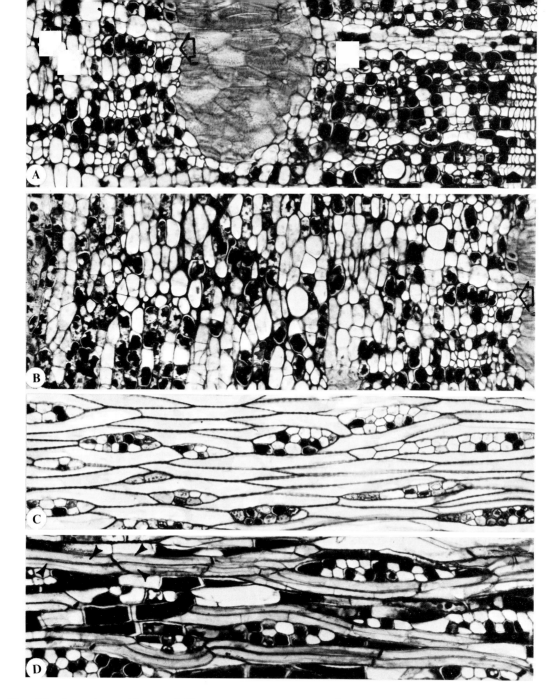

Plate 11. Secondary phloem of *Neocinnamomum delavayii* (Lauraceae). (A and B) transections of younger (A) and older (B) phloem. (A) cambium to the right; group of large sclereids in the middle. (B) in left half only parenchyma cells are intact; the others are obliterated. (C) tangential section of cambium including fusiform and ray initials. (D) tangential section showing results of secondary partitioning of phloem initials. Details are: large, open arrows in (A) and (B) indicate evidence of secondary partitioning; arrowheads in D, individual sieve elements. (A–C) × 130; (D) × 156.

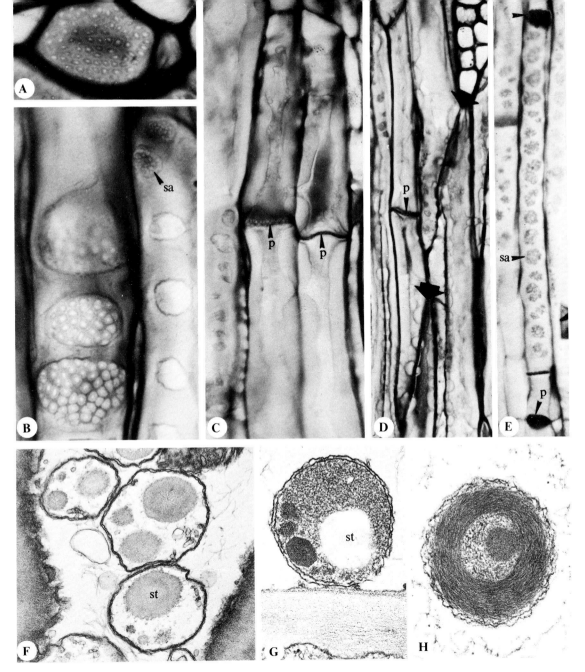

Plate 12. Sieve plates (A and B), sieve elements (C–E), and sieve-tube plastids (F–H). A, *Laurelia novae-zealandiae* (Monimiaceae); B–D, *Magnolia kobus* (Magnoliaceae); E, *Ocotea catesbyana* (Lauraceae); F, *Nicotiana tabacum* (Solanaceae); G, *Mimosa pudica* (Fabaceae); H, *Tetragonia expansa* (Tetragoniaceae). A, simple transverse sieve plate in face view. B, compound inclined sieve plate in face view. Lateral sieve areas to the right. C, D, radial and tangential sections showing transverse and inclined sieve plates. At left in D, the transverse sieve plate (*p*) was formed by secondary partitioning. E, lacmoid staining of sieve plate is more intense than that of lateral sieve areas. F, S-type of plastid. G, alpha P-type of plastid. H, beta P-type of plastid. Details are: *p*, sieve plate; *sa*, sieve area; *st*, starch grain; large arrows in D, inclined sieve plates. A, B, × 1300; C, × 625; D, E, × 390; F, × 30,000; G, H, × 48,000.

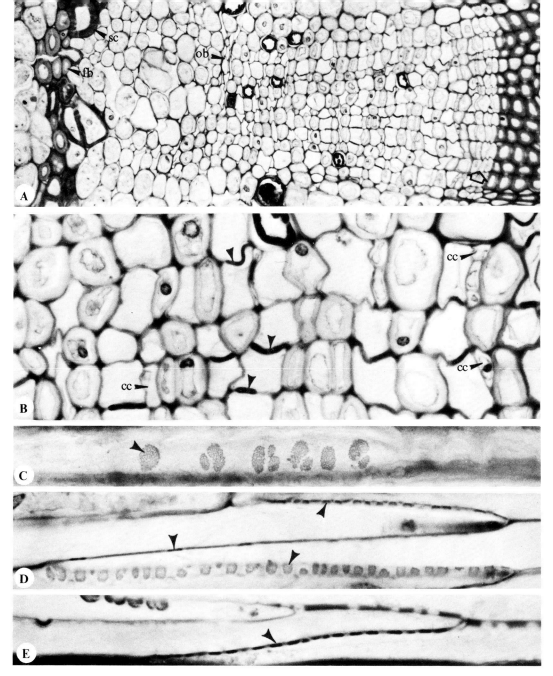

Plate 13. Secondary phloem of *Austrobaileya scandens* (Austrobaileyaceae) in transverse (A, B) and longitudinal (C–E) sections. A, sclerenchyma (left) delimits nonfunctioning part of the phloem from cortex. Some xylem to the right, cambium and functioning phloem next to it. B, sieve areas appear black because of lacmoid stain. Parenchyma cells with thickened walls, one contains tannin. C, sieve areas on radial face of end wall. D, E, tapering ends of sieve elements with numerous small sieve area. Details are: *cc*, presumed companion cell; *fb*, fibre; *ob*, cells in obliteration stage; *sc*, sclereid; open arrow in A, cambium; arrowheads in B–E, sieve areas. A, × 165; B, C, × 600; D, E, × 413.

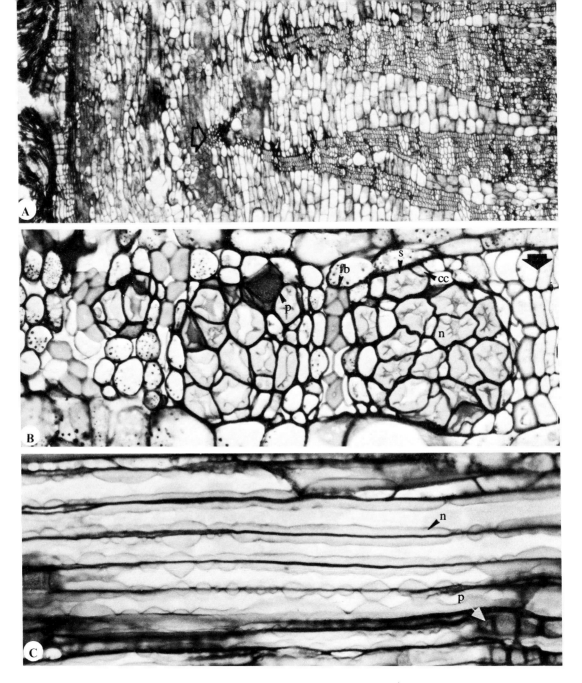

Plate 14. Secondary phloem of Magnoliaceae, *Magnolia grandiflora* (A) and *M. kobus* (B, C) in transverse (A, B) and radial (C) sections. Sieve elements with thick nacreous walls. In A, radial panels of axial system extend to the cortex where some sclereids are present. Sieve elements appear gray because of nacreous walls; parenchyma cells are clear. B, C, show extent of nacreous wall and its crenulated inner margin. Details are: *cc*, companion cell; *fb*, fibre; *n*, nacreous wall; *p*, sieve plate; *s*, sieve element; open arrow in A, protophloem; large arrow in B, cambium. A, × 60; B, C, × 375.

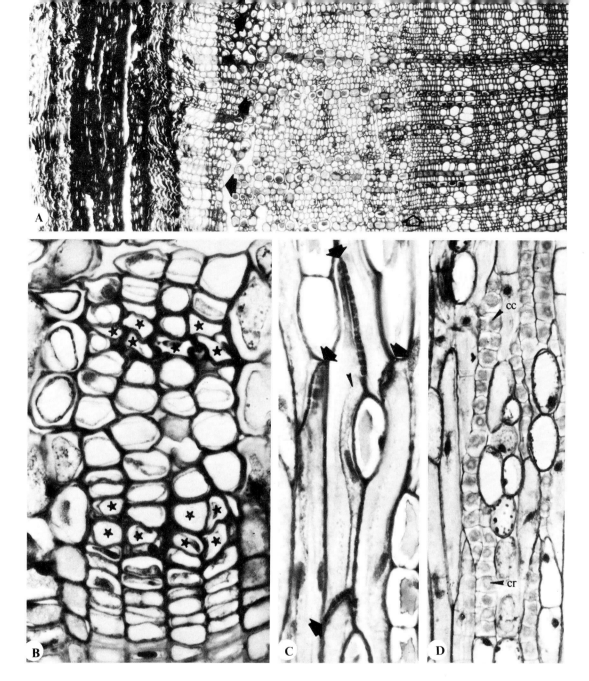

Plate 15. Secondary phloem of *Paeonia suffruticosa* (Paeoniaceae) in transverse (A, B) and tangential (C, D) sections. A, deep-seated periderm is located in the secondary phloem. Homogeneous rhytidome. No fibres in phloem. B, sieve elements marked with asterisks have thin nacreous walls. Details are: *cc*, companion cell; *cr*, crystals in parenchyma strands; solid arrows in A, phellogen; open arrow in A, cambium; arrows in C, sieve plates. A, × 72; B, × 520; C, × 390; D, × 247.

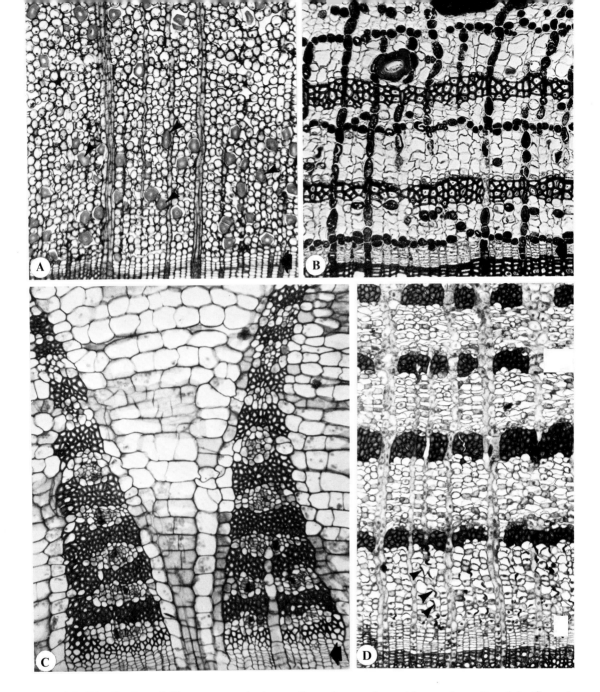

Plate 16. Distribution of fibres as seen in transection of secondary phloem. A, scattered in *Campsis radicans* (Bigononiaceae); B–D, in tangential bands in *Castanea dentata* (Fagaceae), *Cananga odorata* (Annonaceae), and *Fraxinus americana* (Oleaceae). In C, fibres are seen maturing near, the cambium. In D, the functional phloem increment shows darkly stained sieve plates in radial walls. Details are: arrow-heads in A, fibres; arrowheads in D, Sieve plates; large arrow in C, cambium, in D, fibre primordia. A, × 160; B, × 130; C, × 100; D, × 80.

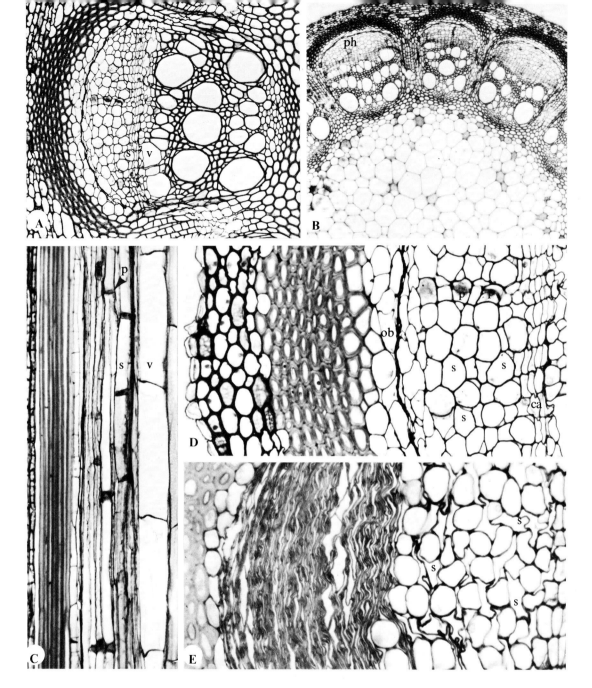

Plate 17. Secondary phloem of *Menispermum canadense* (Menispermaceae) in transverse A, B, D, E, and radial longitudinal C, sections. A, B, complete vascular bundles. Sclerenchymatous pericyclic region of protophloic origin is composed of fibre-sclereids. C, septate fibre-sclereids at left, xylem vessel at right, sieve elements with transverse end walls at *s*. Sieve elements more prominent than parenchyma cells in D, partly collapsed at right in E. All phloem cells collapsed at left in E. Details are: *ca*, vascular cambium; *ob*, obliterated phloem cells; *p*, sieve plate; *ph*, phloem; *s*, sieve element; *v*, vessel. A, C, × 130; B, × 52; D, E, × 325.

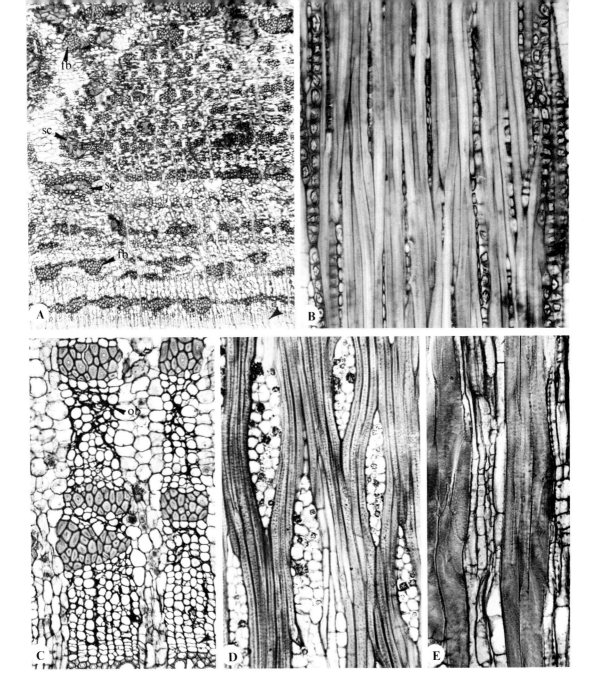

Plate 18. Secondary phloem of A, B, *Populus grandidentata* (Salicaceae) and, C–E, *Bourreria ovata* (Boraginaceae) in transverse (A, C), tangential (B, D), and radial (E) sections. Sclereids associated with fibres in A. Crystals in phloem parenchyma in B. Fibre-sclereids in C–E are wider in radial than in tangential direction and are profusely pitted. Details are: *sc*, sclereid; *fb*, fibre; *ob*, obliterated phloem cells; arrowheads in A and C, cambium. A, × 38; B, × 195; C, D, × 104; E, × 136.